Schrödinger

life and thought

A portrait by Lotte Meitner-Graf, made in London (c. 1935).

Schrödinger
life and thought

Walter Moore

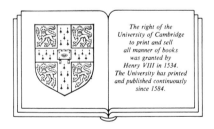

The right of the
University of Cambridge
to print and sell
all manner of books
was granted by
Henry VIII in 1534.
The University has printed
and published continuously
since 1584.

Cambridge University Press

Cambridge

New York   Port Chester

Melbourne   Sydney

Published by the Press Syndicate of the University of Cambridge
The Pitt Building, Trumpington Street, Cambridge CB2 1RP
32 East 57th Street, New York, NY 10022, USA
10 Stamford Road, Oakleigh, Melbourne 3166, Australia

First published 1989

Printed in Great Britain by Bath Press, Bath

British Library cataloguing in publication data

Moore, Walter.
Schrödinger.
1. Science. Schrödinger, Erwin – Critical studies.
I. Title.
509′.2′4

Library of Congress cataloguing in publication data

Moore, Walter.
Schrödinger, life and thought.
Bibliography: p.
Includes index.
1. Schrödinger, Erwin, 1887–1961. 2. Physicists –
Austria – Biography. I. Title.
QC16.S265M66 1989 530′.092′4 [B] 88-25807.

ISBN 0 521 35434 X

Contents

Preface

In this book I have tried to tell something of the life of Erwin Schrödinger in such a way that even those who are not scientists may be able to understand the greatness of his work, the range of his ideas, and the kind of person he was. Montaigne wrote that 'our life is divided betwixt folly and prudence: whoever will write of it only what is reverend and canonical will leave about the one-half behind'. In order to come closer to the whole of Schrödinger, I have included some account of his personal life. It is not thirty years since his death and thus many personal memories and sensitivities have not yet receded into history, a situation that is both a help and a hindrance for a biographer. It is possible to interview former friends and colleagues, but some important collections of letters and memoirs remain closed to research.

My book was made possible by the generosity with which Schrödinger's eldest daughter, Mrs Ruth Braunizer, made available many of the archives concerning her father's life.

Special thanks are due to Professor Ludvik Bass of the University of Queensland, Schrödinger's last postdoctoral student, who reviewed the entire manuscript; his scientific and historical comments have been invaluable.

Dr Linda Wessels of Indiana University provided many helpful suggestions. Professor James McConnell of the Dublin Institute for Advanced Studies never failed to answer my seemingly endless questions about Schrödinger's Dublin years. Dr Anthony Bracken reviewed several theoretical sections; Professor David MacAdam clarified some of the difficulties of color theory. My friend and neighbor, Robert Halporn, assisted in the decipherment of Gothic script and in explaining life in Vienna and the Austrian educational system. Dr John Fletcher helped with the study of Schrödinger's poetry. Professor Sture Nordholm provided translations of the Nobel Archives, and Dr Kirsten Streib translated some of Niels Bohr's letters from the Danish. Professor Karl v. Meyenn of Barcelona made available unpublished material about Schrödinger in Spain, and Professor

Kenji Sugimoto shared his collection of photographs. My family was enlisted in various ways: Professor Joseph Wiecha as consultant on German language and literature, my sister Mary Wiecha for comments on style and content, and my nephew Dr John Wiecha for one of the interviews.

So many others provided help with interviews and research material that I can only mention their names and the relevant section, even omitting their often distinguished titles.

Dublin. Charles Acton, Ronald Anderson, Otto Bergmann, Hermann Bruck, Basil Clancy, Ernesto Corinaldesi, Brigid Dolan, Eithne Dunne, Nuala Feric, Stephen Feric, Gerald Gardner, Herbert Green, Jim Hamilton, Vincent Hart, Margaret Holmes, Fred Hoyle, Ray Kavanagh, John Lewis, Alexander Lieven, A.J. McConnell, William McCrea, Lena McManus, Ernan McMullin, Des McNamara, Harry Messel, Lillian Murnane, Maire Cruise O'Brien, Sean O'Cinneide, H.W. Peng, Jim Pounder, John Ryan, Brendan Scaife, Alfred Schulhof, Neville Symonds, John Synge, Louis Werner, Nicole Whitney, Evelyn Wills.

Austria. Maria Bertel, Bruno Bertotti, Johanna Bianchi, Arnulf Braunizer, Manfred Breitenecker, Ferdinand Cap, Erika Cremer, Auguste Dick, Dieter Flamm, Wilhelm Frank, Leopold Halpern, Otto Hittmair, Berta Karlik, Wolfgang Kerber, Ernst Kohlrausch, Emmy Krafack, Kurt Mählberber, Hermann Mark, Rena Mayer-Rieckh, Suzanne Nemenz, Heinz Oberhammer, Gerhard Oberkaufler, Heinz Pelzer, Kurt Polzer, Mario Rella, Roman Sexl, Ulrike Smola, Walter Thirring, John Ullmann, Paul Urban.

Britain. D.J. Allen, Hansi Bauer-Böhm, Gustav Born, Duncan Davies, Margit Dirac, Patrick Dolan, Margaret Farley-Born, Charles Frank, Fanchon Fröhlich, Herbert Fröhlich, David ter Haar, Barry Hickman, Paul Hoch, Patrick Johnson, Nicholas Kemmer, Heinrich Kuhn, Nicholas Kurti, Catherine Macmillan, Penelope Hughes Minney, Karl Popper, Lucie Rie, N.H. Robinson, Robin Schlapp, Charlotte Simon, Geoffrey Stephenson, Leslie Sutton.

Germany. Ella Ewald, Armin Hermann, Dieter Hoffmann, Dieter Hummel, Theodor Laue, Walter Ledermann, Erika Regener, Fritz Schweighofer, Viktor Weisskopf, K. Wiecke, Eugene Wigner.

Switzerland. Valentine Bargmann, E. Peter Fischer, Beat Glaus, Hans Herwig, Hedwig Herwig, Itha Jennings, Res Jost, Miguel Jungen, Alexander v. Muralt, Gunter Rasche, Michael Weyl.

The custodians of various scientific archives always responded generously and efficiently to requests for research documents and permissions to quote from them: Dr Wolfgang Kerber of the Zentralbibliothek für Physik in Wien; Professor Bengt Nagel for the Nobel Committee; Dr Volker Wahl of the University of Jena; Dr Christa

Kirsten of the Academy of Sciences of the D.D.R.; Dr I. Stolzenberg of the Prussian State Library; Paul Schuchman of the Princeton Institute for Advanced Study; Dr Earle Coleman of Princeton University; Hofrat Dr Waissenberger of the City Museum of Vienna; the staffs of the Library of the American Philosophical Society in Philadelphia, the Bohr Library in New York, the Fisher Library in Sydney, the Deutsches Museum in Munich, and the Indiana University Library in Bloomington.

Special thanks are due to Mr Ehud Benamy of the Einstein Archives for his assistance, and to the Hebrew University of Jerusalem, Israel, for permission to quote from the letters of Albert Einstein.

Dr Simon Capelin, Senior Physics Editor of Cambridge University Press, provided not only valuable editorial comments and criticisms but also unerring judgment in matters of design and production.

The book could not have been written without the devoted assistance of my wife Patricia, who arranged and recorded the many interviews, managed an extensive correspondence, and suggested the most tactful pathways for research.

Chronology

1939.10.7   Arrival in Dublin
1940–1956   Senior Professor, School of Theoretical Physics, Dublin Institute for Advanced Studies
1956.3.28   Return to Vienna as Professor of Theoretical Physics (personal chair)
1958.9.30   Emeritus Professor, Vienna
1961.1.4   Death in Vienna
1961.1.10   Burial at Alpbach, Tirol

Introduction

When the Great War began in August, 1914, Erwin Schrödinger at the age of twenty-seven was a *Privatdozent* at Vienna University, at the outset of a promising career in physics. He served throughout the war as a commissioned officer in the Austrian fortress artillery, first on the Italian front and later assigned to the imperial meteorology department in Vienna. On October 7, 1915, his revered professor of theoretical physics, Fritz Hasenöhrl, was killed by a grenade in an Italian attack on Mount Plaut in the South Tirol.

Early in 1918, Schrödinger received word that he was being seriously considered for the position of *Professor Extraordinarius* [associate professor] in theoretical physics at the University of Czernowitz. 'I made up my mind to lecture there on honest theoretical physics, first of all according to the pattern of the splendid lectures of my beloved teacher Fritz Hasenöhrl, fallen in the war, but besides to concern myself with philosophy, deeply immersed as I then was in the writings of Spinoza, Schopenhauer, Mach, Richard Semon, and Richard Avenarius. My good angel intervened, since soon Czernowitz no longer belonged to us. So nothing came of it. I had to stick to theoretical physics, and, to my astonishment, something occasionally emerged from it.'[1] Even on active duty, Schrödinger found time to read deeply in philosophy, and indeed it was the only thing that kept him from despair during the long years of senseless killing and destruction.

A serious interest in philosophy was not unusual among German and Austrian physicists, but with Schrödinger at that time it was so important that he was even tempted to sacrifice scientific research to devote his life to philosophical study. Much later, in 1963, Max Born wrote: 'I am convinced that theoretical physics is actually philosophy. It has revolutionized fundamental concepts, e.g., about space and time (relativity), about causality (quantum theory), and about substance and matter (atomistics). It has taught us new methods of thinking (complementarity), which are applicable far beyond physics.'[2] By his discovery of wave mechanics, Erwin Schrödinger

1

contributed in a major way to this revolutionary change in our view of the world [*Weltanschauung*]. Now, however, a paradox appears. Schrödinger was more deeply versed in and more sympathetic with philosophic ideas than any other scientist of modern times, yet he rejected the philosophic conclusions that others were so willing to draw from his work, or, to be more precise, he rejected the idea that such far-reaching conclusions could be drawn from any work in theoretical physics.

Arnold Sommerfeld, second to none in his broad understanding of the physics of the first half of our century, said of Schrödinger's discovery of wave mechanics in 1926, 'It was the most astonishing among all the astonishing discoveries of the twentieth century'.[3] This remark might apply also to the discovery of quantum mechanics at almost the same time by Werner Heisenberg. Why were these theories so revolutionary in their philosophical implications? In summary, they had the following results: (1) They devalued the philosophy of materialism that had prevailed in science since the time of Newton. (2) They showed that everything in the world is part of everything else; there are no boundaries and no isolated parts in the world. (3) They contradicted the idea of strict determinism and predictability in nature. (4) They raised deep questions about the role of mind in nature, and restored the human observer to a central position in natural philosophy. As a consequence of all these overturnings of conventional ideas, the influence of the new quantum theory has spread far beyond physics, leaving the philosophy of science, and indeed philosophy in general, in a state of flux whose final resolution can not as yet be discerned.

In his book *The Structure of Scientific Revolutions* (1962), Thomas Kuhn argues that there are two kinds of science, normal science and revolutionary science.[4] In any particular field, normal science is conducted in accord with a set of rules, concepts and procedures, called a *paradigm*, which is accepted by all scientists working in that field. Normal science is similar to puzzle-solving: interesting, even beautiful, solutions are obtained but the rules are never changed. In this normal scientific activity, however, unexpected discoveries sometimes are made that are inconsistent with the prevailing paradigm. Among the scientists, a tense situation then ensues, which increases in intensity until a scientific revolution is achieved. This is marked by a *paradigm shift*, and a new paradigm emerges under which normal scientific activity can be resumed.

This analysis of scientific progress may be less convincing now than when it first appeared; it tries to force the history of science itself into a paradigm with too few variables. Nevertheless, it does fit the work of Erwin Schrödinger. His scientific papers before 1925 were typical products of 'normal science'; they were competent but certainly not remarkable extensions of subjects that were being pursued by his teachers and coworkers at the University of Vienna. In 1926, his *annus mirabilis*, he published at the 'advanced age' of thirty-eight his four great papers on wave mechanics, than which there is nothing more beautiful in theoretical physics. This was revolutionary science by any standard, but when did the revolution begin? Was it in 1913 with Niels Bohr and the quantum theory of the hydrogen atom? Was it in 1900 with Max Planck and the introduction of the quantum? Or was it as long ago as 1861 with James Clerk Maxwell and electromagnetic waves? Has the history of science become a permanent revolution?

After his great discovery in 1926, Schrödinger was unwilling to retreat to normal science. In his later years, he tried, often desperately, to achieve a second revolutionary discovery, attempting to unify gravitation and electromagnetism in a generalized field theory, but he was under no illusions about the likelihood of success. He told a young Irish scholar at Dublin: this problem is suitable only for old men who have already done something in science.

The Kuhnian analysis of scientific revolutions emphasizes the stress that precedes the breakthrough into a new world view. Schrödinger was a passionate man, a poetic man, and the fire of his genius would be kindled by the intellectual tension arising from the desperate situation of the old quantum theory of Planck, Einstein and Bohr. It seems also that psychological stress, particularly that associated with intense love affairs, helped rather than hindered his scientific creativity. In his short autobiography, he wrote: 'To create a true sketch of my life, for that I am lacking a narrator's turn of mind – and furthermore it is impossible, since leaving out relations with women in my case creates on the one hand a great void, and on the other hand seems to be required, firstly on account of scandal, secondly because they are hardly of sufficient interest, thirdly because in such matters no man is really completely sincere and truthful, nor ought he to be.'[5]

Fortunately Schrödinger published a volume of poetry, and a short poem usually unveils more of the psyche than a long scientific treatise. As Leon Edel remarked, a whole new world of biography has been opened by 'confessional poetry'.[6] Personal letters also provide valuable insights into the psychology of the individual, since they paint a portrait midway between the public and the private persona. 'The public facade is the mask behind which a private mythology is hidden', but in a personal letter the mask often slips appreciably. Schrödinger also kept extensive diaries. Orson Welles once contended that 'addiction to diaries, a secret vice like the eating of hashish, degrades the diarist himself to something like the moral status of a drama critic'.[7] I think, however, that Schrödinger wanted us to know that a theoretical physicist need not be a man without qualities or passions and left us his diaries to that end.

Erwin Schrödinger had the most complex personality of all the great creators of modern physics. He was a passionate foe of injustice yet he regarded any political activity as contemptible. He despised pomp and circumstance yet he took a childlike pleasure in honors and medals. He was devoted to the vedantic concept that all men are members of one another, yet he eschewed cooperative work of any kind. His mind was dedicated to precise reasoning, yet his temperament was as volatile as a prima donna's. He claimed to be an atheist, but he always used religious symbolism and believed his scientific work was an approach to the godhead. In many respects he was a true son of Austria. 'In this country one acted – sometimes indeed to the extreme limits of passion and its consequences – differently from the way one thought, or one thought differently from the way one acted. Uninformed observers have mistaken this for charm, or even for a weakness in what they thought was the Austrian character. But that was wrong. It is always wrong to explain the phenomena of a country simply by the character of its inhabitants.'[8]

Hans Thirring, one of his closest friends at the University of Vienna, thought that Erwin had a *schizothymic personality*.[9] This is a category invented by Ernst Kretschmer, a German psychiatrist, who believed with Schopenhauer that 'genius is nearer to madness than to the ordinary intellect'.[10] He classified geniuses into two types, the schizothymic adjacent to schizophrenia and the cyclothymic to manic-depressive psychosis. Furthermore, he found evidence that psychological type can be related to bodily structure: schizothymes have long noses, irregular profiles, long oval facial contours, are often small, thin, wiry men. They have hyperesthetic temperaments, subject to 'tense moods, between which the affect jumps . . . The schizothymic soul, worn out by the noisy pathos of the heroic battle, suddenly feels the need for something diametrically opposed, for tearful tenderness, and dreamy bucolic stillness . . . It is just the same with the style of their private lives.'

Kretschmer found that scholars with schizothymic physiques have been devoted to work of an abstract, metaphysical, or theoretical nature, whereas the cyclothymes tend towards experimental and observational research. Among the markedly schizoid types he included Copernicus, Kepler, Leibniz, Newton, and Faraday. He did not discover any cyclothymic theoreticians, but we should have to place Schrödinger's 'scientific grandfather' Boltzmann in that class, both in bodily build and psychopathology, yet Boltzmann, with his devotion to materialism and his contempt for metaphysics, was an unusual theoretician in other respects also.

Psychology teaches that during his life an individual tends to repeat patterns of interpersonal relationships that were learned in childhood; thus the form of living and even the sexual politics of the man are often determined by his life as a child and by the family in which he was brought up. As the only child in an indulgent family, Erwin might easily have come to believe that the world revolved around his center.

The social milieu of Schrödinger's formative years in Vienna must have influenced his philosophy of life, and through this the way in which he conceived his scientific work. It is not difficult to become bedazzled by the brilliance of Vienna at the turn of the century, as one regards the enchanted characters who share the stage: authors, musicians, artists, philosophers, and scientists. At that time the intellectual world of Austria was not fragmented into academic disciplines and artistic circles that rarely communicated with one another. For example, Robert Musil wrote a thesis on Ernst Mach, and Ludwig Wittgenstein was an excellent architect. But the gleaming of Vienna was superficial, like the phosphorescence covering decay. While Schrödinger was attending the university and enjoying the delights of theater parties, new wine, and excursions to the mountains of nearby Semmering, young Adolf Hitler, twice rejected as an art student at the Academy, having pawned his overcoat to buy bread and milk, was prowling about the snowy streets in his shirtsleeves, with uncut hair and a straggly beard, trying to sell a few pictures of Viennese churches. He told a friend, 'I don't want to seem arrogant, but I think the world really lost something worthwhile when I was rejected by the Art School.'[11]

Everything official in the Austro-Hungarian monarchy was referred to as K & K, *kaiserlich* and *königlich* [imperial and royal]. Musil, Austria's greatest novelist, called the country '*Kakania*', which might be translated politely as 'Poopooland'. 'By its constitution it was liberal, but its system of government was clerical. The system of government was clerical, but the general attitude to life was liberal. Before the law all citizens were equal, but not everyone, of course, was a citizen. There was a parliament, which made such vigorous use of its liberty that it was usually kept shut.'[8]

Yet, after all, the essence of the life of a creative person is expressed in his work: the music of Mahler, the poetry of Hofmannsthal, the politics of Hitler, the physics of Schrödinger. As Ernst Cassirer wrote in regard to Kant: 'In the last analysis the essential task of every biography of a great thinker is to trace how his individuality blends ever more closely with his work and seemingly vanishes entirely, and how its spiritual outlines yet remain embedded in the work and only thus become clear and apparent.'[12] For a philosopher, such an analysis is obviously reasonable; its possibility is less obvious for a theoretical physicist. We may feel intuitively that the physicist and the philosopher cannot be essentially different kinds of human being, yet a causal chain between a scientific discovery and the structure of a personality is not so easy to discern. So far as any individual is concerned, the great discovery itself may be to some extent accidental, the product of circumstance falling into a fertile field of intellect. Schrödinger touched upon this point in one of his poems:[13]

Parabel	*Parable*
Was in unserem leben, freund,	My friend, what in this life
wichtig und bedeutend scheint,	Weighty and important seems,
ob es tief zu boden drücke	Whether causing dark depression
oder freue und beglücke,	Or gladness and rejoicing,
taten, wünsche und gedanken,	Deeds, thoughts and wishes
glaube mir, nicht mehr bedeuten	Believe me, means no more
als des zeigers zufallschwanken	Than a pointer's fluctuations
im Versuch, den wir bereiten	In an experiment that we design
zu ergründen die natur:	To fathom Nature:
sind molekelstösse nur.	Merely molecular collisions.
Nicht des lichtflecks irres zittern	Nor does the light spot's crazy flutter
lässt dich das gesetz erwittern.	Let you smell out the basic law.
Nicht dein jubeln und erbeben	It's not your joy and trembling
ist der sinn von diesem leben.	That makes sense of this life.
Erst der weltgeist, wenn er drangeht,	The World Spirit, if it goes about it
mag aus tausenden versuchen	May from a thousand experiments
schliesslich ein ergebnis buchen. –	Enter finally a result –
Ob das freilich uns noch angeht?	Is it really any of our doing?

A great discovery may be partially a matter of good luck, but it is undeniable that the making of a great discovery has a considerable effect upon the subsequent life of the discoverer. As Franz Grillparzer, Schrödinger's favorite dramatist, wrote:[14]

> Let him who once has won it scoff at fame
> It is no empty echo without meaning
> Its fingers are endowed with divine power.

There is no evidence that fame made any marked change in the personality of Erwin Schrödinger, but it probably made it possible for him to express himself more freely. Before his great discovery, he would appear at scientific meetings in a conventional starched collar; after he achieved fame, he would often turn up looking like a Tirolean mountaineer. Max Born, who had many a fierce scientific argument with him, and who was shocked by some of the things he did, nevertheless wrote: 'His private life seemed strange to bourgeois people like ourselves. But all this does not matter. He was a most lovable person, independent, amusing, temperamental, kind and generous, and he had a most perfect and efficient brain.'[15]

1 Family, childhood and youth

Heredity or environment, nature or nurture, there is no general solution to the problem of how much each contributes to the structure of a personality or the achievements of a person. It has been said that any medical student can become an adequate surgeon provided he or she is willing to work long hours, follow the hospital rules, and placate the nursing staff. In other words, this particular skill is not significantly determined by genetic factors. No such prescription can be applied to mathematical physics; original work in this field requires a special kind of native aptitude, the genes must be right. This is a necessary but not a sufficient condition. A potential Schrödinger may be lost to physics if he happens to be born in a Bengali village rather than a university town.

Many theoretical physicists have come from academic families, which, at least in Germanic countries, are considered to be upper middle class on the social scale. The fathers of Max Planck, Niels Bohr, Max Born, Wolfgang Pauli and Werner Heisenberg were all university professors, of Law, Biology, Anatomy, Colloid Chemistry and Byzantine Studies, respectively. Paul Dirac's father taught French at a Bristol grammar school, in the English context, a respected but lower-middle-class profession; Paul originally aspired only to be an engineer, but turned to pure theory after his ineptness in practical work became evident. Albert Einstein's lower-middle-class background was exceptional among German scientists. The Einsteins lived in an industrial suburb of Munich, and the early stages of Albert's scientific career were unusually precarious.

Thus the ancestry and family history of Erwin Schrödinger are relevant to the story of his life, even if the relative contributions of genetics and economics cannot be clearly separated. Fortunately his Aunt Rhoda compiled a history of his mother's side of the family; less is known about the Schrödinger side.[1]

Family history

The story begins with Franziska Zickler, who was a love child of unknown parentage, born about 1780 in a convent in Hungary where her mother had taken refuge. From certain resemblances, it was thought that she may have been related to Josef of Habsburg-Lorraine, a son of Maria Theresia. She grew up to be a temperamental beauty and married a member of the minor nobility, Anton Wittmann-Denglass who was born at St Bernard in Lower Austria in 1771. They had a daughter Josepha, born in 1803 in Joskowicz, Hungary. Although they were a staunchly Catholic family, Josepha fell in love with a Protestant, and was thereupon forced to marry the Catholic family doctor, by whom she had three children. She was widowed and remarried, this time to Alexander Josef Bauer, who had distinguished himself as her father's secretary and later became manager of his estates. Bauer was a native of Bielitz, born in 1802.

Their eldest son was Alexander Bauer, the maternal grandfather of Erwin Schrödinger, born in Unter Altenburg (Magyarovar) in north-west Hungary, February 18, 1836. After graduation from the *Gymnasium*, Alexander studied mathematics and science at the University of Vienna, but transferred to the Polytechnic Institute (later the Technical University) to specialize in chemistry under Anton Schroetter, the discoverer of red phosphorus.

Erwin's maternal grandmother was English. Her ancestry has been traced back to a Norman family, Forestière, whose stronghold was Bamborough Castle near Durham. The name was anglicized to Forster. Thomas, born in 1772, the son of Colonel Forster, the governor of Portsmouth, married Eliza Walker, and they had five children. They lived in Kensington where their eldest daughter Ann was born on June 27, 1816. She was Erwin's great-grandmother, whom he was to meet when he visited England as a child.

Ann married William Russell, a solicitor in Royal Leamington Spa, Warwickshire. She had gone to Leamington to visit a Walker aunt and met him there. He was descended from a moderately well-to-do family who had been concerned with the legal profession in Warwickshire for many years. The Russells had three children, William, Emily (who was called Minnie in the family), and Ann (called Fanny). Emily was born in Leamington on September 14, 1841. She was baptized in All Saints Church, Leamington Priors (the old name of Leamington Spa), which was the first parish of the Church of England in the town. The family lived in a spacious house at 14, Portland Place, with gardens at the back extending to the river Leam.[2]

How did it happen that Emily Russell met and married Alexander Bauer of Vienna? The causal factor was that Emily's brother William was a chemist. He was not a research scientist like Alexander, but he had set up a laboratory for analytical chemistry, specializing in analysis and control of the waters of the spa. He and Alexander became friends in Paris, where they were both studying chemistry in 1859. Bauer was working at the *École de Médecine* on amylenes and polyamylenes as a pupil of Charles Adolphe Wurtz, known to every beginning student of organic chemistry as the discoverer of the Wurtz synthesis of hydrocarbons, whereby two alkyl halides are linked by the agency of metallic sodium. Emily and her mother visited Paris on their way to a holiday at Montreux, and William took his friend Alex to meet them at their hotel. It was love at first sight between the brilliant and serious young scientist and the lovely nineteen-year-old English girl. We can surmise that her brother William did not hesitate to extol the great abilities of his friend, compared to whom he was a mere dilettante in chemistry.

After this romantic meeting in Paris, Alexander had to return to Austria in the summer of 1860 to work in a sugar factory in Moravia. In 1861, however, his abilities were recognized by appointments as an instructor at the *Handelsakademie* [Trade School] and as Privatdozent at the Institute of Technology. The latter position typically paid little but afforded facilities for research, although the laboratory was poorly equipped even by nineteenth century standards, and water had to be carried in canisters from a nearby brook. The teaching post, however, provided some income and the research gave promise for the future, so that the young chemist was able to ask Emily to be his wife, and he was promptly accepted. He journeyed to Leamington Spa and on December 21, 1862, they were married in the parish church where Emily had been christened. The witnesses were William Russell, Ann Russell, and Fanny Russell.

The newlyweds went to live in a charming small apartment at 20 Kärntnerstrasse, in the heart of old Vienna, not far from the house where Mozart had composed *The Marriage of Figaro*. Alexander registered his marriage at his parish church, the ancient and impressive cathedral of St Stephan. In the summer of 1863, the Bauers visited Leamington, where their first child was conceived. Alex returned to Vienna, while Minnie stayed with her family until the spring of 1864, when she traveled to Vienna, arriving with little time to spare before the birth of the baby on Easter Sunday. This was Rhoda, who became Erwin's most doting aunt. As a child she heard only English spoken at home, and she spent much time in Leamington including one stay of almost five years. A second daughter was born three years later, on March 27, 1867. This was Georgine (Georgie), the future mother of Erwin.

In 1866, Alexander lost an eye in an explosion of a glass vessel in which he was carrying out a chemical preparation. This accident considerably lessened his interest in experimental research, and he turned increasingly to teaching, administration, and especially chemical history, for which he became famous as 'the Nestor of Austrian chemistry'. In 1869, he was appointed associate professor of chemical technology at the *Polytechnicum* [later the Technical University] and then to the chair of general chemistry, where he served till his retirement in 1904.[3]

On March 22, 1874, Emily died of pneumonia shortly after the birth of her third daughter, Minnie. Her funeral was conducted in accord with the Anglican rite. Alexander was desolated by the loss of his young wife, but after a year of mourning he revived and married a seventeen-year-old girl, Natalie Lechner. Natalie was not a great beauty, but a young woman of unusual independence and considerable literary ability; she was an excellent musician, playing the viola in a professional string quartet. The position of a young step-mother is always difficult, but it is likely that Natalie, with her strength of character, managed the rather docile Bauer girls without too many problems. She taught Georgie the elements of violin. After ten years, however, Alexander and Natalie agreed to separate, an event which at the time (1885) caused some negative comment. In 1890, Natalie met Gustav Mahler and for twelve years was his constant companion, until he married Alma Schindler in 1902.[4]

After the departure of Natalie, Alexander devoted himself to his daughters and to a steadily increasing variety of professional and civic duties. He became a noted *Salonlöwe* [salon lion] in bourgeois Vienna society. In 1904 he was honored by designation as *Hofrat*, which may be roughly translated as state councilor. In Imperial and Royal (K & K) Austria-Hungary, honorific titles were highly regarded by members of the upper-middle class, as indeed they still are today in modern Austria. Almost all university graduates are addressed as 'Herr Doktor' and with increasing distinction, the titles become increasingly ornate. Alexander once questioned some advice by his old professor Hofrat Schroetter, and was told quite seriously, 'Young man, a Hofrat never makes a mistake.' The interests of Alexander Bauer were not confined to science and technology. He served for twenty years as a curator of the Museum of Art and Industry, and for twenty-one years as a member of the Theater Commission for Lower Austria. Through this work he met the leading directors and actors of the flourishing Viennese theater, and communicated to his grandson Erwin a love for the theater which began in boyhood and continued all his life.

The three Bauer daughters all espoused men in technical professions whom they met through their father. Rhoda married Regierungsrat Dr Hans Arzberger, Director of the *Medikamenten Eigenregie*

[Pharmaceutical Commission], and Minnie married Dr Max Bamberger, who became Bauer's successor in the chair of general chemistry at the Technical University and later also a Hofrat. Minnie had a daughter Helga, Erwin's first cousin, but Rhoda had no children.

On August 16, 1886, Georgie married Rudolf Schrödinger, who had studied with her father at the Technical University, and who had inherited a small but profitable linoleum and oilcloth factory and wholesale business from his father, and thus was acceptable to Bauer as a son-in-law. The marriage took place in the Lutheran *Stadtkirche*, and the registry indicates that the bridegroom was a Catholic and the bride Georgine Emilie Brenda Bauer was a 'convert' to the Evangelical AB Church.

There are two main divisions of the protestant church in Germany and Austria, Lutheran and Calvinistic. The Lutherans subscribe to the *Augsburgische (AB) Konfession*, a document submitted to the Diet of Augsburg in 1530. They form the major division of the evangelical church in Austria. One may suppose that the Bauer girls were all brought up in the Lutheran religion since it would be difficult to find Anglican churches and instruction in Austria. From a theological standpoint, the differences between the Anglican Church and the Lutheran Church seem negligible, and indeed their differences with the Roman Catholic Church are mostly related to administrative questions such as papal authority and clerical celibacy. The rather tangled religious history of the Bauer family explains how Erwin Schrödinger came to be nominally a Protestant, despite his Catholic background in predominantly Catholic Austria.

Rudolf Schrödinger was born in Vienna January 27, 1857, the son of Josef and Maria Bogner Schrödinger. He was baptized Rudolf Josef Carl in the parish church of Sts Peter and Paul in Erdberg, an outer district of Vienna. Both the father's and mother's families had lived in Vienna for three or four generations, but the Schrödingers originally came from the Oberpfalz area in Bavaria. Rudolf's mother, Maria Anna Josepha Bogner was a nineteen-year-old orphan when she was married in the Catholic church of St Carl on May 14, 1853. Her father had been the proprietor of a coffee house in the suburbs. She had three children, a son Erwin, who died as a child, a daughter Marie, and their last child Rudolf. Rudolf's mother died at the age of twenty-four, six days after the stillbirth of a fourth child. The historian Peter Gay has chronicled the 'devastation that the lethal work of sensuality' so often left in its wake in the lives of nineteenth-century families, when early death in childbirth hung 'like a capricious plague' over the lives of 'flourishing and energetic young women'.[5] Rudolf was less than two years old when his mother died, and his bereaved father did not remarry but undertook the upbringing of his two children.

About three months after their marriage, Rudolf and Georgine effected that fortuitous combination of genes that produces an individual of genius, and on August 12, 1887, the boy was born at home, Apostelgasse 15, in Erdberg, Vienna 3. Georgie, who loved the works of Goethe, wanted to name him Wolfgang after the great poet, but Rudolf, not usually given to sentimental considerations, favored the name Erwin, after his long departed elder brother. The circumstances of the baptism of the child were unusual. It took place at the home of his grandfather Alexander Bauer, Kärntnerstrasse 20. On October 17, the minister of the Evangelical Church came to the house and baptized the baby Erwin Rudolf Josef Alexander. The godfather was Alexander Bauer and there was no godmother. Such a home christening was not unusual but the parental home would have been the expected place. A possible scenario is that Rudolf, who was not a practising Catholic, did not wish to have the baby baptized at all, but that Alexander decided to have it done anyway, and out of deference to his daughter's religion, asked the Evangelical pastor to come to his house and administer the sacrament.

Whether through choice or necessity, Rudolf and Georgie had no more children. Erwin considered his mother to be a woman of delicate constitution all her life. As an only child, he received the full attentions of his mother, and for many years hardly less from Aunts Rhoda and Minnie, as well as the services of a succession of young maids and nurses, all of whom considered him to be a budding genius deserving practically constant adulation. Raised in such an atmosphere of tender, loving, feminine care, it is hardly a wonder that Erwin grew to depend upon it and to expect it as his due, such expectations being fully realized throughout his lifetime.

Childhood

In 1884 a splendid town house, five storeys high, was constructed at 3 Gluckgasse, a fashionable street just off the Neumarkt in the center of the first district of Vienna. The house is of pink limestone faced with marble. Above the second-storey central window is a rococo relief of two cherubs holding a bowl filled to overflowing with grapes. The window above on the next storey is framed by two massive classic male torsi and surmounted with a plaque bearing the date 1884. The house was divided into spacious apartments, only one on each storey. The house was purchased by Alexander Bauer not long after its completion, and in 1890 he rented the top-floor apartment to the Schrödingers and this was to be their Vienna home until the year before the death of Erwin's mother in 1921. The front windows look over tiled roofs to the serrated Gothic spire and flying buttresses of St Stephan's Cathedral.

Erwin and Grandfather Alexander Bauer (c. 1890).

Life in the later years of the Habsburgs was pleasant and unhurried for those with ample incomes. Political freedom was not essential to this good life and its absence did not trouble the ordinary citizen, although there were grumblings from subject peoples in distant parts of the Empire. If anything went wrong, there were always Jewish plots to blame, and the endemic antisemitism helped to satisfy both social and religious needs. Until he entered the Gymnasium, however, the interests of a boy like Erwin would be confined almost entirely to the family circle, with a special emphasis on birthdays and holidays.

Erwin's childhood was somewhat different from that in a typical well-to-do Vienna family owing to the absence of siblings and prevalence of aunts. Some recollections of Erwin's childhood have been given by his Aunt Minnie.[6] She was fourteen years older than Erwin but they were great companions. He learned from Minnie to speak English before he ever spoke German properly. She brought him a book of bible stories from England; he was not terribly impressed by the stories but they were his earliest reading in English.

At the age of four, he displayed an early scepticism of hearsay evidence. Sometimes Aunt Minnie would say one thing and his mother would say another. 'This says Mama, and that says Aunt', remarked Erwin. 'They are both only people. They could just as well say the opposite.'

Even before he could read or write, he kept a record of day-to-day happenings, which he dictated to Minnie. This habit of keeping a day book, he maintained all his life. The later books, called *Ephemeridae*, seem to contain everything but scientific material; it is evident that he thought science has a certain permanence, whereas all else is ephemeral. From one of his earliest records, 1891: 'In evening Aunt Emmy cooked a good supper and then we spoke all about the world.' Minnie recalls that 'Despite my small knowledge of the subject, he was especially interested in asking me about astronomy. For example, I would stand and represent the earth and he would be the moon and run around me, and both of us would move slowly about a light that represented the sun.'

A story that gives early evidence of Erwin's appreciation of the powers of money tells of a long walk with Aunt Emmy in the neighborhood of the old castle of Hartenstein, which is about eighty kilometres along the Danube west of Vienna. They went too far and lost their way back. 'Nothing but woods, and the churning stream [the Else not the Danube] and no possibility of finding our way to the village of Hartenstein. For a short time I carried the little fellow on my back, but naturally not for long . . . Just as our predicament

Erwin with Aunt Emily (Minnie) Bauer (c. 1893).

was at its worst, dear Erwin took a copper coin from his pocket, offered it to me, and asked "will this help?" Fortunately a signpost solved the problem and we got back in time for lunch.'

A visit by army officers in uniform awoke his interest, but all attempts to explain to him the nature of their duties and the necessity for wars were useless. 'What would you do if somebody came and took all your clothes, your shoes, and everything away?' 'Nothing', said the child, 'I would keep cool.'

Ever since 1900, when Freud published his first important work, *Die Traumdeutung*, dreams have been anxiously sought by biographers as mirrors of the subconscious mind. The only one of Erwin's boyhood dreams that has been recounted occurred at Semmering, the mountain resort south of Vienna. They were staying at the enormous Südbahn Hotel. One time he awakened in terror, unable at first to speak. He had been in the middle of a bad dream. What was it? 'It was the horrible word *GEFÄNGNIS* [Prison] written in large letters just above my bed.' It is a horrible word, worse in German than in other languages.

Almost every Sunday Erwin would visit his little cousin Dora, the daughter of his father's sister Marie, who was two years older than he, and the children would play together the whole afternoon. One evening, when Minnie brought him home, he was unusually withdrawn, something seemed to be bothering him. He went quietly to bed but did not fall asleep. 'After a while he called me in. "Yes, what is the matter?" He put his arm around my neck and sort of stammered. "Today I gave the little girl, the cousin, a box on the ear." So, now it was out. After this confession, he cheered up and went to sleep, since the young aunt did not take it so tragically.'

Erwin always had a temper, and tantrums were not uncommon. He had some noteworthy rows with his father, and sometimes had to be sent from the table for bad manners or unruly behavior. He also had the usual quota of boyhood fights, although as an only child he never had to contend with sibling rivalries for family affection.

Erwin experienced the gamut of childhood diseases and accidents, somewhat more than average since he was a rather delicate boy. While still very small, he had a severe case of measles. He became apathetic and almost comatose, and the nurse could hardly wake him. The doctor ordered heart stimulants immediately. Grandfather brought cognac and strong wine into play, and perhaps a catastrophe was thereby avoided. It is interesting to note that it was the practical old chemist and not the rather dreamy parents who came to the rescue.

We know, however, from Erwin's own testimony that the influence of his father was always very important:

To my father I am thankful for far more than only this, that he gave us a very comfortable life, and assured for me an excellent upbringing and a carefree university education, while till almost the end of his life he carried on with little zeal or talent the inherited, prosperous oilcloth business. He had an unusually broad culture; after his university studies as a chemist, there followed many years of intense concern for Italian painters, accompanied by his own landscape drawings and etchings, giving way finally to pyxides and microscope, from which arose a series of publications on plant phylogenetics. To his growing son, he was a friend, teacher and inexhaustible conversation partner, the court of appeal for everything that might sincerely interest him. – My mother was very good, cheerful by nature, sickly, helpless in the face of life, but also undemanding. Besides sacrificial care, I have her to thank, I believe, for my regard for women.[7]

When Rudolf and Georgine Schrödinger were married, she was a sheltered girl of nineteen and he was an experienced man of twenty-nine with a comfortable income from a business that he despised. The manufacture of linoleum involves impregnating a burlap backing with gelled linseed oil and pigments and curing the aggregate to produce the finished product. The process provides limited scope for an artistic temperament. If anything, the skills involved in merchandising the products of the factory would be even more demeaning to an amateur artist and botanist, with a love for mountain landscapes, wildflowers and Oriental ceramics. At this time Vienna surpassed all other cities of the world in its relentless *Drang zum Kulturellen* [drive towards the cultural] and Rudolf was carried along strongly by this current.

Minnie provides only glimpses of the married life of the Schrödingers. 'Doubtless both were a loving couple.' Here the adverb immediately engenders doubt; she will not testify from her own close observations of the family. Minnie continues her portrait of her sister: 'Without being a special beauty the little woman [*kleines Fräuchen*, a double diminutive, which Freud might have underlined] was full of charm, with a lovely expression and an attractive coiffure.' On the honeymoon, as he was downstairs smoking his cigarette, Rudolf once overheard a group of guests wondering 'Is this child really his wife?' Did he ever wonder this himself in the course of his marriage? According to Minnie, her sister concentrated all her thoughts on the welfare of her son, but Erwin was never a mother's boy, an aunt's boy possibly as a child, but soon his father became his principal guide and mentor.

Georgie had an interest in playing the violin. 'As a young bride, she did not find so much need for technique as would be the case today but stressed the winning expressiveness of her playing. After she

Erwin with his parents Rudolf and Georgine on holiday in Kitzbühel.

unfortunately found no response from her husband, her art wasted away.' Almost uniquely among theoretical physicists, Erwin not only did not play any instrument himself, but even displayed an active dislike for most kinds of music, except the occasional love song. He once ascribed this antipathy to the fact that his mother died from a cancer of the breast, which he thought was caused by mechanical trauma from her violin. More likely he learned this distaste for music as a child, echoing his father's lack of response to his mother's art.

A contemporary of Schrödinger, the novelist Robert Musil, also expressed an intense dislike for music, which he called 'the churned-up somatic sediment of the psyche'. Ulrich, 'the man without qualities', the alter ego of the author in *Der Mann Ohne Eigenschaften*, 'had never been able to stand the sight of this always open piano with its bared teeth, this big-mouthed short-legged idol that was a cross between a dachsund and a bulldog, which had subjugated his friends' lives to itself'.[8]

Eithne Wilkins and Ernst Kaiser, who made the superb English translation of *Der Mann Ohne Eigenschaften*, suggest that music symbolized for Musil the mystical aspect of his personality, which he had to repress and deny as a good positivist and intellectual disciple of Ernst Mach. In the last years of his life, Musil was going through a spiritual crisis in an effort to escape from the cold prison walls of the solipsistic philosophy of Mach and Avenarius. Erwin Schrödinger struggled with this same problem all his life: 'One recalls in memory the feeling of anxious, heart-constricting solitude and emptiness that I daresay has crept over everyone on first comprehending the description given by Kirchhoff and Mach of the task of physics (or science in general): a description of facts that is as far as possible complete and as far as possible economical of thought.' This was written in his first philosophical statement, in 1925, before his discovery of wave mechanics.[9] He was not able to lighten his solitude with music, which was the solace of the great Jewish physicists especially, and was compelled to turn to other stratagems.

While Erwin was just beginning to talk, a clever maid began to teach him to read the street signs. His father was somewhat anxious lest he be pushed too quickly. He was not sent to the elementary school, but received lessons at home two mornings a week from a private tutor. This arrangement was not unusual in upper-middle-class families. Their principal goal was to ensure that their boys passed the entrance examination for the Gymnasium, usually taken when they were nine or ten years old.

Instead of taking the examination [*Aufnahmsprüfung*] at the usual time, Erwin had a long holiday and was taken to England by his mother and Minnie to visit the Russells at Leamington Spa. This was in the spring of 1898 when he was ten years old. He saw his great-grandmother in her home Modena Villa, on Russell Terrace named after his great-great-grandfather. He visited his mother's aunt, who had married Alfred Kirk, and who had six Angora cats which soon increased to twenty. One was called Thomas Becket. They visited Kenilworth and Warwick castles. He rode the donkeys on the wide beach at Ramsgate and thought it a great sport. He also learned in England to ride a bicycle, a skill he was to practise all his life, for both pleasure and utility.

Minnie reports that they went by horse and carriage for a 'so called picnic, where one takes a spirit stove to make tea and has all sorts of piquant and sweet baked goodies'. Erwin was laughing happily in the middle of the picnic, but, 'Oh Weh!', he wandered away to a little pond and fell in, which seems a bit clumsy for a ten-year-old boy. He was dragged out, but once he was dry, he did not seem at all sorry about it. Considering what a good swimmer he later became, he probably could swim well enough at ten.

On the trip back to Austria, they crossed the channel from Dover to Ostende, and then visited the beautiful medieval city of Bruges. Proceeding to Cologne, they boarded the Rhine steamer and traveled through the lovely country of wine and castles via Koblenz and

Rudesheim to Frankfurt-am-Main. Here they took the train to Munich and thence to Innsbruck.

Here Erwin had his first experience of school. The parents were worried about the entrance examination and 'afraid that I might have forgotten my ABC's'. Therefore he went to the St Niklaus School for a few weeks. His mother tried to preserve an English atmosphere. When they walked in the park, 'Now we are going to speak only English to each other.' She also tried to observe the English custom of half-holidays, but amid all his other holidays these may have been difficult to distinguish. During all his life Erwin allowed nothing to interfere with his vacations and holidays.

Akademisches Gymnasium

As was to be expected, he passed the examination easily, and entered the Akademisches Gymnasium in the fall of 1898, having just turned eleven. He was about a year older than most of the other students in his class. He had never experienced the rough and tumble socialization of the schoolroom and playground, nor the competition with siblings at home. He had been raised almost exclusively in the company of adults, and it was perhaps only to be expected that he would retain something of the enfant terrible when he reached manhood.

His Gymnasium was the most secular, the least religiously oriented, of all those in Vienna. Ludwig Boltzmann, Arthur Schnitzler and Stefan Zweig had formerly been pupils there. It was, and still is, situated on Beethoven Platz, just off the Schubert Ring, about ten minutes walk from Erwin's house on Gluckgasse. Djuna Barnes once pictured the schoolboys of Vienna as 'flocks of quail, taking their recess in different spots in the sun, rosy-cheeked, bright eyed, with damp rosy mouths, smelling of herd childhood, facts of history shimmering in their minds like sunlight, soon to be lost, soon to be forgotten, degraded into proof.'[10]

As summarized by Friedrich Paulsen, a contemporary authority, the ideal of education in the Gymnasium was to graduate 'a human being whose faculties enable him to form a clear and definite conception of the actual world, and who, by virtue of his will, is able to follow his original bent, whose imagination and fine emotions are trained to the perception of the beautiful and the heroic. This is a man in the full sense of the word; this is true humanistic culture.'[11]

The education designed to achieve this admirable result was based on an intensive study of the Greek and Roman classics. 'The classical literature is, and will continue to be, the source of all our culture. It must remain therefore . . . the most important study in our high schools.' This principle, laid down in 1805 by the first superintendent of the Prussian high schools, had remained in force for at least a hundred years. Actually, the emphasis on Latin and Greek was historical in origin, since all the higher schools had originally been founded by the Church for the education of priests, who required these languages in addition to Hebrew, which was formerly also part of the curriculum, but had become an elective by the late nineteenth century.

Since virtually the only way to enter the university was through the Gymnasium, all university students began with a shared background in humanistic studies and mathematics. Such a preparation for science was welcomed by many who later became eminent in this field. Max Laue, for example, wrote 'I doubt that I should ever have devoted myself entirely to pure science if I had not at that time come into that inner harmony with Greek language and culture, which the humanistic Gymnasium and no other kind of school provided . . . If you wish to bring out scientific development later, I have a recipe: send the youth to a Gymnasium and let him learn the ancient languages.'[12] Of course, there is no single recipe for the making of scientists: compared to Schrödinger, Laue, Born, or Heisenberg, most American physicists have been illiterate in ancient tongues, yet the best of them have also done great physics.

For the first three years at Gymnasium, Latin was the most important subject, eight hours a week of it. When they began Greek, the Latin was reduced to only five hours. First came the grammar and then the classic authors. They started with Nepos, *De Viris Illustribus*, short lives designed to bring out the characters of illustrious men, not only of Rome but also of other nations. Next came Caesar, *Commentarii de Bello Gallico*, then selected parts of Livy's *History of Rome*, and to complete the historical works, something from Tacitus, though perhaps not his penetrating criticisms of the imperial system. From Cicero, they read about the Cataline conspiracy and its ruthless suppression. With eight years available for Latin studies, there was ample time left for the poets: Ovid, selections from *Metamorphoses* and *Tristia*; Horace, the *Odes*; Virgil, some of the *Aeneid*. Among the Greek authors studied were Plato (*The Apology of Socrates*), Sophocles (*Oedipus Rex*), and Homer (*The Iliad* and a little of the *Odyssey*). All these classics had been carefully expurgated to be sure that no intimations of erotic life remained. Some classes on elementary biology were given in the first two years.[13]

Besides the Latin and Greek classics, there were classes each year in German language and literature. The sixth-year pupils studied the literature of Middle High German from its inception to Luther, with emphasis on the poetry of the *Nibelungenlied*; the next year was

Dressed for a special occasion at the Akademisches Gymnasium (1900).

devoted to New High German, with an in-depth study of Goethe and Schiller; in the final year they read literature from the romantic period to the present, with of course considerable attention to the great Austrian dramatist Grillparzer. Only three hours a week were spared for mathematics and science. Mathematics consisted of algebra and geometry; they did not get so far as calculus.

Erwin was reasonably happy with the Gymnasium. 'I was a good student, in all subjects, loved Mathematics and Physics, but also the strict logic of the ancient grammars, hated only the memorization of incidental dates and facts. Of the German poets, I loved especially the dramatists, but hated the pedantic dissection of their works.'[14] Boltzmann once said 'Without Schiller, there might have been a man with my nose and beard, but it would not have been me.'[15]

About five years ahead of Erwin on the academic ladder, Stefan Zweig found the same school a boring treadmill, monotonous, heartless and spiritless. The classrooms were uncomfortable and poorly lighted. 'We sat pairwise like slaves in a galley, on low wooden benches that bent the spine.'[16] Stefan did not like school, but he did not exaggerate. A government inquiry was undertaken to find the reason for the increasing incidence of poor eyesight among the students; there was fear that their eventual use in the army might be adversely affected. Some attempts were then made to improve lighting in the classrooms, notably the introduction of the Welsbach gas mantles. Like many an earnest student, Erwin had to wear increasingly strong glasses from about the age of twelve to correct his nearsightedness. But he was inspired by the joy of intellectual progress, and neither prosaic teaching nor dismal surroundings affected his spirits.

The schoolwork was easy for him and he had ample time for other things. From entry till graduation in 1906, he was always first in his class. Later a schoolmate wrote:

I can't recall a single instance in which our Primus ever could not answer a question. Thus we all knew that he took in everything during the instruction, understood everything, he was not a grind or a swot. Especially in physics and mathematics, Schrödinger had a gift for understanding that allowed him, without any homework, immediately and directly to comprehend all the material during the class hours and to apply it. After the lecture of our professor Neumann – who taught both subjects during the last three Gymnasium years – it was possible for him to call Schrödinger immediately to the blackboard and to set him problems, which he solved with playful facility. For us average students, mathematics and physics were frightful subjects, but they were his preferred fields of knowledge.[17]

Actually on one occasion he did fail to answer a question in class: 'What is the capital of Montenegro?' He became terribly embarrassed, his face turned brick red, and the whole class suffered with him. (The answer was 'Podgorica', now called Titograd.) Also he did not really dispense with homework. If anyone in the house asked where he was, the answer was always 'He is upstairs in his room and is studying.' He had two small rooms of his own overlooking the courtyard at the back of the house.

The classes were held from eight in the morning till one in the afternoon six days a week, allowing twenty-eight hours a week of instruction. When Erwin attended the Akademisches Gymnasium the enrolment was about four hundred students, of whom more than a third professed the Jewish religion, although Jews comprised only about eight percent of the city population. The city government was under the Christian Socialist mayor, Karl Lueger, who had been elected on an antisemitic platform, but there was little evidence of antisemitism in the school. Even though Lueger appealed to the voters with antisemitic speeches, his administration was in fact almost free of overt discrimination and he enjoyed friendly relations with many influential Jews. In general, however, the Christian students in the Gymnasien came from wealthier families than the Jewish ones, and there was little social intercourse between the two groups.[18]

On two afternoons a week, Erwin returned to school for instruction in the Lutheran religion. 'From this I learned many things, but not religion.' He had also received the required religious instruction from his tutor as a preparation for the Gymnasium. His favorite question at the end of a bible story was 'Herr Teacher, do you really believe that?' Thus it was not for lack of formal catechism and bible studies that Erwin became indifferent, sometimes even inimical, to organized religious beliefs and practices. His mother was mildly religious, but he learned his negative attitude from his father, and then reinforced it through his own reading, observation and experience. Among the many anecdotes about Erwin's boyhood, Aunt Minnie recalled not one dealing with a religious observance, not even at Christmas. The Schrödingers never entered a church except to be married or to be buried, or to attend similar rites for their friends.

Nevertheless, Erwin had a great respect for saints and mystics. He became anticlerical rather than basically antireligious, yet he seems never to have made a serious study of Christian theology. It was the history of the Church rather than its beliefs that most distressed him. 'For centuries oppressed in the most shameful ways by the Churches, Science has held up its head, and believing in its sacred right, its divine mission, it has struck powerful, hate-filled blows against its old torturers, not caring that they – even if inadequate, indeed derelict in their duties – for all that were the only appointed guardians of the most holy legacy of the fathers.' The adjectives 'sacred', 'divine', and 'holy' in this example from one of his mature statements evidence a subconscious religious spirit beneath the overt anticlericalism.

During his free time in the afternoons, Erwin continued his study of English and began to learn French, subjects not taught at school. He found the philosophical propaedeutic to be badly taught, and during these classes he used to conceal his French translation in his notebook, and work on it while the teacher thought he was assiduously taking notes.

Another subject not included in the curriculum, except for its cursory denunciation as a heresy in the religious class, was the Darwinian theory of evolution. He was able to go into this subject at considerable depth during walks and botanical excursions with his father. 'On the basis of his botany, my father advised caution. The melding of natural selection and the survival of the fittest with the De Vries mutation theory had not yet been made. I don't know why zoologists were more ardent Darwinians than botanists. My father's friend, Hofrat Anton Handlirsch, a zoologist at the Natural History Museum, taught that development is causal not "final", there is no entelechy.' Erwin was soon converted to this view. 'Naturally I was an enthusiastic Darwinian, which I still am today' [1960].[14]

His best friend in high school was Tonio Rella, who later became professor of mathematics at the Technical University in Vienna. Tonio was always second and Erwin first in class during all their eight years at the Gymnasium. In the last days of the war, in April, 1945, Rella was killed by a stray Russian shell, so that the reunion that Schrödinger had anticipated when he returned to Austria never took place.[19]

The Rella family owned a country inn, Kastell Küb, in Semmering, and Erwin used often to spend holidays with them there. The inn, surrounded by extensive wooded grounds with fine views of the Schneeberg, could accommodate 120 persons, and was called a 'rest and recreation home' [*Erholungsheim*]. Both boys were ardent hikers and mountain climbers. Grandfather Bauer had been one of the discoverers of the beauties of Semmering and one of his most loved pursuits was the ascent of 'Sonnenwendstein', the mountain that sheltered the town. Erwin and Tonio often followed the trails that he had blazed forty years earlier.

In the course of time, there came to be another attraction at Semmering, more powerful than the lure of the mountains. This was Tonio's sister Lotte, called 'Weibi', with whom, Erwin later said 'I was fairly permanently in love' [ziemlich dauerhaft verliebt].[20] Weibi was an Italianate beauty with dark brown eyes and a generous figure; in an earlier time, she might have been a model for a Caravaggio portrait.

Erwin used almost always spend his Christmas holiday from about December 22 till the new year at Semmering with the Rella family, but for Christmas eve he would have to return to Vienna to be with his own family, including grandfather and aunts. He was always unhappy to leave his sweetheart, and anyone who can recall an adolescent first love will understand exactly how he felt, a reluctant member of the family circle whose thoughts were wandering elsewhere. In those days, such a teenage romance would be limited to worship from afar, holding the beloved for a Viennese waltz, perhaps a gentle kiss if rare opportunity occurred, but the sister of a school-friend would be inviolable and a more intimate relationship only a subject such as dreams are made of. Lotte's parents would have been happy to encourage a match with Erwin. His devotion to her was more than a mere *Kinderliebe*, and it would leave an indelible impression on his soul, but he was not prepared at this time to consider any enduring emotional ties.

Tonio Rella, Erwin, Mother Rella, and Friend – at Semmering.

A world of Theater

For the students, the years at the Gymnasium spanned the onset of puberty and the pangs of adolescence. They entered as ten-year-old children and left as sexually mature young men of eighteen. Sexuality was an anarchic factor and bourgeois society protected itself by establishing a dual system of behavior. School, family, popular literature and newspapers formed a world in which sex hardly existed outside the privacy of the conjugal bedroom. Parallel with this conventional world of bourgeois morality, there was a flourishing underworld of pornography and prostitution. Sex for pay was not, however, confined to casual pickups and brothels. For the wealthy, there was a demimonde of music halls and artists' studios where attractive mistresses awaited those able to support them, but such pleasures were outside the financial range of most students.

Karl Marx accused bourgeois men of preying on the daughters of the proletariat, but it was often an unresisting prey. In many autobiographies, the sexual initiation of the hero is provided by the family maid, usually portrayed as a fresh country girl whose upbringing on the farm had left her innocent of bourgeois morality and susceptible to the sophisticated urban student. There was also said to exist a charming institution known as *Das süsse Wiener Mädel*, perhaps a shopgirl whose greatest ambition was a sacrificial affair with a university student, to whom she could devote her generous love with never a thought of complications like pregnancy or marriage. Arthur Schnitzler was particularly expert on the subject of the 'sweet Vienna lass'.[21]

The marvelous Vienna theater, then at one of the high points in its history, provided Erwin with at least vicarious experience in the world of romance, in the works of the great dramatic poets at the Hof-Burg Theater on the Ringstrasse.[22] Josef Kainz, believed by many critics to be the greatest actor of all times, was creating some of his most famous roles, Hamlet, Cyrano, Tasso, Orestes, Mephisto, Don Carlos. The emperor's favorite, Katherina Schratt, was one of the leading ladies. Vienna audiences and critics were the most knowledgable in the world; no *Schlamperei* [sloppy work] was tolerated. Special matinees were held on Sunday afternoons for students and workers' alliances. As Stefan Zweig remarked, this theater was more than a mere stage for plays, it was a microcosm in which Austrian society could see itself. Arthur Schnitzler's *Liebelei*, the story of Christine, the most famous süsse Wiener Mädel of literature, was first produced in 1895. Sigmund Freud wrote to the author: 'Your fascination with the subconscious, with the instinctual nature of human beings, your destruction of the conventional certainties of civilization, the attachment of your ideas to the polarities of love and death – all this affects me with an uncanny familiarity.'[23]

Erwin was a lover of the theater, going as often as he could, sometimes more than once a week, and keeping an annotated record of the performances. As a student, his favorite playright was Franz Serafikus Grillparzer, who was born in Vienna in 1791, the year Mozart died. A shy man, torn all his life between the demands of his work and romantic loves for beautiful but inaccessible women, he was a master of erotic drama.[24] An example from his *Sappho*:

> He who knows what love, what life is, man and woman,
> Does not weigh man's love 'gainst woman's passion.
> Most fickle is man's fitful disposition,
> Subservient to life, most fickle life.

From the later *The Waves of Sea and Love*, the story of Hero and Leander,

> And he, my prince of swimmers and of love,
> Found in the cosmos neither love nor pity.
> He raised his eyes, beseeching to the gods,
> In vain! They did not hear – or were they sleeping?

Sometimes Grillparzer provided simply a great adventure story, as in *The Argonauts*:

> Who sets his heart on snatching the golden fleece
> That death envelops and most potent perils? . . .
> In the cave it lies concealed, defended by all the horrors
> Of cunning and violence . . . daggers beneath each footstep,
> Death by inhalation! Murder in a thousand shapes.
> And the fleece hangs on a tree, besmeared with poison,
> Guarded by the serpent who never sleeps,
> Unrelenting, Unapproachable.

These were stirring dramas for an adolescent student. They must have helped to convince him that there is more to life than books and theory. And should he wish a change from the splendors of the Hof-Burg Theater, there were many lesser companies always playing in Vienna, the Volkstheater, which could seat 1900 spectators in its red, white and gold auditorium, the Carl Theater which might offer a Strauss operetta or an Ibsen play, or the Josefstadt Theater which might be showing a Parisian farce, an avant-garde piece of Strindberg, Hungarian vaudeville, or even a visiting production by Max Reinhardt.

Some pages from Erwin's *Theater Notizbuch* have been preserved, dating from the autumn and winter of 1904/05, his next-to-final year in high school.[25] The first notice is of one of the greatest dramas in German literature, Schiller's *Wallenstein*, which he found most impressive with Sonnenthal in the principal role at the Burgtheater.

On September 3, 1904, the play was at the Raimund Theater, *Herodes and Mariamne*, a tragedy in five acts of blank verse by Friedrich Hebbel (1813–1863). King Herod was played by Wiecke, and Erwin thought that he surpassed even Kainz, having more temperament and more warmth, 'a thin figure of unbelievable elasticity'. The grim story is derived from Josephus, and details the hopeless efforts of the faithful queen Mariamne to convince the vicious and brutal king of her loyalty. Finally he orders her execution, and driven by demonic fear and suspicion proceeds to the slaughter of the holy innocents.

On October 8, Erwin got a seat in the uppermost gallery, 'with the perspective of a bird', to see Hebbel's *Gyges and His Ring*, with Kainz as Kandaules, Reimers as Gyges, and Rompler-Bleibtreu as Rhodope. The story is based on Herodotus and set in prehistoric times. A handsome young Greek, Gyges, presents Kandaules, king of Lydia, with a ring that makes its wearer invisible. The king then allows him, while wearing the ring, to enter the royal bedchamber and gaze upon the beauty of the queen, Rhodope. His sigh of admiration reveals his presence to Rhodope, and since only one living man is allowed to see her unveiled, Gyges and Kandaules fight a duel to the death in which the king is killed. Gyges must then marry the queen, but as soon as the ceremony is performed, Rhodope kills herself to purify her soul. Erwin commented that 'I do not think that a young person of my age can understand this play, not so much the external plot, but just as one can imagine the emotional processes of others when one has experienced them even once in any form, so also with the stage one can enter into an emotional feeling, if one has even once felt something similar, and that obviously does not hold true for the emotional feelings of Rhodope.' Who is the hero of the play, he wonders; from the title, it seems to be Gyges, but actually Rhodope is central. 'The ring, like the golden fleece, is a symbol of superhuman power.' These are perceptive comments: Hebbel originally planned to call the play *Rhodope*. Even mature minds have been perplexed by its psychology and its symbolism; it was rejected by producers all over Europe, and finally performed at the Burgtheater long after the death of the author.

Erwin attended Schiller's *Wilhelm Tell*, a new production, and found Reimers in the title role to be 'much better than I had expected, not so much by what he does as by what he has left undone'.

On January 26, 1905, he saw Schiller's *Don Carlos*, a marathon production, which began at 6:30 p.m. and ended at 12:30 a.m. The title role was played by Kainz, not entirely to Erwin's satisfaction. 'I have read Schiller's *Letters about Don Carlos* with great interest and believe that I shall gradually come to a complete understanding of the play.' Three nights later, he was at the Deutsches Volkstheater to see *The Brothers of Saint Bernard* by Anton Oborn. 'It is a drama with a purpose – not a realistic picture of life.' 'Realistic', however, is not an appropriate adjective for any of the plays described by Erwin in this brief cross-section of his youthful theater-going in Vienna. The classical German drama takes a serious, heroic view of human existence, with powerful men and sacrificial women engaged in matters of momentous portent, usually trapped and destroyed by failures of communication, but going to their destruction with sound and fury. Schrödinger did see a production of *Hamlet* about this time, but one imagines that the subtle and ambiguous psychology of this play was hopelessly weighed down by the German translation and overpowered by the ornate style of the Burgtheater company.

The art world of Vienna at the turn of the century was even more flagrantly erotic than the avant-garde theater, since it had less need to appeal to a large bourgeois audience. In 1897, Gustav Klimt led the revolt of the young artists in founding the movement known as the Secession. Their magazine was called *Ver Sacrum* [Sacred Spring], to recall a pagan Roman ritual in which youths were consecrated to the nation at times of danger. As Carl Schorske aptly said, 'Where in Rome the elders pledged their children to a divine mission to save society, in Vienna the young pledged themselves to save culture from their elders.'[26] In 1894, Klimt, then a brilliant but fairly conventional artist, received a commission to paint three heroic ceiling paintings for the great hall of the new university, to represent philosophy, medicine, and jurisprudence. By the time the first two paintings were completed, in 1900 and 1901, he had fervently embraced the ideas of Wagner, Nietzsche, and Schopenhauer, and translated them into powerful symbolic pictures, filled with disturbingly erotic female figures who seemed to drown the masculine arts of philosophy and medicine in a sea of hopeless sexuality. This was not what either the university professors or the liberal government had anticipated, and a storm of public protest and vicious personal abuse broke over the unrepentant artist. He was accused by the gutter press of being a mere pornographer, by the clericals of being in league with seditious Jewish philosophers, by the academics of being anti-intellectual, and even the social democrats bemoaned his lack of optimism for the future of society. The controversy raged for some years; the government tried to stand firm but the paintings had to be removed to the Modern Art Museum and never graced the university. In 1906 Klimt's friend Egon Schiele was thrown into jail for twenty-four days for painting a lewd picture of a young model, and one of his pictures was publicly burned. It would be interesting to know what Rudolf Schrödinger thought of all this. His own pictures were pleasantly academic and untouched by any of the influences sweeping through contemporary European art.

How was Erwin affected as an adolescent by the erotic art of avant-garde Vienna? Among the papers that he kept all his life were copies of the journal of the university Science Club. The issue for 1909 has an erotic drawing more or less copied from Klimt's 'Fish Blood', which appeared in the first issue of *Ver Sacrum*.[27] Like so many of Klimt's drawings and paintings, it is a desperate effort 'to capture the feeling of femaleness'. Much of Erwin's erotic life would be devoted to a similar effort, not to master or dominate women, but to capture the essence of their sensuality by experiencing it with them and through them.

All this was for the future. The avant garde was certainly not the world of Erwin Schrödinger and Tonio Rella when they entered the university together in the fall of 1906. Erwin's romance with Lotte Rella continued, but it remained within the bounds set by conventional morality. We know this from his *Ephemeridae*, where he carefully recorded the names of all his loves with a code to indicate the denouement.

2   University of Vienna

Vienna is the second-oldest German-speaking university, founded in 1365 by Pope Urban V, seventeen years after the University of Prague. Albert of Saxony, formerly rector of the University of Paris, brought the hitherto secret statutes of that establishment to Vienna as a basis for the new university, and he became its first rector. Students came from all over Europe, but especially from the four nations, Austria, Bohemia, Saxony, and Hungary.[1]

In the sixteenth century, the university joined the battle against Lutheranism, but not without misgivings, since most of the Viennese had become convinced Protestants. In 1550, the Jesuits were called into Vienna, where they established a number of excellent lower schools. The counter-reformation now began in earnest, with many expulsions and a few burnings at the stake, until Catholicism was firmly re-established in the Habsburg lands after a struggle of one hundred years. The professors for the most part resisted the clerics until 1617, when the university was formally united with the Society of Jesus. The Society was a militant order, and no sooner had it achieved its aims, than it began to lose interest in the university and to look for other enemies, which it found in the Dominicans who had refused to swear an oath to the Immaculate Conception of Mary.

In 1683 the Turks appeared in force at the gates of the city. The court fled but the burghers and soldiers defeated the Turkish army, and this siege marked the high point of the Moslem invasions, as the power of the crescent gradually receded from Europe. One thing Vienna learned from the Turk was a love for coffee, and coffee houses soon rivaled university class rooms as centers of intellectual activity.

Charles VI made extensive reforms in the university studies, which he directed more toward the service of the state and less to that of the church. His daughter, the Empress Maria Theresia, who gave Francis I five sons and eleven daughters, embraced the spirit of the Enlightenment with increasing fervor and encouraged the university to develop in new directions. She also provided funds for a magnificent new building, which was opened in 1756 with great celebrations. In 1780,

32

Josef II issued a patent of tolerance, which abrogated the Catholic oath; Protestants and even Jews were admitted to degrees, but not women of any persuasion.

The first professorship of physics [natural philosophy] was established at the University of Vienna in 1554, but no equipment for experimental work was available till 1715 when a collection of apparatus was made by the Jesuits. Andreas Baumgartner, who became professor in 1823, was the first to give physics lectures in German instead of Latin. He helped to found the Vienna Academy of Sciences and served as its president for fourteen years. Through his efforts, the teaching of physics in Austria was modernized, and brought up to the standards of Germany, France, and Italy.

In the spring of 1848, revolutions swept through Europe. In Vienna, workers led by students of the 'Academic Legion' marched through the streets with red banners, hammers and scythes. A unit of the Viennese Grenadiers joined the rebels and the entire city was soon in their hands. Metternich, after forty years of power, announced his retirement and the Habsburgs fled to Innsbruck. By autumn, the forces of reaction were again in control, and Vienna was recaptured by a large Austrian army brought from northern Italy. The imperial army did not forget its ignominious defeat by the 'intellegentsia in arms' and it seized the university and dispersed the staff into makeshift buildings in outer districts. Some concession to the prevailing discontent seemed to be required, however, and it was decided that the half-witted Emperor Ferdinand must abdicate in favor of his eighteen-year-old nephew Franz Joseph.

University physics[2]

In 1850, Christian Doppler, a native of Salzburg, was appointed to the chair of physics. He was famous for his discovery that if a source of light is in motion relative to an observer, the wavelength as received by the observer is different from that emitted by the source. The effect on sound waves is familiar to everyone who has listened to the mournful notes of a passing locomotive, but Doppler analogically applied the same principle to light waves moving through 'the aether'. Schrödinger, who was always interested in anything related to the Vienna school of physics, published a paper on the Doppler effect many years later.

When Doppler retired in 1853, Andreas Ettinghausen, who had served from 1835 to 1848, was reappointed. He supervised the move of the Institute to new quarters, an inadequate building at Erdbergstrasse 15, in the Third District of Vienna, actually not far from where Erwin Schrödinger was born.

The problem of a suitable home for the university could not be solved until a liberal government came to power in 1868. Even then the army clung to 200 hectares of prime building site on the Ringstrasse, which it was using as a parade ground. In 1870, Mayor Felder finally won the approval of the emperor for a grandiose building plan, which accommodated the University, the Parliament, and the City Hall [*Rathaus*] on the old Champ de Mars. The university building, large and ornate in the style of the Italian Renaissance, was completed in 1884, but it did not help the physicists, who were still exiled in Erdberg, although the experimental laboratories had been moved in 1875 to another temporary building on Türkenstrasse, not far from the new university site.

The successor of Ettinghausen to the chair of physics was Josef Stefan, in 1863. His background was unusual, in that his parents were country folk who could neither read nor write, yet his academic brilliance had surmounted all handicaps. Stefan's most famous discovery was concerned with the properties of what was then called 'black radiation'. All bodies are continuously absorbing and emitting radiation. When a body is in equilibrium with its environment, the radiation it is emitting must be equivalent in wavelength and energy to the radiation it is absorbing. It is possible to imagine a body that is a perfect absorber of all the radiation incident upon it, an 'ideal black body'. Stefan, in 1879, discovered experimentally that the energy of this radiation per unit volume (density of radiation) depends upon the fourth power of the absolute temperature. A few years later, a former student of Stefan's, Ludwig Boltzmann, then at Graz, derived the law theoretically, and it became one of the corner stones of radiation physics, usually called the Stefan–Boltzmann equation, $\sigma = aT^4$. The interpretation of the wavelength distribution of the black radiation by Max Planck led directly to the quantum theory.

Another important scientist at the university at this time was Josef Loschmidt. While an assistant to Stefan, he made a major contribution to science, one that required neither experimental nor mathematical skill, but simply the ability to see an important relation that others had missed. As a result of the work of James Clerk Maxwell and Rudolf Clausius, the kinetic molecular theory of gases was becoming generally accepted: a gas consists mostly of empty space in which tiny molecules are flying about at a great rate, colliding with one another and with the walls of their container. The average effect of the collisions with the walls causes the pressure exerted by the gas. A difficulty that prevented universal confidence in this model was that nobody knew either how many molecules are in a given container of gas or what is the size of the molecules of any particular gas. Loschmidt made the necessary calculations of these quantities and reported the results at a meeting of the Vienna Academy. This was the most exciting thing that had ever happened to molecules – for most physicists, they ceased being hypotheses and became realities. Loschmidt, in rapid succession, was awarded an honorary doctor's degree, appointed to an associate professorship in physics, and then to a professorship of physical chemistry.

Like anyone else, a scientist can have only one biological father, but he can have several scientific fathers, whose influence on his life may be even more important. Thus the history of physics in Vienna traces the scientific ancestors of Schrödinger. He had two scientific fathers, Franz S. Exner in experimental physics and Friedrich (Fritz) Hasenöhrl in theoretical physics. Loschmidt and Stefan were scientific grandfathers of Schrödinger on the Exner side.

The most important influence on the development of Exner as a physicist was, however, Viktor Lang. Lang was another distinguished graduate of the Akademisches Gymnasium. He studied at the Vienna University but took his doctorate at Giessen in Germany, and then worked in Paris with the great experimental physical chemist Henri Regnault, and in Heidelberg with Robert Bunsen and Gustav Kirchhoff on spectroscopy. At the early age of twenty-eight, he was called to a professorship in Vienna, which he held for forty-four years. His principal work was in the field of crystal physics. In 1874, he obtained an assistantship for the Institute, to which he appointed Franz S. Exner.

Exner was appointed associate professor in 1879, and in 1891, when Loschmidt retired, he became professor and also director of the Physical Chemistry Institute. At that time he had a choice among Graz, Innsbruck and Vienna, but he chose to stay in Vienna because he was looking forward soon to the building of a new Institute of Physics to replace the inadequate facilities at Erdberg, a hope that was not to be realized for twenty years. Exner worked on many different physical problems, and he inspired his students with an enthusiasm for these problems, so that the Exner Circle, as they came to be called, shared not only a common university background but also common interests in their research fields. These included electrochemistry, atmospheric electricity, radioactivity, crystal physics, spectroscopy, and the science of color [*Farbenlehre*]. It would not be possible today for one person to do experimental physics in so many different fields, for one thing the cost of equipment would be prohibitive, but Exner had practically no equipment except what he made himself, and consequently his scope of activity was less restricted. Erwin Schrödinger was to work in all the Exner fields except electrochemistry, but usually on theoretical problems, where the only equipment needed was pencil and paper.[3]

Professor Franz Exner.

In 1898 the Vienna Academy, on the recommendation of Exner, arranged for the gift to Marie and Pierre Curie of 100 kg of pitchblende residues from the St Joachim uranium mines. From this material, the Curies achieved the first isolation of radium, and they sent Exner a small sample of highly enriched material. With this the Austrian scientists accomplished some noteworthy researches, so that in 1901 the Academy established a Committee for Radioactive Substances, with Exner as chairman, and Suess, Boltzmann, Lang, and Welsbach as members. Welsbach was a brilliant inorganic chemist who had worked with rare earths and made a number of very profitable inventions, including the Welsbach mantle based on thorium oxide which provided the beautiful illumination of the gaslight era. The Academy bought 10 000 kg of residues from St Joachim, and the Welsbach factory processed them to yield 4 g of pure radium chloride, the analytical controls being done at Exner's Institute on the Türken-strasse by Stefan Meyer and Egon Schweidler. In 1907 the Academy sent Rutherford and Ramsay in Manchester 400 mg of radium bromide, and their great discoveries were all made with Viennese radium. In 1908, Karl Kupelwieser, son of a famous painter, gave the Academy 500 000 crowns to construct a building for an Institute for Radium Research to be located next to the new Institute of Physics.[4]

Ludwig Boltzmann[5]

We have described the scientific ancestry of Erwin Schrödinger on the experimental side, and now must go back to consider his even more important scientific forebears on the theoretical side. Here the father was Friedrich (Fritz) Hasenöhrl and the grandfather was one of the greatest theoretical physicists of all times, Ludwig Boltzmann. A scientific family tree is not subject to biological constraints, so that Hasenöhrl was not only Schrödinger's father but also his elder brother, having also taken his first degree under the supervision of Exner, in 1896.

Boltzmann was born in Vienna on February 20, 1844, the son of a 'K & K Cameral Concipist', in other words, a taxation official. The date was Shrove Tuesday, and half seriously he used to ascribe the sudden changes in his spirit between happiness and affliction to the fact that he was born during the dying hours of a gay Mardi Gras ball. He was a typical cyclothyme, short and stout with curly hair; his fiancée used to call him her 'sweet fat darling'. He received his doctorate from the University of Vienna in 1866 for work under Stefan on the kinetic theory of gases. His genius was soon recognized and at the early age of twenty-five he became professor of mathematical physics at the University of Graz.

Boltzmann's temperament was such that he never stayed very long in one place, but wherever he was, he continued to make major advances in the kinetic theory of gases. Quite independently of each other, Boltzmann and Willard Gibbs of Yale created statistical mechanics. This is the science that forms the connecting link between the small-scale world of atoms and molecules and the large-scale world of gases, liquids, and solids. The laws of the large-scale world, for example, the dependence of the pressure of a gas on its volume and temperature, can be derived as the statistical consequence of the mechanical behavior of the enormous number of molecules in any macroscopic volume. So far as the large-scale world is concerned, the positions and velocities of individual gas molecules are *hidden variables* which underlie the ordinary laws of gas behavior but do not explicitly appear in the gas laws. A considerable part of Schrödinger's research was devoted to statistical mechanics, his publications in this field filling the 514 pages of Volume I of his *Collected Works*.

In Botzmann's time, chemists and physicists were much concerned with the question of the 'reality' of atoms and molecules. This concern was not the old philosophic conflict between 'realism' and 'idealism', although this too was mixed up in the scientific question. Many scientists, however, then believed that diamonds and carbon atoms are entities of a qualitatively different kind. Diamonds 'really do exist' and can be bought and sold, but carbon atoms are only theoretical concepts that are useful to derive mathematical equations. Most would admit that a tiny diamond, invisible to the naked eye but observed with a microscope, is just as real as a five-carat gem, but beyond that the fundamental building stuff of nature was energy, which could not really be broken up into little particles called 'atoms' or 'molecules'. For Boltzmann, however, atoms were every bit as real as diamonds.

In 1895, at a conference in Lübeck, an attempt was made to resolve these conflicting views of the fundamental structure of the world. The report in favor of energetics was given by Georg Helm of Dresden; behind him stood Wilhelm Ostwald of Leipzig, the leader of physical chemistry, and behind both was ranged the powerful positivist philosophy of the absent Ernst Mach. The leading opponent of energetics was Boltzmann, seconded by the mathematician Felix Klein. Arnold Sommerfeld reported that the struggle between Boltzmann and Ostwald equaled outwardly and inwardly 'the struggle of the bull with the supple matador. But this time the bull conquered the matador despite all his finesse. The arguments of Boltzmann drove through. All the young mathematicians stood on his side.'

When Stefan died in 1894, Boltzmann was appointed to his chair at Vienna University. The next year Ernst Mach was appointed to the chair of history and philosophy of science. Boltzmann was a good friend of Ostwald, despite Ostwald's opposition to atomism, but he could not stand Mach. Thus in 1900, he accepted a chair of theoretical physics at Leipzig. In 1901, however, Mach retired because of ill health and Boltzmann returned to Vienna, since his former chair there had not been filled. Rather ironically, considering his dim view of most philosophers, he was also assigned the course in philosophy that had been given by Mach. He was a realist and a materialist in philosophy, but in personal relationships he was very soft-hearted, he never failed a student. An ardent supporter of Darwinism, he saw in evolution the mechanism by which inanimate atoms had evolved into mechanical structures called human brains, capable of love, pity, and artistic creativity. Modern molecular biologists like Francis Crick and Jacques Monod would have felt perfectly at home with Ludwig Boltzmann. His philosophical lectures were greeted with great public enthusiasm, the audience overflowing the largest lecture hall available. Even Franz Josef heard about them and he invited Boltzmann for an interview at his palace.

The attacks on atomism continued, however, and Boltzmann was called 'the last pillar of that bold edifice of thought'. As his health declined, he became more depressed, feeling each tremor of what he began to believe was a tottering edifice that would collapse with all his life's work under the rubble. In the summer of 1906 he went for a holiday to the beautiful Bay of Duino near Trieste and committed suicide by hanging while his wife and daughter were enjoying a swim. As Rilke wrote in one of his Duino Elegies: 'Aber die Liebenden nimmt die erschöpfte Natur in sich zurück, als waren nicht zweimal die Kräfte, dieses zu leisten.' [But exhausted Nature takes lovers back into herself, as though there were never the powers to create such a thing twice.]

Although it was sometimes said that suicide was a way of life in Vienna, the loss of Boltzmann was a terrible shock to all the members of the physics department. Erwin Schrödinger was personally broken hearted, for he had expected to begin his studies in theoretical physics within a few months under the great master. He recalled his feelings of that autumn, when he entered the physics building: 'The old Vienna Institute, from which shortly before Ludwig Boltzmann had been torn away in a tragic fashion, the building where Fritz Hasenöhrl and Franz Exner worked and many another of Boltzmann's pupils went in and out, engendered in me a direct empathy for the ideas of that powerful spirit. For me his range of ideas played the role of a scientific young love, and no other has ever again held me so spellbound.'[6]

Fritz Hasenöhrl[7]

Lectures in theoretical physics at the university were suspended for eighteen months, until Hasenöhrl was appointed to the vacant professorship. Like most of his colleagues he was from the comfortable middle class; he had gone to the conservative Theresianum Gymnasium and graduated from the university in 1896. His thesis with Exner was on the temperature dependence of the dielectric constant of liquids and solids. In 1898, Kammerlingh-Onnes, founder of the famous low-temperature laboratory at Leyden, had asked Boltzmann to recommend a research assistant for a year, and Boltzmann sent him Hasenöhrl as his 'best student'. At Leyden he also came to know the great Dutch theoretician Hendrik Lorentz, who has been called the spiritual link between Maxwell and Einstein. He returned to Vienna in 1899 as Privatdozent and in 1904 published his most notable paper, which in some ways anticipated the famous equivalence of mass and energy, $E = c^2 m$, derived from Einstein's special theory of relativity [1905]. Hasenöhrl's theory had nothing to do with relativity; it was based upon the idea that inertial mass has an electromagnetic origin. He showed that a moving charged particle has an inertial mass that depends upon its velocity and that a volume [*Hohlraum*] of black radiation also has a mass proportional to its energy. Unfortunately he did not get the correct proportionality factor c^2, otherwise he would have achieved greater renown. He was thirty-three years old when he succeeded Boltzmann as professor of theoretical physics in 1907.

His inaugural lecture was a masterly synthesis of the statistical theories of Boltzmann and an exposition of the philosophy of that great master. Erwin Schrödinger listened with wonder and excitement. He resolved to make mathematical physics his life work, following in the footsteps of these great Austrian scientists.

Social relationships tended to be quite close between families in the academic and professional world of imperial Vienna at the turn of the century. Doctors, lawyers, government officials, and professors belonged to the comfortable upper-middle class, with spacious but not elegant housing, at least one but usually two servants, money for occasional travel to visit the art centers of Italy, and time for leisurely vacations in the Tirolean mountains. The men had known one another in high school and at the university, where their lifelong friendships were made. They often married daughters of their father's confreres. This was not the world of the haute bourgeoisie like the Wittgensteins, with their country estates and great town house with seven grand pianos. Sometimes a professor sported a 'von' in front of his name, but this was usually no indication of landed aristocracy; it meant only that some ancestor had been useful to the government. This was the world that Alexander Bauer, Erwin's grandfather, belonged to, and the world that Erwin was destined for, once he graduated first in his class at the Academic High School and entered the University of Vienna to study physics and mathematics.

Ernst Mach

Although Boltzmann and Mach engaged in a hard-fought battle over the reality of atoms, most of the physicists seemed to have little difficulty in accepting the philosophical ideas of the latter about the basic nature of science, while using the methods of the former in their daily work in theoretical physics. As Schrödinger explained:

Filled with a great admiration of the candid and incorruptible struggle for truth in both of them, we did not consider them irreconcilable. Boltzmann's ideal consisted in forming absolutely clear, almost naively clear and detailed 'pictures' – mainly in order to be quite sure of avoiding contradictory assumptions. Mach's ideal was the cautious synthesis of observational facts that can, if desired, be traced back till the plain, crude sensual perception . . . However, we decided for ourselves that these were just different methods of attack, and that one was quite permitted to follow one or the other provided one did not lose sight of the important principles . . . of the other one.[8]

The most important scientific research of Schrödinger in the period 1919–24 was in the field of color theory [*Farbenlehre*], and the subject matter and method of approach appear to have been motivated by Machian principles.

In July, 1898, Ernst Mach was taking the train to Semmering when he suffered a stroke that paralyzed his right side and prevented him from speaking clearly. He was then at the height of his influence in Vienna, both in the university and with a wider audience reached through his public lectures and more popular writings. Unable to lecture, he retired from his professorship and early in 1901 was appointed by the emperor to the upper house of the Austrian legislature, a position he held till 1913. He spent most of his time in seclusion at home, and probably Erwin never met him personally.

Ernst Mach was born in 1838 in Chirlitz, in Moravia; when he was a small child the family moved to Vienna, where his father made a poor living as a tutor. At the age of nine, he entered the Benedictine Gymnasium, but he was dismissed at the end of his first year as being 'completely without talent'. He was tutored by his father and did not challenge the K & K educational system again until he entered the sixth form at the age of fifteen and satisfactorily completed the *Matura* that allowed entry to Vienna University, where he studied physics and mathematics under Petzval.[9]

His father was a 'free thinker' and Ernst, though nominally a Catholic, never had much interest in organized religion, although he once acted as godfather to Wolfgang Pauli. He had a quasimystical experience at about the age of seventeen: 'On one bright summer day in the open air, the whole world with my ego suddenly appeared to me as a coherent mass of sensations, only more strongly coherent in the ego. Although the actual working-out of this thought did not occur until a later period, yet this moment was decisive for my whole view of the world.'[10]

What was Mach's view of the world? John Blackmore has given a philosopher's answer to this question.[9] The first basic component of a philosophical view is epistemology [*Erkenntnistheorie*], what and how does one know? The second basic philosophical component is ontology, what exists? Although Mach's views developed and changed somewhat in the course of his life, his mature position was that all we can know of the world are what he called *elements*. The elements include sensory perceptions but in addition other things that can be reliably measured, such as space and time. Elements that refer to the human body are called *sensations*. The world is directly *presented* to us in terms of these elements. Thus the epistemology of Mach is called *presentational phenomenalism*. It is important to distinguish this from *representational phenomenalism*, which states that the external world cannot be directly sensed, but is *represented* to consciousness, so that all appearances are actually mental, but a real external world can be reliably known through inference from empirical evidence of the senses. Mach was not only a presentational phenomenalist in his epistemology, but also in his ontology. A world consisting of the elements is what actually exists; there is no other 'real world' hiding behind the phenomena, nor is there any ego or self hiding in the consciousness of human beings. The ego is simply a group of sensations like an apple or an iceberg. The ego of an individual is part of an inclusive ego of all the individuals in the world. We know by 'irresistible analogy' that other persons have thoughts and sensations similar to our own. Mach thus believed that he could escape from the charge of solipsism by denying the existence of an individual self. 'The philosopher who is a solipsist seems to me to be like the man who gave up turning around because whatever he saw was always in front of him.'

Mach believed that the aim of science was the ordering of the elements in *the most economical way*.

The world consists of colors, tones, warmths, pressures, spaces, times, etc., which now we do not want to call 'sensations' or 'phenomena', because in both names there lies a one-sided arbitrary theory. We call them simply 'elements'. The comprehension of the flux of these elements, whether directly or indirectly, is the actual aim of science. So long as we, not considering our own bodies, concern ourselves with the mutual dependence of certain groups of these elements, which include other bodies, men and animals, we remain physicists. We investigate, for example, a change in the red color of a body through changes in its illumination. As soon, however, as we consider the special influence of those elements that make up our own bodies . . . we are in the field of physiological psychology . . . That the world in this sense is our sensation is not subject to doubt.[10]

In 1860, Gustav Fechner published a two-volume treatise, *Elements of Psychophysics*, which taught that reality has a physical outside and a psychical inside, related to each other by a parallelism of events.[11] He invented the word 'psychophysics' to designate the study of this parallel reality. Mach adopted the idea of psychophysical parallelism, but he denied the existence of a separate outside and inside. 'There is no rift between the psychical and the physical, no "inside and outside", no sensation to which an external "thing" different from sensation corresponds. There is but one kind of element out of which this supposed inside and outside are formed – elements which are themselves inside or outside, according to the aspect in which, for the time being, they are viewed.'[12]

Mach recognized the utility of theories in suggesting connections between phenomena, but he did not think that they had any per- manent value: 'Theories are like dry leaves which fall away when they have long ceased to be the lungs of the tree of science.' In particular he distrusted the idea that atoms are anything but convenient symbols for summarizing experience. 'Certainly one must wonder how colors and tones, which are so close to us, can suddenly appear in a physical world of atoms, how we can suddenly be astonished that what outside so drily knocked and clattered, inside the head can sing and shine. How, we ask, can matter feel, which also means how can a thought symbol for a group of sensations become itself a sensation?'

Richard Avenarius (1843–1896), at the University of Zürich, had independently devised a philosophy called *empirio-criticism*, which was closely similar to Mach's phenomenalism, but which was pro- pounded in an exceptionally opaque style. After reading Mach, Schrödinger also studied in detail the work of Avenarius especially his books *Philosophy as Thinking about the World in Accord with the Principle of Least Energy* (1876), *Critique of Pure Experience* (1888),[13] *The Human Concept of the World* (1891).[14] Much of this reading was done during his long, boring months of wartime service in the fortress artillery. The most readable parts of these works are the introductions, for example:

Is the world really in such a state that only its superficial reflection shows itself with consistency and without contradiction, while it leads into error anyone who wants to comprehend it more precisely in its totality? . . . Or is the world basically consistent and noncontradictory, and is it really some evil spirit that leads in a circle precisely the determined thinker who makes every effort to be consistent? In the first case what is the origin of the inevitability of contradiction to which, up till now, every truly general consideration of the world appears to have led? In the second case, what is the evil spirit that leaves unsatisfied those who hunger and thirst after a true knowledge of the world?

Avenarius answers his first question 'no' and his second 'yes'. Pure experience gives us a natural concept of the world, and the evil spirit is the *introjection* of metaphysical elements into the act of knowing.

In Vienna, poets, artists, and even journalists found the ideas of Mach surprisingly congenial. It was partly a matter of politics. In 1897, Karl Lueger, 'beautiful Karl' [*der schöne Karl*], became mayor of Vienna, a position which he held until his death in 1910.[15] He was one of a few students from lower-middle-class families who managed to attend the aristocratic Theresianum Gymnasium. When he was forty-four years old, he promised his mother on her deathbed never to marry. Originally a liberal, he found the way to political power through antisemitism and a sort of Catholic anticapitalism in the Christian Socialist party. The avante-garde intellectuals mostly followed the socialist Social Democratic Party, which also had its antisemitic elements; Viktor Adler, founder of the party, was an antisemitic Jew, not unusual in Vienna. Mach was not active in politics, but when he spoke on the subject, he spoke as a socialist, and was opposed to militarism, economic injustice, and any form of racism. One of Mach's students was Viktor Adler's son Friedrich, who in 1916 assassinated the prime minister Karl Sturgkh; he was condemned to death but pardoned by the emperor Karl two years later. Of course this crime bears no obvious relation to Mach's teaching.

Another reason for Mach's popularity was that he was not a professional philosopher, and his ideas were expressed in clear and understandable language. Mach was a famous physicist, at a time when members of that profession were regarded as harbingers of progress. Also his philosophy came to be known as 'positivism', which has a pleasant connotation, and its later development was 'logical positivism', which sounds even better.

A final reason for the success of Mach in Vienna may be found in the persistent search of the romantic German mind for a monistic or unitary picture of man and nature. William Johnston has rather convincingly shown that the 'Austrian Mind' was passive, receptive and impressionistic, in short, it was 'feminine', having all those qualities that men at that time professed to find in women.[16] In Viennese society, appearances counted for everything, a doctor was judged by the cut of his suit, and having a beautiful corpse [*schöne Leich'*] was a major concern of the dying. The phenomenalism of Mach was well adapted to the superficial upper-middle-class society of Vienna.

After he retired, Mach became especially interested in the theories of gravitation and electromagnetism, since they both involve fields that vary inversely as the square of the distance from a source. He thought that it might be possible to devise a unified field theory that would supplant current atomic theories. In 1909, he was attracted to the recent work of Einstein and to a book by Paul Gerber, *Gravitation and Electricity*. He asked Professor Wirtinger at the university to find out from some of the theoretical physicists what they thought of the researches of Einstein and Gerber. Wirtinger passed this request along to Schrödinger and wrote to Mach on July 28, 1910. 'I have given Gerber's paper to a young electron man, who otherwise seems quite reasonable, and he offered to give me a detailed opinion in return. Dr E. Schrödinger has now written that detailed letter, and what seems striking to me is his objection that the whole thing [the relation between gravity and electromagnetism] is quite different when another kind of radioactive material is taken into serious consideration.'[17] We may never know what Erwin meant by this mysterious observation. He found parts of Gerber's treatise very obscure: 'Regrettably, I could not carry out my task in a very satisfactory way, since some of the essential points in Gerber's paper remain completely unclear.' Schrödinger and Einstein were both to devote many of their mature years to a fruitless search for the unity of electromagnetism and gravitation.

The ideas of Mach have had a great influence on the development of the philosophy of science in general and on the interpretation of quantum mechanics in particular. From 1920 to 1950, logical positivism, which was formulated by the 'Vienna Circle', originally the 'Mach Circle', reigned almost unchallenged as *the* scientific philosophy. Yet there are a number of fairly obvious defects in presentational phenomenalism. For instance, it fails to explain the close relationship between mathematical reasoning and theoretical physics; mathematical operations and symbols do not denote empirical sensations, and yet one cannot do science without them. Also, experiments are planned *interactions* of the scientist with the environment; how can they be explained as mere collections of sensations? Mach fails to explain the enormous predictive power of physical theories: how can it be that Dirac predicts a positive electron and Anderson finds it in a cloud chamber? Einstein and Planck, who were disciples of Mach in their younger days, eventually cast him aside. As a young man, Schrödinger was able to maintain a pragmatic balance between Machian positivism and Boltzmannian realism, but as he grew older this compromise no longer satisfied him and he sought but probably never achieved a synthesis of the two.

The university student

When Erwin entered the university in the fall of 1906, he brought with him his reputation from the Gymnasium as an outstanding student, in fact something of a genius, and this reputation was soon confirmed by his brilliant performances in mathematics and physics. Hans Thirring, who entered in 1907, recalls his first encounter with Erwin in the library of the mathematics department. A student entered, steely gray-blue eyes and a shock of blonde hair. Another student nudged Hans and said *Das ist der Schrödinger* – the Schrödinger. All the students regarded him as something special, but he was not cold and aloof, he often helped them with difficulties in maths or physics. He became a good friend of Hans, almost like a big brother. 'We saw in him a fire spirit at work, always breaking through to something original in every research.'[18]

His closest friend, indeed his only really close friend, at the university was not one of the physics students, but a student of botany, Franz Frimmel, whom he called Fränzel. They used to spend hours at a time talking together about philosophical questions, which Erwin at the time thought were quite original, but later realized were the same as those exciting the mind of every adolescent student. Sometimes in the evenings they took long walks about the city, engrossed in conversation, discussing their ideas about the meaning of life, not going home till the early hours of morning. His friend had a great reverence for religion, whereas Erwin at this time had a vehement dislike for it, and thought that Fränzel's 'religious teachers had high-handedly torn him from his true way of life', presumably by inhibiting his approach to biological problems.[19]

Fränzel had two brothers, both of whom became doctors. The younger, Silvio, his favorite, lived later in Krems. The elder, Egbert, who settled in Klagenfurt, was an accomplished mountaineer and once acted as a guide for young Erwin in an ascent of the Einser in the Sexten Dolomites, and brought him safely down again. For a reason about which one can only speculate, Erwin in his autobiography does not mention his name but refers to him only as 'E'. Erwin was always a great hiker, but he did not develop an interest in serious climbing in snow and ice. Like most of his university companions, he was a competent skier.

Erwin and Fränzel read together and discussed in detail the book of Richard Semon, *Die Mneme als erhaltendes Prinzip* [*The Mneme as Conservative Principle*], first published in 1904.[20] This is the only time that Erwin ever read a book in this way with another person. The book had a major influence on the development of his philosophical ideas, but it is difficult to discern all the values he must have found in it.

Perhaps his friend Fränzel was keen on the book and communicated his enthusiasm to Erwin. Years later, when he was reading philosophy on the Italian front, Erwin mentions Semon again as one of the important influences on his thinking. His book *What is Life?*, which had such a seminal effect on molecular biology, probably had its distant origin in those midnight discussions with Fränzel about the theory of living organisms.

Richard Wolfgang Semon was born in Berlin in 1859 into a wealthy Jewish family, but he later became a convert to Lutheranism, more as a tribute to German nationalism than from religious conviction. In 1883, he received a doctor's degree in anatomy from the University of Jena as the favorite pupil of Ernst Haeckel, and later he also obtained a doctorate in medicine.[21] At Jena, Semon learned the hodgepodge of darwinism, materialism and German *Naturphilosophie* to be found in Haeckel's book *The Perigenesis of the Plastidule and the Wave Generation of Life Particles.*[22] The plastidules were postulated molecules of basic living matter, which could be decomposed further into special kinds of atoms with souls capable of love and hate, memory and amnesia. 'The whole process of development of living organisms arises from a coordinated ramified undulation of the plastidules, in the course of which the individual waves become more and more differentiated.' It is difficult to do justice to some of Haeckel's bizarre theoretical ideas, but they did not prevent him from doing some excellent experimental work in comparative anatomy and embryology.

From Haeckel, Semon learned the fascination of the wide-ranging theory that tries to explain everything, but the basic philosophy of *Die Mneme* was derived from Lamarck. Erwin sold almost all the books inherited from his father, an action motivated by financial need, but one that he came to regret deeply. Among the few that he kept was *Philosophie Zoologique* by Jean Baptiste Lamarck, originally published in 1809. The author is famous as the founder of the doctrine of inheritance of acquired characteristics, although he was never dogmatic about this. Followers of Lamarck were divided into mechano-lamarckists and psycho-lamarckists. A modern example of the former was Trofim Lysenko. The *Just-So Stories* of Kipling include several lamarckian fables, such as how the elephant got its trunk after a baby elephant had its nose pulled by a crocodile. The question is to what extent changes in the phenotype produced by environmental conditions can affect the genotype, the hereditary information transmitted to the offspring. The consensus today is that such effects have at most a minor role in the hereditary process. In *Die Mneme*, Semon appears to accept lamarckism, but in a final chapter he praises the darwinian selective mechanism of evolution.

Psycho-lamarckism presents a more mysterious theory, which may be why it was so interesting to Erwin and Fränzel as late adolescents. When an organism reacts with its environment, a memory trace is produced in its mind, and this memory can be transmitted to its descendants, either through an effect on the germ cells or directly through some form of psychic inheritance. Instinctual behavior is said to be an example of this inherited memory. A similar idea had been adumbrated earlier by Ewald Hering, professor of physiology at Prague and a colleague of Ernst Mach, and it was given some currency in England by Samuel Butler, the author of *Erewhon*, in his book *Unconscious Memory*.[23] An eloquent modern exponent of psycho-lamarckism was the Swiss psychiatrist Carl Jung. As Einstein remarked in his obituary of Mach: 'Even those who think of themselves as Mach's opponents, hardly know how much of Mach's views they have, as it were, imbibed with their mother's milk.'[24]

In considering the first edition of Semon's book, the one available to Erwin and Fränzel as beginning students at the university, one must remember that Gregor Mendel's marvelous 1866 paper on the mechanism of inheritance, published in the *Transactions of the Brno Society for Natural Sciences*, was unknown to geneticists till 1900, when it was rediscovered by Hugo de Vries and Carl Correns. Charles Darwin himself never learned of Mendel's discovery of the hereditary units which are the basis of the evolutionary mechanism of natural selection. There is no indication that Semon when he wrote his book was aware of contemporary work on genetics, and thus he could let his fancy wander freely over the wildest analogies between memory and inheritance.

Semon's basic idea was that all biological phenomena can be ascribed to the existence in living cells of a sort of memory record called the *mneme*. Thus heredity, differentiation, regeneration, development, instinctive and learned behavior, conscious recall of past events, motor skills – are all due to the *mneme*. Both the inherited and the acquired characteristics of an organism are controlled by *engrams* located within its cells. Thus the regeneration of the limb of a newt is due to the engrams for limb formation in its cells, and the memory of Capri evoked in a traveler by the smell of hot olive oil is due to the corresponding engrams in his brain cells. The sum of all the engrams that an organism has inherited or acquired is called its *mneme*. The production of an engram is caused by a stimulus [*Reiz*], which can be envisaged as an alteration of the energy state of the organism.

Semon's book mentions some interesting examples of experimental work, almost all of which are naively interpreted with little regard for scientific consideration of alternative hypotheses. Thus he calls attention to the behavior of the planarian worm which when cut in two regenerates into two worms having identical characters.

Why did Erwin find this book so fascinating? One can suggest two reasons. Firstly, he had no formal training in biology; it was a *terra incognita* in which bizarre concepts might be encountered without dismay. Secondly, and perhaps more important, all German-speaking scientists have been imbued with the spirit of Goethe, the greatest *Naturphilosoph* of all time. No matter how cold and abstract their university training, they have absorbed in their youth Goethe's feeling for the unity of Nature: 'Faithful observers of Nature, even if they think very differently in other things, nevertheless all agree that everything that appears, everything we meet as a phenomenon, must mean either an original division that is capable of reunion or an original unity that can be split and in this manner display itself. To sever the united, to unite the severed, that is the life of Nature, that is the eternal drawing together and relaxing . . .'[25] Thus as Erwin and Fränzel walked about the streets of Vienna, the gentle gas lights sliding their shadows along the pavements, we can imagine that they were inspired by more than Semon's far-fetched analogies, one boy very religious, the other very agnostic, they were seeking together an answer to the riddle of existence.

The study of Physics

The main focus of Erwin's interest at the university, however, was the course of Hasenöhrl on theoretical physics. This extended over eight semesters with five hours a week of lectures. Hasenöhrl was youthful, full of energy, and a brilliant lecturer. He lectured without notes, but he had not memorized anything, he simply relied on the strong logic of the science and developed it as he went along. Nor did he lose sight of his aim in a thicket of mathematical details, a fault of some theoreticians. His enthusiasm for the beauty of his subject inspired his students; even if Erwin were not already dedicated to theoretical physics, these lectures would have converted him. Erwin always said that he had difficulty learning from books, and to have his subject presented with deep understanding in such wonderful lectures was an intellectual joy of the highest order. 'No other person has had a stronger influence on me than Fritz Hasenöhrl, except perhaps my father.'[26] There was a certain air of chivalry about Hasenöhrl, and his friendliness overcame any barriers of formality or seniority between him and his students. He often had groups of them to his house, where his beautiful wife Ella presided and his small son and daughter added to the happy atmosphere. He was a strong mountaineer and expert in skiing and other winter sports. He organized expeditions with the students and took an interest in student affairs and, as Hans Thirring reported, 'wherever he went he acted as an energizer and brought good fellowship'.

Professor Fritz Hasenöhrl in 1915.

Hasenöhrl lectured in the old rented building on Türkenstrasse. The lectures covered the foundations of analytical mechanics, the dynamics of deformable bodies with special attention to the solution of partial differential equations and eigenvalue problems, Maxwell's equations and electromagnetic theory, optics, thermodynamics and statistical mechanics. The excellence of the lectures made the students forget the dilapidated lecture room in which they were given. There were no proper benches and they had to sit on chairs and hold their notebooks on their knees. 'The floor was an ancient inlaid one, through which gaping crevasses ran, in which even today untold amounts of mercury might remain. Each step made the entire room shake . . . and even the outer walls trembled when a strong wind blew outside or a truck passed by in the street.'[27]

The lectures on mathematics (function theory, differential equations, and mathematical statistics), given by Professor Wilhelm Wirtinger, were uninspired, but the material was essential for any future work in theoretical physics. Some lectures by Gustav Kohn on projective geometry, algebraic curves, and continuous groups were 'strong and clear', and introduced the unifying work of Felix Klein on group theory. Other mathematical topics were treated in lectures by Gustav Escherisch (calculus of variations), and Franz Mertens (calculus in the first year and algebra in the second).

The reason for the close connection between mathematics and science is a mystery. Why are the laws of nature written in the language of mathematics? One answer was given by Einstein, who said that the aim of mathematics is the creation of beauty, and the only physical theories we are willing to accept are beautiful ones. This may be true, but it does not explain why mathematical theories allow us to calculate experimental results about certain small parts of the world with uncanny accuracy. The discoverer of a new physical theory must have available the particular mathematics that is needed. When Newton found his law of gravitation, he needed to invent the differential calculus and the concept of a second derivative; when Einstein found the general theory of relativity, he needed Riemannian geometry; when Schrödinger found wave mechanics, he needed partial differential equations and eigenfunctions.

The theorist who is working at the scientific frontier can never be sure of finding what he needs in the mathematics that has already been applied to previous problems. Schrödinger spent many hours of many days working on mathematics at the university, and his mathematical ability achieved the first rank among contemporary theoretical physicists, perhaps surpassed only by that of Arnold Sommerfeld. Nevertheless there were some branches of mathematics, especially in algebra and group theory, that he neglected. Fortunately

Table 1. *Schrödinger's courses at University of Vienna*

Academic year 1906/07	
V. Lang	Experimental Physics
F. Mertens	Differential and Integral Calculus
G.Kohn	Solid Geometry and Spherical Trigonometry
G. Escherisch	Probability Theory
Academic year 1907/08	
F. Exner	Physics *Praktikum* I & II
F. Hasenöhrl	Mechanics
F. Hasenöhrl	Mechanics of Continua
E. Ozuber	Differential Geometry
G. Escherisch	Function Theory
G. Kohn	Analytic Geometry
F.Mertens	Higher Algebra; Probability Calculus
J.F. Hann	Meteorology
J. Hepperger	Spherical Astronomy; Astrophysics
L. Schrutka	Algebra
Academic year 1908/09	
F. Hasenöhrl	Electrodynamics I & II
A. Lampa	Acoustics
W. Wirtinger	Function Theory I & II
G. Kohn	Group Theory; Number Theory
Z.H. Skraup	General and Inorganic Chemistry; Organic Chemistry
Academic year 1909/10	
F. Hasenöhrl	Thermodynamics and Heat Theory
F. Hasenöhrl	Optics
G. Escherisch	Calculus of Variations
J. Nabl	Screw Geometry
W.Wirtinger	Differential Equations; Mathematical Statistics
G. Kohn	Algebraic Curves

what he studied turned out to be what he would need for his great discovery. The 'bible' of mathematical physics at that time was *The Partial Differential Equations of Mathematical Physics*[28] by Georg Reimann and Ernst Weber, which he mastered in detail as a student.

The course that Schrödinger took in Meteorology with Julius Hann (1839–1921) was to prove an unexpected blessing during the 'Great War', when he was transferred in 1917 from the artillery on the Italian front to the Military Meteorological Service in Vienna. The chemistry course, which included practical exercises in qualitative analysis, was well taught, but he never showed any interest in this subject, and he found organic chemistry particularly tedious.

Table 1 summarizes the courses that Erwin attended during his four years as an undergraduate student.

Several members of the staff of Professor Exner were among Erwin's teachers at the Physics Institute. Egon Schweidler was a Privatdozent, who had taken his Habilitation with Stefan Meyer and Marian Smolu-chowski. Most of Schweidler's research was devoted to atmospheric electricity, a field to which he had been introduced by Exner. Hans Benndorf, another Exner student who worked in this field, was Dozent in Vienna from 1899 to 1904, when he left for a position at Graz. Friedrich Lerch was assistant to Exner from 1904 to 1908, working on the electrolytic separation of radioactive elements.

Erwin's closest friend among the physics staff was Karl Wilhelm Friedrich Kohlrausch, always called Fritz, who came from a family of notable German scientists. He moved to Vienna in 1903, and was just completing his first degree, the doctorate, while Erwin was a second-year student. His thesis was an experimental study of the fluctuations in the rate of disintegration of a radioactive element, a phenomenon discovered by Schweidler the year before.[29] In the winter of 1907/08, with the support of the Academy of Sciences, Kohlrausch went to the West Indies to study atmospheric electricity, and in 1910 he investi-gated the mobility of radium-A atoms in air, a problem that Schrödinger was to take up later.

In 1910 Fritz married Vilma Norer and they had three daughters, Charlotte, Renate and Erika. Erwin was always a welcome visitor at their home, and he remained a lifelong friend of the family.

On the sixtieth birthday of Hans Benndorf, Stefan Meyer and Fritz Kohlrausch, who were then both in Graz, had the happy idea of commemorating the Exner circle with a beautiful silver loving cup, which would be passed from one member to the next as each had his signature inscribed on it on his sixtieth birthday. It bore the invo-cation: 'Travel O Beaker and carry much gladness from one to another. Whoever drinks from you, let him think of the sexagenarian who came before and the one who comes after him.' Benndorf was the oldest and Schrödinger the youngest, and after his death the cup was given into the keeping of the Academy of Sciences.[3]

The world of Vienna outside the university continued in its usual fashion, but the turn-of-the-century flowering of artistic and musical talent was beginning to fade. On November 28, 1905, there was a massive demonstration in which 200 000 persons marched for the right to universal suffrage and this was followed in May, 1907, by the first general election in Austria in which all male subjects were allowed to vote, electing the first 'people's parliament', which convened on June 8. The parliament had little power and the empire continued to be, in the words of Victor Adler, 'a despotism mitigated by slovenliness'. On December 30, 1905, Franz Lehar's *The Merry Widow* opened at the Theater an der Wien, and its lilting waltz became the theme song of pre-war Vienna. In 1906, Enrico Caruso came to the Opera for the first time; Musil published his novel *Young Torless*, a dire premonition of the sadism and perversion of the Nazi era; and Buffalo Bill Cody brought his wild west show to the Rotunda of the Prater. In 1907, Gustav Mahler was dismissed as director of the Opera, and a daring Kabaret, *Fledermaus*, was opened at Kärntnerstrasse 33, a few doors from Grandfather Bauer's house. The next year was marked by the release of the first Austrian feature movie, *From Step to Step*, produced by the photographer Anton Kolm; Moritz Benedikt became the sole proprietor of the *Neue Freie Presse*; the Wiener Werkstätte published Oskar Kokoschka's *The Dreaming Boys*; and Erwin Schrödinger became twenty-one years old and experienced his first mature love affair.

Unfortunately little is known about this early love, except her name, Ella Kolbe, and the fact that Erwin fell in love with her that year and a short but intense affair followed. He was still living in the family home, but another physics student, Jakob Salpeter, with whom he shared a laboratory, had a small apartment near the university where Erwin sometimes stayed and where the lovers might meet.[30] He also continued to visit the Rella family occasionally, but he made it rather clear to Lotte that he was not interested in marriage at this time, despite the hopes her mother may have had for their engagement.

Doctoral dissertation

In the Austrian universities, the first degree in physics was the Doctor of Philosophy, Dr.Phil. A student was not required to take examinations at the end of each course, but at the end of each semester there were optional examinations, called *Kolloquia*, and passing these resulted in a reduction in fees. In addition to completing the necessary courses, the candidate was required to present an original dissertation to the faculty. After the dissertation was accepted, there were final exams called *Rigorosa*, one of which was in some topic in philosophy, as a concession to the name of the degree. The Dr.Phil. degree was not equivalent to a modern Ph.D. in physics, which typically requires at least three years of advanced study and research beyond the first degree. The Austrian Dr.Phil. was at about the academic level of a Master's degree at an American university or an Honour's degree in Britain. Schrödinger was *promoviert* in Physics at the University of Vienna, i.e., awarded the Dr.Phil. degree, on 20 May, 1910.[31]

His dissertation was 'On the conduction of electricity on the surface of insulators in moist air'.[32] The problem was motivated by the importance of electrical insulation in instruments for the measurement of radioactivity and ionization. He thanked Professor Franz Exner in whose department the work was done and acknowledged the support of [Associate] Professor Schweidler. His thesis work was done in the small laboratory shared with Jakob Salpeter, who was working on a thesis problem suggested by Schweidler, the equilibrium of radon between aqueous solution and vapor phase. Towards the end of the term, somebody broke Salpeter's flask containing the radioactive material, and a general evacuation of the laboratory became necessary.

Schrödinger's first paper displays considerable experimental ability but, surprisingly, it lacks any theoretical content. It is a straight-forward, rather routine, set of electrical measurements designed to show the effects of moist air on the conductivity of several solid insulators: ebonite, amber, glass, sulfur and paraffin wax. He assumed that the effects are restricted to the surface but neglected to make the obvious control experiment to prove this, namely, variation of the surface area at constant volume.

He took duplicate thin oblong pieces of each material and mounted them in holders with tinfoil electrodes attached to both ends. One terminal was attached to the high-voltage pole of a series of storage batteries (160 cells). The other electrode was attached to an electroscope. From the charging rate and capacitance of the electroscope the current I could be calculated. Except in some experiments with amber, I was proportional to the voltage V (Ohm's Law) and thus the resistance R could be calculated from $R = V/I$. The samples were kept in atmospheres of various constant humidities, provided by water vapor in equilibrium with aqueous solutions of sulfuric acid. His measurements cover a restricted range of relative humidities from about 80 to 100% for amber and ebonite, and 50 to 100% for glass. No conductivity could be observed for sulfur and paraffin wax, even when droplets of water formed on the surface of the sulfur. He does not mention the temperature at which the measurements were made, an omission that reduces the quantitative value of his data. The increase in conductivity with increasing humidity was not reversible when the humidity was reduced, but sometimes heating the specimens in dry air restored their original insulating state.

Schrödinger did not address any of the interesting theoretical questions that could be raised concerning these observations. His research appears to have been a perfunctory satisfaction of the requirements for the degree. His practical experience with these insulating solids (called *dielectrics*) did have one important consequence: it was the genesis of the masterly survey of dielectricity that he completed four years later. Nevertheless, the dissertation itself, even considering that it was a first effort in research, was not a remarkable performance, and one cannot discern any sparks from the 'fire spirit' portrayed by Hans Thirring. The work was of sufficient interest, however, to be presented at a meeting of the Vienna Academy of Sciences on June 30, 1910. This was the first occasion at which Schrödinger delivered a report on his own research work to fellow scientists. It was published a few weeks later in the *Sitzungsberichte* [Proceedings] of the Academy.

Military training

Austria-Hungary required universal military service, three years from all able-bodied young men. Since the large number of reserve formations needed more officers than could be provided by the military academies and cadet schools, the Empire adopted the Prussian system of 'one-year volunteers' [*einjahrige Freiwillige*]. Men of adequate education and social standing were allowed to volunteer for one year's training as officer candidates, and after passing an examination, were commissioned as reserve officers. The divisions of the army differed in social prestige, the infantry ranking lowest, the cavalry highest, and the artillery in between, the fortress artillery being more snobbish than the field artillery.[33]

On October 1, 1908, Erwin enrolled for military service, and on June 6, 1910, soon after he was *promoviert* at the university, he was accepted as an Einjahrig Freiwillige in the fortress artillery. On October 1, 1910, he presented himself for active service. The personal description entered on the first page of his record book noted that he had blonde hair and blue eyes and blonde eyebrows, but somebody later crossed out 'blue' and wrote 'green?' His height was given as 167.5 cm (5 ft 6 in).

Volunteers were responsible for their own living expenses. The first two months of training were spent in barracks, but after that they were free to find board and lodging in accord with their means in the towns in which they were stationed. Life in the garrison towns could be boring, but there were also happy evenings of wine, women and song for those so inclined. Although the dress uniforms of the officers were elegant indeed, the reserve commissions were not accorded the prestige in Austria-Hungary that they enjoyed in Germany, and many of the one-year trainees never bothered to learn enough military science to pass the final examination. For many young intellectuals, however, the year of outdoor exercise and good fellowship was a welcome change from a sedentary life amid dusty books.

Erwin had leave over Christmas and he went with Hans Thirring for a ski course at Mariazell, in the mountains southwest of Vienna. On December 30, Hans broke his foot while ski-jumping. He was in considerable pain, and Erwin stayed with him all night, moving the foot to a more comfortable position whenever the pain became intolerable. Erwin had planned a long ski tour for the next day, but he did not consider his own need for sleep, only the care of his friend.[18]

A one-year volunteer at his own expense, 1911.

Schrödinger's military service had an important influence on his career. When he graduated there was an assistantship in theoretical physics available with Hasenöhrl, and he would naturally have been appointed, being first in his class. Since, however, he was not available, the appointment went to Thirring, who was excused from military service as a result of his ski accident.

Erwin completed his military training, passed the examination, and received his commission as *Fähnrich* in the reserves, a rank of cadet officer just below that of lieutenant. On New Year's Day, 1911, he returned to civilian life. There is no evidence that he did much physics in the army, but his health was undoubtedly improved.

University assistantship

Schrödinger was appointed to an assistantship in experimental physics under the direction of Franz Exner and immediate supervision of Fritz Kohlrausch, and he was put in charge of the large practical class in first-year physics, a sometimes tedious task, but one upon which he embarked with enthusiasm. As he recalled later:

I learned two things during these years: First, that I myself was not suited to be an experimentalist. Second, that the land in which I lived, and the people with whom I lived there, were no more suited than I to achieve experimental progress along major lines. This was mostly a consequence, among other things, of the tendency of the golden Vienna heart to place amiable nincompoops in key positions (often only on the basis of seniority), where they blocked progress, while active personalities were needed there, who would have had to be brought in from outside. Thus atmospheric electricity and radioactivity, which really had their beginnings in Vienna, were taken out of our hands, and anyone who felt inspired to work seriously in these fields had to go abroad, as for example Lise Meitner from Vienna to Berlin.[26]

Yet Schrödinger afterwards was glad that he had the experimental assistantship with Exner rather than the theoretical one with Hasenöhrl. He was happy to have free access to the excellent collection of instruments. The optical and spectroscopic instruments were kept in his laboratory, so that he was able at any time to make observations of spectra and interferometric measurements. He also had facilities for mixing light of different colors and recording the resultant hues, intensities and saturations. In this way he accidentally discovered that his own color vision was that of a deuteroanomalous trichromat; the cone cells of his retina included those for the three colors red, green, and blue, but the relative intensity of his red perception was enhanced. This anomaly in color vision occurs in about two percent of all those tested.

From his laboratory experience, Schrödinger was able to say, 'I belong to those theoreticians who know by direct observation what it means to make a measurement. Methinks it were better if there were more of them.' When the 'measurement problem' became the subject of a great controversy in quantum mechanics, Schrödinger was able to hark back to his youthful days in the physics laboratory, where he had learned to believe that physics is not based upon mathematical fantasies but on a solid ground of experimental observations. Thus his work as a laboratory assistant helped to determine the philosophical framework that he was willing to accept for a physical theory.

Habilitation: first theoretical papers

As soon as his teaching duties were well organized, Schrödinger began to consider the original research that would allow him to ascend the next step of the academic ladder. This was the *Habilitation* or admission to the *Venia Legendi*, which allows one to serve as Privatdozent at the University. The Dozents were paid only nominal salaries, but could offer courses for which they received the fees. Habilitation required the demonstration of ability to do original research of an acceptable standard, as evidenced by the publication of papers. These papers were considered by a committee of professors, along with the overall record of the candidate and his plans for courses. He was also expected to give a special public lecture attended by members of his committee.

Even for those who could fulfil these requirements, there were few good academic opportunities in Austrian physics. Hans Thirring recalled one discussion with Erwin. 'In contrast with other young titans, he maintained an astonishing inner modesty. Once, in 1911, we were talking about the not so rosy career prospects for physicists. I mentioned that one might perhaps find a position in Germany if all the local ones were filled. Schrödinger stood still for a moment, shook his head, and said "Yes but then one must already have accomplished something quite special".'[18] Those who knew Schrödinger divide into two distinct groups, those who considered him to be a person of amazing modesty, and those who thought he was one of the most conceited men they had ever met, the majority being of the latter opinion.

For his Habilitation papers, Schrödinger chose theoretical problems related to the interests of Franz Exner. His first theoretical paper was presented before the Vienna Academy on June 20, 1912, 'On the Kinetic Theory of Magnetism'.[34] When a substance is placed between the poles of a magnet, two effects can occur. The electric currents due to motions of electrons in the atoms or molecules are altered so as to produce an internal magnetic field that opposes the external field. This effect, called *diamagnetism*, always occurs. In some instances, the atoms themselves act as little magnets, which line up so as to reinforce the external field, an effect called *paramagnetism*. According to the theory worked out by Paul Langevin in Paris in 1905, diamagnetism does not change with temperature, whereas paramagnetism is inversely proportional to temperature. This theory was in good agreement with experiment, except for metals. Schrödinger set out to devise a theory for the diamagnetism of metals.

He based his work on a 1905 paper by the Dutch theoretician Hendrik Lorentz, which pictured a metal as an array of positively charged ions permeated by freely mobile electrons, called an *electron gas*, a theory due to the German physicist Paul Drude (1863–1906). Schrödinger's idea was to calculate the diamagnetism of such an electron gas. At the very beginning, he fell into an error fatal to the final result, by assuming that the electrons would follow the Maxwell law for their velocity distribution and thus would have the same average kinetic energy as a gas molecule at the same temperature. Should Schrödinger have known in 1912 that the 'electron gas' in a metal cannot obey the Maxwell law? In 1905 Einstein had published his theory of the heat capacity of solids, showing that the vibrational motion of the atoms accounts for virtually all the thermal energy of a metal, so that there is nothing left for the electron gas. In 1911, Niels Bohr, two years older than Schrödinger, in his doctoral thesis on the electron theory of metals at the University of Copenhagen rather grudgingly admitted that the assumption of a Maxwellian distribution for the electrons was not tenable.[35] Thus Schrödinger might have been more sceptical about the applicability of the Maxwell law.

Although based on a false premise, his mathematical theory is presented clearly and elegantly. The equation derived for the diamagnetic susceptibility of a metal was in poor agreement with the available experimental data. The odds against Schrödinger in this work were too great. Even if he had suspected that electrons in metals do not follow the Maxwell distribution, neither he nor anyone else at the time had any notion of the electronic basis of paramagnetism. Only after 1921, when Arthur Compton suggested that the electron itself has an intrinsic angular momentum or spin, and consequently acts as a little magnet, would any detailed understanding of the magnetic properties of solids become possible.

When Schrödinger presented this work to his Habilitation committee, Fritz Hasenöhrl, as *Berichterstatter* [reporter], gave it a somewhat equivocal commendation. 'Schrödinger investigates the action of an external field on the free conduction electrons . . . He comes to the remarkable conclusion that this action must appear as a diamagnetism, which, however, far exceeds in intensity the observed diamag-

netism; to explain this, he assumes that this actually occurring diamagnetism is counterbalanced by a strong Langevin paramagnetism.'
The professor did not commit himself on the likelihood that the theory
was correct.[31]

For his formal *Habilitationschrift* Schrödinger chose his paper
'Studies on the Kinetics of Dielectrics, the Melting Point, Pyro- and
Piezo Electricity' which was presented to the Academy on October 17,
1912.[36] He was attracted to the fundamental problem of why and how
a solid melts to a liquid. As he wrote, 'While the very significant
decrease in volume with liquefaction of a gas allows one to understand
the change in mechanical properties of the substance as the result of
the much closer approach of its molecules, crystallization of a liquid
occurs with a much smaller change in volume, which does not even
have the same sign for all substances.' It is not easy to devise a
satisfactory atomic theory for the melting transition from an ordered
anisotropic crystalline state to a disordered isotropic liquid state, and
even today available theories leave much to be desired. Schrödinger
approached the problem with confidence but his underlying model
and the conclusions he derived from it cannot be sustained. 'How do
real solid substances get into this well ordered structure? Why does it
suddenly disappear at a definite temperature? I believe I can give a sort
of atomistic answer to these questions. It is still in very unfinished
form with many and great gaps. Above all it lacks the most important
and the most difficult parts: the equations of state . . . But that is
unfortunately future music. [Aber das is leider Zukunftmusik].'

His starting point was a recent paper by Pieter Debye on the subject
of *polar molecules*. In many molecules the center of positive charge does
not coincide with the center of negative charge, so that the molecule
behaves as a *dipole*, a positive charge Q separated from a negative
charge $-Q$ by a certain distance R. A dipole is measured by its *dipole
moment*, a vector of length QR drawn as an arrow from the negative to
the positive charge. The molecules in a gas or liquid are in constant
motion, colliding with one another and causing the relative orientations of the molecular dipoles to be random and disordered. In the
absence of such thermal motions the dipoles would tend to line up
with each other owing to the attractions between positive and negative
charges, but as soon as any such alignment occurs, a molecule is
violently hit in a collision with another molecule and the orientation is
destroyed. As the temperature is lowered, the thermal motions of the
molecules become less active and the tendency toward alignment of
the dipoles becomes more effective. Now Schrödinger had the ingenious though physically naive idea that the freezing point of a liquid is a
sort of critical point at which the thermal motions of the molecules can
no longer destroy the alignment of the dipoles, so that they suddenly

pass over into a completely aligned arrangement and this ordering of the dipoles is what causes a liquid to freeze to a crystal. He did not realize that the van der Waals attractive forces, which cause nonpolar liquids to freeze, may be more important than the dipolar forces even in polar liquids. Also, if the dipoles in a crystal were really aligned in the way suggested, the crystal would be *ferroelectric*, having a large electric moment even in the absence of a field, but this effect is not generally observed. Schrödinger anticipated this objection, however, by assuming that an actual crystal is made up of microscopic domains within which the orientation of dipoles is the same, but which differ in their prevailing orientation. Ferroelectric crystals with such a structure are in fact known, but they are rare.

In judging the quality of this early paper, we must remember that the structures of crystals were poorly understood in 1912. Just at that time, at the University of Munich, Max Laue was beginning his pioneer work on the study of crystals by X-ray diffraction. In this paper, as in his other early work, Schrödinger displayed considerable mathematical facility combined with a lack of insight into the physical realities of a problem. He also at times neglected to make even rough estimates of the numerical magnitudes predicted by his models, and thus did not discover that they might be physically unsound. Enthralled by his ability to set up a model for a physical situation and to discuss it in mathematical terms, he was apt to proceed with the complexities of the mathematics while ignoring the underlying physics. Although he was working closely with experimental physicists, neither he nor they had yet mastered the basic approach of modern science, the design of experiments with the specific purpose of testing a theoretical model. To them, experimental physics consisted in making careful measurements, a collection of data which would later be subjected to theoretical analysis. Theoretical physics consisted in making a model, usually based on electromagnetic concepts, and then using applied mathematics to derive relations between certain properties. The idea of a close coordination in which theory guides experiment and experiment tests theory was not generally understood at this time.

In the brief report of the *Habilitation* committee on this work, Hasenöhrl did not deal with it critically.

In his work on the Kinetics of Dielectrics, Schrödinger extends the Debye theory, in contrast with Debye, by identifying the critical point that corresponds to the Curie point in the theory of magnetism with the melting point, which yields many interesting consequences, namely for the theories of piezo- and pyroelectricity . . . In the opinion of the committee, all the works of Schrödinger demonstrate a very well founded and broad scholarship and a significant, original talent.[31]

There is evidence here of the parochial character of Viennese physics. If the *Habilitationschrift* had been sent for comment to Debye himself, a less congratulatory critique would undoubtedly have been obtained. Schrödinger's early scientific development was inhibited by the absence of a group of first-class theoreticians in Vienna, against whom he could sharpen his skills by daily argument and mutual criticism.

The committee was not unanimous in recommending acceptance of Schrödinger's application for Habilitation. Professor Rudolf Wegscheider wrote that: 'My opinion is inclined against this premature Habilitation on the grounds of the youth of the candidate and the small amount of research work completed. At the faculty meeting of June 14, the committee report was presented by Hasenöhrl, and approved by 38 votes to 2, with 5 abstentions.

The next step was the lecture or *Kolloquium*. As required, the candidate had suggested three topics: (1) Anomalous Dispersion in the Electric Spectrum, (2) The Magneton, and (3) The Significance of the Quantum of Action in the Theory of Heat Radiation. The committee chose to hear about the magneton, and the meeting took place on June 26. The formalities were not yet completed. There was an oral examination and discussion on the subject of the Kolloquium before several members of the committee on October 15. The matter was then referred again to the professorial board for a final vote, which was 39 in favor and 4 abstentions. Only now could the Dean send the nomination to the Ministry for Culture and Instruction, which he did on November 10. On January 9, 1914, the Minister sent his approval to the Dean, and Erwin Schrödinger became a Privatdozent for Physics at his Alma Mater. He was standing on the first rung of the academic ladder.

Felicie[37]

Just as he was taking his first forward step in his academic career, Erwin was strongly tempted to abandon it altogether for the sake of a young woman.

Among the friends of Rudolf and Georgine Schrödinger for many years were Karl and Johanna Krauss. They were a fairly wealthy family with connections to the minor nobility and a long tradition of strict observance of the Catholic faith. Karl was trained as a lawyer and worked in the *Obersthofmarshcallamt*, a division of the imperial archives. Their daughter Felicie was born in 1896, and was therefore about eight years younger than Erwin. As a boy, he used to be told at children's gatherings, 'now look after little Felicie', a task that he undertook with considerable reluctance. When her father died in 1911, Felicie was not yet fifteen years old. She was growing up to be a

Felicie Krauss (c. 1913).

beautiful teenager, and Erwin was now delighted to be with her as much as possible. He was always captivated by girls who were just at the onset of womanhood. Felicie's mother, however, wanted to discourage any closer relationship with the Schrödingers and viewed Erwin as a quite impossible match for her daughter. With the support of her relations, she issued an edict that Erwin and Felicie would be allowed to meet only once a month. As might be expected, such abstinence made their hearts grow fonder, and they were soon deeply in love. They wanted to get married and considered themselves to be informally engaged.

Erwin's prospects as a physicist were dismal from a financial point of view. At best he could hope to spend several years as a Privatdozent with no assured income or to have an assistantship with a miserable salary. He was desperately in love with Felicie and went to his father to ask if he could give up the university and take a leading role in the family linoleum business. He wanted to get married and saw no likelihood of being able as a physicist to support a wife. With great wisdom, Rudolf said no, he would not allow such a thing. He had sacrificed the scientific work he loved, to go into business, and had regretted it all his life; he did not intend to see his son make the same kind of mistake. Erwin must stay with his university work in physics. Besides, the family factory was not prospering and would not really provide for Felicie to the extent her family would think appropriate.

Erwin was now twenty-five and Felicie seventeen. Baronin Krauss became alarmed at their developing relationship. Her most serious objection to Erwin as a son-in-law was probably not that he was poor. The social standing of his parents was a greater obstacle, for the Krauss family as members of the minor nobility could place a 'von' before their surname, which was important in imperial Vienna. Even worse was the fact that Erwin was nominally a Protestant, but actually a *Freidenker*, a free-thinker who probably did not even believe in God. A noble Protestant might have been acceptable but a poor atheist dabbling in arcane mathematics was unthinkable as a son-in-law.

It would have been a most unsuitable match for many reasons, and Felicie was persuaded (or forced by her strong-willed mother) to break the informal engagement. This happened about the middle of 1913. In the context of Austrian society at that time, it would have been virtually impossible for Felicie to withstand the pressure of her family and marry Erwin despite parental disapproval. Yet the result was an incalculable loss for Erwin. His earlier loves had not been superficial, but he did not think of them as permanent commitments, whereas he was willing to dedicate his emotional life to Felicie. Probably at this time he began to form his negative view of bourgeois marriage. Since he was prevented from a dedication of his spiritual, romantic and sexual longings to the one person of his choice, he would henceforth look with disdain upon the institution of marriage and attempt to construct his emotional life outside its rigid framework. It is doubtful that he could have made a happy marriage with any woman, for he would have demanded complete devotion from his wife and complete freedom for himself, while at the same time his concern for truthfulness would have made the conventional secret adulteries of Vienna society quite impossible.

After the breaking of his engagement to Felicie, many years would pass before Erwin again fell in love with a young woman who belonged to a social class equal or superior to his own. In view of the shallowness of most Viennese religious convictions, he may well have ascribed his rejection by the Krauss family to social rather than ecclesiastical factors, and perhaps inwardly he resolved never again to suffer such a blow to his pride. This interpretation would help to explain the kind of woman he decided to marry, and his subsequent treatment of his wife as a sort of superior domestic servant. He would never have treated Felicie in such a way.

It must be admitted, however, that the harsh policy of Felicie's mother probably prevented future unhappiness for both the young lovers, but one cannot be sure – if Felicie had married Erwin and given him a son, he might have been satisfied to recreate the pattern of his own family circle. One side of his personality was seeking a tranquil life of teaching and philosophical meditation. Yet he may have needed the excitement of tempestuous sexual adventures to inspire the ardent creativity that produced his great discoveries.

In 1917, Felicie married a young lieutenant in the Austrian army, Ferdinand Bianchi. He was a baron like her father and came from an old imperial military family, which at that time was quite wealthy. He was thus eminently agreeable to Felicie's mother on religious, social, and financial grounds. They had one child, a daughter Johanna, who took her doctorate in literature, taught at a Gymnasium, and became a writer and an excellent photographer. Felicie never lost her interest in Erwin's career and later became a good friend of Schrödinger's wife Anny. Felicie sent Erwin a long poem for his seventieth birthday, including the stanza,[38]

> Mir scheint, es ging bei ersten Hahnenschrei
> Grad unsre Kindheit eben erst vorbei,
> Mir scheint, als war es nur ein Augenblick
> Seit unsrer Jugend Übermut und Glück.*

---

* It seems to me, by the first cock's crow / Our childhood time is only just gone by, / It seems to me, as if but a moment has passed / Since the playfulness and joy of our youth.

Atmospheric electricity and radioactivity

One of the major research interests in the Physics Institute II was atmospheric electricity. Professor Exner had encouraged Egon Schweidler to work in this field, and in 1899 there appeared the first in a long series of papers on 'Contributions to the Understanding of Atmospheric Electricity', which eventually extended to seventy-eight publications, twenty of them by Schweidler. In 1911 he became an associate professor and the next year received a call to a chair at Innsbruck. With the support of the Vienna Academy, he established a number of observational stations to study the electrical conductivity of the atmosphere, on the lakesides at Attersee, Ossiachersee, and the principal station at Seeham on the Mattsee near Salzburg. Perhaps there was a scientific reason for favoring such lakeshore resort towns, but they were also exceptionally pleasant places to work during the summers. At that time it was usual for the younger staff to follow the research lines of the professor, and thus it was natural that Schrödinger became interested in atmospheric electricity.

His first paper in this field was a theoretical one, published as No. 48 of the series from the Physics Institute. It was presented at a meeting of the Vienna Academy on December 5, 1912.[39] Paper No. 47 in this series was by Schweidler, and dealt with observations of atmospheric electricity at the experimental station at Seeham in the summer of 1911.

For many years physicists had been puzzled by the origin of the background radiation that caused a slow discharge of electroscopes both in the open atmosphere and in closed vessels. Two sources of radioactivity were found, one in the surface layers of the earth due to radium, thorium, and their decay products, and the other in the atmosphere due mainly to radium emanation (now called radon) and its decay products. Schrödinger's Vienna colleague, Viktor Hess, was on the track of a much more exciting possibility: an extraterrestrial source of high-energy radiation. Hess was an ardent balloonist and in April, 1912, he began a series of ascents with precision electroscopes. On August 7, a flight reached 4800 m where the discharge rate was about three times that at ground level. Hess published his data in the November, 1912, issue of the *Physikalische Zeitschrift*, with the conclusion: 'The results of my observations are best explained by the assumption that a radiation of very great penetrating power enters our atmosphere from above.' Thus the discovery of cosmic rays was recorded, although they were so named by Robert Millikan only some years later.

The paper of Schrödinger can be considered as a calculation of a background with which the observations of Hess can be compared. As had been his wont in previous papers, he chose a mathematically tractable model. He wanted to derive an expression for the variation with altitude of the atmospheric radioactivity, and he assumed that this activity was due to 'suspended material', which was mixed by air currents so as to have an altitude distribution the same as an isothermal variation of barometric pressure, i.e., a simple exponential law. Actually the temperature of the atmosphere varies with altitude and the exponential law is therefore not valid, as Schrödinger knew very well. His final conclusion was that the atmospheric radioactivity should be practically constant from 1000–20000 m if the simple exponential density law is correct. The decrease in density of radioactive material with altitude is compensated by a decrease in the absorption coefficient of the atmosphere. His final statement is: 'In this account extraterrestrial radiation sources are of course neglected.' The following year, at the meeting of the Academy on June 5, Hess presented a definitive paper 'On the Origin of the Penetrating Radiation'. Neither in this paper nor in any of his subsequent publications does he refer to Schrödinger's paper. Hess received a Nobel prize in 1936 for his discovery of cosmic rays.[40]

In August, 1910, Schrödinger's friend and colleague Kohlrausch had made a series of measurements of the radium-A content of the atmosphere at Seeham. Professor Exner suggested that it would be a good idea to repeat the Kohlrausch measurements at the same place in the summer of 1913. Erwin set up his equipment near the window of his hotel room, about 200 m from the location used by Kohlrausch.

The uppermost layer of the earth, the alluvium, contains on average about one part in 10^{12} of radium. The radium decays with a half-life of 1850 years, emitting an alpha-particle and leaving the rare gas radon, an alpha-emitter with a half-life of 3.85 days. The unit of radioactivity, the *curie* was originally defined as the radioactivity of the amount of radon that is in a steady state with one gram of radium. Thus the gases in all soils contain some radon, on average 2×10^{-7} curies per cubic metre. This radon continuously seeps out of the soil into the atmosphere, where it is distributed by diffusion and air currents. When it emits an alpha-particle, the product is radium-A, an isotope of polonium, a semimetallic element in the same family as selenium and tellurium. Radium-A and similar solid decomposition products of the radioactive elements occur in air as positive ions and hence can be separated by deposition on a negatively charged electrode. Schrödinger measured the activities by the aspiration method, in which air is pumped through a cylindrical tube with a coaxial rod which is at a high negative electric potential relative to the walls of the tube. After a measured time, about five minutes in these experiments, the central rod is removed and the charge collected on it measured with a sensitive electroscope. Between July 24 and September 5, he made 229 determinations, certainly earning his spurs as an experimental physicist in this painstaking, rather tedious work.

It is interesting to look at the day-to-day record of his working schedule. Usually he would make several runs in the morning, take time off for some hours at midday, and return to his bench for several more runs in late afternoon before supper. Usually he worked through the weekends, taking only three whole days off during his stay at Seeham. He rarely worked at night, but sometimes would make a few runs between 10 p.m. and 1 a.m. His enthusiasm for data-collecting did not extend to the small hours of the morning, and despite his interest in the diurnal variation of the activity, no data were ever recorded between 1 and 6 a.m.

On average his activity values were only about one-fifth those found by Kohlrausch two years previously. He states 'I have often explored this remarkable result in detail with my friend Kohlrausch . . . There remains no other explanation than that the yield of emanation from the pores of the soil was strongly reduced as a consequence of the repeated and thorough soaking of the soil in the rainy summer of 1913, which seldom left us even one week long without rain.'

This work comprised publication No. 51 of the investigations of atmospheric electricity by various members of the staff of the *II Physikalische Institut*.[41] Schrödinger, though primarily a theoretician, was able and willing to contribute to this communal effort. The subject was more important than its devotees probably realized, since we now know from the work of Hermann Muller and others that an important factor in determining the mutation rate of genetic molecules is the intensity of high-energy radiation from both terrestrial and extraterrestrial sources.

Erwin loved the outdoors and the Austrian countryside. As a compensation for slaving away at his aspirator and electroscope, he was able to enjoy hiking and climbing in the surrounding hills and, when it was not raining at least, swimming in the cool water of the lake.

The Kohlrausches came with their children to spend a holiday there and they had many happy excursions with Erwin. More importantly they introduced him to a pretty and jolly teenage girl, Annemarie Bertel from Salzburg, who was helping take care of the children. Dressed in the local costume, she looked like a peasant girl, but her father was court photographer and a man of some substance in Salzburg. Anny's birthday was New Year's Eve [*Sylvesterabend*] 1896. She had a sister Irmgard, who was three years younger, and an elder brother Erich. Anny was enormously impressed by the young scientist and thought 'he was very good looking.'[42] At that time her meeting with Erwin did not blossom into a serious love affair, since Anny was a country girl, 'practically a child', in dirndl and pigtails, while Erwin was a serious academic, with all the sophistication of the capital city. Besides, he had not yet recovered from the loss of his true love Felicie. Yet it was more than a holiday romance, as the future would tell.

On October 1, 1919, Erwin gave Anny a copy of the 1913 paper with an addendum written in a parody of his scientific style:

As I now first discover, there must have been at that time something else of a different nature in the air at Seeham, besides Radium A, B, and C, of which, however, my electrometer did not detect a trace. The credit for the discovery is due exclusively to Fräulein Bertel (Salzburg) who drew the attention of the author to it . . . A joint publication of the above-mentioned discoverer and the author will follow in another place in the near future.[43]

When his last measurements were made, Erwin quickly packed his equipment, and rushed back to Vienna to take part in preparations for an important international scientific congress, the 85th Meeting of German Scientists and Physicians.

The 1913 Congress of Vienna[44]

Before the previous Vienna meeting of this important society in 1894, Franz Exner had left the city because he was so ashamed of the miserable quarters of the physics institutes. Now he was proud to welcome his confreres to his beautiful new building. Twenty years had elapsed since he had accepted the Vienna chair with the understanding that a new building would soon be provided to replace the rickety quarters on Türkenstrasse, but it was not completed until the spring of 1913. The site faces Boltzmanngasse, next to the Institute for Radium Research and about a kilometre from the main university building on the Ring. The elegant five-storey limestone structure has a classically austere style. It includes a large lecture hall, extensive laboratories and workshops, many offices and seminar rooms, and one of the best physics libraries in Europe.

More than 7000 participants attended the meeting, which included lectures on a wide range of scientific and medical subjects. The Imperial City had never looked more splendid than in these last autumn days of its glory. The festivities included a reception at the imperial court and a banquet given by the city of Vienna in the Rathaus. In addition, there were social activities for the physicists and their wives, including a party held Tuesday evening, with music and dancing in the rooms of the new Physics Institute.

The main physics lectures were given in the large hall of the Institute. The first session on Monday afternoon was chaired by

Privatdozent (c. 1914).

Hofrat Viktor Lang, the emeritus professor and predecessor of Hofrat Ernst Lecher. This program included the following reports: Robert Pohl of Berlin on the photoelectric effect; James Franck and Gustav Hertz, also from Berlin, on collisions between electrons and atoms in gases (in the following year these experiments resulted in the discovery of excitation potentials, for which they were awarded the 1925 Nobel prize); Karl Herzfeld (Vienna) on free electrons in metals.

The most important lecture for the physicists was scheduled for Tuesday morning, by Albert Einstein on 'The Present Status of the Problem of Gravitation'. In April, 1911, Einstein had resigned his associate professorship at Zürich to become professor at Prague, but in August, 1912, he returned to Zürich and a professorship at the Technical University (E.T.H.). In the spring of 1913, Planck and Nernst visited him to ascertain what it would take to attract him to Berlin. In collaboration with Marcel Grossmann, professor of mathematics at the E.T.H., he had made considerable progress during the past two years on the theory of gravitation. He was already widely recognized, for his earlier work in special relativity and quantum theory, as the greatest theoretical physicist in the world, although he had not yet received a Nobel prize. The lecture room was filled to capacity when he began to speak at 9 a.m.

Einstein did not overwhelm his listeners with awesome mathematics. He began with the most simple and basic ideas and developed an analogy between electrical theory and gravitation. The former began with electric charges and the inverse square force between them given by Coulomb's law. Later, electric currents were discovered and it was found that they could be generated by moving magnets. Then, with the work of James Maxwell and Heinrich Hertz, electromagnetic waves were discovered and optical and electromagnetic phenomena became parts of a unified field theory. Our understanding of gravity, he said, is now only at the stage of Newton's law of inverse square attraction between two masses. It is necessary to modify Newton's theory of gravity based on instantaneous action at a distance in order to exclude any physical influences that move faster than the speed of light. Einstein then mentioned that some of the mathematical ideas he was using had already been discussed by a young Viennese theoretician, Friedrich Kottler, whom he had never met personally, and he asked this young man to stand up so that the audience might see him.

Einstein continued with a masterly review of other theories of gravitation and what he perceived to be their shortcomings. He then outlined the work that he and Grossmann had been doing. Their theory, unlike others, requires a definite amount of bending of light rays in a strong gravitational field, and this prediction can be checked quantitatively by observing stars appearing close to the sun during a solar eclipse. 'Let us hope that the eclipse of the year 1914 will allow us to obtain this important decision.'

Einstein's theory was based on the identity of gravitational and inertial mass, and the doctrine of Ernst Mach that inertia is also due to the effect of other masses in the universe. While in Vienna, Einstein visited Mach, who seldom left his apartment. Phillip Frank recalls that on entering Mach's room, 'one saw a man with a gray, unkempt beard and a partly good-natured, partly cunning expression on his face, who looked like a Slavic peasant'. Einstein asked Mach to suppose that he could predict by atomic theory an observable property that could not be predicted by continuum theory, would he then accept that it was 'economical' to assume the existence of atoms? Mach admitted that the atomic hypothesis would then be economical, and Einstein went away quite satisfied.[45]

Schrödinger's interest in gravitation was definitely aroused by this brilliant lecture, but he must have filed it away at the back of his mind until Einstein's great papers on general relativity were published in 1915, and he had time to study them while in charge of an artillery battery on the Italian front. Also the idea of a unified field theory, which would include both electromagnetism and gravity, must have occurred to some of the young scientists listening to Einstein's outline of the analogies between the two fields. In their mature years, both Schrödinger and Einstein would devote themselves to the quest for this unification.

The next physics session was concerned with radioactivity, a field in which the Vienna physicists were particularly strong. Hans Geiger came from Berlin to report on his excellent device for counting alpha-particles. This Vienna meeting marked the general acceptance of the existence of 'cosmic rays', which were to provide one of the most important research fields of modern physics.

The Wednesday session included exciting reports on X-ray interference and crystallography. Max Laue, from Zürich, presented the theoretical foundations of the subject and concluded with the prophetic remarks:[44]

With crystals we are in a situation similar to an attempt to investigate an optical grating merely from the spectra it produces . . . But a knowledge of the positions and intensities of the spectra does not suffice for the determination of the structure. The phases with which the diffracted waves vibrate relative to one another enter in an essential way. To determine a crystal structure on the atomic scale, one must know the phase differences between the different interference spots on the photographic plate, and this task may certainly prove to be rather difficult.

The determination of the phases soon became the crucial problem of X-ray crystallography. A distinguished British visitor was Charles Barkla, who discussed his discovery of characteristic X-rays.

Schrödinger did not lecture at the meeting; he had not yet become a Privatdozent and besides he had nothing special to report. The brilliant array of German physicists impressed him strongly. The X-ray work made a particular impact, for he had begun some work on the effect of temperature on X-ray diffraction. When he received his appointment as Dozent a few months later, his first course, offered for the summer term of 1914, was on 'Interference Phenomena of X-Rays'.

X-ray and dielectric studies

Earlier in 1913, Pieter Debye had published two papers on the effect of the thermal vibrations of the atoms in a crystal on its X-ray diffraction pattern. Although he must have known that Debye was still working on the problem, Schrödinger took it up also. As had been typical of his theories so far, his mathematical treatment was highly polished, but his physical model was less adequate than that used by Debye. The importance of this work for Schrödinger's future research was that it directed his attention to problems of atomic vibrations and scattering of radiation, in which the quantum theory must have an important role. These problems led him away from the special interests of the Vienna School and brought him closer to the mainstream of European atomic physics.

One of the subjects suggested by Schrödinger for his Habilitation colloquium was the theory of anomalous electric dispersion. He wrote this up as a 'Note' for the Discussions [Verhandlungen] of the German Physical Society and it was received 8 November, 1913.[46] This paper marked his first appearance in print outside Austria. Electric dispersion refers to the dependence of the dielectric constant ϵ on the frequency of the electric field at which it is measured. At high frequencies ϵ decreases from its static value to reach its optical value of n^2, where n is the refractive index. In polar liquids, ϵ is observed to decrease markedly at relatively low frequencies, and this dispersion was termed 'anomalous'.

Early in 1913 Debye had published in the Verhandlungen an explanation of this anomalous dispersion based on the inability of dipolar molecules to follow the change in polarity of a rapidly alternating field. Schrödinger pointed out that the mathematical form of the dispersion law obtained by Debye is exactly the same as that of the formula obtained by Drude from a model of aperiodic damped electronic oscillators. He reasoned (correctly) that one cannot, therefore, deduce the correctness of the Debye model merely from the mathematical form of the dispersion law, although the effect of temperature may allow a decision in favor of the Debye model.

During 1913, Schrödinger devoted considerable time to the compilation of an extensive chapter on 'Dielectricity' for the five-volume *Handbook of Electricity and Magnetism* being edited by Leo Graetz, professor of physics at Munich.[47] In this 75-page review, he made an exhaustive survey of all the literature on the subject through the end of 1912. Hundreds of references were cited and the many tables listed every measurement of dielectric constant that had ever been made. These encyclopedic surveys have been peculiarly dear to the hearts of German scientists; they are usually out-of-date by the time the last volume appears, at which point the editor begins a series of supplements, so that involvement with such an enterprise often becomes a lifelong occupation. It was nevertheless a considerable honor for a physicist as inexperienced as Schrödinger to be entrusted with a section of such a treatise. The professional advantage was that after such a publication, one would be recognized as a European authority in the particular field.

Schrödinger reviewed the classical Maxwell theory and the Drude electron theory, but he made no mention of the Debye theory of polar molecules, presumably because it was published just after his cut-off date of 1912. From his concise yet practical account of experimental methods, it is evident that he was familiar with much of the laboratory equipment. The finished article was sent to the editor early in 1914, but it was not published until 1918, after the end of the war. By that time Erwin's interests had turned to other fields, and he published no further papers on dielectrics.

The first outstanding paper

Early in March, 1914, Schrödinger sent to the *Annalen der Physik*, the most important German physics journal, a purely theoretical paper, 'On the Dynamics of Elastically Coupled Point Systems', which is the most significant of his prewar publications.[48]

This paper is especially interesting in that it harks back to one of the persistent themes in the scientific life of Erwin's intellectual grandfather, Ludwig Boltzmann: the absolute necessity of atomistic models in physical theory. In 1897 Boltzmann published a paper 'On the Indispensability of Atomistics in Natural Sciences' [*Wied. Ann.* **60**, 231]. Much of theoretical physics is expressed in the form of differential equations, which prescribe the trajectories in time and space of quantities such as temperature, electric-field strength, concentration, etc. The equations are completely general, and to apply them to specific systems, they must be integrated subject to certain initial conditions (at time zero) and boundary conditions (values of the quantities of interest at the boundaries of the system). The differential equations try to represent the macroscopic behavior of a system without reference to its underlying atomic structure. For example, one might consider an idealized string with a certain elasticity and mass per unit length, and calculate how it vibrates when subjected to some initial disturbance. Advocates of strict phenomenalism, such as Ernst Mach and Wilhelm Ostwald, contended that such differential equations simply represent in mathematical form the world of direct experience.

Boltzmann controverted this view with a number of arguments. In the first place, he said, the differential equations themselves constitute a model or thought picture severely abstracted from observations. 'One ought not to say, with Ostwald, "Thou shalt not make any image", but only "Thou shalt admit in it as little of the arbitrary as is possible".' To suppose that differential equations do not go beyond experience is to suppose that our perceptions are founded on continuous images, but sensory input travels to the brain along a finite number of nerve channels and thus any impression is more like a mosaic. (Artists of the Seurat school of *peinture optique* who based their works on the color theory of Chevreuil would have supported Boltzmann on this point.) Moreover, when we speak of a continuum in physics we do not imagine something truly continuous, but rather a finite number of elements allowed to increase in number until further increase no longer has any effect. 'May I be excused for saying with banality that the forest hides the trees for those who think they can disengage themselves from atomistics by the consideration of differential equations.'[49]

As an ardent student of Boltzmann's work, Erwin would have been well aware of the above remarks of the master, when he began his own extension of the subject:

It is so to speak part of the creed of the atomist that all the partial differential equations of mathematical physics . . . are incorrect in a strictly mathematical sense. For the mathematical symbol of the differential quotient describes the transition in the limit to arbitrarily small spatial variations, while we are convinced that in forming such 'physical' differential quotients we must stop at 'physically infinitely small' regions, i.e., at those that still always contain very many molecules; if we were to push the limiting process further, the quotients concerned, which up to then really were proceeding nearer and nearer toward a definite limit . . . would again begin to vary strongly.

Thus the limit used for the differential equations is only a 'pseudo-limit' and is not the same as the limit conceived by pure mathematics.

If we really wish to take this view seriously, we are confronted with a double task. Firstly, all those differential equations first derived . . . by consideration of a continuous medium, now must instead be derived as difference equations on the basis of a model constructed of molecules . . . Atomistics also calls for a second task, through whose solution it first establishes its supremacy over the phenomenological theory. This consists in finding and stating those conditions under which the differential equation based on a continuum actually leads to an incorrect result because of the truly atomistic structure of matter.

Schrödinger proceeds to investigate a model that had recently been applied by Max Born and Theodore Karman to the specific heat of solids: an infinite linear array of mass points separated at equilibrium by a distance a, each one of which is subject to an elastic restoring force $f\xi$ when displaced from its equilibrium position by a distance ξ. As the most general solution to this problem, he obtains for the displacements ξ or velocities $\dot{\xi}$ an infinite series of Bessel (cylinder) functions of the first kind $J_{n-k}(\nu t)$, where $\nu = 2\sqrt{f/m}$ is a vibration frequency and t is the time,

$$ x_n = \sum_{-\infty}^{+\infty} x_k^0 J_{n-k}(\nu t), \qquad n = -\infty, \ldots, +\infty. $$

Here $x_{2n} = \sqrt{m}\,\xi_n$ and $x_{2n+1} = \sqrt{f}(\xi_n - \xi_{n+1})$.

As the simplest example of the sometimes astonishing results which this formula yields, he considers the following case. At $t = 0$ all the mass points are assumed to be at rest except one, which is given an arbitrary displacement ξ_0. All the mass points then begin to move, each with a motion corresponding to one of the even Bessel functions. There are two astonishing things about this result: it is nothing like the motion of a continuous string as derived from the partial differential equation, and all the mass points begin to move instantaneously, no matter how far they are from the original disturbance. The first surprise is analyzed and explained by Schrödinger in the rest of the paper, but he does not exorcise the second, which, though a valid consequence of Newtonian mechanics, is physically impossible. The future quantum theory would show that the atoms in a crystal (or on a line as in this case) can never be at rest; they always have a 'zero-point motion' that is much greater than any small signal instantaneously caused by the initial disturbance of one distant atom in the sequence.

Schrödinger then asks the question, 'What must be the character of the system of initial values x_n^0 in order that the motion of our discrete model should really display a certain similarity to the motion of a continuous elastic medium?' He recasts the solution in the equation above into the form of a pair of Maclaurin series in powers of νt

$$ \xi_n = \sum_{k=0}^{\infty} \frac{(a\nu t/2)^{2k}}{(2k)!}\, \frac{\delta_n^{(2k)}}{a^{2k}} + \frac{2}{a\nu} \sum_{k=0}^{\infty} \frac{(a\nu t/2)^{2k+1}}{(2k+1)!}\, \frac{\delta_n^{(2k+1)}}{a^{2k+1}} $$

Here the δ_n are functions of differences between coefficients in the expression for x_n and its derivatives with respect to t. In the limit as $a \to 0$, while m/a and fa remain finite, this solution becomes that of the wave equation for longitudinal waves in one dimension, as found by d'Alembert.

Such a limit, however, is physically meaningless since the original solution for mass points on a line specified a fixed value of a. Thus one must define two analytic functions $\xi(x)$ and $\eta(x)$, which for $x = na$ have the values ξ_n^0 or $\dot{\xi}_n^0$, and which also have these values for $x = nb$ at which no mass points exists. The real question, however, is what must be the initial conditions so that for one fixed value of a, the solution of the atomic problem is virtually the same as that for the continuous string. This will be the result if both systems of quantities ξ_n^0 and $\dot{\xi}_n^0$ have sufficient similarity with the distribution of values of two continuous functions of x that detectable differences arise only at distances large compared to the site separation a. At sufficiently long times, nevertheless, there will always be deviations when a is finite.

The result is that one can apply the differential equations to such molecular models if one does not consider the variables themselves, but averages of the variables over regions that contain many atoms. For example, slow acoustic waves can be considered without reference to effects due to rapid thermal motions of the individual atoms.

This paper is undoubtedly the most interesting of all those written by Schrödinger before he was called into military service in 1914. It carries forward one of the basic problems of 'grandfather Boltzmann' by its penetrating analysis of the specification of initial values in a system based on an atomic model, and it forms a bridge to his later revolutionary applications of the differential equations of wave motion in his wave mechanics. In this paper also, for the first time, we hear the authentic Schrödinger style, with its urbane confidence and its ability to relate the question in hand to deeper philosophic concerns of mathematical physics.

3   Schrödinger at war

Erwin began his career in physics during the last peaceful years of the Danube monarchy. Old Franz Josef still retained the respect and affection of the lower and upper classes among his subjects, but his control over the bourgeois politicians and their nationalistic tendencies began to falter. The Hungarian leaders consistently placed the privileges of the great Magyar landowners above all else, and extorted special concessions as the price of loyalty to the throne. The annexation of Bosnia-Herzegovina in 1908 created a large population of disaffected Slavs and Muslims in the southern borderlands. The K & K structure was becoming more unstable, and as for a radioactive atom, the question was not if but when it would disintegrate.[1]

Before the war

The chief of the general staff, Franz Conrad, was a persistent warmonger, urging the emperor to embark upon 'preventive war', one year against Serbia, the next against Italy. At the end of 1911, he was dismissed, not so much for meddling in foreign policy as for his affair with a married woman, Gina Reininghaus, which offended the religious sensibilities of the old emperor. After a year in disfavor, he was called back, but soon thereafter Franz Ferdinand, heir to the throne, was appointed General Inspector of all the Armed Forces. In May, 1913, the war machine was shaken by the revelation that Colonel Alfred Redl, one of the chiefs of military intelligence, was a Russian spy and a member of a homosexual coterie among the army officers. Since he had sold the Russians all the plans for a prospective eastern theater of war, a hasty revision of K & K strategy was required. Meanwhile, the two Balkan wars of 1912–1913 had left Serbia as a much stronger military force in the region. Even Conrad realized that it was now too late for preventive war.

The musical, literary, and dramatic life of Vienna still made the city the cultural center of the German-speaking world, but this eminence was increasingly a creation of the Jews. In October, 1912, the Censorship prohibited the Wiener Deutsches Volkstheater from presenting Schnitzler's play *Professor Bernhardi*, which deals with antisemitism in the medical profession. Many Jews became converted to Christianity, and a few even to antisemitism. The leading antisemite in the Faculty of Medicine was Professor Karl Herzfeld, according to Schnitzler 'not yet baptized'.[2] His son, who became a distinguished physicist, had been one of Schrödinger's students in the practical classes.

As in other large European cities, there was a great contrast between the luxurious life of the upper classes and the struggle for existence of the poor. The typical apartment in central Vienna had seven rooms, whereas less than five percent of flats in the workers' districts had more than one room. The unemployed suffered from cold and hunger during the winters, but the city did provide several goulash kitchens to help prevent starvation; still, the death rate in the outer districts was twice that in the inner city.

Leon Trotsky lived from 1907 to 1914 in Vienna, where he published a paper, *Pravda*, which was smuggled into Russia. His opinion of the Austrian socialists was not complimentary.

In informal talks among themselves they revealed, much more frankly than in their articles and speeches, either undisguised chauvinism, or the bragging of a petty proprietor, or holy terror of the police, or vileness towards women. In amazement, I often exclaimed 'What revolutionaries!' I am not referring here to the workers . . . No, I was meeting the flower of the prewar Austrian Marxists, members of parliament, writers, and journalists . . . In the old, imperial, hierarchic, vain and futile Vienna, the academic Marxists would refer to each other with a sort of sensuous delight as 'Herr Doktor', and workers often called the academics 'Comrade Herr Doktor'.[3]

Erwin took no serious interest in the political situation of the empire. His physics, his intensive reading of philosophy, the theater, and his love affairs left no time for politics. His friend Hans Thirring, however, was becoming a convinced socialist and pacifist, and was fortunately exempt from military service owing to his ski accident.

As was customary at that time for unmarried sons and daughters, Erwin still lived at home with his parents. After the break with Felicie, he fell in love again, with Irene Drexler. Once again it was a romantic love that was never to achieve fulfillment – perhaps future research will elucidate its importance to his life. The fact that during his middle twenties Erwin fell in love with three young women but slept with none of them suggests that Vienna maidens from good middle-class families did not then indulge in love affairs, and that Erwin's other amorous pastimes were enjoyed without romantic complications.

The drums of August

On June 28, 1914, Franz Ferdinand, resplendent in a white tunic and a general's headdress of green cock feathers, and his beloved but morganatic wife, formerly Sophie Chotek lady in waiting, in a picture hat and a high-necked white summer dress, were riding in a fine open car down the main street of Sarajevo in Bosnia. A few hours previously they had narrowly escaped assassination from a bomb thrown by Nedjelko Cabrinovic. Another youth from the same group, Gavrilo Princip, stepped up to the side of the car and fired two shots, one through the neck of the Archduke and the other into the abdomen of his wife. She died almost immediately and he died a few minutes after the cars reached the Residency. Gavrilo and Nedjelko were nineteen years old, part of a group of conspirators who had been provided with weapons by the Black Hand, a powerful secret society with head-quarters in Serbia.

The news of the assassination was received calmly in Vienna, for Franz Ferdinand had been generally unpopular. The old emperor shed a tear but was rather relieved at the removal of a somewhat untrustworthy heir, for now his great-nephew Charles, a saintly but predictable person, was heir apparent.[4] The court was mainly concerned to humiliate Sophie in death as it had enjoyed doing in life. Conrad, however, knew that war was inevitable, and he wrote to Gina on the evening of the assassination: 'It will be a hopeless struggle, but nevertheless it must be, because such an ancient monarchy and such an ancient army cannot perish ingloriously.'[1]

On July 31, 1914, his father brought Erwin's mobilization orders to his office in the new Physics Institute on the Boltzmanngasse. They went together to buy two pistols, one very small and one very large. 'Luckily', recalled Erwin, 'I have never fired them at man or beast.'[5] Before leaving for the front, he sent Anny Bertel a book of essays by Felix Salten, *Gestalten und Erscheinungen*, which he inscribed 'In grateful remembrance of the beautiful long days'.

His orders were to report to Predilsattel, near the Italian border, a fortified position on a mountain pass about 1000 m high, commanding the Seebach valley that opens onto the Venetian plain. This was an excellent place to be at the beginning of the war, well away from the murderous battle of Lemberg on the Russian front, where Conrad lost one-third of his effective troops in the first three weeks of fighting, 250 000 killed or wounded and 100 000 prisoners. The general, of course, practically never ventured near the battle front; he was ensconced in his staff headquarters in a castle at Teschen, in Austrian Silesia, enjoying candle-lit banquets with a succession of lovely lady guests.

On one occasion at Predilsattel, Erwin's commander Captain Reinl sounded an alert but it was a false alarm. Columns of smoke were detected along the shores of the Raiblersee, and he was convinced it signaled an Italian attack, but some peasants were either burning weeds or roasting potatoes. There were observation posts on elevated positions at both sides of the fort, and Erwin spent some time in one of them.

In the approximately ten days before someone thought to fetch us down, I learned two things: First, that one can sleep very well on springy boards with only a sleeping bag and blanket instead of a mattress, much softer than on really hard ground. Second, one night from our observation post, we saw a number of lights moving up the slopes at the head of the lake, where there were no paths, apparently coming towards our position. I sprang up . . . and went through the connecting passage to the post to survey the situation. The observation was correct, but the lights were St Elmo's fire on the points of the barbed wire entanglements only one or two metres away, the displacement onto the background being caused by parallax as a result of the movement of the observer himself. I saw this really enchanting phenomenon another time on the points of grasses on the clods of earth that covered our roomy dugout, once when I went out for a certain purpose. I don't remember ever seeing St Elmo's fire before or since.[6]

Research in uniform

During the first months of the war Erwin was able to complete some scientific work. The advantage of being a theoretician was that he could make calculations even in a dugout, although he missed the possibility of consulting the scientific journals. On October 27, he posted to the *Annalen* from Raibl, a few kilometres from the Predil Fort, a short paper on 'Capillary Pressure in Gas Bubbles'.[7] He had been working on this problem at the outbreak of the war; it was probably suggested by a standard student experiment in the practical class, an experiment that is still popular since it yields good results with a minimum of special apparatus. The method had been improved by Gustav Jäger who was professor at the Technische Hochschule in Vienna.

Three different authors had derived three different expressions for the surface tension and Schrödinger showed that all three were wrong. He gave the correct expression and also a new formula for the Jäger method. He had begun some experimental measurements with wide-bore capillaries (for which the third term in his theoretical formula becomes more important) and they agreed well with his equation. He concluded by saying, 'I intend to continue the measurements, which had to be interrupted at the end of July, so as to obtain more extensive data for testing the formula.' Obviously he did not think that the war would continue four more years.

Erwin's next assignment was to Franzenfeste, a fortress just north of Brixen in South Tirol, which was built in 1835–38 to dominate the Brixener Klause, a gorge at the entrance of the Brenner Pass. Here he was in the mountain country that he loved, surrounded by snowy peaks. It was a perfectly peaceful sector during the winter of 1914/15. Meanwhile the German offensive towards Paris had ground to a halt and the war of attrition had begun, which would take millions of lives in the muddy trenches of the western front. The Central Powers had lost their chance to win the war, but the slow process of their destruction was only beginning.

A most unwelcome surprise to the forces of the Empire was a humiliating defeat by Serbia. The Austro-Hungarians, mostly reserve troops, were under the command of Oskar Potiorek, who had been in charge of security at Sarajevo. His first attack was repulsed by the Serbs, but a second effort captured Belgrade. Then the Serbs under Radomir Putnik counterattacked, recaptured their capital, and drove into Hungary. An alarmed Conrad planned a defensive line from Lake Baloton to Budapest, but the exhausted Serbs could go no farther and all became quiet on the southeastern front.[8] At about this time, Erwin was transferred to Komaron, a Hungarian garrison town between Vienna and Budapest.

While there he was able to write a paper for the *Physikalische Zeitschrift*, 'On the Theory of Experiments on the Rise and Fall of Particles in Brownian Motion',[9] which was received on July 26, 1915. This theory was related to work in progress in Vienna on the charge of the electron.

One of the most important experiments in twentieth-century physics was the determination of the electronic charge by Robert Andrews Millikan at the University of Chicago in the spring of 1909. Earlier attempts at this measurement had collected the total charge on a cloud of suspended droplets, but Millikan was able to isolate an individual droplet of oil or water and to measure its charge by its motion in the gravitational and electric fields between a pair of metal plates. He found that the charge on a droplet was always a small integral multiple of a unit charge $-e = 4.774 \times 10^{-10}$ esu (modern value, 4.8032). Felix Ehrenhaft at the University of Vienna, who in 1907 had been the first to observe the Brownian motion of particles suspended in gases, repeated the Millikan type of experiment but used metal particles. He believed that his results indicated the occurrence of charges less than $-e$ and thus cast doubt upon the atomic nature of electricity. Millikan pointed out errors in this work, and Ehrenhaft was eventually shown to be wrong.[10]

Owing to Brownian motion, the time required for a suspended particle to traverse a fixed vertical distance in a Millikan-type experiment varies considerably from run to run, and it is necessary to average the times over a large number of observations to obtain a reliable result. The unusual statistical problem that thus arises had not been correctly solved before Schrödinger took it up in one of his most elegant early works, one that is particularly relevant to his later research because it is his first publication in statistics, a subject that lies at the heart of interpretations of quantum mechanics.

In his paper, Schrödinger first carefully explained two mistakes in previous work, and then gave the correct solution. The problem is to derive the probability $p(t)dt$ for a first passage time t across an imaginary barrier at a distance l from the origin of particles moving with a velocity v between the plates in a Millikan-type experiment. He called particles that have never crossed the barrier 'white' and those that have crossed at least once, 'red', so that the problem was to calculate the decrease in the number of white particles in a time t. If $v = 0$, the number of red particles would have a gaussian distribution about $x = l$, and the number of whites would be simply the total number minus the number of reds. Schrödinger transformed the distance coordinate $[x' = -vt]$ so that the barrier moved with velocity $-v$. By considering the Brownian motion as equivalent to a diffusion process, he obtained the required solution in the transformed coordinates, and then transformed back again to get it in stationary laboratory coordinates. He calculated the mean square deviation of the first-passage times, showing that it is proportional to the square root of the distance traversed. The relative error in a set of n measurements of the particle velocity varies approximately as $(2/n)^{1/2}$ and a considerable number of measurements is needed for reliable results, about 70 in a typical case.

Millikan's work was statistically valid because he always used a large number of measurements. Surprisingly, however, he did not quote this important Schrödinger paper in the 1922 edition of his book *The Electron*, perhaps because he was an old-fashioned experimentalist who would have found the mathematics incomprehensible. Almost simultaneously with Schrödinger's paper, Marian Smoluchowski obtained the same result by solution of a diffusion equation containing the velocity term. Erwin was very interested in this work and he made a detailed study of the Smoluchowski papers, and thus became conversant with the important subject of fluctuation phenomena.

On the Italian front

On April 26, 1915, the secret 'Pact of London' was signed between the Triple Entente and Italy, promising her large slices of Austrian territory, control of Albania, and some colonies in Africa and Asia

Minor. Britain and France promised Russia the remnants of Turkish lands in Europe, including Constantinople and control of the Straits. Italy hesitated no longer and on May 23 declared war on Austria-Hungary. This action aroused great popular feeling in the Empire. Fritz Hasenöhrl was so eager to fight the Italians that he arranged to be transferred immediately to the Italian front, where on July 20 he sustained a severe wound in the shoulder in an assault on Monte Piano. He was sent to the hospital at Salzburg, but returned to the fight before he had fully recovered.

On July 26, Schrödinger's unit was ordered from Komaron to install a new 12-cm marine battery on the Italian front. For the next two months, he kept a detailed diary of his activities. The marine battery was a great puzzle, for nobody seemed to know exactly where it was supposed to go, only the name of the locality, a place called Oreia Draga on the Istrian peninsula south of Trieste. He set off at 10 a.m. from the railway station at Wiener Neustadt. Irene Drexler was there to wish him farewell from the train platform. By noon next day he was in Marborg [Maribor] trying to find out about transport for the gun. He met the railway officer from Leibach [Ljubljana] and more or less by chance they got the necessary instructions, but the whole project was delayed since his unit did not know where they were supposed to report.

Early next morning the second Italian attack on Görz [Gorizia] began. A battle raged around the Karst plateau, where the barren limestone rocks were pulverized by artillery fire. A major Italian offensive, the first of many battles of the Isonzo, had opened on June 23. In repeated assaults the Italians failed to make significant gains through the Austrian defenses, but losses were heavy on both sides. A realistic picture of this section of the front as seen from the Italian side was given by Ernest Hemingway in *A Farewell to Arms* [1929]. He did not arrive on the Italian front till June, 1918, and he was blown up by an Austrian mortar on July 8. Yet his reporting was so accurate that Mussolini banned the book in Italy. As his protagonist Frederick Henry recalled:

We crossed the river in August and lived in a house in Gorizia that had a fountain and many thick shady trees in a walled garden and a wistaria vine purple on the side of the house . . . The river ran behind us and the town had been captured very handsomely but the mountains beyond it could not be taken and I was very glad that the Austrians seemed to want to come back to the town some time, if the war should end, because they did not bombard it to destroy it but only a little in a military way.

When Erwin was there about two years earlier, the military situation was about the same except that the Austrians still held Gorizia and about half a million young men had not yet been killed. Erwin noted that 'we stood about for hours on the Karst plateau, and about midday had the foresight to take a meal including green bean salad and a passable beer. Towards evening we went through Görz and reached Oreia Draga about 8 p.m. The men had eaten nothing all day, because the supplies had not arrived although the transport officer had ordered them from Ogenia.'[6] They were able to cook some canned goods and a good-natured regimental doctor provided some wine.

On the morning of the 29th, they marched to St Peter, with almost unbroken cannonading beyond Sessana. In St Peter they were met by Captain Novak and Colonel Heckler. Their battery was already in place and had already fired. The men were quartered in a farmhouse and the officers in the castle of the local count. 'Very fine except that all the windows had been blown out by our 30.5 cm Mortar, which had flattened almost everything in the park.' This was the best of the K & K artillery weapons, known as a 'wonder gun'; it had been introduced in 1912, but unfortunately there were still only a few of them.

After a short sleep, Erwin went by motorcar to Görz, finding the town half deserted because of the bombardments. Nevertheless he enjoyed the unusual comfort of the best hotel. That afternoon he went to help direct some artillery that was firing from a suburb. The next day he went to San Marco as an observer, and at noon took over the direction of the firing. The commander of this sector was Major Schuster, for whom he had great respect. The next morning he continued to direct the battery. On August 1, he observed the first firing of the marine battery, which was blasting away at church towers in the distance.

On August 2, he noted that 'we are firing badly. I was balled out. I took control of all the operations myself, which slowed down the action. I was balled out some more. By afternoon it was going well. It is incredible that we were not fired upon at all or only insignificantly since we were using strongly smoky powder. Airplanes were looking for us.' On August 4, the shooting was again poor; Erwin suggested what to do but they did not believe him. 'Thank God! The other battery is also shooting poorly. The emplacement is already quite crooked.' On August 6 the battery was for some time solidly placed, and the shooting went well. The foundation blocks were inclined at 11°. 'Heckler, deputy commandant, is unwilling to move himself to take a look at it. I go out in the evening. The cannon doctor repairs the springs which were already quite slack because of the poor mounting, and he pours in half a litre of glycerine.'

On August 7, troubles with the mortar continued. After six shots, the block elevation increased and Heckler racked his brains as to what was happening. They managed some good shots, but again the block elevation rose and finally the gun dug itself a hole in the earth. They came under heavy fire from at least two enemy batteries, first with shrapnel and then with heavy shells. 'The men were very brave; we are generally well protected and if a shell doesn't hit the roof directly in the middle not much can happen.'

On the 8th, the captain came back.

It was not appreciated that I had taken the responsibility to keep the gun firing . . . That my previous advice had not been taken was forgotten. He says 'You can't picture the situation if you just hear a lot of talk – you've got to see to it yourself. That's what an artillery officer is for.' Exactly the same thing I said yesterday when I begged him to come see it for himself. Next time I'll do much better! Anyhow, I was glad that we had fired again.

A new gun emplacement was set up only 100 m from the old one. With enormous effort they built a great base of 38 m^3 concrete, secure against bomb shells; Erwin doubted that they could cause the enemy as much trouble as they expended in building the emplacement. They fired only on direct orders from corps command, which was at Dornburg, 14 km away.

The news from the eastern front was 'splendid': combined German and K & K forces had driven the Russians out of Galicia and Poland and pushed the attack deep into Russia. For the birthday of the Emperor on August 28, flags flew everywhere, and the officers celebrated with 'bad champagne'. Weigl was in charge of the company, and Novak was put in charge of two companies and three guns, which to Erwin seemed 'merely comical', since neither one, but only Schuster, knew how to shoot.

Every second afternoon, Erwin could go to Görz to stroll about the town and enjoy a visit to the coffee houses. On the 20th, he went for an inoculation; he had the beginnings of an intestinal flu, but was worried by a report of cholera in a nearby small town. Any losses in the artillery companies due to enemy action would, he believed, be only a matter of bad luck. Normally nothing happened but once in a while a shell hit an observer, or a dugout where four or five men were blown apart. 'Otherwise one should have no misfortunes.'

On August 21, there is an abrupt change in the tone of the diary.

Dreamed again of L[otte] the whole night long. 'The whole night' was probably five minutes, but such is the subjective feeling as one awakes. I cannot get free of her, even when I know she is not worth it, but it is as if I were still under the enchantment of that evening when for the first time I held her hand in mine, of the walks in the snow, those few happy days, which no greater favor could, or will, put in the shade. I knew long ago that the structure would tumble down, when she stretched out her hand to me from the railway platform on September 12, 1912. I had no more to hope for, might wish for no more, than the fleeting beauty of the instant. I was a child with the child, and to reawake

Lotte Rella in Semmering.

the vanished *Angst* of a childish eroticism was for the 25-year-old man a source of infinite charm.

For all that, I still hope, I still believe, as I curse her another thousand times, that I can call myself lucky to have escaped the foolish creature.

How deeply I now feel the humiliations which I, — which she imposed upon me, uncalled for. I was foolish, I alone was foolish. I gave the reins that I could have controlled, that I disdained to control, to the child in whom I saw a goddess, which she was to me, not the child that she remained longer than many others. But even so I could have led her, must have led her.

Could have led – I? Not I, but another in my place. And yet I cannot reproach myself.

This confession shows how permanently Erwin's first love affected his soul. Although since then he had experienced a passionate affair with Ella and romance with Irene, and also the desire to make a married life with Felicie, still, when he was in the midst of the cannon fire and random death and mutilation of senseless warfare, it was his high-school sweetheart Lotte who came from his subconscious mind into his dreams.

There was something infinitely appealing to Erwin in the budding sensuality of immature girls. To try to understand this aspect of his personality, which would be disclosed more than once during his life, we might consider the work of Balthus, one of the greatest and most mysterious of modern painters. His eldest son once wrote that 'the fabled theme of the young adolescent girl, which Balthus has treated repeatedly, has nothing whatsoever to do with sexual obsession except perhaps in the eye of the beholder. These girls are emblematic archetypes belonging to another higher realm . . . Eroticism is nowadays confused with libidinousness thus obscuring the true intelligence of esoteric works ultimately pertaining to the divine cosmic mystery of love and desire.'[11] From his early days Balthus, like Schrödinger, was attracted to the wisdom of the East. This oriental interpretation of a sexual pattern is obviously contrary to the pervasive doctrine of psychoanalysis, which seeks always to explain the spiritual in terms of the sensual.

The war diary continued on September 6, with the notation that firing was resumed then after a long break on orders from Army Command. Probably they were short of ammunition, for at the end of August Erwin went to Leibach to collect some. There was a six-hour stop at Opčina, where he took the opportunity to visit the obelisk erected in 1830 in memory of a visit by Emperor Franz I; it is at the edge of the Karst plateau, near a famous observation point with a magnificent view over Trieste and the gulf. Erwin reported it to be 'unforgettable'. On August 24, he was thirteen hours in Leibach; the town was overflowing with about four times its normal population. 'The streets were crowded with an unbroken stream of officers and whores. Most of the female refugees seem to live that way.' The throng at the park concert was like Gastein or Abbazia, two noted spas. 'Next day to Oberleibach, a lovely little village. A dimwit of a transport officer. One must extract every hobnail from him with wind from his backside. A moving scene of farewell of a gypsy from his black pony.'

At the evening mess in the Castle, there were interesting accounts from some officers who had taken part in the debacles in Serbia and Galicia. In Serbia, they all lost their heads entirely in the precipitate retreat. The Serbians appeared astonished and did not push the pursuit. Typical of the panic was the blowing up of the Belgrade bridges with the Austrian troops still on them, and long before they had all crossed. No attempt at all was made to hold the city. Major Schuster was there with the heavy mortars. It took a long time to bring them forward, and then they had to be immediately brought back again. 'Later they were ordered to cover the retreat. Heavy mortars!' The Serbs were depicted as extraordinarily gallant and chivalrous enemies. During their control of Semlin, not a house was destroyed and not a nail was stolen.

On 19, September, Erwin wrote:

Interesting events zero. For a long time it was absolutely quiet here, now at night there are front line attacks on the plateau. It is utterly boring. When I have otherwise nothing to do, I fill my mind with the psychology of the fundamentals of consciousness (memory, association, the concept of time).

Tonight again dreamed very vividly of L[otte], a remarkable thing that. With my own conduct in the matter, the more I review it, the more satisfied I am with it. Especially that I was so clever at that time – long ago – to write the letter in question to Johanna [a first cousin and good friend of Lotte]. Instead of a shackle, it gives me now the most complete freedom to maintain my dignity, in which I can be silent as long as I want to. And that was urgently necessary for me. I make for myself, so to speak, a paper lock, allow nobody to enter, until the ban is broken in the right way. And then I shall have the situation that I need, which could never have been created in any other way.

I do not need to speak, because I have already spoken. Like a coat of mail this one word guards me against every reproach, ensures me against every disaster from either one side or the other.

I will have only one thing three times for every argument from her side: peace, peace, peace. Be silent and do not allow yourself to be duped. Silent as the *earth*, as a buried stone.

What are we to make of this – one of the greatest intellects of his time talking to himself about an affair of the heart as if he were a shyster plotting a defense against a lawsuit? Many years ago he must have written to Johanna to explain his intentions, or lack of intentions, towards Lotte. Why does she now come back to haunt his dreams?

The final entry in the diary is made on September 27:

It is really horrible and I have a home-sickness for work. If this goes on for long I shall be a wreck physically and mentally. I am no longer accustomed to work or to think for half an hour. Every rational thought is entangled with another one: what's the use of it all if the war is not finally finished. The last two months here were the best of it. But I have outgrown the whole thing. Is this a life: to sleep, to eat, and to play cards? Through incessant military service, every bodily activity is turned into an illusion.

Many days I lie in bed till the midday meal, merely to avoid facing the question of what I should begin to do in the morning. After eating, naturally a siesta. To stand up for five minutes is too wearying, so I sit down. I scarcely notice the names of my acquaintances.

Why is it that I am the only one with nothing to do? Is it my fault? Is it the standard that I apply? Or the special conditions of my demand for activity – especially because of the futility in which I now live.

I am not completely desperate about it – much too apathetic to be that. I think: it continues in its idiotic way, nothing can be done. Frightful.

Remarkable: I no longer ask when will the war be over? But: will it be over? Not naively: hopefully. Are 14 months so horrible? That already one actually begins to doubt the end.

It is really to be doubted: people are becoming accustomed to the war. I really don't know whether there are not people today for whom the burning question is: how will we endure the peace? as others ask: how the war?

For now a revolution of an entire people must be reversed. Revolutions, however, never become simple retrogressions.

What fears for the future the war already raised in us even at its beginning. We set them aside when we saw that the war might be endured more easily than we had thought. Was that not a great fallacy? Is therefore the same true of the coming peace?

We have war! [*Wir haben Krieg!*] The word sounds playful. Truly, I swear, the word sounds playful. For it sounds as though this were some out-of-the-ordinary state.

At this point the war diary breaks off. Schrödinger began to receive some books of philosophy and copies of scientific journals, and he was able to rouse himself from the desperate boredom and futility of military life and to turn his mind to the kind of problems that he loved.

Military citations

By the end of 1915, a combined German, Bulgarian and Austrian offensive had finally eliminated Serbia from the war, and some reinforcements could be sent to the Italian front. The third great battle of the Isonzo opened on October 18, with a seventy-hour Italian artillery barrage of unprecedented ferocity. Fortunately the K & K Fifth Army was in well fortified positions and when the attack was broken off early in December, the Italian gains had been limited to a

Schrödinger as Leutnant in the Fortress Artillery, at the front 1916.

small area of the west border of the karst plateau. The Italians lost 125 000 killed and wounded, compared to the K & K loss of 80 000.

Schrödinger was awarded a citation for his outstanding service during this battle:

In the battle of Oct. 23 to Nov. 13, while acting as a replacement for the battery commander, he commanded the battery with great success. During the preparations as well as in many engagements, he was in command as first officer at the gun emplacement. By his fearlessness and calmness in the face of recurrent heavy enemy artillery fire, he gave to the men a shining example of courage and gallantry. It was owing to his personal presence that the gun emplacement always fulfilled its assignment exactly and with success in the face of heavy enemy fire . . . He has been at the front since July, 1915.

This was signed by his division commander on November 12, counter-signed by divisional and corps commanders, and entered in his record in the War Archive.[12]

In the Tirolean sector of the Italian front, Fritz Hasenöhrl had been ordered back to duty with a military service cross III Class and his shoulder wound not yet completely healed. At Vielgeruth [now Folga-ria] in South Tirol, he was killed by a grenade while leading his batallion of the 14th infantry regiment in an attack. Hermann Mark, then twenty years old, was a corporal in a nearby company, and he recalls the great excitement at the news that the famous physicist had been killed. Hans Thirring later commented that 'the moral qualities of duty and sacrifice stand higher than the intellectual qualities of critical thinking and foresight. In those days before Hiroshima there were few intellectuals and researchers who would apply the scientific scep-ticism and criticism which they used in their work also to the questions of power politics . . . We must have even more respect for his spiritual qualities than we have for his bravery.' Hasenöhrl was posthumously awarded another decoration and the emperor sent the widow a telegram of condolence, which was unprecedented for a mere *Ober-leutnant* in the reserves.[13]

On May 1, 1916, Schrödinger was promoted to *Oberleutnant*, and May 15, after a long winter break, fighting was resumed. The Italians were able to advance about 10 km in a salient around Gorizia and then captured the town. These were the bloodiest of the Isonzo battles, with Italian losses of 286 000 men and the Austro-Hungarian less than half that many. The Italian general Cadorna was impervious to the slaughter of his army for slight tactical gains, and although his grand plan to capture Trieste never seemed likely to succeed, full credit must be given to the tenacious defense by the imperial Fifth Army.

There is no detailed record of Erwin's military service in this battle, but at about this time he was given charge of a battery which had been installed at Prosecco, about 300 m above Trieste, and somewhat to the north of the city. He called this a 'still more comical naval gun in an extremely boring but beautiful lookout spot'. He spent the rest of 1916 here, either by good luck or because, after the death of Hasenöhrl, somebody in Vienna did not want to see his possible successor meet a similar fate.

On November 21, 1916, Franz Josef died, in the eighty-sixth year of his life and the sixty-seventh year of his reign. The last news he heard from the war had been encouraging – the capture of Bucharest. Both he and Kaiser Wilhelm had made an effort to stop the war in 1916, but the Kaiser's manner was so arrogant that these peace overtures were rejected by the Entente. The new emperor Charles was that rarest of species among the powerful, a sincere Christian. Much to the distress of the military commanders, he forbade any attacks on civilian targets and the use of poison gas. He immediately sent the brother of Empress Zita, Sixtus of Bourbon-Parma, to Paris, to begin negotiations for peace. Sixtus was an officer in the Belgian Army, since as a Bourbon he was legally forbidden to serve in the French armed forces.

Early in 1917, Schrödinger was surprised to find Sixtus making a tour of inspection of the K & K forces around Trieste. He did not know until later of the peace negotiations. Sixtus persuaded Charles to agree to a separate peace with the Entente, territorial concessions to Italy, and eventual restoration of Alsace Lorraine to France. As Schrödinger remarked later, this was 'a treacherous betrayal of Germany, which unfortunately did not come to pass'. The Italians, however, demanded everything promised in the secret Treaty of London, and when the Germans learned of the Sixtus affair, they were able to tighten a virtual noose around the neck of the hapless Charles, who lacked the stamina to resist the military dictatorship that now controlled Germany.

Erwin was now more or less resigned to the continuing war, having recovered from the deadly period of idleness and depression of 1915. He must have been cheered also by a visit from Annemarie Bertel. It is interesting that none of his Vienna girl friends came to see him, but only the country girl from Salzburg, now a delightful young woman of twenty. According to Anny's account, they did not become lovers at this time, but this denial is somewhat hard to believe. She was much in love with Erwin and the idea of a sisterly visit to a gallant soldier facing possible even if not probable death for his country would not have been consistent with her passionate nature. On the other hand, Erwin's record of his loves lists Anny with the date 1919. Thus it would appear that Erwin's sexual experience before his marriage was restricted to one love affair, with the probability, however, of many casual encounters that he did not record. It is difficult to imagine that he spent all his time in Leibach drinking coffee with other subalterns, although it is also doubtful that his sex life was as uninhibited as that of Lieutenant Henry on the Italian side. The truth of the matter may be more complicated and more interesting.

Meteorology service

In the spring of 1917, Schrödinger was transferred back to Vienna. He was assigned to teach an introductory course in meteorology at a school for anti-aircraft officers at Wiener Neustadt [50 km south of Vienna] and also to teach a laboratory course in physics at the University, similar to his former practical course but more elementary. He was stationed for a while at the *Sternwarte* [Observatory] outside Vienna, where Hans Thirring was also involved in meteorological work; this was the principal weather station for all of Austria.

Schrödinger's lectures on meteorology were similar to those he had heard from Julius Hann ten years previously. In his notebook on this course, the topics include the composition of the atmosphere, the variation of temperature with altitude and its diurnal and seasonal variations, areas of high and low pressure and how they move, types of clouds, thunderstorms, wind currents and how they are affected by bodies of water and mountains.

Hermann Mark was one of the technically qualified soldiers who were given special furloughs to take some science courses. He first met Schrödinger as his instructor in the physics laboratory. They were both in uniform. As Mark recalled, Schrödinger 'was already very impressive to all of us because of his kindness. Of course there was not much physics involved in these experiments that we carried out. We would weigh something out and put it in water and weigh it again and determine the volume, in other words, make very simple experiments. But he was infinitely patient with us although we apparently did nothing for months – well that lasted a few months and I had to go back to the front.'[14]

Schrödinger's first paper after his return to Vienna was sent to *Physikalische Zeitschrift* the last week in July, 1917.[15] It seems to have been inspired by his experience as an artillery officer, for it deals with the problem of the so-called 'outer zone of abnormal audibility' of large explosions. As one moves laterally away from the point of the explosion, the sound level is first attentuated but then may rise again in a zone 50–100 km distant, before finally dying out at longer distances. In 1911, G. Borne had suggested that the effect was due to reflection of the sound waves from layers of the atmosphere at heights of 70–100 km. This hypothesis was attacked in 1915–16 in a number of papers by Wilhelm Schmidt, who had studied with Exner but later turned to meteorology and geophysics. Schmidt showed that, according to the usual equation for soundwaves, the intensity (energy) of the wave would be directly proportional to the atmospheric density, and thus the wave would be so attenuated at high altitudes that the reflection mechanism of Borne could not hold.

Schrödinger suspected that Schmidt had made an error in his analysis, since it appeared to contradict the law of conservation of energy – where did the supposedly lost energy go, he asked? He therefore decided that the error must lie in the wave equation used by Schmidt, which did not include the gravitational effect responsible for the variation of atmospheric density with altitude. He first considered the equation of motion for a sound wave in an isothermal column of air, and found for waves of small amplitude the differential equation,

$$\frac{\partial^2 \eta}{\partial t^2} - v^2 \frac{\partial^2 \eta}{\partial x^2} + g\gamma \frac{\partial \eta}{\partial x} = 0$$

where η is the compression, x is the height above ground level, and γ is the ratio of specific heats c_P/c_V.

The solution of this revised wave equation showed that the amplitude varies inversely as the square root of the density, so that the increased amplitude of the waves at high altitudes just compensates the effect of density on intensity derived by Schmidt. Thus there is no extra attentuation of the wave due to altitude. Schrödinger next showed that this conclusion is not altered for sound waves moving in three dimensions in the isothermal atmosphere. The effects of variations in temperature and composition could be estimated by considering successive layers of the atmosphere to be approximately homogeneous. Indeed the refraction at the interfaces between such layers was used to calculate the bending of the sound waves in the original Borne theory. There might be a little attenuation associated with this refraction, but not nearly so much as suggested by Schmidt.

Having disposed of Schmidt's objections to the Borne theory, however, Schrödinger raised some new objections of his own, which made it seem unlikely to him that the sound could be reflected with appreciable intensity from the upper layers of the atmosphere. In air at ground-level pressures, the damping of sound by thermal conduction and viscosity is practically negligible, except for high frequencies near the limit of hearing, but at the lower pressures of the upper atmosphere, these damping factors become appreciable at all frequencies. Ordinary experience demonstrates this effect, for example, the weakness of calls shouted between mountain climbers or balloonists. Schrödinger used formulas from Rayleigh's famous *Theory of Sound* to calculate the damping as a function of altitude for sound of various frequencies. As soon as the damping becomes appreciable, it increases steeply with the altitude; it is also very dependent on frequency, the higher the frequency the greater the attenuation. His final conclusion is that reflection from the upper atmosphere is unlikely unless it can occur from a level considerably lower than the 70–100 km that had been suggested as the likely location of appreciable changes in atmospheric composition. He suggests that new measurements to determine the frequency spectrum of the sound in regions of abnormal audibility would help to establish its mechanism.

Even though it did not solve the problem of abnormal audibility, this paper was an excellent example of classical applied physics. It is written with the clarity and attention to style and organization that already can be recognized as Schrödinger hallmarks. The harvest of long Gymnasium years devoted to the classical authors can be seen in the graceful prose of his scientific papers. If it were not for the mathematics, they could be read with pleasure as literary essays.

Schrödinger's stint in the army obviously had not dulled his theoretical skills, yet neither had the fallow period led to an outburst of original thinking about problems in the forefront of physics, in particular the problems of quantum theory and atomic structure that were stretching the fabric of classical physics to the breaking point. He was still reacting to various concerns of the somewhat isolated Vienna school, still using his great mathematical facility to make improvements in structures built by others, although he was now thirty years old, an age by which most great theoretical physicists have been prepared to rebel against the paradigms received from their university teachers.

First paper on quantum theory

About midway through 1917, Schrödinger sent an article on 'The Results of New Research on Atomic and Molecular Heats' to *Die Naturwissenschaften* [*The Sciences*].[16] This journal, published by Springer Verlag, was founded and edited by Arnold Berliner, who encouraged authoritative reviews of the status of important fields in addition to brief reports of original research. Schrödinger's article appeared in two parts in successive issues.

Most of the paper is concerned with the heat capacity of solids. In 1819 Dulong and Petit had noted that molar heat capacities at constant volume of the elements are almost all close to a value of 6 cal/deg mol. This value is explained on the basis that the energy taken up when a solid is heated is distributed uniformly among the vibrational motions of its atoms, with vibrations in three dimensions each having energy kT, amounting to $3\,kT$ per atom or $3\,NkT = 3\,RT$ per mole, so that

$C_V = (\partial E/\partial T)_V = 3R = 6$ cal/deg mol. There are some exceptions to the rule, notably light elements like diamond [1.5 cal/deg mol]. Below room temperature, C_V decreases and ultimately approaches 0 as $T \to 0$. In 1909, Einstein explained this behavior on the basis of quantization of the vibrational energy; he assigned a characteristic vibration frequency v to each element, and the solid can take up energy only in increments hv. The equation he derived for C_V represents the data quite well, the low value for elements such as carbon being explained by the high vibration frequency of the light atoms, but his calculated C_V went to zero more rapidly than the experimental value. The experiments were mainly the work of Walther Nernst and Frederick Lindemann in Berlin and Willem Keesom in Leiden. In 1912, Debye, then in Zürich, realized that the Einstein model, by taking a single v to fit the data at high temperatures, fails to consider the vibrations of low frequency, which are important for the uptake of energy at low temperatures. On the assumption that the vibrations are distributed over a range of frequencies, he derived a new equation that gave an excellent fit with the data over the entire temperature range. Another approach to the vibrations of solids was that of Max Born and Theodor Karman, based on the dynamics of an atomic lattice. In his review article, Schrödinger presented these developments in a way that would be readily understood by a nonspecialist in the subject.

He then turned to a brief discussion of the heat capacities of gases. The state of understanding of C_V for the diatomic gases like H_2, O_2, and Cl_2 was much less satisfactory than that for solids. Schrödinger was not able even qualitatively to disentangle the contributions due to translational, rotational and vibrational motions, and he did not consider the detailed data on temperature dependence. He was apparently unaware of the important work in Berlin of the Danish physical chemist, Niels Bjerrum, in 1912, which set forth the quantization of rotational and vibrational energies in molecules. In general the approach to quantum theory in Vienna at this time was through the applications to thermodynamics as made by Planck and Einstein, and not through the spectroscopic studies emphasized by the Copenhagen school of Niels Bohr. Nevertheless, this essay of 1917 marks the first time that Schrödinger wrote anything about the quantum theory, and two years later it was followed by a much more extensive review of the same general area.

Smoluchowski and statistics

Among Schrödinger's notebooks from 1914 to 1918 are three devoted to the work of Marian Smoluchowski. The notebooks are undated, and it has been suggested that the first one, entitled 'Fluctuation

Opalescence', may have been written late in 1914, but other historians place it several years later. The other two, entitled 'Discussion of the Last Works of Smoluchowski', were written late in 1917 or early in 1918. They deal with Brownian motion, fluctuation phenomena, diffusion, and the statistical basis of the Second Law of Thermodynamics.

Marian Smoluchowski was born in Vorderbruhl near Vienna in 1872, and attended the upper-class Theresianum Gymnasium at the same time as Fritz Hasenöhrl, with whom he formed a lifelong friendship. He completed his physics studies at Vienna, worked in Glasgow and Berlin, and received an appointment in Lemberg in 1899. In 1913, he became professor of experimental physics at Cracow. Schrödinger did not have an opportunity to know Smoluchowski as a coworker in Vienna, but he came indirectly under his influence through Hasenöhrl, and came to recognize him as the Vienna physicist who was the most worthy successor to the heritage of Boltzmann. Smoluchowski died in an epidemic of dysentery in 1917. He is especially remembered for his theory of Brownian motion.[17]

In 1827, shortly after the invention of the achromatic lens, the Scottish botanist Robert Brown observed under his microscope a curious perpetual motion of pollen grains suspended in water. 'These motions were such as to satisfy me . . . that they arose neither from currents in the fluid, nor from its gradual evaporation, but belonged to the particle itself.' In 1888, Georges Goüy suggested that the particles were propelled by collisions with the rapidly moving molecules of the suspension fluid. In the phenomena of Brownian motion, we can see with our own eyes events occurring at the borderline between the macroscopic and the molecular worlds. The perpetual Brownian motion does not contradict the First Law of Thermodynamics, for the source of energy that moves the particles is the kinetic or thermal energy of the molecules surrounding them. We may assume that in any region where the microscopic particles gain kinetic energy, there is a corresponding loss in energy of the surrounding molecules, which thus undergo a localized cooling. Brownian motion thus reveals that the Second Law of Thermodynamics is a statistical law: in sufficiently small regions, the entropy does not remain absolutely constant but fluctuates about a mean equilibrium value. In 1905/06 Einstein and Smoluchowski independently provided mathematical theories of Brownian motion, in several of the most important papers of twentieth-century physics. Schrödinger recognized the basic importance of fluctuation phenomena and he plunged into a detailed study of this field, mastering all the papers, and then making original and important contributions to its development. Fluctuations would become one of the leit-motifs of his future scientific work.

Schrödinger studied carefully the paper that Smoluchowski had contributed to the *Festschrift* for the sixtieth birthday of Boltzmann in 1904, 'On Irregularities in the Distribution of Gas Molecules and their Influence on Entropy and the Equation of State'.[18] In this paper he ascribed the critical opalescence, which appears in a dense gas in the immediate neighborhood of its critical point, to the large fluctuations in density of the gas that occur under these conditions. Smoluchowski also gave important lectures at the Münster meeting of the German Society of Physicians and Scientists in 1912, and at Göttingen in 1913 under the auspices of the Wolfskehl Foundation. In these lectures he explored the problem of the statistical nature of the Second Law and the limitations on its validity. He was not troubled by the apparent paradox that natural processes are irreversible, always increasing the entropy of the universe, whereas molecular processes are essentially reversible, and he believed that: 'irreversibility is only a subjective concept of the observer, the applicability of which does not depend upon the type of natural process, but rather upon the position of the initial point and the duration of the observation . . . Processes will appear to us to be irreversible, if their initial points lie far beyond the average range of a fluctuation, and if they are observed only for a period of time that is short compared to the time of recurrence.'

Schrödinger's first paper related to Smoluchowski's work was the solution of the diffusion equation in the presence of a gravitational field, which he completed while on active service at Komaron. Smoluchowski independently solved this same problem and published it at almost the same time, but this coincidence did not bother Erwin; he was in fact almost delighted that his result was confirmed by the acknowledged master of the subject. The equation solved by Schrödinger and Smoluchowski was an example of a general type applied by Adriaan Fokker (brother of the aircraft designer) and Max Planck to the time development of stochastic processes [a stochastic variable is one that does not depend in a completely definite way upon the independent variable (e.g, the time t) but is subject to random effects that can only be defined in statistical terms].

Problems in statistical dynamics

During 1918, Schrödinger wrote two long papers on statistical dynamics, devoted to a complete analysis of the random fluctuations in the rate of radioactive decay, the so-called *Schweidler fluctuations*. We are today so familiar with radioactivity that it is impossible to recapture the wonder and mystery that surrounded its manifestations at the time of its discovery. The idea that an atom of an element may exist for a period of time and then suddenly emit a particle and change into an atom of a different element was in itself strange enough, but the fact that the process cannot be influenced by any external factors and appears to be a completely random event subject to the laws of pure chance was even more remarkable. The best criterion for the randomness of the decay process is the occurrence of fluctuations in the rate of decay when measured over successive short time intervals; these fluctuations follow the laws of probability theory. Schweidler was the first to report quantitative measurements of the fluctuations, at the First International Congress of Radiology at Liège in 1905.[19]

In his first paper, presented at the Academy Meeting on March 14, 1918, Schrödinger gave a complete statistical analysis of a convenient method for measuring the fluctuations, which had been devised by his friend Fritz Kohlrausch in 1906.[20] Two almost equal sources of α-radiation are incident upon two separate ionization chambers with high tensions of opposite signs, and the difference between the two ionization currents is applied to one pair of quadrants of an electrometer, which are connected to ground through a high resistance, the other pair of quadrants being directly grounded. A potential difference between the quadrants is measured by a torque on a suspended vane, the displacement being measured with a lamp, mirror and scale arrangement.

To analyze the fluctuations in such measurements, Schrödinger first wrote down the Fokker–Planck equation in its most general form. From this general equation, he could obtain the generalized diffusion equations with external forces, for example, the case with elastic forces treated by Smoluchowski, and the case with gravitational forces that he, and also Smoluchowski, considered earlier. The present application to the quadrant electrometer is somewhat analogous to the former case. Schrödinger says that he has a double goal in this paper: 'on the one hand to show a pretty, sufficiently complicated example of the general statistical theory, which is relatively easy to realize, even as a demonstration in a large auditorium; on the other hand, to construct a rational, secure theory for an experimental procedure suitable for deciding important fundamental questions, the lack of which has been felt to be very disturbing.' He intends to confine this paper mainly to the basic theory, and in a later paper to apply it to the Schweidler fluctuations.

The three parameters chosen to describe the experiment are the scale displacement s, the velocity of the pointer $ds/dt = v$, and the voltage φ on the quadrants. His special Fokker–Planck equation for $W(s, v, \varphi, t)$ is as follows: if W is the probability that at time t the parameters lie between values s and $s + ds$, v and $v + dv$, φ and $\varphi + d\varphi$, then

$$\frac{\partial W}{\partial t} = -v \frac{\partial W}{\partial s} + \frac{\partial}{\partial v}\left[(ks + \mu v - K\varphi)W\right]$$

$$+ \frac{\partial}{\partial \varphi}\left[\frac{1}{C}\left(\frac{1}{\omega'} + 2nE\eta'\right)\varphi W\right] + \frac{\partial^2}{\partial \varphi^2}\left[\frac{nE^2}{C^2}W\right]$$

Here k is the elastic force constant of the torsion wire $[(d^2s/dt^2)_{el} = -ks]$; μ is the frictional constant $[(d^2s/dt^2)_f = -\mu v]$; K is the voltage response constant $[(d^2s/dt^2)_{el} = K\varphi]$; C is the capacitance of the quadrants; ω' is the resistance to earth; E is the quantity of electricity transferred to the quadrants for each α particle passing through an ionization chamber; n is the mean number of α-particles that pass through one ionization chamber in unit time; η' is the limiting value of the variation of the 'degree of saturation' of the electrometer with applied voltage.

Schrödinger calculates the mean and mean-square values of s, v, and φ, from the stationary-state equation obtained by setting $\partial W/\partial t = 0$. The result he derives for $\overline{s^2}$ agrees with that obtained in 1909 by a British worker using less elegant methods.

He next examines the properties of some nonstationary solutions of his Fokker–Planck equation, for the important case of a strongly damped aperiodic electrometer, but although he obtains a solution, it is in terms of variables that cannot be observed, and hence has no practical application.

Finally, therefore, he considers briefly a measurement that employs a different kind of electrometer, a rapidly reacting type, for which an approximate solution of the general case is adequate. The simple equation so obtained yields the well known result, $s = (K/k)\varphi$. Schrödinger himself admits that 'This result may appear trivial, everyone would have expected it under the given conditions without any lengthy calculation.' He believes, however, that the rigorous derivation is valuable since it allows an exact determination of the effect of the various approximations on the final result.

One might view this paper as an exercise in resharpening his theoretical tools after the wasted years in military service. One sees again his deep immersion in the subjects cultivated by the older generation of Vienna physicists, and the choice of problems governed by a reaction to local, almost parochial, interests. Even if the subject matter is classical and unexciting, one must admire his ability to write papers with such style and elegance while he was undernourished and cold and beset by grievous family troubles.

Schrödinger's second paper on the statistical dynamics of radio-active decay was entitled 'Studies in Probability Theory of Schweidler Fluctuations, especially the Theory of their Measurement'.[21] It was presented before the Academy on January 16, 1919, and has the distinction of being the longest paper he ever wrote, occupying sixty pages in the *Proceedings*.

Part One of the paper was concerned with several statistical questions of particular interest to the workers at the Institute for Radium Research.

The more important Part Two is an extension of his previous theoretical treatment, based on the Fokker–Planck equation, to three different experimental procedures for measuring Schweidler fluctuations. In each case he derives the stationary probability distribution of the pointer readings, the fluctuation of the pointer displacements for the given time, and the fluctuation in the times required for a given pointer movement. He at first neglects the inertia of the electrometer, so that the three parameters that determine the 'Smoluchowski motion' of the pointer are the position of the neutral point σ, the fluctuation range b, and the time constant of the decay λ_3. The three procedures are: (1) the positions of the pointer are recorded for a large number of random trials; (2) from an arbitrary starting point, the positions of the pointer are recorded after a fixed time; (3) the times are recorded for the pointer to move from a fixed starting point to another fixed point.

The first procedure had been used most frequently and its statistical theory was already worked out by N. Campbell in 1909. The theory for the second procedure was given in his previous paper, except for the inertial effect to be considered later.

The theory for the third procedure is then considered. He remarks that 'it is scarcely to be believed how much more complicated the analysis and the calculation turn out to be in this case compared to the preceding ones'. Let $W(t)\,dt$ be the probability that the pointer, starting from s_0 at time 0, reaches for the first time the position s_1 between t and $t + dt$, in short, the probability of a *first passage time*. Then the mean value of any function of t will be

$$\overline{f(t)} = \int_0^\infty f(t)\, W(t)\,dt.$$

Schrödinger confessed 'I could not calculate $W(t)$, and I doubt that it can be expressed in terms of known functions.' Nevertheless he was able to devise a method for calculating mean values such as $\overline{f(t)}$ even without having an expression for $W(t)$. The truly ingenious method by which he does this is the major accomplishment of this long paper. He first notes that

$$W(t) = -\, d/dt \int_{s_1}^{\infty} w(s, s_0, s_1, t)\, ds$$

where w is a solution of the appropriate Fokker–Planck equation,

$$\frac{\partial w}{\partial t} = \lambda_3 \frac{\partial (s - \sigma) w}{\partial s} + \lambda_3 b^2 \frac{\partial^2 w}{\partial s^2}$$

He transforms this into the familiar form of a diffusion equation,

$$\frac{\partial^2 u}{\partial x^2} - \frac{\partial u}{\partial y} = 0$$

through the substitutions,

$$x = (s - \sigma) \exp(\lambda_3 t); \qquad y = \frac{b^2}{2} \exp(2\lambda_3 t); \qquad u = w \exp(-\lambda_3 t)$$

For some reason, he did not choose to use the solution of the diffusion equation for the first passage problem, which is equivalent to having an absorbing barrier at $s = s_1$, but instead calculated the desired averages by a clever application of Green's theorem and the properties of Hermite polynomials. His final results for the mean and mean-square first passage times are given in the form of transcendental equations:

$$\overline{\exp(\lambda_3 t)} = \frac{s_0 - \sigma}{s_1 - \sigma} \qquad \text{and} \qquad \overline{\exp(2\lambda_3 t)} = \frac{(s_0 - \sigma)^2 - b^2}{(s_1 - \sigma)^2 - b^2}$$

When these formulas are applied in at least two long sets of observations with different s_0 or s_1, the values of σ, b, λ_3, can be calculated. 'It is certainly not to be denied that for numerical calculations they are extraordinarily impractical. The difficulty lies in the fact that the unknown λ_3 occurs under the mean-value bar, and can be found only through systematic trials.'

The final sections of the paper are concerned with the inertial effects in the electrometer. 'All the previous considerations were utopian, inasmuch as they related to a completely inertia-free electrometer; or if you wish: they did not actually relate to the indications of the electrometer, but to the potential itself, which, as it continuously changes, is not synchronous with its measurement.'

The inertial effects in Method I had been already given by Campbell, and Schrödinger therefore considers Method II only. He goes back to the general equation of $W(s, v, \phi, t)$, which he obtained in his previous paper, and calculates the mean and mean-square deviation of s. The result obtained for the latter is a formidable expression with twelve terms, which is of doubtful practical value, but which yields several approximation formulas of more use to experimenters.

This long paper is essentially an exercise in applied mathematics. It shows that Schrödinger was a master of statistical theory, who was able to fill with distinction the vacancy left by the untimely death of Smoluchowski. Nevertheless the more exciting problems of physics appeared to lie elsewhere, in the theory of atomic structure and in relativity theory. Schrödinger in 1918 made two valuable contributions to the latter subject, but at this time he displayed little interest in the former, in this respect sharing the prevalent attitude in the Vienna school of physics.

Two notes on general relativity

Schrödinger first learned of Einstein's theory of general relativity while he was stationed at the front at Prosecco. He immediately recognized the great importance of this theory, which by representing the gravitational field as an aspect of the geometry of space–time, opened new vistas for the world view of physics. When he returned to Vienna in 1917, he found that the university physicists were also excited by the Einstein work. Ludwig Flamm, who had married Boltzmann's youngest daughter Elsa, had already published a paper in the field, and Hans Thirring was working on another application of the theory. Unfortunately Friedrich Kottler, an early enthusiast for relativity theory, was still on active service in the army. For the first time in his scientific life, Erwin was able to enjoy critical discussions with knowledgeable colleagues about a subject at the forefront of theoretical physics. His own intensive studies of the mathematical foundations of general relativity are recorded in three notebooks (undated as usual) entitled *Tensoranalytische Mechanik*. Notebook III also included an outline of the analogies between mechanics and optics, such as the relation between Huyghens' principle and Hamilton's equations, a subject that was to play a leading role in his development of wave mechanics in 1925.

The short paper that he sent to *Physikalische Zeitschrift* in November, 1917, went to the heart of a fundamental question, how to express the total energy and momentum of a closed system in terms of the formulas of general relativity.[22] In 1916, Einstein had introduced an entity t^μ_ν, which he called the energy components of the gravitational field. This entity, however, is not a general tensor density and hence is not invariant with respect to transformation of the coordinate system. Schrödinger considered a special system for which solutions to the general-relativity equations had been obtained by Karl Schwarzschild in 1916, the surroundings of a stationary sphere of incompressible gravitating fluid. He calculated the sixteen t^μ_ν for a coordinate system 'almost equivalent' to a cartesian one. Since the t^μ_ν have tensorial character with respect to *linear* transformations, it was sufficient to calculate the values for one point on the x_1-axis; he showed that they were all equal to zero at this point, and therefore at all points in the coordinate system. His final comment was: 'The result which one gets in this special case – exact identical vanishing of all the t^μ_ν – seems to me for all that so strange that I believe it deserves general discussion. It appears relevant to the physical nature of the gravitational field.'

The problem raised by Schrödinger is part of a complex of questions relating to the localization of gravitational energy, which continue to be studied even today. It is remarkable that in his first note on general relativity he was able to uncover such a deep problem, but it is typical of an approach that became increasingly evident in his work: a disinterest in facile applications and a concern for fundamental principles, often in frontier areas where theoretical physics meets metaphysics. There is no doubt, however, about the local validity of the principle of conservation of energy, which can be shown by the powerful symmetry theorem of Emmy Noether (1918), who is generally considered to have been the greatest of female mathematicians.

A week after his first note to *Physikalische Zeitschrift*, Schrödinger sent a second one (received November 30), 'On a System of Solutions of the General Covariant Gravitational Equations'.[23] In this, his starting point was a recent paper by Einstein which gave a system of energy components and gravitational potentials that exactly integrated the field equations and provided an approximation to a possible large-scale structure of space and the distribution of matter in it. The model consisted of a resting fluid of uniform density which forms a closed spatial continuum of finite volume having the metrical properties of a hypersphere. Einstein had included a 'cosmological constant' in his solution, but Schrödinger showed that there was another solution without this constant. The Schrödinger solution had some curious properties: the fluid universe was under a considerable tension and its mass density was zero. The latter property was explained away by a Machian suggestion that nonvanishing mass arises only in the form of mass differences in a nonuniform universe. Einstein replied immediately to this note, saying 'The path taken by

Schrödinger does not seem to me to be an accessible one, since it leads too deeply into a thicket of hypotheses.'[24] Nevertheless this note is interesting since it marks Erwin's first essay in cosmology, and nineteen years later, after he had studied the work of Eddington, he added a handwritten footnote to calculate the value of the tension on the universe.

The end of the war

During 1917, both Kaisers, Karl and Wilhelm, as well as Lloyd George in Britain, tried to bring about negotiations for peace, but by now the generals were in command of the warring nations. From their comfortable headquarters, they lusted for the one final offensive that would complete the slaughter of their enemies. Hindenburg and Ludendorff assured Wilhelm that unrestricted submarine warfare would bring Britain to its knees within a few months. Karl warned against this suicidal policy, but Wilhelm dragged him, still pleading for peace, into the final debacle. On April 6, 1917, the United States declared war on Germany – they did not declare war on Austria-Hungary till eight months later.

The multinational K & K Army had fought courageously for four years. Of the eight million men mobilized, more than one million were killed and almost two million taken prisoner. The generals on both sides of the war were devoted to the doctrine of attack, which consisted in sending masses of soldiers against machine-gun emplacements. When Pershing arrived with the American forces in France, he delegated one man in ten to the military police, whose duty was to follow the attackers and shoot any soldiers who lagged behind.[25]

The supply situation in Austria-Hungary was deteriorating. The soldiers lived mostly on soup made of dried vegetables. The meat ration was 200 g a week in the front line and 100 g in the rear echelons, less than a McDonald hamburger, but usually lean horse meat. The meat was often full of worms, but the government informed the troops that, though unappetizing, they were not dangerous to health. In some divisions, only front-line troops had uniforms; the reserves waited in underwear until it was their turn to be thrown into the battles. Under these conditions, Conrad persuaded Karl to launch a 'final offensive' on the Italian front; when it foundered with heavy losses, he was finally retired.

The situation in Vienna was better, since a flourishing black market kept those who could afford it fairly well fed. In January, 1918, the bread ration was cut from 200 to 160 g a day, and workers went on strike in the armaments factories. It was necessary to detach seven divisions from the front to suppress the strikes.

Schrödinger remained on active service till the end of 1918. Among his random jottings from that year a curious fragment has been preserved:

> Da sind zwei Armeen ausmarschiert seit 1914
> Die eine kämpft noch
> Die von der andern haben Frieden gemacht [unter] der Erde
> Wahle! Zu welcher willst du?*

* Two armies have marched out since 1914 / One is fighting still / Those of the other have made peace underground / Choose! Which one do you want to belong to!

This may have been written by Erwin himself or copied from some underground peace propaganda, which was beginning to appear more openly. His grandfather's estranged wife, Natalie Lechner-Bauer, became so active in the peace movement that she was arrested and convicted of treason.

During these times, life was particularly difficult in the large fifth-floor apartment which the Schrödingers rented from grandfather Bauer. They had never installed electric lighting, partly because Alexander did not wish to pay for it, and partly because Rudolf preferred the mellow white light of the Welsbach gas mantles to the reddish-yellow of the available electric bulbs. For convenience ('servant miseries') they had removed the great tiled stoves from several rooms and replaced them by gas fires with copper reflectors. Also they cooked with gas, although a huge slow-combustion stove stood like a venerable monster in the middle of the kitchen. Thus when word came from City Hall that every dwelling was to be allowed only one cubic metre of gas per day, no matter how it was to be used – any violation to be punished by complete cut-off – the Schrödinger household had serious energy problems in addition to the terrible shortage of food. Erwin's mother appeared to be recovering from a radical operation for breast cancer, which had been performed in 1917, but she was still weak and in pain. Erwin's own health was far from adequate: in August, 1918, an inflammation of the apex of the lung was diagnosed, which may have been tuberculosis, since this disease was epidemic among the undernourished population of the city.

Rudolf's business had been destroyed by wartime shortages, and in 1917 Groll Brothers, his oilcloth and linoleum company on Stephans-platz, was closed. When the war ended, he no longer had either the money or the strength to try to salvage it. Thus the family for the first time was facing serious financial difficulties, since Erwin's military pay stopped at the end of 1918 and he had barely enough income from the university to support himself. During this winter the Schrödingers often ate at a community soup kitchen.

After the Armistice, as the empire disintegrated, the situation in

Vienna became much worse than during the actual war, for food supplies from Hungary were cut off, and the Entente maintained its blockade. Thousands of people in Vienna were starving and freezing in the winter of 1918/19. The streets were filled with beggars, ex-soldiers in various extents of mutilation, with decorations pinned to their rags. The worst were the *Zitterere* [tremblers] who owing to nerve damage exhibited grotesque facial tics and continuous jerky movements.[26]

The women assumed most of the responsibility for foraging for their families; the farmers in outlying areas had large stocks of food, but it was necessary to seek them out and barter precious possessions for a few eggs, vegetables, and other edibles. Even to obtain these, ladies had to importune and beg in ways that they found humiliating, and the farmers drove hard bargains with the starving city dwellers. The official meat ration was 100 g a week, but even this was often not available. Thousands of women queued overnight, and when the rations appeared, they rushed forward and seized them, leaving nothing for those at the rear. Mounted police were called to disperse the women, some of whom collapsed in the street from exhaustion. During one disturbance, a horse fell and in an instant the mob was upon it, slicing it up and bearing off the meat within a few minutes.

Schrödinger recalled that 'In 1918, we had a sort of revolution; Emperor Karl went and we became a republic, without changing much in our lives. For me personally the dismemberment of the country had one result, since I had a call as Extraordinarius to Czernowitz, where I intended to lecture on theoretical physics, but in private life to concern myself more with philosophy (I was just now with great enthusiasm becoming familiar with Schopenhauer and, through him, with the doctrine of unity taught by the Upanishads).' Czernowitz, however, had become a part of Romania and the new government would not tolerate foreign teachers.[27]

The Austrian government tried to make the country part of Germany, but this action was blocked by France. Italian troops occupied Vienna and the Czechs cut off the export of coal. Housewives went to the Vienna woods to gather fuel, and soon whole slopes were denuded of trees and bushes. Many lives were saved by the smuggler [*Schleichandler*] who sneaked across the border from Hungary and called once a week with food supplies; paper money was useless, but everything else from carpets to grand pianos was carted away. Karl Renner, the socialist president, implored the British and French to raise the blockade so as to allow some food to enter the country. The blockaded Germans sent a little dried fish, Sweden sent rice, and the Swiss and Italians did what they could to provide some relief. On Christmas day, the official toll of war dead on both sides was announced as seven and a half million.

Herbert Hoover came to Vienna on January 17, and arranged for some American relief through Quaker agencies. His motives were as much political as humanitarian; food was used as weapon to stop the spread of bolshevism. In March, as the blockade continued, it became a crime to forage in the countryside around Vienna. Finally, on March 22, the blockade of Austria was lifted, although it continued to be applied to Germany. It is difficult to find a parallel in history to this deliberate starvation of a defeated enemy. On March 23, the emperor left the country, to die in exile after one final typically inept attempt to regain his throne.

4　From Vienna to Zürich

In the midst of all the postwar turmoil and suffering, Erwin took no respite from his intensive research at the Physics Institute. He also filled notebook after notebook with commentaries based upon his reading of European and Eastern philosophers. It was in these dying days of the Danube Empire that he formed the foundations of his philosophy, which was to remain remarkably constant all his life. His interest in philosophy at this time was so intense that he even considered the possibility of devoting his intellectual life to that field. It would have been an inopportune decision, since philosophers now began to consider their subject as a self-contained discipline pursued by specialists for academic ends. Wittgenstein's *Tractatus Logico-Philosophicus* became the bible of the Vienna School of logical positivism. 'What is history to me?', he wrote, 'Mine is the first and only world'. The task of trying to repair the wreckage of European civilisation may, however, have been hopeless even if help from the philosophers had been forthcoming.

Schopenhauer

Erwin read everything written by Schopenhauer. The luminous words of the philosopher of pessimism were perfectly suited to the world of 1919, and in that context may even have had a consoling effect, making some sense of Europe's four years of self destruction.

Arthur Schopenhauer (1788–1860) was born in Danzig, the son of Heinrich, a rich Hanseatic merchant, and Johanna, a romantic novelist. As a youth, he traveled widely, becoming fluent in English and French, so that his prose style acquired a lightness and clarity quite unlike the murky philosophic German of his times. His first education was that of a man of the world, only later did he obtain the usual academic credentials. He became a friend of Goethe, and in 1816 wrote a small book on color theory inspired by the ideas of the older man.

Schopenhauer regarded himself as the true spiritual descendant of Kant, and he did not hide his view that Hegel, then at the height of his

111

fame, and other university philosophers were charlatans who had perverted the Kantian gospels. The first edition of his major work, *The World as Will and Representation* appeared in 1818. He followed Kant in the belief that the mind is not merely a passive recipient of sense impressions, but takes an active role in fitting the phenomena into the categories of space and time, the principle of causality being the necessary method for creating this representation of the world. Kant taught that the real world, the *noumenon*, the thing-in-itself [*Ding an sich*], can never be accessible to human thought or experience. Schopenhauer did not agree: he believed that the thing-in-itself can be identified as *will*. Every person experiences himself in two different ways, as an object like any other, and through self-consciousness as a *will*. The will is neither a phenomenon nor a representation, it is a directly experienced reality. What is true of the microcosm of man, is also true of the world: its thing-in-itself is will. On the foundation of this primary intuition, which of course can be neither proved nor disproved, Schopenhauer constructed a philosophy that has continued to fascinate and influence thinkers of all kinds: Wagner, Nietzsche, Thomas Mann, Freud, Klimt, and Schrödinger, to mention a few.[1]

Schopenhauer's views on the arts and sciences were unusual. Nature is not a rational order but a chaotic struggle for existence, and the analysis of nature by science in terms of causality is directed towards making use of things to gratify the will. Artistic experience, on the other hand, is 'a sabbath from the penal servitude of willing'; the contemplation of a work of art allows a person to transcend the will by direct experience of the archetypal Platonic forms. (Any inconsistency here may be overlooked as deference to the beauty of great art.) Music is a different kind of thing altogether: its subject is the will itself, and thus music is closest to the ultimate reality.

Schopenhauer kept a faithful dog called 'Atman' [Sanskrit: the soul], a copy of the Hindu scriptures at his bedside, and an ancient statue of the Buddha dressed as a beggar and thickly covered in gold leaf. His philosophy is closely related to the ancient wisdom of the East, and many westerners first learned of Vedanta and the Upanishads through his writings. His direct influence on Schrödinger was considerable, but equally important was the introduction he provided to Indian philosophy. Schopenhauer often called himself an atheist, as did Schrödinger, and if Buddhism and Vedanta can be truly described as atheistic religions, both the philosopher and his scientific disciple were indeed atheists. They both rejected the idea of a 'personal God', and Schopenhauer thought that 'pantheism is only a euphemism for atheism'. Yet Schopenhauer claimed kinship not only with the Buddha but also with the Christian mystics, and although Meister Eckhart may have been a heretic, he was certainly not an atheist.

Vedanta

The books on Indian philosophy that Schrödinger studied at this time included:

Henry C. Warren (1896) *Buddhism in Translation*. Harvard.
Max Walleser (1904) *The Philosophic Foundations of Older Buddhism*. Heidelberg.
Richard Garbe (1896) *Samkhya and Yoga*. Strassburg.
Richard Garbe (1894) *Samkhya Philosophy, a Presentation of Indian Rationalism according to its Sources*. Leipzig.
Paul Deussen (1906) *The System of Vedanta*. Leipzig.
F. Max Muller (1880) *On the Origin and Development of Religion with Special Consideration of Religions of Ancient India*. Strassburg.
T.W. Rhys Davids (1877) *Buddhism*. London.
Richard Pichel (1910) *Life and Teaching of the Buddha*. Leipzig.

He did not merely read these books and make notes on them. He thought deeply about the teachings of the Hindu scriptures, reworked them into his own words, and ultimately came to believe in them. Possibly his half-famished state at this time was an involuntary mortification of the flesh conducive to religious experience.[2]

In July, 1918, he wrote:

Nirvana is a state of pure blissful knowledge . . . It has nothing to do with the individual. The ego or its separation is an illusion. Indeed in a certain sense two 'I's' are identical, namely, when one disregards all their special content – their *Karma*. The goal of man is to preserve his Karma and to develop it further. The goal of woman is similar but somewhat different: namely, so to speak, to create an abode that accepts the Karma of man . . . When a man dies, his Karma lives and creates for itself another carrier.

This survival of Karma is not orthodox Hinduism, and it appears that Erwin was unwilling to give up his Ego completely, or even his male supremacy, at least at this stage in his studies.

A few days later: 'No self is of itself alone. It has a long chain of intellectual ancestors. The "I" is chained to ancestry by many factors . . . This is not mere allegory, but an eternal memory.'

In August: 'The stages of human development are to strive for: (1) *Besitz* (2) *Wissen* (3) *Können* (4) *Sein*. [(1) Possession (2) Knowledge (3) Ability (4) Being.]'

Erwin said that he was 'under the very strong influence' of Lafcadio Hearn (1850–1904). Hearn was born in the Ionian Isles of Irish-Greek parentage, went to America at the age of nineteen, and from the age of forty lived in Japan, where he immersed himself in Japanese Buddhist culture, taking the name Koizumi Yakumo. In his essay on 'The Diamond Cutter', he wrote:[3]

The ego is only an aggregate of countless illusions, a phantom shell, a bubble sure to break. It is *Karma*. Acts and thoughts are forces integrating themselves into material and mental phenomena – into what we call objective and subjective appearances . . . The universe is the integration of acts and thoughts. Even swords and things of metal are manifestations of spirit. There is no birth and death but the birth and death of Karma in some form or condition. There is one reality but there is no permanent individual . . .

Phantom succeeds to phantom, as undulations to undulations over the ghostly Sea of Birth and Death. And even as the storming of a sea is a motion of undulation, not of translation, – even as it is the form of the wave only, not the wave itself, that travels – so in the passing of lives there is only the rising and vanishing of forms, – forms mental, forms material. The fathomless Reality does not pass . . . Within every creature incarnate sleeps the Infinite Intelligence unevolved, hidden, unfelt, unknown, – yet destined from all eternities to waken at last, to rend away the ghostly web of sensuous mind, to break forever its chrysalis of flesh, and pass to the extreme conquest of Space and Time.

Perhaps these thoughts recurred to Erwin when he made his great discovery of wave mechanics and found the reality of physics in wave motions, and also later when he found that this reality was part of an underlying unity of mind. Yet, in the course of his life, belief in Vedanta remained strangely dissociated from both his interpretation of scientific work and his relations with other persons. He did not achieve a true integration of his beliefs with his actions. The *Bhagavad-Gita* teaches that there are three paths to salvation: the path of devotion, the path of works, and the path of knowledge. By inborn temperament and by early nurture Erwin was destined to follow the last of these paths. His intellect showed him the way, and throughout his life he expressed in graceful essays his belief in Vedanta, but he remained what the Indians call a *Mahavit*, a person who knows the theory but has failed to achieve a practical realization of it in his own life. From the *Chandogya Upanishad*: 'I am a Mahavit, a knower of the word, and not an Atmavit, a knower of Atman.'

Death of the Father

The first postwar years were filled with difficulties and sorrows for Erwin, the loss of father, mother and grandfather within a span of less than twenty-one months.

In the summer of 1919, the family was able to afford a lakeside holiday at Millstatt. While there, Rudolf, then sixty-two years old, showed the first signs of the hypertension and atherosclerosis that already must have been advanced. On the walks and excursions which he always loved, he fell behind the others and had difficulties in climbing hills. He tried to hide his debility by saying that he was collecting botanical specimens for his research on morphogenetics. When Georgie called 'Rudolf, come along now', he did not react, but that caused no surprise because when something interested him he always lost track of time. When they returned to Vienna, there were more ominous symptoms, which at first they failed to understand, nosebleeds, retinal hemorrhages, and oedema. The climb to their fifth-storey apartment became an ordeal for him. With the severe gas shortages, they faced both cold and darkness. He bought some carbide mine lamps to illuminate his library, but they pervaded the whole house with a frightful stink.

Rudolf was the only one truly aware of their precarious financial situation. With the failure of his business, they were dependent upon income from investments, and many of these were in government bonds. Late in 1919, the cost of living began to rise steeply, although rampant inflation did not begin until about two years later. Rudolf worried that Erwin had practically no income, only 2000 crowns that year from the university, at a time when about 3000 crowns a month was needed to provide food for an average family. Nevertheless, the university was a refuge from the troubles at home, and Erwin spent every day there, managing to accomplish a remarkable amount of work under the most difficult circumstances.

Rudolf's strength declined steadily, and he died on December 24, peacefully resting in a grandfather chair. Erwin later reproached himself for not securing better medical care for his father, for example, by asking for help from specialists on the university faculty whom they knew. In 1930, he published a short contribution in the Christmas-eve number of a German paper, under the heading 'Neglected Duties':[4]

Thus there was no longer much more to destroy in the sentimental complex of Christmas eve, when, on 24 December 1919, between 6 and 7 in the evening, my father died. And strange. Also on this evening, a beloved person, who knew nothing of the change for the worse in the last days, was not with me but in the circle of her family. From there, sent with loving, unsuspecting intentions, an hour later there stood at the door a messenger with a basket of presents. – With all that, it served me right. For however reluctantly I confess it to myself today: had the immediate crisis not come, I would also on this evening, as so often in the previous weeks, have left my mother alone for several hours with the desperately sick man; and this, although I could understand that the holy day meant something to her, and that my father would never experience another one. So to me Christmas is a feast which I am not fond of, from which I expect nothing good, and which more than any other reminds me of neglected duties.

The Christmas hamper had undoubtedly been sent by Anny from Salzburg. During the summer, Erwin had been able to see her quite often, since Millstatt is only about 120 km from Salzburg. During the autumn, he made a special visit to Salzburg, during which they became lovers (if they were not already) and also became engaged, but which event came first is not known. Anny's sister Irmgard was a devout Catholic, but Anny herself, though still practising her religion at that time, was less strict.

Anny came to Vienna and obtained a position as a secretary to General Direktor Fritz Bauer of the Phoenix Insurance Company. The Bauers were a wealthy Jewish family (not related to Erwin's grandfather). Anny's monthly salary was more than Erwin's annual income, a situation that caused him embarrassment and bitterness. He used to spend almost every evening with her, but later, in a rather irrational way, reproached her for keeping him away from his dying father.

Anny once took her fiancé to visit the Bauers in their magnificent house in the suburb of Sievering. He met the family, including fourteen-year-old Hansi, who found him to be 'stiff and suffering' during afternoon tea.

At the time of his death, Rudolf was vice-president of the Zoological–Botanical Society of Vienna, and his old friend Professor Anton Handlirsch delivered a eulogy at its first meeting in the new year:[5]

His outstanding personality is still so fresh in the memory of all of us. He was always with us and we saw him always at work, which he carried out voluntarily in a completely unselfish way. And it was no small work! . . . When Brunnthaler, then general secretary, became ill, Schrödinger took care of his duties for several years and thereby protected his sick comrade from material loss. Elected in 1907 to the board of directors, Schrödinger, perhaps without wanting to, soon became the administrative soul of the society . . . He was a man of rare versatility, always ready to give, always full of humor and animation, incisive but just [*scharf aber gerecht*] in his judgements, and responsive to all that is beautiful and noble.

This is how friends and confreres saw the father. Erwin did not take after his father in all these ways. He was never elected to office in a scientific society and his skill as an administrator was a negative quantity. He was not a person eager to spend much time in helping others, and he had few students, although he was considerate of these few. But also unlike his father, he was a very great scientist.

In retrospect, Erwin thought that perhaps his father died at an opportune time, for during the next year runaway inflation wiped out his savings and he would have been truly destitute. The proceeds from the sales of his Persian rugs, his scientific instruments, and most of his extensive library – all became worthless paper money. Erwin wrote:

I very seldom remember my dreams and (except perhaps in early childhood) I have very seldom had bad dreams. After the death of my father, and indeed for quite a long time after, with certain intervals, this anxiety dream often recurred: my father was still alive, and yet I knew that I had sold off all his beautiful instruments, and disposed of his books, especially the botanical ones. My God! What is he expected to do now, I have recklessly scattered to the four winds the essentials of his intellectual life, beyond recall! – I believe this anxiety dream was caused by a bad conscience due to my less than honorable conduct towards my parents in the years 1919/21. This explanation is consistent with the fact that I have usually been exempt from nightmares because of course I very seldom, as they say, 'have a bad conscience about anything' [aus etwas eine Gewissen mache].[6]

Coherence experiments

To do any scientific research in Vienna in 1919 must have required unusual powers of concentration. Early in the year Erwin decided to spend some time on an experimental project, perhaps because it was a refreshing change from theoretical work, or perhaps because he thought his position in the Physics Institute II implied an obligation to work in the laboratory. He always enjoyed precise optical measurements, and the problem he chose required interferometric techniques, which illustrate so beautifully the wave properties of light.

The immediate motivation for his work was a recent paper by Einstein which discussed the emission of light as particles or quanta. Schrödinger found it difficult to reconcile the quantum theory of light, based on the original work of Planck, with its well known wave properties.

Max Planck [1858–1947] belonged to a scientific generation just prior to that of Schrödinger. As a young man, his view of science was completely different from that which would prevail thirty years later. He believed that the outside world is something independent from man, something absolute. 'The quest for the laws that apply to this absolute appeared to me to be the most sublime scientific pursuit in life.'[7] When Planck became professor of physics at Berlin in 1892, his subject did appear to be firmly based upon a number of absolute principles: the absolute three-dimensional space and linear time of Newtonian mechanics; the laws of thermodynamics, conservation of energy and continuous increase of entropy; Maxwellian electrodynamics which unified electromagnetism and light as wave phenomena in an all pervading aether. By the time Schrödinger succeeded to the Planck chair in 1927, all these absolutes had been destroyed or

substantially modified. Planck himself, a most reluctant revolutionary, began the upheaval with his work on the quantum theory of
energy and radiation.

The first conspicuous failure of the wave theory of light was found
in the experimental data on the wavelength distribution of black-body
radiation. The results of measurements by Otto Lummer and Ernst
Pringsheim, at the Reichanstalt Laboratory in Berlin-Charlottenburg,
were in flagrant disagreement with the predictions of electromagnetic
wave theory.

Planck used a statistical thermodynamic approach to the problem of
black radiation, similar to the treatment by Boltzmann of the distribution of kinetic energy among molecules in a volume of gas. On
October 19, 1900, he announced to the German Physical Society the
mathematical form of a law that was in good agreement with the
experimental data over the whole range of frequencies,[8]

$$u_v = \frac{8\pi h v^3}{c^3} \left[\exp(h v/kT) - 1 \right]^{-1}$$

Here u_v is the density of radiant energy and h is a constant having the
dimensions of *action*, energy × time. On December 14, he presented a
theoretical basis for his distribution law:

We . . . consider – and this is the essential point – the energy E to be composed
of a determinate number of equal finite parts, and employ in their determination the natural constant $h = 6.55 \times 10^{-27}$ (erg × sec). This constant multiplied by the frequency v of the resonator yields the energy element $h v$ in ergs,
and dividing E by $h v$ we obtain the number of energy elements to be
distributed over the N resonators.

Planck did not intend to imply by this statement that energy is really
restricted to discrete quanta. His 1900 papers were certainly not
immediately recognized, even by himself, as containing a discovery
that would remake the foundations of physics. He was a deeply
conservative man, and for some years he was content to regard his
quantization of energy $E = Nhv$ as simply a convenient assumption
that allowed him to carry through an application of Boltzmann statistics to the radiation problem.

In 1905, Einstein, then an Examiner Third Class at the Patent Office
in Bern, published a paper that boldly accepted the full consequences
of Planck's hypothesis by stating that radiation acts as if it consists of
quanta of energy $h v$, which occur not only in emission and absorption
but can have an independent existence as particles in empty space[9].
He applied these ideas to explain the photoelectric effect, discovered

in 1887 by Heinrich Hertz. When light of suitable frequency strikes the surface of a solid, electrons are emitted. Lenard had shown, in 1902, that the kinetic energy of the emitted electrons depends on the frequency of the light and not at all upon its intensity. The intensity, however, determines the number of electrons emitted. Einstein showed that these results are readily understood if the elementary process of photoemission is a collision between a particle of light with energy $h\nu$ and an electron in the solid, which overcomes the potential energy holding the electron and imparts to it an excess kinetic energy $\frac{1}{2}mv^2$, so that

$$h\nu = \tfrac{1}{2}mv^2 + e\Phi$$

The Einstein concept of light quanta was taken up enthusiastically by Johannes Stark, who applied it in 1908 to ionization of gases by ultraviolet light and to photochemical reactions. At that time, however, it is doubtful if anyone except Einstein and Stark believed in the reality of light particles. [They were first called *photons* by G.N. Lewis in 1926.]

The motivation of Schrödinger's work was the idea that a suitable experimental test should be able to decide between the particle theory and the wave theory of light emission. Thus if an excited atom emits a photon, the emitted light should travel along a path of narrowly defined angular width, whereas if the atom emits a wave, the light should constitute a spherical wavefront spreading uniformly in all directions from the point source. For Schrödinger, and for almost all other physicists in 1919, it seemed obvious that these two pictures cannot both be true.

Schrödinger thought at first that if the particle theory is true, the degree of coherence between two 'light bundles' emitted from a small source should be less at large than at small angles of separation, and that such an effect might be detected by examining the interference between the two bundles. No sooner had he started to test this idea experimentally than he realized it was invalid, since in the interferometer the two light sources are defined by two pinholes or slits, which act as spherical sources in accord with the Huygens–Kirchhoff construction. His experimental approach had been based on the use of extremely small light sources, but he also soon realized that even the smallest laboratory source is enormous compared to the dimensions of an atomic or molecular emitter of light. Thus, as he admits in the introduction to his paper, the original purpose of the experiments soon vanished, but he persisted anyway, simply to study the effect of source dimensions on the clarity of the interference fringes.

His light source was an electrically heated fine wire; the rays to be compared were selected by a pair of fine slits, the separation of which could be varied. The rays defined by the slits were brought together at a small angle by an objective lens, and the image of the source with its interference fringes was examined through a microscope objective and ocular. He employed thinner filaments than had ever been used before, 2–4 μm platinum wires. These could not be used in a vacuum lamp, but it was possible to maintain a small section of the wire, about 1 mm long, at red heat in air for long enough to make his observations, provided he worked quickly. With this arrangement, he was able to detect interference fringes at angular ray separations of up to 56°, much larger than ever seen previously. He concluded that he had clearly demonstrated the increase in the limiting angle for well-defined interference (and hence coherence) as the diameter of the light source is decreased, which is the result that he expected from a purely mathematical analysis of the problem.

This research was published in the *Annalen der Physik*, the most prestigious German physics journal, the third paper by Schrödinger to appear there.[10] Except for a few observations on color vision, it was his last work as an experimental physicist, and it demonstrated laboratory skill of a remarkable delicacy and precision. The long hours spent sitting in front of his observation microscope in the dark, cold laboratory, watching the elusive interference fringes come and go, must have left him with a firm conviction that light waves have an incontrovertible physical reality. This research was also important for his future in that it made him study and ponder the strange ambiguities of quantum theory, and at least temporarily diverted his thinking from the more classical physics so popular in the Vienna School. Later that year, however, he began to work intensively on an entirely different Viennese specialty – color theory.

Color theory

Schrödinger's most important research at the University of Vienna from 1918 to 1920 was in the field of color theory [*Farbenlehre*]. He continued to publish occasional papers on this subject between 1921 and 1925, but the basic papers appeared in the *Annalen* in 1920. As with most of his work since graduation, the origin of his interest in color theory was research in progress by his colleagues at the University. From 1917 to 1920, his friend Fritz Kohlrausch held a lectureship in color theory at the School of Applied Arts, during which he investigated the colors of artists' pigments in terms of their perceived hue, saturation, and brightness (or lightness). These are the three most distinctive perceptual qualities of color. The term *hue* specifies the spectral color (wavelength) perceptually most similar to an ordinary broad-band color; *saturation* specifies the amount of white that seems to be mixed with a spectral (monochromatic) color in an ordinary color (the less whitish it appears, the more saturated it is); *brightness* is related to the perception of the intensity of light contained in, or reflected by, a color.[11] In 1922, Kohlrausch received the Lieber prize of the Vienna Academy for his work on artists' colors. Erwin's old professor, Franz Exner, also presented important experimental papers before the Vienna Academy in 1918 and 1920 on the comparison of brightness of different artists' colors.

In 1919, Exner published the first edition of his influential book, *Lectures on the Foundations of the Natural Sciences*, which included four chapters on color vision.[12] He took a catholic view of the epistemology of color theory: 'To be sure, pure phenomenalism has a claim to the greatest interest – in the field of *Farbenlehre*, the description of our sensations; but the aim of research still remains the genesis of these sensations, the knowledge of the physical and physiological processes that determine the psychic ones.'

The basic inspiration of Erwin's work on color theory, however, seems to have been closer to the philosophy of Ernst Mach, which he had continued to study intensively during his tedious years of service with the fortress artillery. Human color vision provides one of the best examples of the Machian elements of sensation. Schrödinger based his color theory on direct observations in which he matched areas of colored light for hue, brightness, and saturation. He divided the subject into 'elementary color theory' in which colors are matched in all three perceptual respects, so that the sensations become indistinguishable, and 'advanced color theory' in which the sensations differ (e.g., comparison of brightness between patches of different hue).[13] Sensations of color can become extremely complex, and all the theories considered by Schrödinger are based upon comparisons between isolated patches of color, so that effects due to contrast with different backgrounds, temporal after-images, etc. are all excluded.[14]

Schrödinger's first paper on color was 'Theory of Pigments of Maximal Luminosity', which was submitted to the *Annalen* just before Christmas, 1919, only a few days before his father died.[15] It recalls the great interest of grandfather Bauer in artists' pigments, and Erwin would certainly have discussed this paper with the old man.

The color of light that is reflected from a streak of pigment never reaches the degree of saturation of a pure spectral light, but always appears more or less whitish compared to a pure light of the same hue . . . The impossibility of realizing in pigments colors of spectral saturation is not a technical problem but a matter of principle . . . To achieve the full saturation of a spectral color, a pigment can actually reflect only an infinitesimal range of wavelengths while completely absorbing everything else. In this case, as Helmholtz has already remarked, it would appear to be extremely dark, in the limit, black.

Given this limitation, what is the greatest brightness that can be achieved with mixtures of pigments?

Schrödinger introduced the concept of *ideal* colors. These are colors for which the spectral reflectance has values only of either 0 or 1. In practice, of course, the reflectance does not jump abruptly from 0 to 1, but some actual pigments approximate this behavior quite closely. Schrödinger proved that if a color is obtained by mixing pigments having the characteristics of ideal colors, it will have the maximal brightness obtainable for the color. The *ideal* colors are therefore also *optimal* colors. These results provided a theoretical understanding of some empirical rules that had been given by Wilhelm Ostwald a few years previously. Schrödinger's paper was an important contribution to color theory and the concept of ideal colors has since been used in the analysis of many color problems.

Schrödinger's next work on color theory was a long three-part paper, submitted to the *Annalen* in March, 1920, on 'Fundamentals of a Theory of Color Measurement in Daylight Vision'.[16] This was his major contribution to the subject, and it was honored by the prestigious Haitinger prize of the Vienna Academy of Sciences. It was in part a 'tutorial paper', in which he aimed to present a logical account of the state of the subject as well as the original advances he himself had made. He first presented the results in three lectures before the newly established Vienna section of the German Physical Society, delivered on February 26, March 4 and March 11. The restriction to daylight vision allowed him to consider color vision independently of light intensity; in terms of physiology it is almost exclusively *cone vision*.

Erwin based his analysis of color vision on the three-color theory of Thomas Young (1806), surely the most prescient work in all of psychophysics, which was rediscovered, developed, and extended by Hermann Helmholtz in the latter part of the nineteenth century. The Young–Helmholtz theory is based on the hypothesis [since proven] that the normal [trichromat] human retina contains three types of receptors, each with a particular spectral response curve; these may be called *red, green, and blue* receptors on the basis of their spectral response curves. Any spectral color (light) F or any mixture of such colors can be matched by a linear combination of the three basic colors, R, G, B, so that one can write

$$F = x_1 R + x_2 G + x_3 B$$

The colors present a manifold of things for which equality, addition, subtraction (with limitations), and multiplication by a nonnegative number are defined in an unambiguous manner . . . ; the manifold (in the normal case) is of dimension three; and all of this is connected with experience uniquely by judgements of the equality of adjacent color fields and is firmly based thereon. . . . If one now compares these empirically established laws of the colors with the *axioms* that must be postulated for vectors drawn from a point, in order to establish the affine geometry of such a bunch of vectors, complete agreement is found. Thus the manifold of colors, or the *color space* as we wish to call it, is a three-dimensional system with, from the standpoint of the equality relation, only an affine structure.

In tribute to Machian principles, Erwin re-emphasizes the primacy of the sensation:

The color space owes its existence as well as its affine structure to the equality relation quite without reference to the vectorial or point space which serves for its elucidation.

The geometry of color space is not the ordinary Euclidean variety that we learn in high school. It is a more general geometry, called *affine geometry*, of which the Euclidean variety is a special restricted case. Affine geometry deals with those properties of figures which are unchanged when the original coordinates of their points x, y, z, are transformed to new coordinates x', y', z', by a system of linear equations,

$$x' = a_{11}x + a_{12}y + a_{13}z$$

$$y' = a_{21}x + a_{22}y + a_{23}z$$

$$z' = a_{31}x + a_{32}y + a_{33}z$$

This set of linear equations with constant coefficients a_{ik} defines an *affine transformation*, which plays the role in affine geometry that the concept of congruence has in Euclidean geometry. A general affine transformation corresponds to a displacement (e.g., translation, rotation, reflection in a plane) plus a *dilatation*, i.e., an expansion or contraction of space in three mutually perpendicular directions. A dilatation transforms each line into a parallel line. The importance of affine geometry is greatly enhanced owing to the fact that more general transformations become linear in the limit of very small displacements. Thus any geometry that deals with infinitesimal displacements, i.e., a *differential geometry*, is necessarily affine.

In affine geometry, the basic elements are points A, B, C, etc., segments AB, BC, etc., and the idea of intermediacy, e.g. of B in a segment ABC. In affine geometry, lengths of segments can be

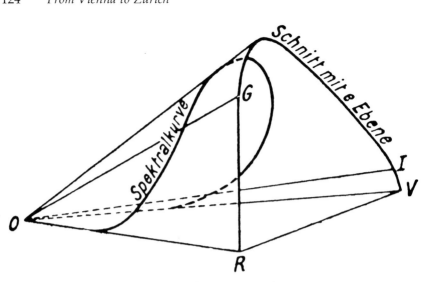

Figure 4.1. The color 'cone'. The surfaces ORG *and* OVI *are planar. A spectral curve is shown.*

compared only if they are collinear or lie on parallel lines. Thus, in color space, the length of a vector from the origin O to color A cannot be compared to the length of a vector from O to B unless OB is in the same direction as OA. In terms of colors, one cannot directly compare the lightness of colors having different hues, for example, by saying that a patch of green is lighter than a patch of yellow.

Schrödinger pointed out that the empirical data of elementary color theory are derived exclusively from sensations of equality between color samples, which are best compared by presenting two adjacent color areas to the observer. It is possible to match one of the qualities of hue, brightness, or saturation, when the other two are kept the same. When one of these qualities is altered continuously, the observer does not perceive a change until a certain minimal difference has been presented; this is called the *threshold of distinction*. All colors that are at the same threshold of distinction from a given color are said to be at the same *distance* from it. Thus the difference in stimulus required to reach the threshold of distinction defines a unit length along any vector in color space. By proceeding with stepwise matches it is thus possible to compare lengths along collinear vectors by the number of thresholds required to cover the distance in question.

Elementary color theory is not so simple as it may seem. There is an infinity of different spectral distributions of energy (or of reflectances or transmittances) which can match any given color in the visible range. Helmholtz was the first really to understand this fact. The visual system performs a formidable job of reduction of physical data before it presents a color sensation to the mind.

In his second paper Schrödinger developed a geometrical representation of elementary color theory based on an arbitrary choice of three noncoplanar basis vectors, **A**, **B**, **C**. Any other color is specified by its components in the directions of the basis vectors. The manifold of vectors which correspond to the various colors then defines a color 'cone', and any given spectral distribution is represented by a curve on the surface of this 'cone' (Figure 4.1). The most desirable basis-vector set would be that corresponding to the three different color receptors in the human eye and Schrödinger shows how this can be obtained by experimental measurements with subjects who have dichromatic color vision, lacking either the red or green receptors. Blue color-blindness appears only in individuals suffering from other eye disorders and the corresponding basis vector cannot be exactly defined.

Advanced color theory is concerned with questions such as how to measure a difference of brightness between two colors that have different hues. Consider two arbitrary colors **X** (coordinates x_1, x_2, x_3) and **Y** (y_1, y_2, y_3), and let a function $s(x_i, y_k)$ be a measure of the 'distance' between them. Instead of trying to match two closely neighboring colors exactly, Helmholtz introduced a 'principle of greatest similarity'; all those colors that appear equally most similar to a given color are said to be at the same distance ds from it. He wrote ds because the colors are very close together and the finite distance is approximated by the differential. The differential distance or *line element* is expressed as

$$ds^2 = a_{ik}\, dx_i\, dx_k \qquad [a_{ik} = a_{ki}]$$

[The usual convention of summation over repeated subscripts is followed, with the sums from i, $k = 1$ to 3.] In advanced color theory, therefore, a *metric* has been introduced, and the geometry is no longer affine, but Riemannian. It is interesting that this is the same kind of geometry used by Einstein in his general theory of relativity, although his space is four-dimensional (space-time) whereas the color space is three-dimensional.

The difference between any two colors **X** and **Y** can now be calculated as the integral of ds along the shortest path between them, $\int_X^Y ds$, provided this integration can be carried out. The shortest path or *geodesic* is the one that requires the least number of steps of greatest similarity. The coefficients a_{ik} can in principle be determined experimentally by measurements of ds when the dx_i are varied independently. Note that the geodesic in Riemannian space is not a straight line but in general a curved path.

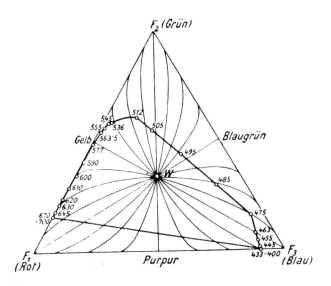

Figure 4.2. A sheaf of geodesics at constant brightness. Schrödinger defines colors of constant hue as those lying on the geodesics between the fully saturated color and white. Only the geodesics from the basic colors and their complementary colors are straight lines. For all intermediate colors, admixture with white causes an apparent shift in hue. (Thus Machian purity has been sacrificed to theoretical expediency.)

Thirty years after the work of Helmholtz, Schrödinger took up the problem of specifying a line element in color space. The differential ds is not in general an exact differential of a function s, but it may be possible to find an integrating multiplier such that ds becomes integrable. Schrödinger derived the condition for the existence of such a multiplier and proposed an expression for ds^2 based on experimental work of Abney, Exner, and Kohlrausch:

$$ds^2 = (a_1 x_1 + a_2 x_2 + a_3 x_3)^{-1} \left(\frac{a_1 dx_1^2}{x_1} + \frac{a_2 dx_2^2}{x_2} + \frac{a_3 dx_3^2}{x_3} \right)$$

From any color of fixed hue and brightness, he can calculate a sheaf [*Büschel*] of geodesics in color space as shown in the example in Figure 4.2. Later work, however, has shown that these theoretical geodesics, calculated from the Schrödinger line element, are not in very good agreement with experiment, but the concept that loci of constant hue are geodesics in color space was an important contribution to color science.

This *Annalen* paper, which has been called 'masterly' by modern experts in color theory, is an impressive example, probably the most impressive in all the scientific literature, of how the philosophy of Mach can be applied to an actual problem. The data are not pointer readings or mechanical measurements of any kind, but rather the direct results of human sensations. Yet, despite its profundity and elegance, this work has had little consequence in practical color measurement, which still relies on empirical tables and color charts. In real life, color vision does not involve simply the comparison of isolated patches of color. It may not be possible to fit its complex reality into the framework of Mach's 'elements of sensation', since color vision is determined by the total Gestalt of visual perception, which does not contain 'elements' at all in the Machian sense.

Schrödinger did not publish any further work on color theory till 1924, when, in a brief article in *Die Naturwissenschaften*, he considered the question of why the color sensitivity curve of the human eye is shifted towards the blue compared to the spectral distribution curve of sunlight.[17] He suggested that the reason was that man's color vision originally evolved in fish, which experienced a bluish light filtered through water. There does not seem to be any way to test such an hypothesis, but considering the fact that most mammals are color blind, it seems rather unlikely.

On May 1, 1925, he published another article on color in *Die Naturwissenschaften*, 'On the subjective Colors of the Stars and the Quality of Twilight Sensitivity'.[18] The problem was to explain the discrepancy between the colors seen for fixed stars and their temperatures, as compared with the colors and temperatures of light sources on earth. For example a source that appears white on earth appears yellowish for a star of about the same temperature. This paper includes some experimental observations made with a simple device that enables an observer to compare daylight colors (cone vision) with dark-adapted colors (rod vision). Rod vision is not completely colorless but has a bluish tinge. Four normal trichromats matched the twilight color as a pale unsaturated violet similar to the color of the common lilac. Thus Schrödinger explained the yellowish appearance of the white star as the consequence of contrast against the lilac background of the rod vision. A similar explanation cannot hold, however, for the subjective shift toward the red in stars whose temperatures correspond to orange. To explain this effect, he calls upon the variation of hue with brightness that becomes apparent at low brightness (Bezold–Brücke phenomenon). With the two effects mentioned, he finds all the subjective star colors to be explainable. As a deuteranomalous trichromat he matched the twilight color with a pale greenish-blue.

At the very end of this paper he included a remark that must have made Mach turn over in his grave. 'The remarkable difference of twilight colors for normal and anomalous trichromats . . . can, I believe, be explained by differences in the daylight system alone, while the rod color itself is "in reality" the same for both – and apparently for all – types of eyes.'

On December 17, 1925, Schrödinger presented before the Vienna Academy of Sciences his last paper on color theory, 'On the Relation of the Four Color to the Three Color Theory'.[19] It is interesting to note that this paper was published with the aid of a grant from the Jerome and Margaret Stonborough Fund; Margaret was the youngest sister of Ludwig Wittgenstein, for whom he designed a beautiful and luxurious house. For many years the four-color theory of Ewald Hering was a strong rival of the Young-Helmholtz theory. The Hering theory relied on the obvious psychological fact that 'for the naive observer, besides red, green, and blue, pure yellow is also a psychologically unitary color, in which with the best will in the world an equal mixture of green and red cannot be seen'. In 1905, the *zone theory* of Johannes Kries provided a synthesis of the two rivals, in which Young–Helmholtz accounted for the interaction of light with the visual receptors, while Hering operated at the cortical stages of color processing in the brain. Schrödinger liked the turn that Kries gave to the discussion but in his paper he wished only to display the formal relation between the three- and four-color theories. Thus he defined a set of basis colors for the Hering theory and showed how the Young–Helmholtz color space can be transformed into the Hering color space. Thus both theories are equally capable of representing color mixtures at the level of simple sensations. The three-color theory has been used by almost everyone ever since biochemical research identified three cone pigments as red, green, and blue receptors. Recent electrophysiological studies support a modified form of the zone theory, in which the 3-to-4 transformation occurs in or prior to the lateral geniculate nucleus, so that Hering's ideas are still very much alive.

A fair amount of emotion was often displayed in the defense of favorite color theories, but scientists in the twentieth century have been less prone than those of earlier times to discover moral depravity in their intellectual opponents. Erwin rarely indulged in polemics, although he sometimes allowed himself a mild irony in describing

what he believed were dubious propositions. Thus the communication to the *Physikalische Zeitschrift* on March 16, 1925, in which he
demolished a paper by T. Oryngs on 'The Physical Definition of Bright
Bodycolors', was exceptional, possibly because Oryngs had slighted
the work of his friend Kohlrausch.

In the early 1920s Schrödinger became recognized as the world
authority on color theory. Accordingly he was asked by the editor of
the new eleventh edition of Muller-Pouillet, *Lehrbuch der Physik*, to
write the section of 'The Visual Sensations'.[20] The *Lehrbuch*, which
originally appeared in 1868, was one of the best of those many-
volumed scientific encyclopedias which have been faithfully produced
by German scientists for many generations. They define in authoritative terms the contemporary status of a subject and provide a solid
foundation for future advances. Erwin produced a 104-page article
with hundreds of references, which for many years was the standard
work on the subject. Nowadays such articles are prepared with the
help of computer searches and skilled assistants, but for him it was a
lonely and time-consuming task. The article displays his mastery of
not only the mathematics of color measurements but also the optics
and physiology of the eye. Schrödinger was a worthy successor to
Helmholtz in this field. Although color theory was a field attractive to
psychologists and physiologists, it had few apparent liens to basic
physical problems. The clue to Erwin's great interest in it was probably
that he was still tempted by his wartime vision of himself as a
philosopher, still immersed in Mach, Schopenhauer, and by now also
in eastern mysticism. Color theory stands at the crux of the ancient
mind-body problem. By working in this field Erwin could be both a
philosopher and a scientist.

Marriage and departure

From 1917 through 1920, the Institutes of Physics of Vienna University
were kept in turmoil by intrigues and arguments about the appointments of professors to succeed Hasenöhrl and Exner. The decisions
were complicated by political and racial alignments, and constrained
by the ingrown character of Vienna physics which tended to prevent
serious consideration of candidates from the world outside of Austria.
The main contenders for chairs were the theoretician Gustav Jäger,
who belonged to the German Nationalist party although his wife was
Jewish, and the Jewish experimentalist Felix Ehrenhaft, whose wife
was a friend of an influential politician in the Social Democratic party.
The leading candidate to succeed Hasenöhrl was obviously Smoluchowski, but Ehrenhaft and Jäger formed an unlikely alliance to
prevent his nomination on the basis that selection of a Pole would be

Erwin and Annemarie, wedding picture, March 1920.

damaging to German interests in Vienna. The issue became moot with the untimely death of Smoluchowski in 1917. At the end of June, 1918, Jäger, then fifty-three years old, was named to the theoretical chair as of October 1 and given the title of Hofrat. He was the last student of Loschmidt, Stefan and Boltzmann still in Vienna. He was a good administrator and a competent but uninspired theorist who worked on kinetic theory of liquids; he admitted that he knew little about modern theories and requested that he have Hans Thirring at his side to lecture in this area.[21]

In 1920, Exner retired from the Physics Institute II after a final *Ehrenjahr*. Jäger and Ehrenhaft both wanted this chair, but the enmity between Exner and the latter was serious, and the chair went to Jäger, after some perfunctory canvassing of foreign candidates. The Social Democrats were then in power, and a new Institute (III) was created for Ehrenhaft.[22] The Viennese physicists had thus managed to achieve mediocrity in all their professorships.

The question of the theoretical chair was still to be settled. On January 17, 1920, the Faculty had proposed Schrödinger for an associate professorship. The salary would still have been inadequate to support a wife and he was anxious to marry Anny as soon as possible, and without having to rely on her income as a secretary. He had been offered an assistantship with Max Wien in Jena, with an assignment to give some theoretical lectures. The salary of 2000 marks per annum appeared to be satisfactory – the runaway German inflation had not yet begun. Hans Thirring, who had some outside income from consulting, accepted the associate professorship in Vienna and was ultimately [1927] named professor of theoretical physics there, after he had virtually ceased any research in that subject. In retrospect, one can say that Schrödinger was fortunate to escape from Vienna when he did, for it is unlikely that his great work could ever have been accomplished there.

Erwin Schrödinger and Annemarie Bertel were married twice, the first time in a Catholic ceremony on March 24 in the rectory of the church of St Leopold, in the parish where Anny was living in the xviii District. The second ceremony was a more formal wedding on April 6 in the Evangelical Church on Martinsgasse in the xviii District. This duplication of the sacrament must have been a good omen for the duration of their marriage, which despite all trials and temptations, endured until they were parted by death.

Anny was twenty-three years old and Erwin thirty-two when they were married and they were in some ways an incompatible couple. Anny had little academic education and limited intellectual interests. She was an amateur of arts and crafts, but her tastes in music and the fine arts were strictly middle-class – as indeed were Erwin's. Hansi Bauer would shudder when she regarded their domestic interiors. Anny played the piano not badly and she had a pleasant singing voice, but these talents would have little appeal for Erwin, and he never allowed her to have a piano. She was considerate and kind-hearted, with an outgoing personality of the cyclothymic type. She was not extremely feminine, a bit of a tom-boy as a girl, and becoming distinctly mannish in appearance as she grew older. Although Erwin probably did not suspect this at the time, she was either unable to have children, for reasons that are obscure, or was deterred from having any by the history of mental instability in her family. One of her greatest attractions was her fervent admiration for everything about Erwin, his looks, his personality, and his intellectual brilliance. She once told Hans Thirring, 'You know it would be easier to live with a canary bird than with a race horse, but I prefer the race horse.'[23] She entered the marriage with the hope and expectation that it would be a true union of minds and bodies, in which she would achieve happiness through boundless submission to her brilliant and beautiful lover. These illusions may have lasted for at least a year.

Considering the relatively high frequency of his postmarital affairs, it is noteworthy that Erwin did not record any serious loves between the time he was dismissed by Felicie and the time he became engaged to Anny. During these eight years, he must have had casual sexual encounters, with sweet Vienna maids and the like, and during the war both professional and amateur sex were readily available. After the war, he had no home of his own and the appeal of domesticity must have been considerable. In this regard, Anny became an excellent wife, relieving him of all everyday concerns, providing the kinds of food and wine that he preferred, nursing him when he was ill, and after his interest in her as a sexual partner disappeared, remaining his friend and even helping him to find other feminine companionship. From his viewpoint, it was hardly an ideal marriage, but it had many compensations. In her case, it eventually led to emotional frustration, only partly compensated by her satisfaction in being the wife of a great man. A neutral observer must find them to be a strange couple, surrounded by a certain mystery, which could only be solved by a much more intimate understanding of their psychosexual histories.

Jena, Stuttgart, Breslau

Erwin and his bride arrived in Jena in April. It was a charming small city of about 70 000 nestled in the valley of the River Saale. The university, founded in the middle of the sixteenth century, had enjoyed its greatest fame from 1790 to 1805, when Friedrich Schiller was professor of history and his friend Goethe lived only a short distance away at the petty court of the Duke of Weimar. More recently it had become an important center for scientific research, gaining support from the large Zeiss Optical works established nearby.[24]

The Schrödingers received a hospitable welcome from Max Wien and his wife, whom, however, they found to be 'antisemites, but more as a matter of custom, not very bad'. Max Wien's brother Wilhelm (Willy) was one of the most powerful figures in German physics, professor at Munich, editor of the *Annalen*, and a Nobel laureate in 1911 for his discovery of a semi-empirical black-body radiation law. For advice about assistants, Max Wien had written to Arnold Sommerfeld in Munich, who replied 'Take a look at the three young Austrians, Flamm, Thirring, and Schrödinger.'

Schrödinger's assignment was to supplement the lectures of Professor Auerbach with 'something newer in theoretical physics'. Auerbach and his wife were very friendly and helpful to the young couple. They were Jews and later chose to depart this life together when their colleagues and neighbors turned on them with scorn and abuse after the Nazis seized power in 1933. While at Jena, Erwin also came to know the old philosopher Rudolf Eucken, who had received a Nobel prize in literature in 1908. Eucken based his philosophy on the life of man – he said it is partly good and partly evil, but must be accepted in an optimistic way. He rejected socialism as merely living in the present without spiritual depth and he hoped that science would help to regenerate outworn traditional religions.[25]

Schrödinger gave his inaugural lecture on recent developments in atomic theory. He immediately made a favorable impression on the physics community in Jena, both in the laboratory and in his lecturing, for he had been there only a few weeks when the faculty recommended that he be promoted to an associate professorship (without chair). The German inflation was worsening, and even though his total income in Jena would now amount to about 11 000 marks a year, his associate professorship was not a permanent position. Thus when a regular associate professorship was offered at the Technische Hochschule in Stuttgart, he resigned from his various posts at Jena as of October 1, and moved with his wife to the great industrial city and capital of Baden-Württemberg.

The professor of experimental physics in Stuttgart was Erich Regener, an expert in X-rays, who had worked for many years in Berlin with colleagues including Otto Hahn, Lise Meitner, Gustav Hertz, and Albert Einstein. The Regeners, who had two children, Erika and Victor, lived in a spacious apartment attached to the Institute of Physics, where they often entertained friends from the university. Erika, who was a student at the recently established Gymnasium for girls, became a good friend of Anny.[26]

While in Stuttgart, Erwin also often met Hans Reichenbach, who was lecturing in mathematics but even then had a great interest in the philosophy of science. Although he had never studied in Vienna, he became an ardent exponent of logical positivism and later wrote several books on the philosophy of quantum mechanics.

The closest friends of the Schrödingers in Stuttgart were the Ewalds, Paul and Ella. Paul was appointed professor of theoretical physics in 1921 and arrived early that year, but Ella stayed in Munich until her second son Arnold, named after Sommerfeld, was born. Paul was a big, handsome man, and Anny was greatly attracted to him while, perhaps responding to her unfeigned admiration, he in return became fond of her. Anny's women friends always found it difficult to understand her attraction for men.

While Erwin and Anny were in Stuttgart, his mother became seriously ill with a recurrence of cancer. Fortunately she had been able to come to visit them while still able to travel, and Anny accompanied her back to Vienna at the end of February. Her condition worsened, and during the three months before her death in September, 1921, she was unable to leave her bed. After the death of Rudolf, she had been forced to give up her beautiful Vienna home, since her father needed more rent money to pay the heavy postwar real-estate taxes; he was in dire financial straits himself as the inflation had destroyed his pension and savings. Thus the widow was turned out and the Schrödinger flat rented to a 'rich insurance Jew' – this was the rude phrase used by Erwin many years later in recalling the event in his *Autobiography*. It reflects the bitterness he felt at his mother's fate, his frustration that he was unable to do anything about it, and his bad conscience at leaving her in the lurch. (Presumably, however, her two sisters who were comfortably situated were able to take care of her.) Erwin's almost neurotic concern for financial security and widow's pensions were caused by these years of financial destitution, during which he had to leave his native land to secure even a marginally adequate income and was completely bereft of assets that could be used to help his mother. In later years he worried continually about what might happen to Anny as a widow, as he was haunted by the vision of his dispossessed widowed mother. As it turned out, during her few years of widowhood, Anny would be more wealthy than at any other time in her life.

In the spring of 1921, several German universities, Kiel, Hamburg, Breslau, and also Vienna, were seeking professors of theoretical physics, and Schrödinger was being seriously considered for all these posts. The position at Breslau was the most attractive, giving due weight to both the reputation of the university and the level of the appointment. Otto Lummer, famous for the measurements of black radiation that inspired the Planck quantum hypothesis, was professor of experimental physics there; Fritz Reiche was working on quantum theory, and Rudolf Ladenburg, an outstanding spectroscopist, was also deeply interested in this subject.

Schrödinger had serious doubts about a move to Breslau, since the city was in the heart of the Silesian industrial region near the new Polish border. He wrote to his good friend Stefan Meyer in Vienna that he felt a 'certain dread of the mad leftists . . . who are continually gaining more adherents among the workers, and where that can lead has been seen in Halle, where 2/3 of the socialist party have just declared for a union with Moscow, i.e., for the red terror. I often have the feeling: let me get away from this powder keg.' He says that he would not mind going alone to Breslau but even the remote possibility of exposing a wife, 'and perhaps children', to such circumstances without being able to protect them is the most dreadful that can be imagined. In late October, he wrote to Hans Thirring:[27] 'Although now as before I aspire to Vienna, it is clear that it would be for me a really embarrassing thing . . . to refuse the Breslau *Ordinarius* for the Vienna *Extraordinarius*.' He went on to say that he was not confident that he would be promoted to a Vienna *Ordinarius* when it became available since during recent months he had been passed over by the Austrian government for two full professorships in Graz, even though he had been ranked *primo loco* for both appointments. Erwin did not have political influence with either the clericals or the socialists and thus was at a serious disadvantage in the Austrian intrigues for academic preferment. Thus, despite his misgivings, he accepted the Breslau professorship for the beginning of the summer semester and once again Anny packed their few possessions for another move to a new city, the third within eighteen months.

As *Ordinarius* in charge of a department, Schrödinger for the first time undertook the chores of academic administration, which he performed adequately but not enthusiastically.

Old quantum theory – spectra

While in Stuttgart, Schrödinger studied intensively the recent book of Arnold Sommerfeld, *Atomic Structure and Spectral Lines*, the first edition of which had been published in 1919.[28] The book had immediately become the 'Bible' of its rapidly developing field. It is remarkable that Schrödinger, who was new to the subject, was at once able to make an important contribution, which he sent to *Zeitschrift für Physik* in January, 1921. As he explained some years later:[29]

In my scientific work (and moreover also in my life) I have never followed one main line, one program defining a direction for a long time. Although I can work only poorly in collaboration, and unfortunately also not with pupils, my work in this respect is still not entirely independent, since if I am to have an interest in a question, others must also have one. My word is seldom the first, but often the second, and may be inspired by a desire to contradict or to correct, but the consequent extension may turn out to be more important than the correction, which served only as a connection.

Thus, the quantum theory of atomic structure and spectra was already well advanced when Schrödinger wrote his 1921 paper. The first correct application of quantum theory to spectroscopy was not made to atomic spectra but to the infrared absorption spectra of molecules. This work was done in 1912 by a young Danish chemist, Niels Bjerrum, who was at Nernst's laboratory in Berlin. Nernst was able to explain the specific heats of gases on the basis of quantized rotational and vibrational motions of the molecules, and Bjerrum obtained the vibrational and rotational frequencies from the molecular absorption spectra.

The interpretation of atomic spectra was achieved by another young Dane, Niels Bohr, who was working with Rutherford in Manchester. Bohr was only two years older than Schrödinger, and he made his great discovery in 1913 at the age of twenty-eight. He solved the problem of atomic spectra by taking what was correct in older ideas, throwing away what was incorrect, and then adding exactly the necessary new ideas. He based his theory on the Rutherford model of the atom, a central positive nucleus surrounded by negative electrons. He postulated that in the absorption or emission of radiation corresponding to a spectral line, a single electron jumps between two different states characterized by *discrete* values of angular momentum and energy. When an electron makes a transition between two states with energies E_1 and E_2, the frequency of the spectral line v is given by the Planck–Einstein relation,

$$hv = E_1 - E_2$$

Finally Bohr proposed a revolutionary concept, which aroused a storm of controversy among scientists and philosophers: We must renounce any attempt to visualize or to explain the behavior of the electron during a transition of the atom from one stationary state to another.

Having invented a satisfactory model, Bohr found no difficulty in its application to the hydrogen atom, which he simply treated as a microscopic Kepler system, with the electron revolving around the nucleus (the proton) like a planet around the sun. An electron is held in its orbit by the electrostatic force that attracts it to the nucleus; if the nuclear charge is Ze, this force is Ze^2/r^2 from Coulomb's Law. For a stable state, this is the centripetal force mv^2/r, so that

$$Ze^2/r^2 = mv^2/r$$

The magnitude of the angular momentum of an electron of mass m moving with speed v in a circular path of radius r is $L = mvr$, and this is assumed to follow a quantization condition, $mvr = nh/2\pi$, where n is called the *principal quantum number*. Therefore, $r = n^2 h^2/4\pi me^2 Z$. For hydrogen $Z = 1$, and the smallest orbit, that with $n = 1$, would have a radius

$$a_0 = h^2/4\pi me^2 = 0.529 \times 10^{-10} \text{ m}$$

This is now called the *Bohr radius* and it provides a natural atomic unit of length.

Bohr calculated the energy level of a stationary orbit, the sum of the kinetic and potential energies of the electron,

$$E = E_k + E_p = \frac{Ze^2}{2r} - \frac{Ze^2}{r} = \frac{-2me^4}{n^2 h^2} Z^2$$

Finally, the frequency of the spectral line arising from a transition of the electron between levels with quantum numbers n_1 and n_2 is

$$v = (1/h)(E_{n_1} - E_{n_2}) = \frac{2me^4}{h^3} \left(\frac{1}{n_1^2} - \frac{1}{n_2^2} \right) Z^2$$

In 1885, J.J. Balmer had shown that the experimental data on the frequencies of lines in the visible spectrum of atomic hydrogen fit a simple formula, $v = R c(2^{-2} - n^{-2})$, where R, the Rydberg constant, can be obtained to extremely high accuracy. Bohr calculated the theoretical value of the Rydberg constant for hydrogen as

$$R = 2me^4/ch^3 = 109\,737 \text{ cm}^{-1}$$

A small correction is necessary to account for the fact that the electron actually revolves not about the proton but about the center of mass of the electron–proton system. With this correction, agreement between theory and experiment becomes perfect to within experimental uncertainties, a triumph for the Bohr theory. The mathematics of the theory is elementary and once the model is constructed, the equations are child's play. Bohr had a reasonable acquaintance with mathematical physics, but he never employed advanced mathematics in his work.

One of the first applications made by Bohr of his new theory was to explain the lines in a series first observed by Pickering in the spectrum of a star ζ-Puppis. They were commonly thought to be hydrogen lines since they had almost the same series limit as the Balmer series, but Bohr showed that they really arise from singly ionized helium, in which a single electron is held by a nucleus of charge + 2. Einstein was at the meeting in Vienna in September 1913, when he heard about this result from George Hevesy. He was astonished and said: 'Then the frequency of light does not depend at all on the orbiting frequency of the electron. And this is an *enormous achievement*. The theory of Bohr must be right.'[30]

Not all physicists were prepared to leap to this conclusion. The conservative influence of Planck was still pervasive; in 1914 he advanced a 'third theory' in which both absorption and emission of radiation obey the laws of classical electrodynamics, while quantization occurs only in collision processes.

Influential support for the Bohr theory came, however, from Arnold Sommerfeld. A native of East Prussia complete with duelling scar, he had studied with Hilbert and Klein at Göttingen, and was the most skillful mathematician among the physicists. In 1905, Röntgen, who was professor of experimental physics, persuaded the Bavarian government to re-establish the chair of theoretical physics at Munich, which had been vacant since Boltzmann left eleven years previously, and in 1906, after Nernst, Lorentz, and others had turned it down, it was happily accepted by Sommerfeld, then thirty-eight years old and at the height of his powers. As a young man, he had been rather severe in personality and reactionary in politics, but as he grew older he became steadily more genial and more liberal. His ability as a teacher was legendary and he established the best school of theoretical physics in the world, with important help from his faithful assistant Pieter Debye.[31]

Sommerfeld extended the Bohr theory to consider not only circular electron orbits but also elliptical ones. For an allowed orbit, he used the general criterion that the increase in action in going once around the orbit is an integral multiple of the Planck constant, $\oint p\,dq = nh$. In 1915, he introduced two new quantum numbers to distinguish the orbits, n, the radial quantum number and k, the azimuthal quantum number, the ratio of the major to the minor axis of the ellipse being n/k. He applied his theory to the spectra of the alkali metals, and used the value of k to distinguish the different observed series, which had been called by spectroscopists, *sharp, principal, diffuse, fundamental, etc.* Later $l = k - 1$ was used for the azimuthal quantum number and it was denoted by letters s, p, d, f for $l = 0, 1, 2, 3$ respectively. Actually the lines in the alkali-metal spectra are doublets, and Sommerfeld introduced an *inner quantum number j*, to represent the quantization of the unknown kind of angular momentum responsible for this multiplicity.

Sommerfeld also made a remarkable application of relativity theory to the calculation of electron orbits. In an elliptical orbit, the velocity of an electron near its perihelion can become so close to that of light c, that the relativistic increase of mass with velocity becomes appreciable. The ratio of electron speed to that of light is $2\pi e^2/hc$. As predicted by the theory, this dimensionless *fine structure constant* is closely (but not exactly) equal to 1/137.

Penetrating orbits[32]

Schrödinger wished to obtain a satisfactory model, based on Bohr–Sommerfeld theory, for the so-called *sharp series* in the emission spectrum of alkali metals such as sodium. The line frequencies in this series are given by $v = 2P - mS$, where $m = 2, 3$, etc., i.e., the spectra arise when the outermost electron in the atom makes a transition from one of the S levels ($k = 1$) to the lowest P level.

Schrödinger pointed out that elliptical and circular orbits similar to those proposed for the electron in the hydrogen atom cannot occur for

$k = 1$ states in an atom like sodium. The sodium atom has eleven electrons, two in the innermost shell, eight in the next shell, and a sole valence electron in the outermost shell. When $k = 1$, the elliptical orbits according to Bohr–Sommerfeld theory have a perihelion distance of from 0.5 a_0 to 0.54 a_0, depending on the value of the radial quantum number. The radius of the supposedly circular orbits of the eight middle electrons is estimated to be 0.61 a_0. Therefore, it appears that the supposed elliptical orbit of a k electron must sometimes penetrate this shielding shell and experience an effective nuclear charge closer to $Z' = 9$ than to $Z = 1$, and thus the suggested elliptical orbit based on $Z = 1$ is unrealistic. The exact calculation of the true orbit 'appears to be hopelessly complicated', but he made an approximate calculation, based on a model in which the electron passes smoothly from a large outer elliptical orbit to a small inner one as it penetrates the shielding shell. With some clever mathematics, he carried out the integrations $\oint p dq$ for this complex orbit and obtained effective quantum numbers to designate the spectral terms. He did not make any experimental tests of his model, since the Bohr–Sommerfeld theory was quite incapable of any quantitative calculation of the energy levels of even a two-electron atom like helium.

The qualitative concept of penetration of outer electrons through the shielding shell of inner electrons was nevertheless important and Schrödinger's paper was frequently cited in later work on atomic structure. After the advent of quantum mechanics and the consequent demise of all models based on electron orbits, the concept of penetration effects remained valid and was used to help explain how physical and chemical properties of the elements vary with nuclear charge Z through the periodic table.

The University of Zürich[33]

Erwin and Anny had only just returned to Breslau after witnessing the interment of his mother in Vienna's Hietzinger Friedhof, when they prepared to leave turbulent Germany for peaceful Switzerland, in response to a call from the University of Zürich.

The University of Zürich was founded in 1833. It was not organized from above but in response to demands by the people of the district for an institution of higher learning. At that time the only Swiss university was in Basel. Even by 1860, Zürich was still a small town with about 20 000 inhabitants, but it grew rapidly, to 200 000 in 1920. In 1855, to meet the need for technical education, the Eidgenossische Technische Hochschule [always called the E.T.H., ay tay ha] was opened, and Rudolf Clausius was named professor of physics there. In 1857 he became also professor at the University. Clausius was one of the greatest physicists of the nineteenth century. He established the second law of thermodynamics and defined the concept of entropy; he also made important contributions to the kinetic theory of gases.

In 1878, Alfred Kleiner was named as associate professor of experimental physics (*Extraordinarius*). Kleiner's greatest work, as he himself often admitted, was his approval in 1905 of Albert Einstein's thesis, 'Determination of Molecular Sizes', and his influence in calling Einstein in 1909 to a post as *Extraordinarius in Theoretical Physics*, the first appointment in this subject at the University. In 1911 Einstein went to Prague as full professor; he returned to Zürich in 1912, but soon moved to the Kaiser Wilhelm Institute in Berlin, a city that was then the center of world physics.[34]

The Swiss authorities tended to be parsimonious; while the university was able to attract excellent young physicists, they were easily lured away by higher salaries and better research conditions elsewhere. In 1911, the brilliant Dutch chemical physicist Pieter Debye, who had been working with Sommerfeld in Munich, was appointed to the Einstein vacancy, but only as *Extraordinarius*. He did his important work on the theory of heat capacity of solids while at Zürich. When he received an offer from Tübingen, he was promoted to the first chair in theoretical physics at Zürich in 1912, but left soon after to go to Utrecht. To succeed Debye, another brilliant choice was made, Max Laue, also from Sommerfeld's department. He had already made his great discovery of X-ray diffraction and was to receive the Nobel prize a few years later, but the frugal Swiss gave him an associate professorship. Following the now familiar pattern, he left after two years for a professorship at Frankfurt. An attempt was made to get Debye back, but the salary offered was not enough. In 1920, however, the E.T.H. had more success, and he stayed there for seven years before moving to Leipzig.

This musical chairs at the German-language universities was an accepted game. The curricula at all these universities were much the same, and there was a well-established order of prestige, with Berlin, Munich, and Göttingen at the top, and places like Aachen and Innsbruck at the foot. If he received an offer of a full professorship at another university, an associate professor was expected to transfer even if the new place was lower on the prestige scale. The Austrian universities tended to be somewhat chauvinistic, often preferring to promote a professor from within a department instead of seeking the best man from abroad.

The chair of theoretical physics at Zürich remained unfilled all during the war. The department managed to function owing to the devoted services of its Dozents, Paul Epstein, Simon Ratnowsky, and Franz Tank. This arrangement saved the government some money, but in November, 1919, Professor Meyer wrote to the dean of the Philosophy-II Faculty, Hans Wehrli, to request that the position of associate professor of theoretical physics be filled.[35] A selection committee was appointed, including the dean, the alternate dean Wolfer, and professors Meyer, Fueter, Speiser and Karrer. They wrote to Niels Bohr to ask his advice; he said he was busy and would write within a few weeks, but he never did, since at that time he was recuperating from nervous exhaustion probably caused by overwork in establishing his new Institute. At the end of February the committee met, decided it would not be possible to attract an eminent outside candidate, and considered the three Dozents. There were six votes for Ratnowski, five for Epstein, and one abstention. They sent this information to the Education Commission [*Erziehungsrat*] of the Canton of Zürich. On June 3, the Commission informed the faculty that they should consider the matter further, in the order Epstein, Tank, Ratnoswki. The committee thought all three candidates were deserving, but they had some doubts about how Epstein, with his Polish accent, would manage the large class in the elementary lectures. They decided to assign him to teach the course in the winter semester 1920/21 and raised his salary to Sfr 3500.

By the end of January, 1921, some doubts remained, and meanwhile the committee had consulted Sommerfeld for his recommendations concerning a list of names, including Leon Brillouin, Erwin Schrödinger, Paul Ewald, Wilhelm Lenz, Tank, and Epstein. Sommerfeld replied in a detailed handwritten letter.

Epstein has doubtless among all your candidates the greatest name. His Stark-effect work belongs among the really greatest achievements. His general understanding and his mathematical aptitude are also remarkable. Yet it seems that in recent years he has become less productive. But that may be due to his precarious external [Polish] financial circumstances. The fact that he did not receive a call in the last round of German appointments is due to the fact that Germany with its handicaps can no longer afford the luxury of foreign professors. I would really not begrudge Epstein the Zürich professorship. The choice between Lenz and Epstein is difficult. Debye would apparently favor Lenz. I think that Epstein has the prior claims.

Sommerfeld was lukewarm about Brillouin, whom he thought was not highly original as a theoretician, but a long letter from Victor Henri in Paris recommended Brillouin as the best student he ever had.

About Schrödinger, he said only: 'A first-rate head, very sound and critical. *Full* professor at Breslau, and thus certainly not available to you [as associate]'.

On February 2, Laue wrote that he would like to come back to Zürich, but his family had lost all its money in the postwar inflation, and since he would have to rely entirely on his salary, it would have to be an exceptionally good one.

On February 5, a letter arrived from Fürtwangler in Vienna praising Schrödinger highly.

About Herr Schrödinger as a person and as a teacher, there is only the most favorable to report. That he enjoys the greatest esteem here, you can see from the fact that he was ranked *primo loco* for the post in theoretical physics here. Unfortunately the ministry was not sufficiently forthcoming as to salary and he did not accept, Thirring did. He has gone from Stuttgart to Breslau, and I hear he also had a call to Kiel. You see that he is apparently greatly esteemed everywhere.

On February 16, the staff of the physics department presented a petition to the *Regierungsrat*, asking that Paul Epstein be appointed to the position in question. This petition was evidently organized by Richard Bär.

On February 26, Dean Wehrli went to Stuttgart to attend a lecture by Schrödinger and he asked him confidentially if he would accept a professorship at Zürich if it were offered. The answer was positive, and was confirmed in a letter of March 2.

In fact I was appointed at the University of Breslau on January 15 to a scheduled associate professorship with the personal rights (not salary) of a full professor. I take up this post on April 1. I would, however, *absolutely unequivocally give preference* to the professorship mentioned by you at the University of Zürich, if I should be called to it. My Breslau obligations limit me only to the extent that I could not take the new appointment till the end of the following semester, that is at the earliest October 1 of the current year. I might also mention that owing to the unprecedented lack of housing in Breslau, I should certainly not be able to undertake the move of my household there on April 1, but at best several – I'm afraid many – months later. Thus also in this aspect, there is no problem that would impede a possible call to Zürich.

On March 3, the dean wrote to the Education Commission that the first choice of the faculty was Laue and giving the other names in alphabetical order. There is a gap in the record here, but evidently the directorate could not find the funds needed to attract Laue, and they sent the short list back to the faculty with a request that an order of recommendations be provided. Always on the lookout for savings, they also asked whether Professors Debye and Scherrer at the E.T.H. might not also provide for theoretical physics at the university.

Meanwhile, a long letter from Langevin was received, which praised Brillouin in glowing terms. A letter from Erich Regener at Stuttgart (March 17) spoke highly of Schrödinger: 'His scientific work I need not describe. As a teacher, however, he speaks with extraordinary clarity, and everything he says is thoroughly thought out. Also he has a nice wife [*nette Frau*].'

The task of the faculty was made easier by the sudden resignation of Paul Epstein, who notified the Education Commission on March 15 that he had accepted a post with Lorentz at the University of Leiden, and he left the faculty at the end of the month. Evidently his Polish accent and 'too foreign ways' would be less of a problem among the Dutch.

On June 13, the dean sent a thirteen-page letter to the Commission, summarizing the records of all the candidates, and giving the faculty's order of preference as (1) Schrödinger, (2) Brillouin and Lenz, *aequo loco*, and (3) Ewald. He strongly argued that there was no way in which the E.T.H. professors could also provide for the teaching and research in theoretical physics at the university. At its meeting of July 20, the *Regierungsrat* of the Canton approved the recommendation of the Faculty and the appointment of Erwin Schrödinger as full professor of theoretical physics. In its report the Faculty had pointed out:

the versatility of the works of Schrödinger in the fields of mechanics, optics, capillarity, electrical conductivity, magnetism, radioactivity, gravitation theory, and acoustics. The appointment of Schrödinger was also deemed desirable by the Faculty because it would enable the holding of the lectures on biometry long desired by the biologists.

He was thus represented as a man conversant with all fields, but as yet he had not accomplished a truly outstanding piece of work in any particular field. To some extent the appointment was an act of faith that such wide ranging interests and abilities would not go unrewarded by some major discovery. The conditions of the appointment were specified as follows:

(1) A tenure for six years to commence 15 October 1921.

(2) Teaching duties to comprise 8 to 12 hours weekly of lectures and exercises in the field of theoretical physics, including a course of four hours a week on the mechanics of solid bodies every winter semester.

(3) The annual salary would be 14 000 Sfr, to begin on date of resignation from his previous post on October 1, 1921. The appointee will receive 30% of any tuition fees paid for his lectures and courses, after subtraction of the state quota and applicable costs of the institute; the rest will be added to university funds.

(4) The state undertakes to pay the costs of entry into the state widows and orphans pension fund as well as an appropriate part of his moving expenses.

(5) The appointee is obliged to live in the city of Zürich or its suburbs, and to enrol in the widows, orphans and pension funds of the university.

(6) To be transmitted to the appointee, Breslau, Tiergartenstrasse 74 (in duplicate), to the Dean of Philosophical Faculty II, to the Rector and Bursar of the University, and to the Education Directorate.

It is evident that the careful Swiss authorities left nothing to chance. The close control they exercised over the university administration at that time is also noteworthy. The salary offered to Schrödinger, equivalent to about $2500 a year, was at the top of the range for full professors. The country was in the midst of a postwar depression with severe unemployment in Zürich. The cost of living was fairly low since inflation in Switzerland had been much less than elsewhere. The cost of food had increased by sixty percent during the war years, with the best veal, for example, now at four francs [$.80] a kilo.

On September 16, Schrödinger wrote from Vienna to accept the appointment formally. At the age of thirty-four, he had achieved the first ambition of every academic, a full professorship at an excellent university. As the successor of Einstein, Debye, and Laue, he must have felt inspired by a firm resolve to accomplish some physical research of the high standard set by these illustrious predecessors.

5 Zürich

The city of Zürich is divided by the River Limmat into the newer *Grosse Stadt* on the left bank, and the *Kleine Stadt* on the right bank, where the E.T.H. and University are located on heights overlooking the city, the river, and its source in the Zürcher See. The Schrödingers rented a spacious flat consisting of an entire floor of a modern stucco house called *Zu Vier Wachten* [At The Four Guards] at 9 Huttenstrasse, a broad avenue just a few blocks behind the university.

Liegekur

Even before his lectures for the winter term began, Erwin was physically and psychologically exhausted. As he wrote about a year later to Pauli, 'I was actually so *kaputt* that I could no longer get any sensible ideas. Not the least to blame for it were the many complications, the constant decisions about one's own fate, negotiations with ministries, etc., which I was not at all cut out for. Now that's all finished for a long time.'[1] Also he had lost his father, his mother, and a grandfather within the past two years. He had scarcely started his lectures when he was forced by a severe attack of bronchitis to interrupt them in the middle of November. He suffered intermittent respiratory illnesses during the winter, from which he never managed to recover completely; finally a mild case of pulmonary tuberculosis was suspected, probably a recurrence of the infection of the apical lung area that was noted two years previously in Vienna. He was ordered to undertake a complete rest cure [*Liegekur*] at high altitude. The only bright aspect of this miserable winter was that his financial worries were over. Anny was able to engage a wonderful cook from Vienna, who expertly provided all their favorite dishes. This cuisine was a happy change from the near starvation of the postwar years in Germany and Austria.

The place chosen for Erwin's Liegekur was Arosa, an Alpine *Kurort* at about 1700 m altitude, not far from the ski-resort of Davos, and overlooked by the great peak of the Weisshorn. Anny worried about

Erwin like a mother with a sick child; she could not do enough to take care of him. The Viennese cook was there to provide culinary comforts. The theory of the cure was that the high altitude causes proliferation of red blood corpuscles, which somehow help to overcome the deadly bacillus; actually, *myobacterium tuberculosis* likes a plentiful supply of oxygen and any beneficial effect of high altitude may be directly related to the low atmospheric pressure. Erwin did improve remarkably during his nine-month stay in Arosa, and all evidence of active infection disappeared.

They were visited for a couple of weeks by the Ewalds, Paul and Ella, from Stuttgart; the Schrödingers paid all their expenses since inflation had made the German currency worthless. Ella recalls that Anny and Erwin seemed to be a happily married couple at this time.[2] Erwin postponed his return to Zürich till November 1, two weeks after the beginning of winter term; by then he had been pronounced cured, but he still tired easily. He resumed his heavy load of teaching but had little energy left for research.

He had, however, managed to write two short papers at Arosa. The first was sent to the *Zeitschrift für Physik* on September 5 'On the Specific Heat of Solids at High Temperatures and On the Quantization of Vibrations of Finite Amplitude'. It was a comment on a recent paper by Born and Brody on the effect of anharmonic oscillations on the Dulong and Petit law of specific heats.

A remarkable discovery

The second Arosa paper was a much deeper and more original contribution, 'On a Remarkable Property of the Quantized Orbits of a Single Electron'.[3] In the summer term of 1917, Hermann Weyl had given a course of lectures at the E.T.H., which were published the following year as *Space – Time – Matter*.[4] The book gave a complete account of relativity theory including all the necessary mathematical background. It was very popular among physicists, both old and young, and by 1921 was already in its fourth edition. Schrödinger studied this book carefully and used it in several phases of his subsequent work.

Weyl had shown that for a manifold to be a metric space it must have a measure at every point, determined by a quadratic form, $dl^2 = g_{ik} dx_i dx_k$, and also every point must be metrically related to the domain surrounding it by a linear form $dl = -l\varphi_i dx_i$. The geometric meaning of the latter relation is that the magnitude l of a vector does not remain constant in a congruent displacement to an infinitely near neighboring point, but undergoes a change dl. Thus by congruent displacement of a length along a finite stretch of a world line, the measure of the length is multiplied by the factor exp($-\int \varphi_i dx_i$). According to Weyl, the coefficients of the linear form $\varphi_i dx_i$ are proportional to the electromagnetic potentials, so that the multiplying factor becomes

$$\exp[-(e/\gamma)\int (V dt - A_x dx - A_y dy - A_z dx)]$$

where V is the scalar potential and A_x, A_y, A_z the components of the vector potential, and γ has the dimensions of action. Schrödinger pointed out that for an electron in a hydrogen atom, where $A = 0$, the exponent in the Weyl factor becomes $-e\bar{V}\tau/\gamma$, where τ is the orbital period and $e\bar{V}$ the average potential energy. For a Bohr orbit of quantum number n, the term thus equals $-nh/\gamma$, where n is an integer.

This was a far-reaching result, but although Schrödinger said that he believed it must have deep physical significance, he did not pursue it further at this time. Later he would realize that it contained the basis of de Broglie's explanation of Bohr orbits in terms of electron waves.

At the end of his paper, Schrödinger mentioned two possible values for γ: a real number, $\gamma = e^2/c$, and $\gamma = -ih/2\pi$; for the latter case the Weyl multiplying factor becomes unity. As Chen Ning Yang explained at the Schrödinger Centenary celebration (London, 1987),[5]

the importance of the introduction of complex amplitudes with phases into physicists' description of nature was not fully appreciated until the 1970s when two developments took place: (1) all interactions were found to be some form of gauge field; and (2) gauge fields were found to be related to the mathematical concept of fibre bundles, each fibre being a complex phase or a more general phase. With these developments there arose a basic tenet of today's physics: *all fundamental forces are phase fields* (Yang, 1983). Thus the almost casual introduction in 1922 by Schrödinger of the imaginary unit i into [the Weyl factor] has flowered into deep concepts that lie at the very foundation of our understanding of the physical world.

In 1926, Fritz London wrote a remarkably playful letter to Schrödinger about this paper:[6]

Very Respected Herr Professor:
Today I must talk with you seriously. Do you know a certain Herr Schrödinger who described, in the year 1922, a 'noteworthy property of quantum orbits'? Do you know this man? What, you say you know him rather well, you were even with him when he wrote this paper and were implicated in the work? This is truly shocking. Hence you already knew four years ago that one does not possess rods and clocks for the definition of an Einstein–Riemannian measure in the continuous description that occurs in analyzing atomic processes; thus one must see whether perhaps the general principles of measurement that arise from Weyl's theory of distance transfer might help. And you even realized four years ago that they help very well . . . and you

showed that for real discrete orbits the gauge factor reproduces itself on a spatially closed path; and especially you then realized that on the nth orbit the unit of measure swells and shrinks n times, exactly as in the case of a standing wave describing the position of charge. You therefore demonstrated that Weyl's theory becomes reasonable – i.e., *leads to a unique determination of measure* – only if combined with quantum theory; and one has no other choice if the whole world of atoms represents a process in a continuum without any identifiable fixed point. You knew this and said nothing about it . . . Will you now immediately confess that, like a priest, you held the truth in your hands and kept it a secret? . . .

The four years from 1922 to 1926 spanned a revolution in theoretical physics, so that what seemed crystal clear to London in 1926 was only a clouded intuition to Schrödinger as he was fighting the tubercle bacillus at Arosa in 1922. One can speculate that perhaps – had his health been better – he might then have realized all the implications of that strange property of electron orbits. Three years later, in this same mountain retreat, he would make his great discovery of wave mechanics.

Life in Zürich

Schrödinger had a heavy schedule of lectures for the academic year 1922/23. In the winter semester, four lectures a week on Analytical Mechanics, four on Special Problems of Electron Theory, the seminar and proseminar in Theoretical Physics, and one lecture a week for nonspecialists on The Modern Development of the Physical World Picture: Fine Structure of Matter, Atomic Structure, and Relativity Theory. In all, therefore, he had eleven hours a week of lecturing and teaching. Tuesday was an especially arduous day, with lectures from 8 to 10 a.m., and from 3 to 5 p.m. His colleague in experimental physics, Professor Edgar Meyer, had three lectures a week, but also direct responsibility for all the practical classes. In contrast, Privat-dozent Ratnowsky had only two lectures a week, on Relativity Theory. Among the other Dozents, Richard Bär lectured once a week on the History of Mechanics and Thermodynamics; he belonged to a rich Zürich banking family and had no need to teach for a living. Karl Meissner gave an hour on Geometrical Optics, and Mieczyslaw Wolfke two hours on Quantum Theory. Unlike the situation today, the distinguished professors did most of the teaching.

Alexander Muralt was a student in Schrödinger's classes at the University during 1922/23.[7] He recalls that the lectures were:

extremely stimulating and impressive. At the beginning he stated the subject, and then gave a review of how one had to approach it, and then he started exposing the basis in mathematical terms and developed it in front of our eyes.

Sometimes he would stop and with a shy smile confess that he had missed a bifurcation in his mathematical development, turn back to the critical point and start all over again. This was fascinating to watch and we all learned a great deal by following his calculations, which he developed without ever looking at his own notes, except at the end, when he compared his work on the blackboard with his notes and said 'this is correct!'. In summer time, when it was warm enough, we went to the *Strandbad* [bathing beach] on the Lake of Zürich, sat with our own notes in the grass and watched this lean man in bathing trunks before us writing his calculations on an improvised blackboard which we had brought along. At that time few people came to the *Strandbad* in the morning and those who did watched us from a discreet distance and wondered what that man was writing on the blackboard!

Besides his colleagues at the University, Erwin soon developed close relations with the mathematicians and physicists at the nearby E.T.H. Pieter Debye, who had returned to Zürich in 1920, was about three years older than Erwin. Of all the theoreticians, he had the greatest insight into experimental problems and the best intuitive feeling about what was physically reasonable. He often met Schrödinger at the joint E.T.H.–University seminars; they were always cordial but never became intimate friends – solid Dutch virtues and fugitive Austrian charms made a somewhat immiscible combination.[8] Debye had arranged the appointment of Paul Scherrer, a native of Switzerland, to an associate professorship at the E.T.H. Paul was three years younger than Erwin. His wife Ina came from the wealthy Sonderegger family, and her father had objected vehemently to her marrying a relatively poor professor. Paul, however, proved to be a bon vivant and pillar of Zürich society, as well as a competent director of the Physics Institute. Erwin's closest friend when he first came to Zürich was Hermann Weyl, always called Peter by his intimates. Weyl was born in 1885 in Elmshorn, Germany, received the doctorate under Hilbert in Göttingen, and soon displayed mathematical genius of the highest quality. Those familiar with the serious and portly figure of Weyl at Princeton in the 1940s would hardly have recognized the slim, handsome young man of the twenties, with his romantic black moustache. His wife, Helene Joseph, from a Jewish background, was a philosopher and literateuse. Her friends called her Hella, and a certain daring and insouciance made her the unquestioned leader of the social set comprising the scientists and their wives. Anny was almost an exact opposite to the stylish and intellectual Hella, but perhaps for that reason Peter found her interesting and before long she was madly in love with him. They were not a snobbish group – some were rich and some were poor, and their favorite entertainments were not necessarily expensive.

Zürich is one of the oldest cities in Europe, the Roman town of

Turicum having been established in 58 B.C. It became famous during the Reformation under Ulrich Zwingli, a precursor of Calvin, and calvinism allied with capitalism remained the strongest influence in the life of the city. The Zürich *Frauenverein* was a league of right-minded ladies who tried to ward off any threats to the morals of their fellow citizens. During the war years, however, an influx of artistic and literary people formed an avant-garde minority culture, almost an underworld. Its symbolic representation was the Steppenwolf, created by Hermann Hesse, who when in Zürich lived with two dwarves in a house on the bank of the Limmat. A favorite meeting place was the Café Odeon where every table was the territory of some special group: ballet dancers, poets, anarchists, and so on. Students favored the Café Terasse, where they could read *La Vie Parisienne*. At the Cabaret Voltaire, Hugo Ball, Emmy Hennings, Hans Arp, Tristan Tsara and other notables until 1922 presented a magic theater of Dada, a reaction to the brutal horrors of the war. Ball once defined Dada as 'a fool's play from nothing, in which all higher questions are developed, a gladiatorial gesture, a game with shabby relics, an execution of sham morality and sentiment'.[9] It would probably have appealed to the mischievous, anti-bourgeois side of Erwin's character. After the end of the war, as Europe counted its dead and the Swiss banks counted their profits, it was impossible to return Zürich to its prewar calvinistic austerity. The *Frauenverein* erected a solid wooden fence at the *Strandbad* to separate male and female bathers – one night it was torn down by rebellious students and never replaced.

At the traditional *Schauspielhaus*, Shakespeare, Schiller and Ibsen, in that order, remained the most popular playwrights, but innovative productions sometimes appeared. In 1923, Lote Schubert played the role of Gretchen wearing metre-long blonde hair and an amazingly low-cut dress. The painter and architect Otto Zollinger created a colorful stylized magic forest for *A Midsummer Night's Dream*. Zürich provided Erwin ample opportunities to indulge his love of the theater. There was little public for his old favorite, Grillparzer, but *The Jewess of Toledo* was presented in 1923–24. Schnitzler was never played during Erwin's time in Zürich, his iconoclastic bitter-sweetness did not appeal to a people who loved milk chocolate.[10]

Erich Hückel was a student at the E.T.H. during the early twenties. He recalls some of the amusements of the town.[11] There was a small cabaret off the Bahnhoffstrasse, to which he went with Debye. It featured Carl Sellmayer, a bawdy Bavarian comedian, and Anita Berber, who 'danced embracing herself as if she were her own partner, in a frivolous yet elegant way, which, however, one must admit was obscene'. In view of his sensitivity to noise and distaste for music, it is unlikely that Erwin ever wasted any time in Zürich nightclubs. He would prefer to spend a leisurely hour or two sipping wine with friends at the 'Bauschänzli', a tavern located on the little island in the Limmat, just before the Quaibrücke where the river joins the lake. When Erwin wished for a more adventurous outing, he could rely upon Debye to provide an introduction to the less calvinistic circles of Zürich society.

A favorite excursion was to take the steamer to Rapperswil, about 25 km down the lake, where there was wonderful swimming and picnicking on the Island Ufenau. On one occasion, a young lady at the Institute, whose father was fabulously rich, invited all the physicists to a party at the most expensive hotel Baur au Lac.

Erwin was always a good swimmer and he once wrote a sonnet after basking in the lake on a hot midsummer day:

> *Zürich*
>
> Wenn auf dem see die sonne brütend ruht
> und kaum bewegt in schwachen atemzügen
> die wellen leise auf und nieder wiegen
> im heißen glaste schwerer mittagsglut:
>
> wie wohlig fühlst du dann die weiche flut
> geschlossnen augs in ruhigem genügen
> sich schmeichelnd dir um leib und glieder schmiegen
> und sänftend kühlen das erhitzte blut.
>
> Die stunde steht und alle wünsche schweigen
> und lösen sich im weiten stillen raum.
> Doch öffnest du das aug zu fernen grenzen:
>
> dort, wo die himmel sich zur erde neigen,
> aus nebelschleiern steigt ein weißer traum,
> der reinen firne überirdisch glänzen.*

Zürich

* The brooding sun rests softly on the lake / And barely marks its shallow breathing / As gentle waves rock up and down / In heavy glare of midday heat.

Your eyes then close in peaceful satisfaction / What comfort now to feel the loving water / Embrace caressingly your waist and limbs / And sweetly cool the heated blood.

Then time is stopped and wishes all fall silent / Dissolving in the broad and quiet space / But yet you open eyes to distant ends:

And there, where heaven bows to meet the earth / From veil of mist a white dream rises / To gleam unworldly in glacial snow.

Even in summer, the mountain snows never seemed far from Zürich. In winter, when cold fogs swept though the city streets, there was often bright sunlight above the clouds, and one could ascend nearby heights like the Uetliberg and escape from the urban gloom.

When it was sunny on the heights, Zürich was filled with brilliant yellow placards announcing 'Uetliberg Hell'. Visitors sometimes thought they were advertising a light beer. Sunrise from the summit was a sight never to be forgotten, as the distant peaks appeared suddenly one by one, Jungfrau, Eiger, Mönch, and Finsterhorn.

The year 1923 was a low point in Schrödinger's scientific production, for he published no papers at all. He had not completely recovered his health, and Debye recalls that he was 'not sick, not really sick, but very sensitive . . . He had quite a nice wife, but they never had children, for instance, which is also a little bit out of the order . . . He was rather nervous, but not nervous of the type of Zermelo. On the other hand he was quite more like a nice Austrian.'

He did contribute that year a *Feuilleton* [short article] on 'Sound and Color' to the *Neue Zürcher Zeitung*.[12] In this, he pointed out that physically they are both periodic phenomena, but their aesthetic effects through the respective senses of hearing and sight are quite different in modality. Music depends upon a temporal sequence of sounds, whereas painting depends on a spatial arrangement of colors. He did not believe, however, that an abstract arrangement of colors could constitute a satisfactory painting; there must also be a depiction of reality.

Inaugural lecture

On December 9, 1922, Schrödinger gave his inaugural lecture at the University of Zürich.[13] Such lectures were intended for a public audience and hence supposed to deal with comparatively general themes, often designed to show the philosophic view of the new professor towards his subject.

While he was preparing his inaugural lecture, he wrote to Wolfgang Pauli, who was then with Bohr in Copenhagen:

I for my part believe, *horribile dictu*, that the energy-momentum law is violated in the process of radiation. What does the whole $h\nu$ relationship tell us then? (The experimental one, I mean). That an equivalence exists between electrons or atoms of definite energy and radiation of definite frequency, an *equivalence* relative to the achievement of a certain effect in an atomic system. Nothing more . . .

In the case of emission of light from an atom, he held it probable that:

there will be a spherical wave emitted – and is there still a recoil there? Why? Can the conservation of energy-momentum not be merely a macroscopically valid average relation, of which atomic physics knows nothing, like the 2nd Law? At least it can be that way, and I see almost no other way out.[14]

The title of the lecture was 'What is a Natural Law?', harking back to a similar occasion in the autumn of 1907, when Franz Exner gave his inaugural address as Rektor Magnificus of Vienna University, on 'Laws of the Sciences and Humanities'.[15]

Exner's thoughts had been inspired by the magnificent work of the late Ludwig Boltzmann, in which the laws of thermodynamics were derived from the statistical behavior of a myriad of randomly moving and interacting molecules. He explained that:

> every happening in Nature is the result of accidental occurrences. The greater the number of individual occurrences from which a phenomenon is composed, the greater the probability that it has a determinate character, the greater the likelihood that it follows definite laws. If the probability that an event occurs in a given way is so high that from the human viewpoint it has become a certainty, then we speak of a 'law of Nature'. That is possible, however, only if there is an unimaginably large number of individual occurrences, as can happen in molecular processes. In all other fields, there are no laws, only regularities, and these become more doubtful the smaller the number of occurrences from which they are derived, until finally, when the number becomes too small, they pass over into random events.

In 1919, Exner published the first edition of his *Lectures on the Physical Foundations of the Natural Sciences*, in which he gave a detailed account of his views on the statistical foundations of the laws of Nature, and suggested that the individual molecular events which comprise a statistical ensemble may not be themselves subject to determinate laws, but may instead be completely random and devoid of any casual explanation.[16] He had no empirical evidence for such a view, but it appealed to his aesthetic feeling that physics should rest upon a unitary foundation: since the large-scale laws are basically statistical, he did not like to have the small-scale laws governed by a strict determinism.

This was the idea that Schrödinger expounded with enthusiasm in his inaugural lecture. It would have been an unusual notion at that time, since the *principle of causality* was then accepted by most scientists. In the words of Cassirer: 'Every natural event is absolutely and quantitatively determined at the very least by the totality of circumstances or physical conditions at its occurrence.'[17] All physicists would grant that macroscopic laws, like those governing the behavior of gases, are based upon the statistical behavior of the countless molecules in any volume of gas – Schrödinger pointed to the phenomena of fluctuations and Brownian motion as experimental proofs of this hypothesis. Yet most physicists would believe equally strongly that the behavior of each individual pair of colliding gas molecules is governed by basic physical laws such as conservation of energy and momentum. Following Exner, Schrödinger denied that such determinism and regularity are necessary at the molecular level. For example, to maintain conservation of energy for a gas, it is not necessary that energy be conserved in each individual molecular collision; a sufficient condition is that in the average over many such collisions equal amounts of energy are gained and lost.

Even allowing for the fact that he was talking to a general audience, Schrödinger's treatment of the conservation laws was surprisingly superficial. As early as 1904, Georg Hamel had shown the relation of these laws to the fundamental symmetries of space and time.[18] In mechanics, the principle of conservation of energy can be derived from the invariance of physical processes under a displacement in time. This invariance is such a fundamental part of physical reasoning that one is almost forced to infer that neither Exner nor Schrödinger was aware of the deep roots of the energy principle when they offered to eradicate it in such a facile fashion. Yet invariance with respect to temporal and spatial displacements is also the basis of Einstein's work on relativity, and conservation of energy and angular momentum are essential in the Bohr theory of atomic structure. Indeed Schrödinger made a comment in his lecture that contradicted his own thesis: 'I will not deny, however, that it is precisely the Einstein theory that in no uncertain terms makes plain the *absolute validity of the energy-momentum principles.*'

Schrödinger summarized his conclusions in support of Exner.

The contention of Exner amounts to this: It is quite possible that the laws of nature without exception have a statistical character. To postulate an absolute law of nature behind the statistical one, as is generally done today as a matter of course, *goes beyond the bounds of experience.* A dual foundation of this kind for the lawfulness of Nature is in itself improbable. *The burden of proof lies upon the advocates of absolute causality, not upon those who doubt it.* For to doubt it is today by far the more *natural* viewpoint.

It is regrettable that Schrödinger never found time to rework this semipopular lecture into a more philosophical essay, in which he could elucidate his ideas about the subjects here touched upon so briefly. Scientific law, causality, determinism, conservation principles – these concepts need to be disentangled and more carefully defined before one can begin to understand the relevance of Exner's views. What is nevertheless clearly evident in the lecture is Schrödinger's enormous respect for the Vienna tradition of physics, and his desire to bring it to the attention of his new academic colleagues in Zürich.

In the light of subsequent scientific history, these debates about causality and determinism in the early twenties may appear to foreshadow the much more cogent attack on causality derived from quantum mechanics by Werner Heisenberg in his *indeterminacy principle* of 1927.

Some historians of science have discerned a relation between the indeterminism of the new physics and the climate of despair and disillusionment that prevailed in central Europe as an aftermath of the war.[19] In July, 1918, Oswald Spengler [1880–1936] published the first volume of his *Der Untergang des Abendlandes* [*The Decline of the West*].[20] Written before and during the war years, the book appeared just as the war turned conclusively against the Central Powers; its prediction of the doom of western civilization was perfectly attuned to the mood of intellectuals in defeated Germany and Austria. Spengler was not an academic mandarin but a high-school teacher who had studied literature, art, mathematics and the natural sciences. He compared a great civilization to a living organic creature, which passes through well defined stages of development to maturity, decline and dissolution. The characteristic achievements of each stage are determined by the prevailing culture so that the arts and sciences are essentially products of the creative spirit of the age [*Zeitgeist*] and have no absolute or universal significance. 'Nature is a possession that is saturated through and through with the most personal connotations. *Nature is a function of the particular culture.*' In the foreword to the second edition of his *Lectures*, Exner quoted Spengler with approval, especially his remark that the world is the image of human consciousness.

Spengler contrasted direct intuitive forms of knowing with logical scientific forms, usually to the detriment of the latter.

Causality is the reasonable, the law-bound, the describable, the badge of our whole waking and reasoning existence. But destiny is the word for an inner certainty that is *not* describable. We bring out that which is causal by means of a physical or an epistemological system, through numbers, by reasoned classification; but the idea of destiny can be imparted only by the artist working through media like painting, tragedy and music. The one requires us to *distinguish* and in distinguishing to dissect and destroy, whereas the other is creative through and through, and thus destiny is related to life and causality to death.

Outsiders will find it difficult to appreciate the resonances which such sentiments can evoke in a germanic soul, but they will not be surprised to learn that Spengler became an ardent Nazi.

When Schrödinger came to Zürich, he found in Hermann (Peter) Weyl a kindred spirit in many ways, including their mutual distrust of causality. As a Dozent at Göttingen, Weyl had been influenced by Edmund Husserl [1859–1938] who held the chair of philosophy there from 1901 to 1916, and his wife Helene Joseph had been a student of Husserl. Husserl was the creator of a philosophy called 'phenomenology' [not to be confused with 'phenomenalism'] in which psychological introspection largely replaces any empirical study of an external world; all things that appear in consciousness are worthy of attention and analysis, whether or not they are metaphysically real, and perhaps the phenomenological structures of the mind are even more interesting when they happen to be unreal. Weyl became the strongest supporter of the intuitionist theory of mathematics, founded by Luitzen Brouwer [1881–1966] of Amsterdam, in which the concept of number is based on man's intuitive sense of the concept of time. In 1920, Weyl published a paper 'On the Relation of the Causal to the Statistical Consideration of Physics', in which he rejected causality because of its incompatibility with our intuition of the inwardness of time.[21] He would be sympathetic indeed with Erwin's interests in eastern religions, and especially with the concepts of tantrism, in which all creative activities in the cosmos are allied to the creativity of human erotic experiences.

Schrödinger may have had second thoughts about his inaugural address, since he did not allow it to be published until six years later, by which time he had become a famous scientist who could afford to express radical ideas; but also by that time he had become an outspoken opponent of the concept that physical events at the atomic level are random manifestations of pure chance.

Zürich 1923/24

In the summer semester of 1923, Schrödinger lectured on the Theory of Light and on Atomic Structure and the Periodic System of the Elements, in addition to conducting the two seminars in theoretical physics.

After the end of classes, he and Anny went to Rügen, an island in the Baltic Sea, noted for its wild sand dunes and good salt-water swimming. The Ewalds came to spend two weeks with them there. Erwin's health was still in a precarious state, but he had improved considerably, and his lingering cough disappeared in the summer sun and warmth.

Hunger and starvation were still widespread in Germany. The Allied blockade had been maintained after the Armistice, and thousands perished in the cities. In Switzerland, many collections and benefits were organized for German Relief. Zürich citizens contributed to maintain a soup kitchen in Stuttgart that fed thousands of persons every day. Anyone in Germany on a fixed or infrequently adjusted salary was in dire circumstances. A professor paid 2.6 billion marks a year found it was worth only 2.6 gold marks, perhaps enough for several days subsistence. At this time one of the main stories in the *Neue Zürcher Zeitung* was the drawn-out trial in Munich of Hitler and Ludendorff as ringleaders of the abortive Beer-Hall putsch.

Erwin was happy with the natural environment and the intellectual life of Zürich, but he was growing older and still had not made any major contribution to physics. As a schoolboy and university student, he had always been at the top of his class, but now he had to watch younger contemporaries forging ahead of him. He was well aware of sentiments like that expressed by Dirac in a student ditty:

> Age is of course a fever chill
> That every physicist must fear.
> He's better dead than living still
> When once he's past his thirtieth year.

Except for *Farbenlehre*, which was outside the mainstream of physical theory, there was no division of physics in which he was an acknowledged leader. When he ventured to propose a novel theory, as in his early work on magnetism and on melting of solids, the model had usually turned out to be naive and untenable. If Schrödinger had died in 1924 at the age of thirty-seven, his work would have merited only a footnote in the history of modern physics.

He was invited to attend the Fourth Solvay conference, but not to present one of the important papers there. He was as yet an observer and commentator on the progress of physics, not a major contributor to it. The Conseil met in Brussels from April 24 to 29 to consider 'The Electrical Conductivity of Metals and Related Topics'.[22] The conference had been established in 1911 by Ernest Solvay, a wealthy Belgian industrial chemist, after consultation with Walther Nernst. It was intended to bring together about twenty leading physicists to survey and discuss, under luxurious conditions, the status of some important field. Einstein once called it 'that witches' sabbath in Brussels'.[23] Almost six years after the end of the war, it was still the policy of the Solvay committee to exclude all German scientists. An exception would have been made for Einstein but he refused to come under such conditions, and Niels Bohr also would never attend the conference until it put aside political hatreds.

The theory of metals had actually advanced very little since Bohr wrote his thesis in 1911, but it is interesting to read the discussions among the great scientists as they struggled to understand the paradoxical behavior of electrons in metals, where they are free yet bound, carriers of thermal energy yet incapable of acquiring it. Three years later, after the discovery of quantum mechanics, all these anomalies would be resolved. In 1924, however, the savants were like the proverbial blind men examining an elephant.

Percy Bridgman of Harvard outlined a theory in which collisions between electrons and atoms are usually perfectly elastic and exchange of energy occurs only when an electron enters or leaves an atom. Schrödinger politely demolished this idea by pointing out that 'each collision, perfectly elastic or not, will be just as effective in destroying the small component of velocity acquired in the direction of the field, provided that there is an exchange of momentum'. This comment was considered for a while before he made an even more cogent intervention. 'In view of the identity in exterior constitution between metallic ions and rare-gas atoms, according to the theories of Kossel and of Bohr, one could easily understand the perfect elasticity of the collisions as envisaged by Monsieur Bridgman, if it was permitted to assume that most of the atoms have lost their valence electrons. But, since M. Bridgman assumes an equipartition of energy, the specific-heat data permit only a very feeble ionization.' Instead of focusing upon this dilemma, the discussion then wandered to other topics, but Schrödinger had come close to a correct statement of the problem – its solution would come only with quantum mechanics and the Fermi–Dirac statistics.

Abram Joffe from Leningrad recalled that he and his wife had several long conversations with Erwin and Anny, both of whom he found to be not unfavorable to the Soviet political experiment.[24] The conferees had an audience with the Belgian king, who probably thought that Schrödinger was a citizen of Switzerland, for he engaged him in a discussion about fine details of the educational system in that country.

In 1924, Schrödinger regained his research momentum and published six papers. One on the specific heat of hydrogen combined an incorrect value for the moment of inertia of the molecule with an erroneous model for the energy levels to obtain results that, *mirabile dictu*, were in good agreement with available experimental data, illustrating the fact that agreement with experiment is a necessary but not a sufficient condition for the validity of a theory. One cannot blame him for not knowing of the existence of the nuclear-spin isomers of H_2, ortho and para hydrogen, which were not suggested till 1927.

In the summer semester of 1924, Schrödinger gave for the first time the course on Mechanics of Deformable Bodies, which had been mentioned at the time of his appointment. This was an intensive course, consisting of five hours of lectures and a special seminar each week. It provided an opportunity to go deeply into some applications of tensor calculus, which he would later put to good use in his work on unified field theories.

Erwin's personal life was entering a period of turbulence. He and Anny had become members of a close group of friends, almost a clique, that included the Weyls, the Bärs and the Meyers. The families spent many hours together in picnics and swimming parties. Erwin became popular among the children for his ability to stand on his head, but he was unhappy that he had no offspring of his own – he had always wanted a son so that he could re-enact his own close relation with his father. Although he had enjoyed several casual love affairs, they had not been serious, and he was becoming unhappy with Anny's evident interest in other men. They began to consider the possibility of a divorce. Thus, as the year 1925 began, and his physical health approached normality, his psychic state was far from tranquil. It would prove to be a marvelous, almost miraculous, year for his theoretical physics.

The nature of light

As a result of contacts and discussions with European colleagues, Schrödinger was by now deeply concerned with problems in atomic physics and quantum theory, especially the nature of radiation and how it interacts with electrons and atoms. These new interests brought him into closer relations with work in progress in the schools of Sommerfeld in Munich, Born in Göttingen, and especially Bohr in Copenhagen, all of which had close connections with one another.

In 1916, Bohr had returned to the University of Copenhagen to a chair of theoretical physics that had been especially created for him. Although he was still only thirty-one years old, his subsequent influence was exerted not so much through original research as through his inspiration of others and his provision for them of an ideal climate for intellectual work. He also began to delve into the logical and philosophical foundations of physics. Even his first papers from Copenhagen are often essays in search of verbal understanding rather than mathematical analyses of crucial problems. His lecturing style was a discursive mumble, but with small groups and especially in man-to-man discussions he was without an equal in his enthusiasm, his empathy, and his contagious love of his subject. At the end of the war, young physicists from all over the world flocked to Copenhagen, where science could be pursued in an atmosphere unpoisoned by politics. Victor Weisskopf called them 'many of the most active, the most gifted, the most perceptive physicists in the world'. Not everyone, however, fell under the spell of Copenhagen. Einstein was too independent and too interested in relativity theory. Schrödinger also remained an outsider, sustained by his superior mathematical abilities and sceptical of any orthodoxy.

Bohr was at first concerned to develop some kind of liaison between quantum theory and classical electrodynamics. He set forth a general rule called the *correspondence principle*, according to which, in the limit of large quantum numbers, the frequencies, intensities and other properties of radiating systems as calculated by quantum theory must pass over into those calculated classically. Many results that Bohr 'derived' from his correspondence principle appear to have been based on his deep knowledge of the experimental facts of spectroscopy and the chemical properties of the elements. He was able to explain the structure of the periodic table by assigning electrons one by one to orbits specified by their quantum numbers n and k. He used the concept of orbital penetration to explain the properties of the transition elements and even the rare earths. In 1922, he was awarded the Nobel prize in physics 'for his services in the investigation of the structure of atoms and of the radiation emanating from them'.

In June, 1922, Schrödinger sent to the *Physikalische Zeitschrift* a note on 'The Doppler Principle and Bohr's Frequency Condition'.[25] This was his first publication from the University of Zürich. By applying the laws of conservation of energy and of momentum to the emission of a quantum from an excited atom, he calculated the correction to the Bohr formula due to the recoil velocity of the quantum. The calculated effect was too small to be detected even for emission of quanta in the far ultraviolet. In this paper, the independent existence of light particles was taken for granted. If he had extended his considerations to X-ray quanta, he might have made a major discovery, the Compton effect.

As early as 1912 it had been noted that when a beam of X-rays of well defined wavelength is scattered by material of low atomic weight, some of the scattered X-rays have a longer wavelength than before. Several explanations of this effect were proposed, only to be disproved. In October 1922, Arthur Compton at Washington University (St Louis), gave the correct explanation by applying the laws of conservation of energy and momentum to the collision of an X-ray photon with a single electron. The shift in wavelength on inelastic scattering was calculated to be:

$$\Delta\lambda = (h/m_0 c)(1 - \cos \theta)$$

The equation is in excellent agreement with experimental data.

The Compton effect appeared to bring the particle theory and the wave theory of light to a decisive confrontation. Sommerfeld wrote to Compton that his work had sounded the death knell of the wave theory. In the Compton effect, the incident and scattered X-ray quanta have definite linear directions of motion. Bohr, however, knew that interference experiments required spherically spreading waves of radiation. To save the wave model, he too was prepared to abandon the laws of conservation of energy and momentum in interactions between light and electrons. The horns of the dilemma were sharp indeed. It seemed as if every physicist was now summoned to take a stand concerning the nature of light: *either* for waves *or* for particles.

John Slater, twenty-two years old, had just taken his Ph.D. at Harvard. Uninhibited by the factions in European physics, he asked himself, 'Why must it be either–or? Why can't radiation be both wave and particle?' He took this heretical idea to Cambridge, England. Like most ideas, it was not entirely new, but he carried it further. In Bohr's theory the radiation is emitted instantaneously as the atom jumps from the state E_2 to the state E_1.

Any student of physics knows that a wave train of finite length has a frequency spectrum which is not strictly monochromatic, but which instead has a frequency breadth Δv of the order of magnitude of $1/T$, where T is the length of time during which the train is emitted . . . The observed sharpness of spectral lines . . . is consistent with emitted wave trains . . . of the order of 10^5 or more waves in the train. How, I asked myself, could a physicist with the insight of Bohr have suggested that the radiation is emitted instantaneously? Surely it must have taken long enough for 10^5 waves to be emitted.[26]

In his Nobel prize lecture, Bohr had said: 'According to the correspondence principle, it is assumed that every transition process between two stationary states can be coordinated with a corresponding harmonic vibration component in such a way that the probability of occurrence of the transition is dependent on the amplitude of the vibration.'[27] Slater took this quite literally and proposed that all the time an atom is in an excited state, it is emitting electromagnetic waves of all the frequencies corresponding to transitions to lower states which would be allowed by Bohr's theory. These electromagnetic waves, however, are of a peculiar kind; they do not carry energy, but are connected with the probability of finding photons at a given point.

Slater arrived in Copenhagen at Christmas time, 1923. Bohr and his assistant Hendrik Kramers were enthusiastic about his ideas, but they proceeded at once to modify and extend them in ways that made him quite unhappy. Bohr coined the term 'virtual oscillations' for his electromagnetic waves. The dictionary defines 'virtual' as 'existing in essence or effect though not in actual fact'. Slater was willing to accept this term, although he did not in actual fact understand it, but Bohr and Kramers went much further: they did not include the actual existence of light particles in the theory, and they abandoned the conservation of energy and momentum as applied to an individual emission or absorption of light, retaining only a statistical validity for these laws. They may have been encouraged in this adventure by learning of Schrödinger's views from Pauli. The three men worked furiously on a joint paper, which was published in the May, 1924, issue of *Philosophical Magazine*.[28]

After a stylish summary of the paradoxical state of quantum theory, they discussed 'Radiation and Transition Processes'. 'We will assume that a given atom in a certain stationary state will communicate continually with other atoms through a time-spatial mechanism which is virtually equivalent with the field of radiation which on classical theory would originate from the virtual harmonic oscillators corresponding with the various possible transitions to other stationary states'. The spontaneous transitions of the Einstein formulation are due to the virtual radiation field of the atom itself, whereas the induced transitions are due to the fields of neighboring atoms. There is only a statistical connection between actual transitions in different atoms. 'This independence reduces not only conservation of energy to a statistical law, but also conservation of momentum.'

Their attempt to describe the Compton effect without photons predicted that the recoil electron and the scattered quantum would not appear simultaneously in Compton scattering. Experiments by Walther Bothe and Hans Geiger in 1925, however, confirmed the simultaneous process.[29]

Schrödinger's reaction to the Bohr–Kramers–Slater (B–K–S) paper was quite enthusiastic. In a letter to Bohr, May 24, 1924, he wrote:

I have just read with the greatest interest the interesting change in your ideas in the May issue of the *Phil. Mag.* I am extremely sympathetic to this change. As a pupil of old Franz Exner, I have long been fond of the idea that the basis of our statistics is probably not microscopic 'regularity', but perhaps 'pure chance' . . . [30]

In the September 5 issue of *Die Naturwissenschaften* Schrödinger published a paper on 'Bohr's New Radiation Hypothesis and the Energy Law'.[31] He first condensed the B–K–S paper into a three-page summary, and then noted that 'the most exciting thing about it is the fundamental violation of the energy-momentum laws in each radiation process. This violation is not something trivial. For example, an atom has reached its second-lowest energy state with an excess energy ϵ over the normal level . . . Thus every atom experiences an energy fluctuation of mean value ϵ about 10^8 times per second.' On superficial examination, it seems unavoidable that the entire system experiences considerable, completely irregular changes in its energy content, like a gambler who in rapid succession bets a large part of his funds on a game of chance that he has a 50–50 chance of winning, but just this comparison solves the difficulty, since he can bet a lot each time in a huge number of small bets.

To get a relatively large fluctuation in energy, the system must be at high temperature. Schrödinger shows that the B–K–S fluctuations must cause an isolated system to take a random walk through allowed states, leaving its energy indeterminate. The First Law of Thermodynamics then applies to an isolated system only on average over defined times. In the limit of $t \to \infty$, the energy of the system is completely undetermined. The only way to avoid this situation is to immerse the system in an enormous heat bath. The *exact* validity of thermodynamics thus would require a double limit, $t \to \infty$, and heat-bath volume $\to \infty$.

Thus one can also say: a definite stability of the state of the world *sub specie aeternitatis* can only occur through the *connection* of each individual system with the whole rest of the world. The separated individual system would be, from the standpoint of the unity, a chaos. It requires the connection as a permanent *regulator*, without which, energetically considered, it would wander about at random. – Is it an idle speculation, to find in this a similarity to social, ethical and cultural phenomena?

Only a lack of understanding of the foundations of the conservation laws would have allowed Schrödinger (and Bohr) to cast them aside so blithely. Einstein and Pauli, in contrast, were scathing in their criticisms of the B–K–S theory. Bohr became upset and accused Pauli of ridiculing Kramers, but the Schrödinger paper was too much even for Bohr; however it may have been intended, as he pondered its implications and the results of the Bothe–Geiger experiment, he realized that the B–K–S theory could no longer be taken seriously. In retrospect, the only part of the theory that was useful was Slater's original idea of virtual oscillators, which was consistent with Kramers' dispersion equations and would be used again in the Dirac quantum mechanical theory of radiation.

Innsbruck

The 88th Meeting of German Scientists and Physicians was held in Innsbruck, September 21 to 27, 1924.[32] The arrangements were in charge of Egon Schweidler, Erwin's former colleague in Vienna. Arthur March, a Dozent at the University of Innsbruck, was a member of the local committee.[33] It was a large meeting, with important lectures by Sommerfeld, Ewald, Planck, Laue, Born and others. Schrödinger was there but did not present a paper, nor did he contribute to the published commentaries, but he was an enthusiastic participant in the informal discussions and social events. He was happy to be back again in Austria and to renew old friendships with the Austrian physicists.

Pauli had a long discussion with Einstein about the B–K–S paper, and when he wrote to Bohr from Hamburg [October 2] after the meeting, he reported that Einstein found the violation of the First Law of Thermodynamics 'to be disgusting'. Despite its failures, the B–K–S paper was an important precursor of quantum mechanics, both in its emphasis on oscillators and in the way it encouraged the Copenhagen school and its Göttingen affiliate to move away from the old quantum theory and to seek new methods for dealing with the mechanics of electrons in atoms. Born introduced the term *quantum mechanics* in a paper with this title received by the *Zeitschrift für Physik* on June 13, 1924.

Sommerfeld's lecture was on 'Teleological Causality'. An electron in a Bohr transition, he said, must know to which state it is going before it can radiate the quantum equal to the energy difference between final and initial states. Therefore everything must be instantaneous – no time is required for the radiation process. Yet he thought that the difficulty thus posed may be due only to our excessive belief in our own models. An atomic model should be regarded more as a scheme for making calculations and less as state of reality [*Zustandsrealität*].

Early in 1925 the professorship of theoretical physics at the University of Innsbruck became vacant as a result of the retirement of Ottokar Tumlirz. On June 6, the Philosophical Faculty sent to the Austrian Ministry for Education the following short list of recommendations for the Chair:[34]

> *primo loco*: Dr Erwin Schrödinger, ordinary professor of theoretical physics in the University of Zürich.
> *secundo et aequo loco*: Dr Arthur March, Privatdozent in the University of Innsbruck; Dr Adolf Smekal, Privatdozent in the University and Technische Hochschule in Vienna.

Arthur March [1891–1957] was born in Brixen in South Tirol, an Austrian town that was part of the territory given to Italy after World War I. He entered Innsbruck University in 1910 but found Tumlirz to be such a fierce opponent of quantum theory that he went as soon as possible to work with Sommerfeld in Munich, and then with Hasen-öhrl in Vienna, where he became a good friend of Schrödinger and Hans Thirring. After graduation from Innsbruck in 1913, he taught for ten years in the Innsbruck Girls High School, and from 1917 also as Privatdozent at the University.

The nomination of Schrödinger included a reference to his experimental abilities, since a pure theoretician was still regarded askance in official circles. It was said to be desirable 'that the post be filled by an investigator of many-sided abilities, who has an overview of the entirety of his science and therefore also a feeling for empirical research. Our university would not be a suitable base for one-sided specialists, who are devoted exclusively to purely mathematical formulation or – an increasingly common type at this time – who pursue speculations far removed from reality.'

As proof of his versatility, the forty-two publications of Schrödinger were cited: mechanics (4), atomic physics (11), radiation theory (10), color theory (7), geophysics (2), and experimental investigations (4). The faculty was mindful of the situation of its own colleague and requested that if Schrödinger was appointed professor, the appointment of March as associate, which had been pending for four years, should no longer be delayed.

On 21 July 1925, Erwin wrote to Sommerfeld[35]: 'I nearly forgot to tell you – if you don't know about it – that I am expecting to receive a call to Innsbruck. I should like to return home, because the Swiss are far too cheerless [*gar zu ungemütlich*]. However, I must tell myself again and again that Innsbruck still appears gilded by the memory of last autumn.' An indication of Erwin's lack of enthusiasm for Zürich, and perhaps of a growing estrangement from Anny, can be seen in another letter to Sommerfeld that year: 'She is tonight at the ball . . . Frankly I have found them a bore and would any time prefer a student party in the German or Austrian style to such a "fine" Zürich ball, where the local money aristocracy hook up to loges at 300 francs and as a simple professor one is a *misera plebs*. Therefore I let her go along under the wings of Meyer, Bär and Scherrer – besides for me a ticket at 25 Fr times two was too much for the dubious pleasures.'[36]

On August 19, Schrödinger reported to the Dean at Zürich that he had received a call to Innsbruck. He was thinking seriously about accepting, since he loved Austria and especially the Tirol. Willy Wien heard about the situation and wrote on September 16 to Erwin: 'Congratulating you heartily. I can only add that I should enjoy it very much if you moved closer to us. I expect that then you would establish contacts with not too distant Munich, and we would have a chance to talk more often about scientific questions.'[37] Schrödinger replied the next day, thanking Wien for his kind words, and adding 'If I am strongly attracted to Innsbruck, then one of the main reasons is its closeness (in kilometres and in marks) to Munich.' He pointed out, however, that normal salaries in Austria were still very poor, and he would have to receive a special stipend to make it possible for him to accept the position, and this was still a matter under negotiation.

On January 15, 1926, the Innsbruck newspaper reported that 'in times of great financial stringency, the Vienna ministry is preparing to select the most expensive teacher. It is about to call Professor Dr von Schrödinger [*sic*] from Zürich and the ministry is prepared to find an extra salary of 1200 schillings a month to pay him.' A few days later an Innsbruck paper complained that 'five state positions with an average pay of 300 Sch must be abolished in order to afford this costly luxury call to a small provincial university (1600 students). The post could be filled with a local incumbent for 470 Sch a month.' [The salary offered to Schrödinger was 14137.60 schillings a year.]

On January 29, 1926, Schrödinger wrote to Sommerfeld:[38]

Officially I have not yet decided about Innsbruck. I do believe, however, that I shall stay here. It is mainly the fact that Schweidler is leaving for Vienna that decides the case for me . . . Please do not mention my decision as being certain when you talk to anybody. It would be unpleasant for me in dealing with both ministries. On the other hand, the delay is perfectly fine with me, because now I shall obtain here after much effort some little improvements, that is, a new blackboard in the lecture hall and, hopefully, a little larger budget for the seminar library.

Schrödinger officially declined the Innsbruck offer about March 16. By this time, the first of his great papers had appeared and he was about to be recognized as one of the world's most eminent theoretical physicists. In August the Austrian ministry appointed Arthur March to an associate professorship.

Michelson–Morley

In addition to all his other activities in 1925, Schrödinger became involved in a controversy arising from the degenerate state of German politics. Reactionary groups in Germany never accepted the reality of the defeat of the Kaiser's military machine, and they sought to exploit an historical fantasy in which an undefeated army was betrayed by a sinister alliance of Jews and socialists. The reactionary forces included remnants of the military, industrialists and large landowners, and many university professors. They all hated the social-democratic government of the Weimar Republic and were dedicated to its destruction, the generals and business magnates by overt actions and the university mandarins by incessant and insidious propaganda.

An important experimental support for Einstein's special theory of relativity was the null result of the Michelson–Morley experiment of 1887, the failure to detect any motion of the earth relative to a fixed aether. In 1921, Dayton C. Miller, who had been a junior colleague of Michelson at the Case Institute in Cleveland, performed a Michelson–Morley-type experiment at the top of Mount Wilson, and reported a positive result, a small but significant effect of the earth's motion on the velocity of light.[39] He theorized that at sea level the aether is carried along by the earth, while at higher altitudes a small aether wind becomes evident. Einstein was visiting Princeton when he heard of these results and made his famous statement, 'Raffiniert is der Herr Gott, aber boshaft is er nicht' [The Lord God is subtle, but he is not malicious]. The Miller result was received with delight by the enemies of the Einstein theory, but, according to Schrödinger, surprisingly Lenard did not believe it.

On June 22, 1922, Walther Rathenau, the German foreign minister, was assassinated by right-wing gangsters, with the cry 'Rathenau, Rathenau, Judensau' [Jewish pig]. Einstein was threatened with a similar fate. June 27 was proclaimed as a day of national mourning, but at the Physics Institute in Heidelberg, Lenard declared business as usual and refused to lower the flag to half-mast. The institute was invaded by a mob of union workers, and Lenard was hauled off to jail in 'protective custody'. Henceforth his virulence increased. At the next meeting of the Physicians and Scientists Society, September, 1922, in Leipzig, he organized an antisemitic demonstration against Einstein, even though the latter had canceled his speech in fear for his life.

On April 1, 1924, Adolf Hitler was sentenced to prison for his part in the Munich beer-hall putsch, and Lenard and Stark published a statement in his support in the *Grossdeutsche Zeitung*: 'One ponders how much it means that we are allowed to have this type of spirit living amongst us in the flesh . . . such spirits are embodied only with Aryan-Germanic blood.' Henceforth, Lenard's institute became a center for right-wing and Nazi agitation. Willy Wien's son wrote home in 1925, 'I haven't quite figured out yet whether one first becomes *völkisch* and then a doctoral candidate, or the reverse. In any case, the institute appears to be rather homogeneous in this respect.'[40]

Such was the background of the plans for a cooperative effort of some German and Swiss scientists to arrange a replication of the Miller experiment at a much higher altitude. Schrödinger wrote to Willy Wien, a moderate antisemite, on September 17, 1925:[41] 'The result of the Miller experiment is very important, but it has been played down in Jewish circles of physicists. I should like to see the experiment repeated on the Jungfraujoch.' [3457 m altitude]. Wien found that Rudolf Tomaschek from Heidelberg would be willing to do the experiment; he was highly qualified, having already published excellent interferometric studies.

At this time, Schrödinger's experimental colleague at Zürich, Edgar Meyer, intervened, saying that he would not trust any of Lenard's people to be objective about the work. Erwin was incensed, and reported to Wien on December 27:[42]

Please forgive me that I have not answered you before on the Tomaschek matter. I told Edgar Meyer that you had informed me that Tomaschek was prepared to do the experiment. The astonishing answer of Meyer was that he knew this long ago. And now came a bunch of 'buts and howevers' which finally came down to the point that Tomaschek as a pupil of Lenard on account of the antagonism against Einstein (and perhaps also because of the present very outspoken political orientation of the Lenard group) would not be suitable. Besides Joos may be obtained and that would be better yet. I don't like to complain about a colleague, but the whole way in which this affair has been managed doesn't please me at all.

In the event, Georg Joos carried out an excellent series of experiments on the Jungfraujoch, financed jointly by the Germans and the Swiss. The results confirmed the absence of any detectable 'aether wind', and Einstein's theory thus received further experimental support.

In this aspect of the controversy, Schrödinger found himself aligned with the more nationalist and antisemitic wing of the German physics community. Actually, he was basically contemptuous of politics, and he was correct in his view that Tomaschek was a trustworthy scientist, who should not be held responsible for the political activities of his professor.

My world view (Meine Weltansicht)[43]

In the autumn of 1925, Schrödinger wrote an intensely personal account of his philosophy of life. He did not publish it at that time, perhaps because he was just then swept up in the onrush of creativity that led to wave mechanics, but also perhaps because this book revealed his innermost heart in a way that might seem immodest for a physicist yet to gain a place among the immortals. The testament of 1925 must have been the result of many days of meditation: here am I, thirty-eight years old, well past the age at which most great theoreticians have made their major discoveries, holder of the chair that Einstein once held, who am I, whence did I come, where am I going? The time of introspection recorded in *Meine Weltansicht* was also the time during which his psychological resources were subconsciously marshalled for the creation of a new world of physics in the papers of 1926.

The 1925 part of *Meine Weltansicht*, called 'In Search of the Way', consists of ten short chapters. In 1960, the year before his death, he wrote five more chapters, and published all together. The earlier chapters, which are considered here, state his view of the desperate state of metaphysics, provide a means of rescue based upon the philosophy of Vedanta, and conclude with reflections on several psychobiological themes such as inherited memory.

He begins with a consideration of 'Metaphysics in General': Since Kant, it is easy enough to blow away the structure of metaphysics like a house of cards, but having done so, one is left with the empty feeling that art and science are no longer worthy of serious dedication. The task of post-kantian scientific philosophy is to understand the contradiction between the necessity of a metaphysical framework for scientific endeavor, and the flimsy unscientific structure of the framework itself. Metaphysics is like a far outpost in the land of an enemy, it is indispensable for defence of the realm but it is vulnerable and easily demolished.

The 'death of metaphysics' has led to an overdevelopment of technology and a decline of the arts. The churches, guardians of the most holy treasures of mankind, wasted their spiritual resources, and 'gradually and unnoticed, the sparks of ancient Indian wisdom, kindled to a new glow by the wonderful Rabbi by the Jordan, have been almost extinguished . . . The middle classes have become unprincipled and leaderless. They believe in no God and no gods, recognize the Churches only as political parties and morality as an irksome restriction . . . that has lost every support . . . A general atavism has set in . . . crass, unrestricted egoism lifts its smirking head and with its primeval irresistible iron fist seizes the rudder of the leaderless ship.'

This chapter reveals a man profoundly disturbed by the decline of Western civilization – and the worst by far was yet to come. It is a cry of spiritual pain of a soul torn between the need for religious belief and the inability to accept such belief without treason to his intellectual standards.

Schrödinger suggests that the apostles of freedom from metaphysics – Kant and the philosophers of the Enlightenment – would be horrified to experience the present consequences of their ideas in ordinary life. [Other observers, however, such as Thomas Mann, have ascribed the decline of the West and in particular the degradation of Germany to too little rather than too much enlightenment. Nietzsche spoke of 'the profound and comprehensive antipathy of the Germans to the Enlightenment'. The German intellectuals were thus able to accept ideologies filled with contradictory ideas and not susceptible to rational analysis.]

Schrödinger wants to make a statement about the basic concepts of philosophy, but he feels held back because he knows the lack of clarity, the falsity, the one-sidedness of every such statement. Thus the situation calls for the Buddhist wisdom that sought to express the unexpressible through contradictory statements: a thing can be either A or not A, but can also be both A and not-A at the same time.

He found in Epicurus the idea that the origin of philosophy is the feeling of wonder, θαυμάξειν, which arises when an experience departs from the ordinary. Our experience of the world is fraught with wonder, but what causes us to regard with wonder the particular facts of our own existence when we have known no other? Here, he says, is a paradox that would bring metaphysics to life even if we held its death notice in our hands.

Schrödinger now describes how he has achieved a resolution in his own mind of these problems. He rejects as 'naively childish' the idea of a soul dwelling in a body like a tenant in a house, and quitting it on death to take up another existence. Instead he wishes to consider four questions:

(1) Does an 'I' exist?
(2) Does a 'World' exist besides 'me'?
(3) Do 'I' cease to exist on bodily death?
(4) Does the 'World' cease to exist on 'my' bodily death?

He shows that there is no set of 'Yes' and 'No' answers to these four questions that does not lead to an endless circle. He dismisses the answers 'Yes' to all four questions as 'dreadful nonsense', that is, he rejects solipsism, as all philosophers who publish anything must do. He next tries 'No, Yes, Yes, No': i.e., only the material world exists. In this set, he finds a paradox which up till now 'has been fully recognized only in Indian Samkhya philosophy'. Assume two human bodies A and B. Put A in a situation such that a garden is seen; put B in a dark room. Then let A and B change places. On the assumption that there are no distinct 'selves' in A and B, the interchange should make no difference. But if A is *my* body, there is then no view of the garden. Therefore, besides the world, the 'self' must exist, a 'soul-self' inside a particular body. Yet this is the assumption that was rejected earlier as naive and childish.

He next considers the perception by the self of an object like a tree. Here he restates the conclusions of Mach and Avenarius, who were among his favorite philosophers when he was a student. I look at a tree – what I see is the tree of physics *as well as* the tree of psychology. Any 'tree in itself' is inaccessible to us and, as such, is of no interest. Mach's phenomenalism can thus be said to be a 'vindication of naive realism'. [Realism is ordinarily taken to be the doctrine that physical objects exist independently of any experience of them. Naive realism asserts, moreover, that these objects are what they seem to be. Note that Schrödinger (and Mach) actually reject both these doctrines.] Suppose I am standing with a group of companions in front of a tree. The tree must then be simultaneously the content of several consciousnesses, it belongs at the same time to several selves. Mach did not shrink from this conclusion; he made 'no essential difference between my sensations and the sensations of another person. The *same elements* cohere at several points of connection, which are selves'.

Chapter 5 'The Basic View of Vedanta', is the culmination of the book. Here Schrödinger attempts to find a solution to the problem raised by the spatial and temporal multiplicity of observing and thinking individuals. Vedanta teaches that consciousness is singular, all happenings are played out in one consciousness only, and there is no multiplicity of selves. Schrödinger does not believe that it will be possible to demonstrate this unity of consciousness by logical or rational argument; in order to reach any understanding of it, one must make an imaginative leap, guided by communion with nature and the persuasion of analogies. The epigraph to this most important chapter is from Goethe:

> Und deines Geistes höchster Feuerflug
> Hat schon am Gleichnis, hat am Bild genug.*

* And thy spirit's highest fiery flight / Is satisfied with likeness and with image.

Erwin does not mention the arduous spiritual exercises which Indian ascetics pursue to reach contemplative understanding of the unity of the world and consciousness. He does not claim that he himself has achieved this fusion of self with Nature, but the lyrical intensity of this chapter suggests that he may have come close to such a mystical experience in the mountains of his beloved Tirol.

A brief summary of Vedanta was given by George Thibaut who translated the Sutras with the commentary of Śankara in the great series, *The Sacred Books of the East*, edited by Max Muller, which made these scriptures available to those who do not read Sanskrit. The Sutras are brief, often cryptic or incomplete statements, which require interpretation to become intelligible. The interpreters may differ among themselves on important points, but Śankara (~788–820) provides a standard and orthodox view of Vedanta. In his opinion, the *Upanishads* teach the following:

'There exists only one universal being, called Brahman, which comprises all of reality in an undivided unity. This being is absolutely homogeneous in nature. It is pure thought. Thought is not an attribute of Brahman, but constitutes its substance. Thus Brahman is not to be called a thinking Being, rather it is Thought itself. It is absolutely destitute of qualities.'[44] Whence then comes the phenomenal world of ordinary experience, and our own individual selves and qualities? The answer is that Brahman is associated with a certain power called Maya to which is due the appearance of the entire world. Maya is neither being nor not-being, but a principle of illusion. As a magician creates illusions on a stage, Brahman through Maya projects the appearances of the world. Maya is the material cause of the world. In all the apparently individual forms of existence, the indivisible Brahman is present. All that is real in any individual soul is Brahman, the appearances are unreal. External things are Maya, the thoughts we have of them are Maya, neither one is more or less real than the other. Thus Vedanta is not an idealist philosophy in a technical sense.

The unenlightened soul is not able to look through or beyond Maya. It is engaged in an endless *samsara*, a cycle of birth, action, and death. The soul, which in reality is infinite Brahman, pure thought, is enmeshed in the unreal world of Maya. The only way out is provided by the Veda. Devout meditation on the Sutras can finally lead the soul to know that there is no difference between its true self and the highest Self. It will then be released from Maya and will realize its true identity with Brahman.

Other Vedic theologians, if one may call them such, do not agree in all respects with Śankara. For instance, Ramanuja says that Brahman does not have a homogeneous nature, but truly manifests itself in a diversified world. The world is not Maya, an illusion, but a real part of Brahman's nature, although the unenlightened person does not see it as it truly is. All forms of Vedanta, however, agree that things and selves do not have individual existences, but are all joined together as effects of Brahman, or in brief are Brahman itself.

Schrödinger says: 'This is what Brahman expresses with the holy, mystical, and yet actually so clear and simple formula "*Tat Twam Asi* – This Thou Art" or also with such words as "I am in the East and in the West, I am below and I am above, *I am this entire world*".' This is not meant in Spinoza's pantheistic sense, that you are a part, a piece, of an eternal infinite being, since then the old question would arise, what piece are you? 'No, as inconceivable as it seems to ordinary reason: you – and every other conscious being taken in itself – you are all in all . . . it is a vision of this truth that forms the basis of every morally valuable activity.'

Schrödinger is here testifying to a religious conviction. He has wandered far from the perhaps firmer ground of scientific knowledge, and almost as far from the conventional ground of philosophical discussion, and his thought has taken wings into the realm of eastern mysticism. He was obviously strongly influenced by Schopenhauer, of whom he had read every word.

Such a vision of the unity of the world – nature, man and God – has not been confined to eastern seers. From India it spread to the neoplatonic Greek philosophers, notably Plotinus, for whom the universe is a living organic whole, which emanates from the One or the Good, the divine Intellect. 'In this divine Intellect, thought and its content are one, so that it can be regarded either as a unity in a diversity of forms, or as a unity in a diversity of minds, each of which thinks and so is the whole. In terms of our consciousness, Intellect is the level of intuitive thought that is identical with its object and does not see it as in some sense external.'[45]

The ancient Hindu religion was reborn in the West in various forms of Gnosticism. The Sanskrit word 'Veda' and the Greek 'Gnosis' both mean 'Knowledge'. The gnostic doctrine is that salvation can be achieved through secret knowledge. While the apostles were spreading the gospel of Christ, Simon the Magician was wandering through Asia Minor with his faithful companion Helena, whom he had redeemed from a brothel in Tyre. He preached an elaborate gnostic mythology, in which the harlot typified the female holy spirit of wisdom, Sophia or Ennoia. Her representation as a whore showed the depth to which the divine had fallen by becoming involved in the creation of a material world.

The secret knowledge of gnosticism is knowledge of the self. Schrödinger had read the Christian mystics, who found God in the emptiness of their innermost souls. A favorite metaphor was 'the inner eye'. Thus Jacob Böhme, 'The unity is an eye which looks, which sees itself, which creates vision', and Johann Eckhart, 'The eye with which Man sees God is the same as the eye with which God sees Man.' In modern gnosticism, 'The universe is like the visual field of a living brain.'

Schrödinger found no need for God in his view of the world. His system – or that of the Upanishads – is delightful and consistent: the self and the world are one and they are all. He did not consider the basic religious question: why is there something rather than nothing? He rejected traditional western religious beliefs (Jewish, Christian, and Islamic) not on the basis of any reasoned argument, nor even with an expression of emotional antipathy, for he loved to use religious expressions and metaphors, but simply by saying that they are naive.

Vedanta and gnosticism are beliefs likely to appeal to a mathematical physicist, a brilliant only child, tempted on occasion by intellectual pride. Such factors may help to explain why Schrödinger became a believer in Vedanta, but they do not detract from the importance of his belief as a foundation for his life and work. It would be simplistic to suggest that there is a direct causal link between his religious beliefs and his discoveries in theoretical physics, yet the unity and continuity of Vedanta are reflected in the unity and continuity of wave mechanics. In 1925, the world view of physics was a model of the universe as a great machine composed of separable interacting material particles. During the next few years, Schrödinger and Heisenberg and their followers created a universe based on superimposed inseparable waves of probability amplitudes. This new view would be entirely consistent with the vedantic concept of the All in One.

The remainder of Schrödinger's 1925 *Meine Weltansicht* reflects his interests in evolutionary biology and mechanisms of heredity. In the act of reproduction, a genetically related individual is produced that is continuous with its forebears in bodily and mental life. He suggests that the identity of the consciousness of an infant and its ancestors is like the identity of his own consciousness before and after a deep sleep. The argument usually made against this idea is the absence of any memory for ancestral events, but instinctive behavior, as readily observed in animals, can be regarded as a supra-individual memory. Here Erwin's youthful study of the work of Richard Semon is recalled.

In this part of the book, the examples are based on personal experience, although it is reported in generalized terms. Erwin was intensely concerned with sexual experience – one might say that he was devoted to it as the principal nonscientific occupation of his life. Not only did he enjoy making love, but also he consciously conceived of it as a way to achieve transcendence and to perpetuate himself. Human sexual behavior, as he points out, is to a considerable extent unlearned and instinctive: 'All those difficult to describe, partly painful, partly heavenly feelings, but especially the headstrong choice, the falling in love which indicates special memory traces in the individual, not common to the entire species.' This is a fanciful idea, especially for someone like Erwin who might fall in love with any attractive young woman who came within hailing distance. One cannot deny that falling love is a momentous psychological event, even for the tenth time, but Stendhal's analogy to crystallization from a supersaturated solution is more relevant than Schrödinger's evocation of a genetic memory.

His other example of 'ecphoria' (Semon's term) is 'having a row' as it occurs in everyday life. Erwin was an irascible man, given to occasional childish temper tantrums, and this phase of his behavior is the background to the philosophical discussion. He describes the physiological effects of the fight or flight reaction, the result of an outpouring of adrenalin into the bloodstream. A man who behaves in this way, when he later reflects, indeed feels that he has been carried away by instinctive forces and emotions.

In all this discussion, he avoids the question of the nature of consciousness itself. He does not differentiate it clearly from physical or somatic factors. At one point he says: 'No self stands alone. Behind it lies an extended chain of physical and – as a special kind of these – intellectual happenings, to which it belongs as an interacting member and which it carries on.'

Displaying a wide acquaintance with biological phenomena, he illustrates the idea of 'nonplurality' with arguments based on a variety of organisms. He considers, for example, the polyp *Hydra fusca* and the worm *Planaria*. These primitive animals, he says, must have some form of consciousness. If they are cut into two, they will regenerate new complete animals, each of which will then have the same consciousness as that of the original creature. He gives this statement as an argument for the thesis that consciousness is indivisible, but it seems more like an unverifiable deduction from his hypothesis of a singular consciousness. Then there are the Siphonophores, the groups of coelenterates that are composed of a number of differentiated organs, one for movement, one for alimentation, one for reproduction, etc. These organs can function together as a unified organism, but some of them can also break off and as Medusae lead a free swimming existence on their own. We ourselves are actually an aggregation of cells: how is my 'Self' composed of the individual 'Selves' of my brain cells?

Schrödinger mentions the idea that all human beings together constitute a higher unity, analogous to the constitution of an individual man from his somatic cells. Theodor Fechner was a well known proponent of this idea. But this is not what Schrödinger means by the unity of consciousness; he means that each 'I' is the whole, so that in a sense each of us can say 'L'état, c'est moi.'

Considered critically, this 1925 testament, 'In Search of the Way', rises to a peak of interest in about the middle of the book, and gradually dies away into inconsequence in the final chapters. The first half of the book builds up to his vision of the unity of mind in the universe. The second half relies heavily on the work of Richard Semon, and thus harks back to his early days at Vienna University and his long discussions with his friend Fränzel, the religious student of biology. Almost twenty years have passed since then, but maturity and experience have not diluted his enthusiasm for Semon and his Mneme.

Schrödinger is fascinated by the process of inheritance, the genetic line extending through the evolution of man. He exalts the idea of immortality through the continuity of the genetic material. He always wanted a son. He must have dreamed of imparting to a young mind his own wisdom, as his father, everpresent friend and companion of his youth, had passed on his ideas to him.

At the time this book was written, Erwin's own marriage was close to disintegration. After four years together, he and Anny had no children, and their sexual incompatibility was such that it was unlikely that they ever would. The special circle in which they lived in Zürich had enjoyed the sexual revolution a generation before Alfred Kinsey discovered it in Indiana. Extramarital affairs were not only condoned, they were expected, and they seemed to occasion little jealous anxiety. Anny would find in Hermann Weyl a lover to whom she was devoted body and soul, while Weyl's wife Hella was infatuated with Paul Scherrer. Erwin and Anny considered the possibility of a divorce, but the idea did not appeal to Anny. Outside the bedroom, they were a compatible couple; she was an excellent housekeeper and a faithful correspondent. Thus they agreed to continue the marriage, while each remained free to find sexual relationships elsewhere. In view of Erwin's desire for a son, this was a curious arrangement, but Anny was nominally a Catholic and divorce would have left her in a precarious position. Furthermore, divorce is expensive, both at the time and afterwards, and this factor must have had some weight, given Erwin's financial prudence. The principal factor, however, was the romantic tendency of Erwin's character. He was convinced that bourgeois marriage, while essential for a comfortable life, is incompatible with romantic love. On the evidence of his diaries, he was not a libertine for whom sexual conquest was an aim in itself – it was the falling in love that he valued most. Several of his greatest loves never led to sexual union, but the romantic passion was valued for its own sake and as a source of inspiration: this was the pattern that Goethe had made famous and acceptable for the life of a creative person.

This personal background may explain a certain aridity that pervades Schrödinger's 'World View'. There is little joy in it and no love at all. An abstract contemplation of the continuity of inherited behavior is a bleak substitute for the happiness of shared family life, and the unity of consciousness in Brahman, especially as a vicarious experience, provides no guidance through the actual conflicts of egotistical personalities. Even if we share our minds with all other mortals, with our dogs and cats, and even with flowers and crystals, there is limited consolation in such theoretical communion. Erwin, however, never allowed his distrust of Maya to inhibit his enjoyment of its illusory pleasures. He found happiness in his intellectual work, the companionship of friends, good wine, poetry and drama, the love of women, and explorations of mountains and seashores. In these ways also he resembled his mentor Schopenhauer, who was able to combine a pessimism about the world with an indulgence in its pleasures, and an almost paranoid misogyny with an ardent pursuit of fair women.

The philosophy of Schrödinger at this time does not appear to have been influenced by his physics. In fact he often said that one cannot derive philosophical conclusions from physics. In contrast, however, he was willing to admit that philosophy could influence physics. Like Schopenhauer, he accepted an hierarchial view of our understanding of the world, with philosophy above and physics below. The philosopher once said that 'physics is unable to stand on its own feet, but needs a metaphysics on which to support itself, whatever fine airs it may assume towards the latter'.[46] Perhaps this is why, as a young man, Erwin considered giving up serious physics to become a philosopher. Yet, in *Meine Weltansicht* he does not attempt to provide a metaphysical foundation explicitly designed for physics.

At this time, his philosophy of science (as distinct from his philosophy in general) was derived in part from Mach and in part from Boltzmann, and any synthesis of these conflicting elements must have seemed impossible. Moreover, both physics and its philosophy were on the verge of revolutionary changes. The influence of Vedanta on Schrödinger's physics is obscured by the fact that he was a reluctant revolutionary. Thus others would argue that quantum mechanics brought the world view of physics closer to that of Vedanta, but Schrödinger, the most consistent follower of Vedanta among physicists, refused to derive this consequence from his own discoveries.

Gas statistics

Besides his work on color theory, Schrödinger's principal research during his early years in Zürich was on the statistical thermodynamics of ideal gases. This field was part of the great legacy of Boltzmann, which related the macroscopic to the microscopic world. Schrödinger's choice of a statistical theme for his inaugural lecture showed his abiding interest in such problems.

Thermodynamics is based on three fundamental postulates or laws. The First Law is the principle of conservation of energy: energy can be neither created nor destroyed in any process – the energy of the universe is constant. This law, however, tells nothing about the direction of changes in the world, yet natural processes occur in particular directions, for instance, heat flows from hotter to colder bodies, gases mix by interdiffusion. The energy function cannot predict the direction of a change, since it always remains constant in isolated systems or in the universe as a whole. We need a function that changes when the system changes and stays constant when the system rests at equilibrium. Midway through the nineteenth century, Rudolf Clausius in Zürich and William Thompson in Glasgow postulated the Second Law of Thermodynamics, and Clausius used it to define a new function S, which he called *entropy*, from the Greek for 'in change'. In all the large-scale changes that occur in the universe, the entropy always increases.

Why does the entropy of the universe always increase in any naturally occurring process? The greatest achievement of Ludwig Boltzmann was to answer this question in terms of the motions of the enormous numbers of particles that make up any large-scale material system. Boltzmann saw that the increase in entropy of a system means that the particles that compose it are moving from a less probable to a more probable arrangement. The equilibrium state, in which the entropy has reached a maximum, is the state of maximum probability. He set the probability proportional to the number of different complexions, W, of the particles in a system, for example, the number of ways in which a total energy E can be divided among N molecules. By such reasoning, Planck obtained the famous relation

$$S = k \ln W \tag{1}$$

The state of each molecule can be specified by giving the three components of its position q_1, q_2, q_3, and the three components of its momentum p_1, p_2, p_3. One can define a six-dimensional *phase space*, which is divided into small cells, each having the same volume, $\delta v = \delta q_1 \ldots \delta p_3$. The number of ways of arranging N distinguishable molecules among the cells in the six-dimensional space such that N_j molecules are in cell j is

$$w = N!/N_1! \, N_2! \ldots \tag{2}$$

The total number of distributions W is the sum over all the particular distributions w, subject to the condition that the total energy E of the system and the number of molecules N it contains are both held constant. Thus,

$$S = k \ln \Sigma \, w = k \ln \Sigma \frac{N!}{N_1! \, N_2! \ldots} \tag{3}$$

Since the number of molecules in any macroscopic volume of gas is enormously large, it suffices to take the largest term in the sum (3), i.e. the most probable value w^* of w, giving

$$S = k \ln w^* \tag{4}$$

In 1912, Otto Sackur and H. Tetrode[47] independently carried out the calculation of w^* and obtained the entropy of an ideal gas, starting from

$$S = Nk \, \ln[(2\pi mkT)^{3/2} \, V/\delta v] + \tfrac{3}{2} Nk \tag{5}$$

where δv is the volume of a cell in the phase space. Sommerfeld had noted that the product $\delta q \, \delta p$ has the dimensions of action, which according to Planck should be quantized in units of h. Thus Sackur and Tetrode set the volume element $\delta q_1 \ldots \delta p_3$ in phase space equal to h^3.

They further noted that Equation (5) must be wrong since entropy is an extensive state function: if you divide the volume of the system into two equal halves, the S of the whole must equal the sum of the S of the halves. Therefore they also subtracted $k \ln N! \simeq kN \ln N - kN$ from the S in Equation (5) to obtain finally the remarkable formula:

$$S = Nk \ln[(2\pi mkT)^{3/2} V/h^3 N] + \tfrac{5}{2}Nk \qquad (6)$$

The subtraction of $k \ln N!$ is equivalent to dividing w^* by $N!$

To many physicists at the time the Sackur–Tetrode formula seemed to have been produced like a rabbit from a magician's top hat. They had dark suspicions of the derivation and during the next ten years many papers were published discussing various aspects of the $S = k \ln W$ problem.

Another focus of interest in the entropy of ideal gases was the experimental work of Walther Nernst in Berlin. In 1906, he proposed his 'heat theorem', which before long was recognized as the Third Law of Thermodynamics. One statement is that in the limit of absolute zero, the entropy change $\Delta S = 0$ for any process. Furthermore, the heat capacity C_V of any substance goes to zero in the limit of 0 K. At first, Nernst restricted these theorems to liquids and solids, but later he saw that they are valid also for gases. Attention was directed to what was called 'the degeneracy of gases' at very low temperatures, by which was meant the failure of their specific heats to follow the ideal-gas law. The energy of an ideal monatomic gas is just the translational kinetic energy, which in accord with the equipartition principle is $E = \tfrac{3}{2}NkT$ for three degrees of freedom of translational motion. Hence $C_V = \tfrac{3}{2}Nk$. Only at temperatures in the neighborhood of absolute zero would a gas become 'degenerate' and its C_V depart from this value, and such degeneracy had never been observed experimentally.

Early in 1924, Schrödinger published a paper on 'Gas Degeneracy and Free Path Lengths' in *Physikalische Zeitschrift*.[48] He mentioned that many of the theories of gas degeneracy led to a characteristic frequency $v = h/8ml^2$, where l is a characteristic length. He proposed to set l equal to the mean free path traveled by a molecule in a gas between collisions with another molecule. One can trace his interest in this problem in his notebook titled *Chemical Constants and Gas Degeneracy II* (1922–23). From his notes it appears that he was mainly concerned with the problems created by attempts to calculate entropy from $S = k \ln W$, in particular the division by $N!$ and the extensive nature of the entropy function. The German physicists were engaged in lively arguments about these problems, with contributions by Sommerfeld, Planck, Ehrenfest, Stern, Scherrer, and others, flying back and forth in the journals and in personal correspondence.[49]

Schrödinger's calculation of the entropy gave the same result as Equation (6) except for the $k \ln N!$ correction term. (The characteristic length l actually drops out of the formula for S.) Following a paper by Ehrenfest and Trkal, he argued that S need not be an extensive function. Such a conclusion seems today to be absurd, but it shows the inadequate understanding of thermodynamics even among outstanding scientists at that time.

The situation in regard to the $N!$ correction was incongruous – the result obtained by applying it is correct, but the reasons given for it were confused and erroneous. It is not surprising, therefore, that Schrödinger vacillated between disapproval and approval of the tantalizing factor. Moreover, he must have felt under some pressure to get the question settled, since he was lecturing at that time on quantum statistics and even planning a book on *Molecular Statistics*.

There is no satisfactory physical basis for the introduction of the mean free path into the problem of gas degeneracy and Schrödinger abandoned this model a few months later. At the end of his paper, however, he included some pertinent remarks on the degeneracy of the electron gas in metals. He had first broached this subject in his Habilitation paper of 1912 on diamagnetism of metals, in which he was led astray by his use of classical statistics for the electron gas. Now he deduced from his mean-free-path model that the conduction electrons should be almost completely degenerate at moderate temperatures, and their failure to contribute to the specific heat necessarily follows. This conclusion is correct, but the real nature of the degeneracy of conduction electrons was not revealed until the Fermi–Dirac statistics was discovered after the advent of quantum mechanics.

Schrödinger's next attack on the entropy problem was a non-mathematical paper, 'Remarks on the Statistical Definition of Entropy for an Ideal Gas'. Perhaps to make amends for his rejection of Planck's views in the earlier paper, he sent this one to Planck for submission to the *Sitzungsberichte* of the Prussian Academy of Sciences, where it was reported in July, 1925.[50]

In the eighteen months since his previous paper, there had been a major advance in the statistical theory of gases. In June, 1924, Satyendra Nath Bose, a young Bengali physicist at the University of Dacca, sent Einstein a letter with a copy of a paper that had just been rejected by the *Philosophical Magazine*, an English physics journal. The paper contained a new derivation of Planck's radiation law. Einstein was so impressed with the paper that he translated it personally and sent it to the *Zeitschrift für Physik*, with a note saying 'In my opinion, Bose's derivation of the Planck formula constitutes an important advance. The method used here also yields the quantum theory of the ideal gas, as I shall discuss elsewhere in more detail.'

Bose derived the Planck distribution law for thermal radiation without any reference to electromagnetic theory, but simply by applying an appropriate statistics to the photons in a container of volume V. He considered the photons to be massless particles capable of existing in two states of polarization, with momenta given by $p = h\nu/c$. They were assumed to be indistinguishable and their number N was not necessarily conserved. All three of these assumptions are correct; Bose did not justify them, however, by arguments of any kind; they seemed to appear intuitively to his mind, perhaps because his thoughts were uncluttered by the ongoing controversies of the European physicists.[51]

Based on the ideas of Bose, Einstein gave a derivation by considering the number g_k of cells in a one-particle phase space $q_1 \ldots p_3$ that corresponds with a frequency range from ν_k to $\nu_k + d\nu_k$. The number of arrangements of N_k photons in the g_k cells is

$$W_k = (N_k + g_k - 1)! \, / \, N_k! \, (g_k - 1)!.$$

[Think of a linear array of N_k particles and $(g_k - 1)$ partitions between them.] The total number of microstates is then

$$W = \prod_k W_k!$$

Calculating the maximum of W in the usual way, one arrives at the Planck Distribution Law.

Einstein saw immediately that the Bose method of counting can also be applied to the statistics of indistinguishable molecules. The difference between molecules and photons is that for the former the number of particles N is held constant in calculating the maximum of W. Einstein published three papers on the subject, the first in September, 1924, and the last two early in 1925. In his second paper, he wrote 'If it is justified to conceive of radiation as a quantum gas, then the analogy between the quantum gas and the molecular gas must be a complete one.' Einstein believed that the relation was in fact more than a mere analogy, and that molecules as well as photons must have both particle and wave characters. He cited the work of Louis de Broglie at this point, thus making a connection between quantum statistics and the wave properties of matter that would certainly have attracted the attention of Schrödinger during his careful reading of the Einstein papers.

In his first paper, read before the Berlin Academy on September 24, 1924, Einstein showed how to calculate the entropy of a monatomic gas from the Bose distribution law for indistinguishable particles, without relying on the *ad hoc* corrections needed when Boltzmann statistics is used. The Boltzmann statistics then is seen to become valid at sufficiently high temperatures as the limiting form of the quantum statistics. At low temperatures, the correct equations for the degenerate gas are obtained. Einstein's second paper, read on January 8, 1925, discussed gas degeneracy in more detail, and pointed out that the gas molecules at low temperature cannot be considered to be independent of one another, even if the gas is so dilute that conventional intermolecular forces are absent.

The Bose–Einstein statistics was not received with universal approval. Max Planck, in his usual conservative fashion, preferred the Boltzmann statistics, but he devised a new derivation of the thermodynamic formulas, one that is actually used in many elementary textbooks today. This paper was presented to the Academy in July, 1925. He expressed the partition function Z of the entire gas as a product of molecular partition functions $z = \exp - (\epsilon/kT)$, so that $Z = z^N/N!$. This method is not, of course, able to cope with the region of gas degeneracy.

Schrödinger's paper was presented at the same meeting as that of Planck, but published some months later. He gave a review and critique of several equations relating entropy to probability. His main concern was to elucidate the division of the expression for w^* by $N!$. Planck had explained this in statistical terms by saying that the molecules must not be too strongly individualized: they are really indistinguishable, and an exchange of roles by two molecules does not result in a new microstate of the gas. In his previous paper Schrödinger had rejected this explanation, but now he accepted it and wished to show its limitations. It is only an approximate correction for a counting of microstates that was made wrongly in the first place. The Bose–Einstein statistics gives the correct general counting for the microstates of indistinguishable molecules, but when the temperature is not too low and the density not too high, the number of states is so much greater than the number of molecules, that N_i, the number of molecules per state, equals either 0 or 1. States with $N_i>1$ are so rare as to be negligible. Under these conditions, the $N!$ division gives the same results as the Einstein statistics. This paper of Schrödinger, relying entirely on crystal-clear logical analysis, was an important contribution, in a sense the last word on a difficult and controversial subject until quantum mechanics showed the necessity for two different kinds of statistics, depending on the symmetry of the particle wave functions.

When Schrödinger read the first Einstein paper, he evidently did not understand, at the beginning of the derivation, that the particles were considered to be indistinguishable, and on February 5, he wrote to Einstein to suggest a possible error in the probability formula. Einstein wrote back to explain that the 'quanta or molecules are not treated as *independent of one another*'; he even gave a little diagram to show how the counting procedure in Bose statistics differs from the Boltzmann case. He ended his letter with the assurance that 'There is certainly no error in my calculation [Fehler ist gewiss keiner in meiner Rechnung].'

Schrödinger did not reply to this elucidation from Einstein. On September 26, Einstein wrote to him to say that he had enjoyed his Prussian Academy paper on entropy as well as another recent paper on classical relativity theory. Einstein went on to say that Planck's idea of treating the energy levels of the entire gas instead of those of individual molecules could be formulated classically by considering a $3N$-dimensional phase space and equating its volume to $nN!$ h^{3N} to obtain the energy level of the nth quantum state of the gas. Schrödinger was delighted with this idea and used it to write a paper in which he calculated the sum-over-states of the ideal gas and its thermodynamic functions. He sent an outline of the paper to Einstein and asked him if he would be a co-author of a joint publication.[52]

Einstein replied that he did not believe he should be a co-author since Schrödinger had done all the work, and he would not wish to appear as 'an "exploiter" as the socialists so prettily put it'. Schrödinger then sent him the completed paper with the authors' names omitted. 'Even in jest, I would not consider you as an "exploiter". To remain with the sociological framework, one could say instead: When kings build, the hod-carriers have something to do.'

Schrödinger did not learn the final fate of this paper until Einstein sent it to the Prussian Academy in February, 1926. The paper contains explicit formulas for the thermodynamic functions of a sort of semi-classical ideal gas, which, however, are not of any practical use. Since the Boltzmann formula was used to evaluate the sum-over-states, any conclusions about gas degeneracy were invalid. In retrospect, Einstein was wise not to put his name on this paper – it would have added nothing to his three great papers on the subject.

Particles and waves

The discovery that laid the foundation for Schrödinger's revolutionary papers in wave mechanics came from an unexpected source, a thesis presented by a young French physicist for a doctoral degree at the University of Paris.[53]

Louis de Broglie was born on August 15, 1892, in Dieppe, the youngest of five children. His father, a member of the French parlia-

Louis Victor de Broglie (c. 1924).

ment, died in 1906 and his elder brother Maurice, who was seventeen years his senior, became head of the family and responsible for his education. They lived in a mansion in Paris, where Maurice had a private laboratory for research on X-rays. At first, however, Louis was more interested in history than in science, but in 1911 after Maurice served as secretary at the first Solvay conference and told him about the exciting developments in quantum theory and relativity, Louis decided to study physics. In 1913, he received his *Licence ès sciences*, equivalent to a B.Sc. degree, and entered the army for his year of military service, in the 8th Regiment of Engineers. When war broke out, he was assigned to the radiotelegraphy unit in the Tour Eiffel. He remained there for five and a half years, learning an enormous amount of practical electromagnetism.

Thus, like Schrödinger's, his scientific career was interrupted by the war, but his military service was more relevant to his future development as a physicist. After the end of the war, he followed the university courses in theoretical physics and mathematics of Langevin and Borel, and did some experimental work on X-rays in his brother's laboratory.

He read and reread the discussions of the 1911 Solvay Conference and everything published since about the quantum theory. At the Solvay Conference, Sackur had broached his idea that phase space is divided into cells of volume h^3, and in 1922 Louis de Broglie applied this method to a 'gas' consisting of light quanta and was able to calculate the Wien distribution law for the energy density as a function of frequency,

$$du = (8\pi h/c^3) \exp(-h\nu/kT)\, \nu^3 d\nu.$$

Thus he partly anticipated the 1924 paper of Bose, by providing a derivation of the energy-density formula that did not depend on classical electrodynamics.

It is interesting that both de Broglie and Schrödinger were working on the theory of ideal gases just prior to their respective discoveries of wave-particle duality and wave mechanics. The success of the ideal-gas theory when applied to radiation quanta indicated that material particles and radiation are entities of a similar kind.

De Broglie recalled that 'all of a sudden' he saw that the so-called crisis of optics was simply due to a failure to understand the true universal duality of wave and particle.[54] He published the basic elements of his discovery in three short notes in the *Comptes rendus* of the Paris Academy in September and October, 1923, and a more elaborate version in his thesis for the doctorate of science, defended on November 25, 1924. 'Having a very "realist" conception of the nature of the physical world and little inclination toward purely abstract considerations, I wished to represent to myself the union of waves and particles in a concrete fashion, the particle being a little localized object incorporated in the structure of a propagating wave.'

Like Niels Bohr, Louis de Broglie did not rely on advanced mathematical analysis but on innovative ideas about physical reality. He began by considering two simple expressions for the energy of a particle at rest, $\epsilon = m_0 c^2$ and $\epsilon = h v_0$. He accepted Einstein's special theory of relativity, which requires that these expressions be invariant under a Lorentz transformation. Since c and h are invariant constants, m_0 and v_0 must follow the same transformation law. For m_0, this is known to be

$$m = m_0(1 - \beta^2)^{-1/2} \qquad \text{where} \quad \beta = v/c.$$

Hence also,

$$v = v_0(1 - \beta^2)^{-1/2}$$

This was an important result, since it was known that the frequency of a moving clock is slowed (relative to a stationary observer) by $v = v_0(1 - \beta^2)^{1/2}$. Thus the periodic phenomenon associated with the moving particle cannot be an oscillator (or clock) but must be a wave.

A wave moving in the x direction can be represented by a function $\sin \varphi$, where the phase $\varphi = 2\pi(vt - x/\lambda)$, λ being the wavelength. The change in φ when the particle moves a distance $dx = vdt$ is

$$d\varphi = 2\pi(vdt - dx/\lambda) = (2\pi/h)\left[\frac{m_0 c^2}{(1 - \beta^2)^{1/2}} \, dt - \frac{h}{\lambda} \, dx\right]$$

De Broglie postulated a 'theorem of the harmony of phases', according to which the phase of the wave associated with the particle must always remain in accord with the phase of a clock at the position of the particle (as measured by an observer at rest). The phase change of the clock in time dt would be

$$d\varphi_1 = 2\pi v_0(1 - \beta^2)^{1/2} dt = (2\pi/h) m_0 c^2 (1 - \beta^2)^{1/2} dt$$

Thus, from the requirement $d\varphi = d\varphi_1$, he obtained

$$m_0 c^2 (1 - \beta^2)^{-1/2} \, dt - (h/\lambda) dx = m_0 c^2 (1 - \beta^2)^{1/2} dt$$
$$m_0 v^2 (1 - \beta^2)^{-1/2} = hv/\lambda$$
$$m_0 v(1 - \beta^2)^{-1/2} = p = h/\lambda$$

This is the famous Broglie equation which relates the momentum of a particle to its wavelength.

In his thesis, Louis de Broglie also obtained this result by setting forth the close analogy between the Principle of Fermat in optics and the Principle of Hamilton in mechanics. Both assert that certain integrals have stationary values. The Hamilton principle can be written in accord with special relativity as

$$\delta \int J_i dx^i = 0$$

where J_i are the covariant components of a vector, the time component being ϵ/c and the space components being those of momentum p_i. The Fermat principle can be written,

$$\delta \int O_i dx^i = 0$$

where the time component of O_i is v/c, and the space components are proportional to the wave numbers of the wave of length v/u. It is plausible that $O_4 = J_4/h$. If one postulates $O_i = J_i/h$, the dynamically possible trajectories of the particle are identical to the possible rays of the wave, so that again $p = h/\lambda$.

The examiners of the thesis were Jean Perrin, Elie Cartan, Charles Mauguin, and Paul Langevin. They found the idea that particles are also waves too fantastic to be taken literally, but nevertheless accepted the thesis as a virtuoso performance. Langevin spoke to Einstein by telephone about it, and arranged to send him a third copy, which fortunately had been typed. Einstein at once recognized its importance and wrote to Langevin: 'Er hat eine Ecke des grossen Schleiers gelüftet. [He has lifted a corner of the great veil].'

Intimations of wave mechanics

In the winter semester of 1924/25, Schrödinger gave a course of two lectures a week on 'Molecular Statistics' and he prepared a seventeen-page outline for a book on the subject. In the summer semester of 1925 he lectured on Quantum Statistics. Many scientists find the preparation of a course of advanced lectures to be an inspiration for further research in the subject concerned, as they must read all the current literature and in explaining it to students often become aware of gaps in the theory or in their own understanding. Such was the background of Schrödinger's paper 'On Einstein's Gas Theory', submitted to the *Physikalische Zeitschrift* on December 15.[55]

Since this was the last paper that he wrote before the wonderful explosion of creativity in his discovery of wave mechanics, it has attracted much attention as an intellectual precursor of that development. The paper demonstrates how seriously he was thinking about particles as waves. Schrödinger later said that 'wave mechanics was born in statistics', and it may have been conceived during this work on the statistical mechanics of the ideal gas.

This paper is the best of his contributions to quantum statistics, but it still has the character of his earlier works, a critical reaction to an idea proposed by someone else, followed by an attempt to refine its mathematical analysis and to sharpen its theoretical relevance. In this case, he took a method of evaluating statistical probabilities, given in 1922 by Charles Darwin and Ralph Fowler, and used it for a new derivation of the Einstein–Bose gas statistics.

He points out that Einstein applied to molecules the same kind of statistics that yields the Planck radiation law when applied to 'atoms of light', but it is also possible to get the Planck law from 'natural statistics' applied to so-called 'aether resonators', i.e., to the degrees of freedom of radiation. Louis de Broglie had emphasized that every particle has a wavelength and a frequency. In his paper on gas statistics, Einstein had accepted wave-particle duality, and many physicists first learned of this concept through him, and were obliged to take it seriously by the weight of his authority. Schrödinger must have studied the de Broglie papers carefully in the summer and fall of 1925. He says that his approach to gas statistics 'means nothing other than taking seriously the Broglie–Einstein wave theory of the moving particle, according to which the latter is nothing more than a kind of "whitecap" [*Schaumkamm*] on the wave radiation that forms the basis of the world.' This statement goes beyond anything that de Broglie ever wrote about the relation between waves and particles. It is the first expression by Schrödinger of what was to become a major theme in his interpretation of wave mechanics: the world is based on wave phenomena, while particles are mere epiphenomena.

He considers n molecules of a monatomic gas in a volume V. Each molecule has a discrete set of states specified by energy levels, ϵ_1, ϵ_2, ..., ϵ_s, ... At any instant, an arbitrary number of molecules can be in the same state. Each state accessible to an individual molecule can also specify a degree of freedom of the entire system, and if n_s molecules are in the state ϵ_s, one can say that the system oscillates with the energy $n_s \epsilon_s$. This treatment of the gas as a whole is the most important feature of the paper from the standpoint of statistical mechanics. Schrödinger treats the system by Planck's method of the sum-over-states [partition function]. The general term in this sum is

$$[\exp - (n_1 \epsilon_1 + n_2 \epsilon_2 + \ldots + n_s \epsilon_s + \ldots)/kT]$$

If one were dealing with radiation in a volume V, there would be no condition on the n_s, and the sum of all terms of this type would split into a product of simple sums, each of which could be easily summed as geometric series.

$$\prod_s \sum_{n_s=0}^{\infty} \exp(-n_s \epsilon_s/kT) = \prod_s [1 - \exp(-\epsilon_s/kT)]^{-1} = \prod_s (1 - x_s)^{-1}$$

The peculiarity of the molecular system resides in the fact that the n_s are subject to the restriction $\Sigma n_s = n$.

The condition $\Sigma n_s = n$ means that in the development of the infinite product above, only those terms are admissible which are homogeneous of degree n in the variables x_s. As shown by Darwin and Fowler, the task of selecting such terms 'is accomplished by the beautiful residue theorem of function theory. One writes in the right side of the equation instead of x_s the product zx_s, multiplies the whole by z^{-n-1} and seeks the residue of the function of the complex variable z so obtained at the point $z = 0$. This residue is evidently $2\pi i$ times the sum-over-states that is sought, which we shall call Z:'

$$Z = \frac{1}{2\pi i} \oint dz \, z^{-n-1} \prod_{s=0}^{\infty} (1 - zx_s)^{-1}$$

He evaluates Z by the method of steepest descents and obtains an expression which reduces to that of Einstein. The formula is a general one, in which the energy spectrum of the 'oscillations' remains unspecified. It is not in a form convenient for calculation of thermodynamic properties since it involves a parameter r, the value of z for which the integrand is a minimum.

In a section called 'Determination of the Frequency Spectrum', he uses the results of Louis de Broglie to obtain expressions for the energy levels. He correctly excludes the state $s = 0$ because it would correspond to an infinite wavelength. He then, however, makes the mistake of saying that the Einstein 'condensation' will thereby be prevented. Actually the behavior of the gas cannot depend on the choice of the zero level of energy, and the Einstein 'condensation' can occur just as readily into a state with $s = 1$ as into one with $s = 0$.

He spends some time discussing the nature of the particle-waves. His starting point is always the mass-energy relationship of special relativity, so that the molecule has a 'rest frequency' given by $v_0 = mc^2/h$, and the dispersion law [dependence of speed on frequency] becomes $u = cv/(v^2 - v_0^2)^{1/2}$. 'The universal radiation, as "signals" or perhaps singularities of which the particles are supposed to occur, is therefore something significantly more complicated than the wave radiation of the Maxwell theory, and indeed not only because it generally shows dispersion, but especially because the dispersion law for the phase velocity of a wave group depends upon the kind of singularity produced through superposition of the group of waves concerned.' Schrödinger is now taking the wave nature of matter very seriously, beginning to think in detail about the kind of wave that is required and the laws and equations that it must obey. He is formulating the dispersion laws in relativistic terms, and thus it will be only natural that his first efforts to find a suitable wave equation will also be based on the relativistic equations.

The final section of his paper considers 'The Possibility of Representing Molecules or Light Quanta through Interference of Plane Waves'.

It is immediately clear that through the superposition of a great number of plane waves with the same wave normal and closely neighboring frequencies one can produce a 'signal' that is limited almost exclusively to a thin plane parallel 'slice' of the total space. On the other hand, one is perhaps in doubt for a moment whether and how it is possible to restrict the signal to a small region of space in all three directions. According to Debye and Laue, one achieves this by allowing not only the frequencies to vary slightly, but also the wave normals over a small region, a small solid angle $d\omega$, and then integrates together a continuum of infinitesimal wave functions within this range of frequencies and wave-normals.

Of course one naturally cannot guarantee through classical wave laws that the 'model of a light quantum' thus produced – which after all still mixes wavelengths in every direction – will also permanently stay together. On the contrary, it will be scattered into ever greater space after passing through a focus point. If one could avoid this last consequence by a quantum-theoretic modification of the classical wave laws, then a way would be open to escape the dilemma of the light quantum.

When he did achieve his wave equation for material waves, this spreading of the wave packet would turn out to be one of the ineluctable features of wave mechanics and the source of some of his most desperate problems in its interpretation.

6 Discovery of wave mechanics

When Schrödinger returned from his summer holidays for the winter semester of the academic year 1925/26, he would have been quite unaware that he was on the verge of an outburst of creative activity that would change forever the world of physics. He had just reviewed and codified his personal world view and it was unlikely that his philosophical outlook would change in any revolutionary way. He had set down his belief in Vedanta and thus in Schopenhauer, through whom the Kantian *a priori* categories of space and time had been refurbished. During five carefree years in Switzerland, he had been insulated from both the hypocrisy of Vienna and the brutality of German politics. It would not be surprising if Swiss society, with its deep roots in economic security, had made him more conservative in other ways. His more unconventional beliefs, he kept to himself, and neither his inaugural lecture nor *My World View* was published at this time.

Hermann Weyl once said that Schrödinger 'did his great work during a late erotic outburst in his life'.[1] His marriage with Anny was at a high point of disagreement and tension, with constant talk of breakup and divorce. Their Zürich circle provided ample opportunities for amorous adventures, and if Erwin cared to venture outside the usual academic liaisons, both Weyl and Debye were available as competent guides to the uninhibited night life of the city.

Before the breakthrough

The semester began on October 15, and Schrödinger had his usual heavy teaching schedule, five hours a week on Theory of Electricity, two hours on Theory of Spectra, a weekly seminar for the electricity course, and the joint seminar with E.T.H., which met for two hours every fortnight. Pieter Debye suggested that Erwin should give a seminar talk on the de Broglie thesis, which had just been published in *Annales de Physique*.[2]

On November 3, Schrödinger wrote to Einstein:

A few days ago I read with the greatest interest the ingenious thesis of Louis de Broglie, which I finally got hold of; with it section 8 of your second paper on degeneracy has also become clear to me for the first time. The de Broglie interpretation of the quantum rules seems to me to be related to my note in *Zeit. f. Physik 12*, 13, 1922, where a remarkable property of the Weyl 'gauge factor' along every quasi-period is shown. As far as I can see, the mathematical content is the same, only mine is much more formal, less elegant, and not actually shown in general. Naturally de Broglie's consideration in the framework of his comprehensive theory is altogether of far greater value than my single statement, which I did not know what to make of at first.[3]

He was also corresponding with Alfred Landé, professor of physics at Tübingen. On November 16, after discussing derivations of the Planck distribution law, he wrote:

I was especially pleased with your news that your work would be a 'return to wave theory'. I am also strongly inclined that way. I have been intensely concerned these days with Louis de Broglie's ingenious theory. It is extraordinarily exciting, but still has some very grave difficulties. I have tried in vain to make for myself a picture of the phase wave of the electron in the Kepler orbit. Closely neighboring Kepler ellipses are considered as 'rays'. This, however, gives horrible 'caustics' or the like for the wave fronts.[4]

A caustic is a curve that gives the boundaries of an initially plane wave after reflection or refraction. The electron waves of the Broglie theory are traveling waves, and Schrödinger wanted to find the structure of such waves when they are refracted sufficiently to travel in one of the Bohr orbits.

His colloquium report on the de Broglie thesis was most likely given at the meeting on November 23. According to the recollections of Felix Bloch, at the end of this colloquium:

Debye casually remarked that he thought this way of talking was rather childish. As a student of Sommerfeld he had learned that, to deal properly with waves, one had to have a wave equation . . . Just a few weeks later [Schrödinger] gave another talk in the colloquium which he started by saying: 'My colleague Debye suggested that one should have a wave equation; well, I have found one!'[2]

On November 25, Erwin Fues arrived from Stuttgart on a Rockefeller Foundation fellowship to work as Schrödinger's assistant. He had done his Habilitation with Ewald at the Technische Hochschule Stuttgart. The first problem suggested was a theoretical study of the interaction between atmospheric shock waves. This was a curious choice, harking back to Erwin's wartime experience in the artillery, where he had noticed the relatively quiet zone equidistant from two simultaneous detonations. It is likely that in the back of his mind was some analogy to the way in which particles may ride on the crests of

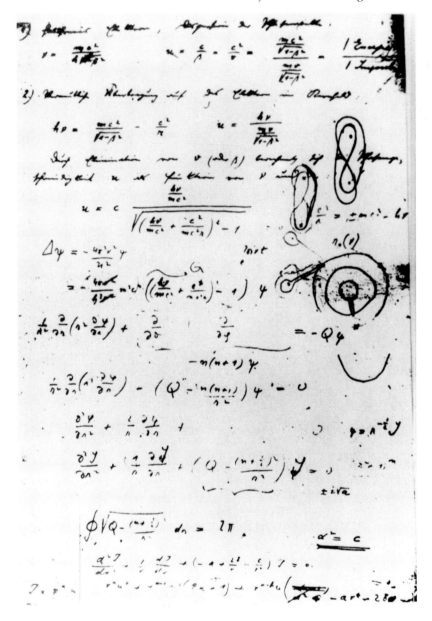

A page from Notebook N1, with the first record of the wave equation.

Broglie waves, but he did not mention this. When Fues returned to Stuttgart for the Christmas holidays, Ewald was not favorably impressed by his research project, and strongly advised him to ask for a new problem, one related to quantum theory, in which great developments were expected.[5]

Schrödinger's first derivation of a wave equation for particles was based entirely upon the relativistic theory as given in de Broglie's thesis, but he did not publish this equation. The crucial test of any theory would be its application to the problem of the hydrogen atom. At the nonrelativistic level, it must at least be able to reproduce the results of the old quantum theory of Bohr for the energy levels and quantum numbers. A relativistic theory should also be capable of explaining the Sommerfeld equation for the fine structure and perhaps improving upon it.

A considerable mystery now obscures the historical record. Schrödinger did not keep systematic, dated research notebooks; he used separate notebooks for different subjects, and entries were seldom dated. The only surviving records from this time (about December 1925) are a three-page set of rough notes (which we shall call N1) titled 'H Atom – Characteristic Vibrations [*Eigenschwingungen*]', and a 72-page research notebook (N2) titled 'Eigenvalue Problem of the Atom I'. N1, which is believed to be the earlier, contains the first written record of a wave equation for the hydrogen atom; it is a relativistic equation, as will be discussed below.[6]

In 1963, thirty-seven years later, Fues recalled definitely that the relativistic equation was never discussed with him: 'Schrödinger's first, relativistic attempt to establish a wave equation was already concluded and filed away [*ad acta gelegt*] when I came there.' If this is exactly true, Schrödinger must have accomplished the work on the relativistic equation before November 25. Hans Thirring recalled that: 'In November 1925 Schrödinger wrote me a four-page letter in which he revealed his wave mechanics. This letter, like many others, was lost when the Gestapo seized my correspondence.'[7] The letter may have contained the relativistic wave equation for the Kepler problem shown in N1 but it is hardly possible that it contained the complete solutions, with the eigenvalues and eigenfunctions. It is more likely that Fues did not have a serious discussion of quantum theory with Schrödinger until they had both returned from their Christmas holidays. While Fues went home to Stuttgart, a few days before Christmas Schrödinger went to Arosa, where he stayed until January 9, 1926.

Christmas at Arosa

Erwin wrote to 'an old girlfriend in Vienna' to join him in Arosa, while Anny remained in Zürich.[8] Efforts to establish the identity of this

*The Villa Herwig, where wave mechanics was discovered during the
Christmas holidays, 1925–26.*

woman have so far been unsuccessful, since Erwin's personal diary for
1925 has disappeared.

Whenever Erwin and Anny visited Arosa they stayed at the Annex
of the Villa Dr Herwig, in a room next to the living quarters of the old
Dr Otto Herwig. This is where Erwin spent his *Liegekur* from July 1 to
October 29, 1922. The more severely ill patients lived in the main
building, and Dr Herwig was always careful to separate those less ill as
much as possible from them. Erwin and Anny also stayed in the same
room during the Christmas holidays in 1923 and 1924, as is attested by
the hotel registry. There is, however, no record of his stay there in
1925, although we know from his letters that he was there, or at least
gave that address to his correspondents.

Like the dark lady who inspired Shakespeare's sonnets, the lady of
Arosa may remain forever mysterious. We know that she was not
Lotte or Irene. In all likelihood she was not Felicie; her husband had
lost his fortune in the postwar inflation and had gone to Brasil, leaving
her with an infant daughter. Whoever may have been his inspiration,
the increase in Erwin's powers was dramatic, and he began a twelve-
month period of sustained creative activity that is without a parallel in

the history of science. When he was enthralled by an important problem, he was able to achieve intense and absolute concentration, bringing to bear all his great mathematical powers.

On December 27, he wrote from Arosa to Willy Wien in Munich.

At the moment I am struggling with a new atomic theory. If only I only knew more mathematics! I am very optimistic about this thing and expect that if I can only . . . solve it, it will be *very* beautiful. I think I can specify a vibrating system that has as eigenfrequencies the hydrogen *term* frequencies – and in a relatively natural way, not through *ad hoc* assumptions. But you don't actually get these term frequencies themselves, i.e., not $-R/n^2$, but $mc^2/h - R/n^2$ (*m* is electron mass) . . . If, say

$$v_n = mc^2/h - R/n^2, \quad v_m = mc^2/h - R/m^2,$$

then

$$v_n - v_m = R(1/n^2 - 1/m^2)$$

I hope I can report soon in a little more detailed and understandable way about the matter. At present, I must learn a little more mathematics in order to survey completely the vibration problem – a linear differential equation similar to Bessel's, but less well known.[9]

This letter shows that Schrödinger now had clearly in hand the way in which quantization conditions with whole numbers arise as eigenvalues of a wave equation. It is highly probable that by now he had separated the three-dimensional wave equation and found how the azimuthal and magnetic quantum numbers (as they are now called) arise from the angular parts of the equation. He has intuitively realized that the principal quantum number will arise in a similar way from the radial wave equation, but he has not yet been able to solve the latter.

The relativistic and the nonrelativistic radial equations are so similar that the solution of one could be followed almost immediately by the solution of the other. Since in his first paper on wave mechanics he thanks Hermann Weyl for indicating the way to solve the equation, he would not have found the method till shortly after his return to Zürich on January 9. The actual solution would then take no more than a day or two, so that by January 11 he should have had the solution to the relativistic wave equation (now called the Klein–Gordon equation). Then followed the decision to present only the nonrelativistic theory. The time schedule is rather tight but quite possible. Unfortunately, we do not have a manuscript for the earlier (relativistic) paper. The details of its content are in Notebook N2. In 1956, he wrote to Wolfgang Yourgrau:[10]

Admittedly the Schrödinger theory, relativistically framed (without spin) gives a *formal* expression of the fine-structure formula of Sommerfeld, but it is *incorrect* owing to the appearance of half-integers instead of integers. My paper in which this is shown has . . . never been published; it was withdrawn by me and replaced by the non-relativistic treatment.

The reason why Schrödinger's relativistic equation did not agree with experiment is of course that it did not include electron spin, which was not known at that time. It is a perfectly good equation for particles of spin zero.

The relativistic wave equation

The steps leading to the relativistic wave equation are sketched briefly in the three pages of notes N1. Schrödinger used a method similar to that found in many elementary textbooks today – substitution of the Broglie relation for the wavelength or frequency of a particle into the ordinary steady-state wave equation of mathematical physics. The procedure is so simple that perhaps he believed it would not be convincing, so that when he published his equation, he gave two different and more difficult 'derivations'.

Of course, one cannot actually *derive* the Schrödinger wave equation from classical physics. The so-called 'derivations' are therefore never logically rigorous – they are justifications rather than derivations. It would be reasonable simply to write down the final equation as a *postulate* of wave mechanics, and then to show that it gives correct results when applied to various calculations. But before you can write it down, you have to have it, and it did not spring fully fledged from Erwin's mind like Venus from the scallop shell. The paths leading to the equation are thus of interest, since they show the process by which the discovery was made. Schrödinger saved correspondence and notes about many uninteresting things, wrote several short autobiographical accounts, and talked on all sorts of subjects to friends of all persuasions, yet he never gave a chronological account of his pathway to the discovery that made him one of the immortals of science.

We shall first consider the nonrelativistic equation, even though Schrödinger started with the relativistic one. The space dependence in three dimensions of the amplitude ψ of a wave motion, independent of time, is given by a linear second-order differential equation,

$$\Delta \psi + k^2 \psi = 0$$

where Δ is the Laplacian operator (sometimes written ∇^2). In cartesian coordinates, this is $\Delta = \partial^2/\partial x^2 + \partial^2/\partial y^2 + \partial^2/\partial z^2$. The parameter $k = 2\pi/\lambda$, where λ is the wavelength, so that

$$\Delta \psi + \frac{4\pi^2}{\lambda^2} \psi = 0$$

When the Broglie relation $\lambda = h/mv$ is inserted, the wave equation becomes

$$\Delta \psi + \frac{4\pi^2 m^2 v^2}{h^2} \psi = 0$$

The kinetic energy of a particle is the total energy E minus the potential energy V, so that $E - V = \frac{1}{2}mv^2$, and the final form of the wave equation is

$$\Delta \psi + \frac{8\pi^2 m}{h^2} (E - V) \psi = 0$$

It is easy enough to follow this almost trivial 'derivation' once somebody has written it down for the first time, but it is quite impossible to recapture the tension and confusion in the minds of the physicists of 1925. For them, hundreds of pathways appeared to be open and there were no signposts on any of them. Only after the trail was blazed, was it easy to follow.

Schrödinger's first derivation, for which the first written evidence is in N1, began with the relativistic equations from de Broglie's thesis,[11]

$$u = E/p = \frac{hv(1 - \beta^2)^{1/2}}{m_0 \beta c} \qquad \text{and} \qquad hv = \frac{m_0 c^2}{(1 - \beta^2)^{1/2}} - e^2/r$$

Here u and v are respectively the phase velocity and the frequency of the wave associated with the bound electron, and r is the radius of its classical orbit. After eliminating $\beta = v/c$,

$$u = \frac{hv/m_0 c^2}{[(hv/m_0 c^2 + e^2/m_0 c^2 r)^2 - 1]^{1/2}}$$

This is substituted into the differential wave equation to yield ($u = v\lambda$):

$$\Delta \psi + \frac{4\pi^2 m_0^2 c^2}{h^2} [(hv/m_0 c^2 + e^2/m_0 c^2 r)^2 - 1] \psi = 0$$

This equation has the same general form as the nonrelativistic equation, and the solution of one leads immediately to the solution of the other.

The first step is to separate the variables. In view of the spherical symmetry of the electric potential, $V = -e^2/r$, it is convenient to work with spherical polar coordinates, r, θ, φ, where φ is a longitude and θ a colatitude. Schrödinger, following a standard procedure, substituted $\psi(r, \theta, \varphi) = R(r)\ \Theta(\theta)\ \Phi(\varphi)$ into the equation, to obtain three ordinary differential equations. The solutions for the angular functions are

$$\Theta(\theta)\ \Phi(\varphi) = \frac{1}{\sqrt{2\pi}}\ \exp(im\varphi) \left\{\frac{(2n+1)(n-|m|)}{2(n+|m|)}\right\} P_n^{|m|}\ (\cos\ \theta)$$

where $P_n^{|m|}\ (\cos\ \theta)$ are the well known spherical harmonics first described by Legendre. The radial equation is:

$$\frac{d^2R}{dr^2} + \frac{2}{r}\frac{dR}{dr} + \left(A + \frac{2B}{r} + \frac{C}{r^2}\right) R = 0$$

where

$$A = \frac{8\pi^2 m}{h^2}\ (E - mc^2), \qquad B = \frac{4\pi^2 mc^2}{h^2}, \qquad C = -l(l+1)$$

This equation is included in the early notes N1, and Schrödinger worked intensively on it during his stay at Arosa. When he returned to Zürich, Dean Schlaginhausen asked him if he had enjoyed the skiing during his vacation. Erwin replied that he had been distracted by a few calculations. This may have been the only time in his life when he allowed anything to distract him from a holiday.

It is surprising that Schrödinger had so much difficulty in solving the radial equation. He was using as a reference the little book of Ludwig Schlesinger, *Introduction to the Theory of Differential Equations*, published in 1900.[12] As the title indicates, it is not a book devoted to the practical problems of solving differential equations of mathematical physics. Schrödinger evidently was not yet using the book that became the *vade mecum* of theoretical physicists, *Methods of Mathematical Physics I*, published in 1924 by Richard Courant and David Hilbert.[13] On page 161 of this book, the equation satisfied by the Laguerre polynomials appears, and its form is very close to that of the radial equation for the H-atom, but the associated Laguerre polynomials, which actually provide the solution for Schrödinger's equation are not cited. The book of Philipp Frank and Richard Mises, *The Differential and Integral Equations of Mechanics and Physics* was published only at the end of 1925; this gave a complete account of the associated Laguerre polynomials and their corresponding differential equations, but it would have been just too late to help Schrödinger.[14]

In November, 1926, he wrote a foreword to a collected edition of his first six papers on wave mechanics[15] (which we shall denote as Q1 to Q6).

With reference to the six papers, whose present republication was caused solely by the strong demand for reprints, a young friend of mine recently said to the author: 'Hey, you never even thought when you began that so much sensible stuff would come out of it.' This expression, with which I fully agree after suitable discounting of the complimentary adjective, may recall the fact that the works united here in one volume emerged one after the other. The knowledge of later sections was often still unknown to the writer of the earlier ones.

The young friend was fourteen-year old Itha Junger, whose acquaintance will be made subsequently. She remembers that when Erwin wished to work without any distraction, he used to place a pearl in each ear to shut out any noise, to which he was very sensitive.[16] During the early months of 1926, as his creativity reached a peak and the papers poured from his pen, he found frequent need for this stratagem.

The first paper

The first communication on 'Quantization as an Eigenvalue Problem' (Q1) was received by the *Annalen* on Tuesday, January 27, 1926.[17] It was less than three weeks since Erwin had returned from Arosa. In that time, he had consulted Weyl more than once, found the solution to the relativistic equation, written a paper about it, then found the nonrelativistic solution, written another paper, and withdrawn the first effort because of its disagreement with the hydrogen-atom data. There is no evidence that the first paper was actually sent to the journal and no manuscript has been found.

In Q1 he gives a 'derivation' for the wave equation which seems to be almost deliberately cryptic. He knew the equation before he devised the 'derivation', and his main purpose in this paper was to show that his equation gives the correct quantization of the energy levels of the hydrogen atom.

In this communication I wish first to show in the simplest case of the hydrogen atom (nonrelativistic and undistorted) that the usual rules for quantization can be replaced by another requirement, in which mention of 'whole numbers' no

Hermann (Peter) Weyl in 1925.

longer occurs. Instead the integers occur in the same natural way as the integers specifying the number of nodes in a vibrating string. The new conception can be generalized, and I believe it touches the deepest meaning of the quantum rules.

The derivation was based on the Hamilton–Jacobi equation of classical mechanics,

$$\partial S/\partial t = -H(q,\ p,\ t)$$

where S is called *Hamilton's principal function*, and p and q are generalized momenta and coordinates respectively, and H is the Hamiltonian,

$$\partial S/\partial t = -H \qquad \text{and} \qquad \partial S/\partial q = p$$

The integral of the Hamilton–Jacobi equation has the form

$$H(q,\ \partial S/\partial q) = E \tag{1}$$

Schrödinger now substituted for S a new function $\psi(q)$, such that

$$S = K \ln \psi \tag{2}$$

He was at this point concerned only with stationary (time independent) systems. Then, from (1),

$$H\left(q,\ \frac{K}{\psi}\frac{\partial\psi}{\partial q}\right) = E \tag{1'}$$

The energy of an electron moving in the field of the proton is

$$E = \frac{1}{2m}\ (p_x^2 + p_y^2 + p_z^2) - e^2/r$$

Or, in terms of S,

$$(\partial S/\partial x)^2 + (\partial S/\partial y)^2 + (\partial S/\partial z)^2 = 2m(E + e^2/r)$$

From (2), since $\partial S/\partial x = (K/\psi)\,\partial\psi/\partial x$,

$$\left(\frac{\partial\psi}{\partial x}\right)^2 + \left(\frac{\partial\psi}{\partial y}\right)^2 + \left(\frac{\partial\psi}{\partial z}\right)^2 - \frac{2m}{K^2}\ (E + e^2/r)\ \psi = 0 \tag{1''}$$

with

$$r = (x^2 + y^2 + z^2)^{1/2}$$

'We seek those functions ψ which are single-valued, finite, and continuously twice differentiable over the whole configuration space and which make the integral of the just mentioned quadratic form, extended over the whole configuration space, an extremum. *We replace the quantum conditions by this variation problem.*'

There are two astonishing things about this procedure. It makes no reference to the wave properties of particles discovered by Louis de Broglie, and it makes no reference to the quantization of energy and Planck's constant, although it does mention that K has the dimensions of action.

The variation condition becomes

$$\delta J = \delta \iiint dx\,dy\,dz \left[\left(\frac{\partial\psi}{\partial x}\right)^2 + \left(\frac{\partial\psi}{\partial y}\right)^2 + \left(\frac{\partial\psi}{\partial z}\right)^2 \right.$$
$$\left. - \frac{2m}{K^2}\left(E + \frac{e^2}{r}\right)\psi \right] = 0 \qquad (3)$$

The Euler–Lagrange equation for the variation problem is then found to be,

$$\Delta\psi + \frac{2m}{K^2}\left(E + \frac{e^2}{r}\right)\psi = 0 \qquad (5)$$

If K is set equal to $h/2\pi$, this is the same equation that Schrödinger obtained earlier by substitution of the Broglie equation into an ordinary wave equation.

He separated the equation as already shown and obtained the same radial equation with different constants,

$$A = 2mE/K^2, \qquad B = me^2/K^2, \qquad C = -n(n+1) \qquad (7)$$

Now, he knew how to solve the equation. He first made the substitution $R = r^\alpha U$, with α chosen so that the term in $1/r^2$ falls out, giving a new equation,

$$\frac{d^2U}{dr^2} + \frac{2(\alpha+1)}{r}\frac{dU}{dr} + \frac{2m}{K^2}\left(E + \frac{e^2}{r}\right)U = 0 \qquad (7'')$$

This equation has exactly the form of the Laplace equation discussed in Section V-49 of Schlesinger's textbook.

One can only speculate about the point at which Erwin found his mathematics to be inadequate and was forced to call Weyl to the rescue. It seems unlikely that he failed to find the substitution that leads to the Laplace equation, and once this equation was in hand, the solution was available in Schlesinger's textbook as an example of the application of the Laplace transform method. Alternatively, as Schlesinger points out, the equation can be integrated by substitution of a series expansion of U and calculation of the coefficients from the recursion formulas. The most probable source of his difficulties was in the mathematical arguments necessary to select the solution that is in accord with the boundary conditions. In the published paper this part of the derivation is the most difficult and it would not be available in any reference book. The first presentation of a new idea is seldom if ever as perspicuous as the later version of the same material, a newborn truth is bound to bear evidence of the pangs of parturition. Thus Schrödinger makes heavy going of mathematical arguments that seem elementary in modern textbooks. One can almost hear an echo of the discussions with Weyl in the terminology of function theory that is employed.

If Schrödinger took Schlesinger's book with him to Arosa, he should have had the general solution before his return to Zürich on January 8, and, with the assistance of Weyl, the fitting of the boundary conditions and derivation of the radial eigenfunctions probably were accomplished within a week at most. At this time, however, neither he nor Weyl recognized the 'completely transcendent solution' as the associated Laguerre polynomials. The final result for the eigenfunctions is:

$$R = f\left(r\frac{\sqrt{-2mE}}{K}\right);$$

$$f(x) = x^n \exp(-x) \sum_{k=0}^{l-n-1} \frac{(-2x)^k}{k!}\binom{l+n}{l-n-1-k}$$

The eigenvalues are:

$$me^2/K \sqrt[+]{-2mE} = l; \qquad l = 1, 2, 3, 4 \ldots \ldots \qquad (15')$$

Schrödinger summarized the result for states of negative energy (bound states of the electron) as follows: *For negative E, our variation problem has a solution if and only if condition (15') is satisfied. The whole number n which gives the order of the spherical harmonics in the solution can then take on only values smaller than l (of which at least one always occurs).* [In present-day notation Schrödinger's *l* is called *n*, the principal quantum number, and his *n* is called *l*, the azimuthal quantum number.]

This paper (Q1) has been universally recognized as one of the greatest achievements of twentieth-century physics. As Dirac was to remark later, it contains much of physics and, in principle, all of chemistry.[18] By 1960, more than 100 000 papers had appeared based on application of the Schrödinger equation. From the beginning, it was accepted as a mathematical tool of unprecedented power in dealing with problems of the structure of matter. And almost from the beginning, scientists began to ponder and worry and argue about what it might be telling them about the nature of the physical world. As J. Robert Oppenheimer said some years later:

Here is this quite beautiful theory, perhaps one of the most perfect, most accurate, and most lovely man has discovered. We have external proof, but above all internal proof, that it has only a finite range, that it does not describe everything that it pretends to describe. The range is enormous, but internally the theory is telling us, 'Do not take me absolutely seriously, I have some relation to a world that you are not talking about when you are talking about me.'[19]

One way in which the range of the theory is limited was of course well known to Schrödinger – his equation is not consistent with the requirements of relativity theory: it is not invariant under a Lorentz transformation.

Schrödinger devoted several pages at the end of his paper to an interpretation of his new theory.

Naturally it is very obvious to relate the function ψ to a vibration process in the atom, to which we can ascribe more than today's doubtful reality of the electronic orbits. Originally I had the intention of establishing the new foundation of the quantum rules in this more pictorial [*anschaulich*] way, but then gave preference to the above more mathematical formulation, because it allows the important points to emerge more clearly. What seems to me to be important is that the mysterious 'whole number requirement' no longer appears, but is, so to speak, traced back to an earlier stage: it has its basis in the requirement that a certain spatial function be finite and single-valued.

I don't want to go into more detail now about possible interpretations of the vibration process until it is clear that the new method can give new results. We don't know for sure that its results may not be merely a rehash of the conventional quantum theory [*ein blosser Abklatsch der üblichen Quantentheorie*]. For instance, the relativistic Kepler problem, when calculated in exactly the same way as given here, strange to say, leads to half-integral quantum numbers (radial and azimuthal).

He now mentions that the inspiration of his work has been the ingenious *Thèses* of Louis de Broglie. The difference is that the latter based his theory on traveling waves, whereas the present theory is based on standing waves. 'I have recently shown that the Einstein gas theory can be derived from such standing eigenvibrations, to which the de Broglie dispersion law is applied. The above considerations for the atom can now be regarded as a generalization of that application to the gas model.' This statement is important direct evidence for the idea that the work on statistical gas theory was indeed the intellectual precursor of the discovery of wave mechanics.

The final two pages of the paper are devoted to an attempt to give a picture of the vibration processes responsible for the Bohr frequency conditions for emission and absorption of radiation. If the ψ-functions describe characteristic vibrations, the energy levels E must be related to the vibration frequencies. Usually, however, in a vibrating system, the energy is proportional to the square of the frequency. For negative energy values, this would give imaginary frequencies. He therefore proposes to add to the negative energies a large constant energy term, which is related to the rest energy mc^2 of the electron. This addition can be justified by the fact that the zero level of energy is anyway arbitrary. He then imagines the atom to be vibrating with a 'potpourri' of very high frequencies corresponding to these high energies. The frequencies given by the Bohr condition $h\nu = E_2 - E_1$ are then visualized as beats between these high vibration frequencies. He seems pleased with this picture and remarks: It is scarcely necessary to emphasize how much more appealing than the conception of jumping electrons would be the conception that in quantum transitions the energy passes from one vibration pattern to another. The change in the vibration pattern can take place continuously in space and time, and can readily persist as long as the emission process does.' He feels that this picture will overcome the difficulties that caused the demise of his old favorite, the B–K–S theory. He even goes so far as to suggest that simultaneous emission of many spectral lines by the same atom can now occur. This anachronistic idea had not been seriously entertained since the work of Arthur Conway in 1907. It was perhaps true, as Schrödinger stated, that there was no experiment explicitly disproving it, but the Franck–

Hertz experiments on the excitation of atoms by electron impact came close to doing so.

Thus in the same paper in which he created a revolution in quantum theory and opened the way to a new era in microphysics, Schrödinger sought to use his discovery as a pathway back to a classical physics based on a continuum undisturbed by sudden transitions, in which quantization was only a misleading approximation. He hoped that the relativistic theory, which he had so far failed to find, would complete this process.

The second paper

Just four weeks after the first paper (Q1) the *Annalen* received on February 23 the second paper (Q2) in the series 'Quantization as an Eigenvalue Problem'.[20] It consists of a detailed exploration of the Hamiltonian analogy between mechanics and optics, leading to a new derivation of the wave equation, an analysis of the relations between geometrical and undulatory mechanics, and applications of the wave equation to the harmonic oscillator and the diatomic molecule.

Schrödinger gives a leisurely and polished development of the analogy between mechanics and optics, written in his most elegant style, informal yet magisterial. It is a masterpiece of scientific exposition, and must have given its author a justified feeling of satisfaction. The connection between Hamilton's theory and wave motion was by no means new. Sommerfeld had called Schrödinger's attention to the work of Felix Klein in this area, dating from 1891. Hamilton's principle can be shown to correspond with Fermat's principle for wave propagation in configuration space (q-space), and the Hamilton–Jacobi equation expresses the Huyghens' construction for wave propagation. The analogy is one between mechanics and *geometrical* optics, not physical or *undulatory* optics. The important ideas of wave theory, such as wavelength, amplitude, and frequency do not enter into the analogy since they have no parallels in classical mechanics.

There should be, however, an essentially classical theory of matter waves that would have the role in mechanics that the Maxwell theory of electromagnetic waves has in optics. Schrödinger remarks that '*we know today in fact that our classical mechanics fails for very small dimensions of the path and for very great curvatures. This failure is so similar to that of* geometrical optics that it becomes a question of searching for an undulatory mechanics by further working-out of the Hamiltonian analogy.'

The image point (or 'particle') of the mechanical system must then be represented by a wave group with small dimensions in every direction (now called a *wave packet*). In such a packet one must be able to neglect any spreading of the waves in comparison with the dimensions of the path in the system. The image point of a mechanical system corresponds to that point where a certain continuum of wave forms coalesces with the same phase. The image point moves with the group velocity of the wave packet, $v = dv/d(v/u)$, where u is the phase velocity.

The true mechanical process is represented by the wave pattern, not by the image point. In the interior of an atom, for example,

no special meaning is to be attached to the electron path itself . . . and still less to the position of an electron in its path . . . The wave group not only fills the whole path domain all at once, but also extends far beyond in all directions . . . This contradiction is so strongly felt that it has even been doubted whether what goes on in an atom can be described within the scheme of space and time. From a philosophical standpoint, I should consider a conclusive decision in this sense as equivalent to a complete surrender. For we cannot really avoid our thinking in terms of space and time, and what we cannot comprehend within it, we cannot comprehend at all. There *are* such things but I do not believe that atomic structure is one of them.

It is evident in this paper that Schrödinger believed that a classical wave picture based upon continuous matter waves would provide the most satisfactory foundation for atomic physics.

The most important advance in Q2 is the demonstration that ψ is a function in configuration space, $\psi(q_k)$, for example for two particles, $\psi(x_1, x_2, y_1, y_2, z_1, z_2)$. This fact establishes the relationships between the states of different particles in a system, and is the basis of all atomic and molecular structure.

Schrödinger mentions that he has read the recently published work on quantum mechanics of Heisenberg, Born and Jordan, but he has not yet been able to discern the relation of their method to his own. He believes the strength of their method is its ability to calculate spectral line intensities, while the strength of his, if he may be permitted to say so, is in its clear physical viewpoint which provides a bridge between macroscopic and microscopic phenomena.

In the rest of the paper several important applications of the wave equation are given. The solution for the one-dimensional harmonic oscillator is readily obtained, since now he has the Courant and Hilbert book (which he did not have while working on the first paper), and the eigenfunctions are the well known Hermite orthogonal functions, and the eigenvalues are $E_n = (n + \frac{1}{2})h\nu$, as found by Heisenberg. He treats the rigid rotor, both with fixed axis and with free axis. The eigenfunctions are the spherical surface harmonics (associated Legendre functions) and the eigenvalues are $E_n = n(n + 1)h^2/8\pi^2 A$, where A is the moment of inertia. Finally he considers a problem important in molecular spectroscopy, a diatomic molecule in which rotational and vibrational degrees of freedom are coupled through variation of the moment of inertia as vibration changes the internuclear distance. He does not consider the anharmonic oscillator, remarking that it is a problem requiring perturbation theory.

Reactions to Q1 and Q2

Schrödinger sent Max Planck a reprint of Q1, and on April 2, 1926, Planck replied to thank him for the paper, which he had read 'like an eager child hearing the solution to a riddle that had plagued him for a long time'.[21] He was delighted by the application of the action function S, the importance of which he had long felt to be underestimated. Erwin at once replied to the 'highly revered Herr Geheimrat' that the card had made him very happy, and he included a brief account of his work on the Stark effect (Q3), in which 'the fundamental assumption is that the space density of electricity is given by the square of the wavefunction'.[22]

On May 24, Planck wrote about Q2: 'You can imagine with what interest and enthusiasm I plunged into the study of this epoch-making work, although I now progress very slowly in penetrating thought processes of this kind.'[23] He was pleased to learn that Grüneisen, the Chairman of the Berlin section of the German Physical Society, had invited Schrödinger to give a lecture so that the Berlin physicists would soon have an opportunity to listen to a first-hand account of his work. He then added a rather curious remark: 'I do not know whether you already know Berlin, but I hope that you will find that in certain respects one lives here more freely and independently than in a smaller city, where each one checks up on the other, and one does not have the possibility of sometimes going into seclusion without anyone noticing it.' Even at this early date Planck wished to acquaint Schrödinger with the attractions of Berlin, and he had already conceived the plan of bringing him there as his successor after his retirement in 1927. Erwin replied, confirming his plans to visit Berlin in July, and asking whether he should give a more general or a more advanced lecture.

On April 16, Einstein wrote to say that Planck had showed him the wave mechanics papers and he thought that 'the idea of your work springs from true genius!' No praise could have pleased Erwin more than these words from the scientist he admired above all others. The schoolboy who was always first in class had at last taken his rightful place at the forefront of physics. Ten days later, Einstein wrote again: 'I am convinced that you have made a decisive advance with your formulation of the quantum condition, just as I am convinced that the Heisenberg–Born method is misleading.'

Paul Ehrenfest wrote from Leiden on May 19: 'I am simply fascinated by the $\Delta\psi + (8\pi^2 m/h^2)(E - V)\,\psi = 0$ theory and the wonderful new viewpoints that it brings. Every day for the past two weeks our little group has been standing for hours at a time in front of the blackboard in order to train itself in all the splendid ramifications.'[24]

A more cautious note had come from Willy Wien: 'How would you derive black-body radiation in your theory? . . . The physical significance of the constant h has not become clear to me . . . In any case I congratulate you on your achievement and wish you good luck with all my heart.'[25]

Matrix and wave mechanics

After finishing Q1 and Q2, Schrödinger had time to consider the relation of Heisenberg's matrix mechanics to his wave mechanics. In a paper received by *Zeitschrift für Physik* in July 1925, Werner Heisenberg, then twenty-four years old, had given a preliminary account of a new and highly original approach to the mechanics of the atom.[26] Having finished his Ph.D. with Sommerfeld in Munich, he was working with Max Born in Göttingen. He had been influenced considerably by discussions with Niels Bohr. Since the B–K–S paper, the Copenhagen physicists had been fascinated by the concept that an atom can, in a sense, be equated with a set of oscillators having frequencies equal to its absorption frequencies v_{nm}. Heisenberg's aim was to develop a mathematical theory based entirely on experimentally observable quantities such as the frequencies, amplitudes and polarizations of spectral lines, without any reliance upon pictorial models of atomic structure based on concepts such as 'the position of an electron within the atom', which he believed to be in principle unobservable. This was a thoroughly Machian (or positivist) program.

In accord with the Rydberg–Ritz combination principle, the lines observable in an atomic spectrum can be represented as a two-dimensional array of differences between spectral terms, $v_{nm} = T_n - T_m$. Heisenberg proposed to substitute for the position coordinate of an electron a similar array, $q_{nm} \exp(2\pi i v_{nm}t)$. As is often the case with the first expression of a new idea, his paper is somewhat obscure, but while he was away from Göttingen on a visit to Cambridge, the concepts in his paper were extended and clarified by Born and Pascual Jordan (September, 1925)[26], and when he returned he joined these two colleagues in a definitive paper on the new quantum mechanics (November, 1925).[26]

Born had recognized immediately that Heisenberg's arrays q_{nm} are infinite matrices, the algebra of which was well known. Multiplication of matrices does not obey the usual commutation law, i.e., $AB \neq BA$.

The classical momentum p was also replaced by a matrix $p_{nm} \exp(2\pi i \, \nu_{nm} t)$ and the q and p matrices were postulated to follow a commutation law,

$$qp - pq = (ih/2\pi) \, |1|$$

where $|1|$ is the unit matrix. The classical Hamiltonian equations of motion were now assumed, with q and p representing the matrix quantities,

$$\dot{q} = \partial H/\partial p = p/m; \qquad \dot{p} = - \, \partial H/\partial q = - \, \partial V/\partial q$$

The conservation of energy, $\dot{H} = 0$, then requires that H be a diagonal matrix in which the diagonal elements H_{nn} are the energies of the various states of the system, so that the mathematical technique of matrix mechanics is the procedure of diagonalization of the H-matrix.

When Heisenberg lectured in Cambridge before the Kapitza Club in July 1925, Paul Dirac, a 23-year-old student, was in attendance, but quantum mechanics was not discussed. Later, however, his professor, Fowler, gave him a preprint of the basic Heisenberg paper, and Dirac independently carried out a reformulation of the theory.[27]

Schrödinger was aware of the first two papers on matrix mechanics when he devised his wave mechanics, but he did not have the definitive third paper. They had no influence on his work since their approach was so different from his. He was surprised to find, however, that when the same problem was treated by both methods, the harmonic oscillator for example, they gave concordant results. Heisenberg described his work as 'the true theory of a discontinuum', whereas Schrödinger's concept of a wavefunction filling all space evoked a continuum even more pervasive than that of classical mechanics. At first he could see no relationship between the two theories, but in the last week of February and the first two weeks of March, he was able to establish the relationship in one of his most brilliant papers. [Carl Eckart at Caltech also did this independently.[28]]

The paper 'On the Relation of the Heisenberg–Born–Jordan Quantum Mechanics to Mine' displays his mathematical skills, as he uses the then new methods of functional analysis to solve the problem.[29] He remarked in a footnote: 'My theory was inspired by L. de Broglie and by brief but infinitely far-seeing remarks of A. Einstein (*Berl. Ber.* 1925, p. 9ff). I was absolutely unaware of any genetic relationship with Heisenberg. I naturally knew about his theory, but because of the to me very difficult-appearing methods of transcendental algebra and because of the lack of *Anschaulichkeit*, I felt deterred by it, if not to say repelled.'

Schrödinger's procedure was to demonstrate the following: (1) To each function of position and momentum one can relate an infinite matrix that satisfies the Born–Heisenberg calculating rules. The relation of matrices to functions is general, the functions forming an arbitrary, complete orthogonal set in configuration (q) space; the particular Hamiltonian function does not enter into the connection between the theories. (2) Heisenberg's so-called 'equation of motion', the system of algebraic equations that relate the matrices of position and momentum with the Hamiltonian matrix, is completely solved if the general functions in (1) are a particular orthogonal system, namely, the eigenfunctions of the Schrödinger partial differential equation for the problem. The solution of the boundary-value problem for this equation is *completely equivalent* to the solution of the Heisenberg algebraic problem.

The result of the remarkable analysis in this paper is thus that the two theories appear to be equivalent from a mathematical standpoint. Here then are two theories, one based on a clear conceptual wave model of atomic structure and the other based on a radical statement that any such model is meaningless, yet both lead to the same final results. Schrödinger was taken aback by what he had discovered: 'There are today not a few physicists who, exactly in the sense of Mach and Kirchhoff, see the task of physical theory to be merely the most economical description of empirical connections between observable quantities . . . In this view, mathematical equivalence means almost the same as physical equivalence.' The spirit of Mach might consider matrix mechanics, with its complete lack of any model [*Unanschaulichkeit*] to be superior to wave mechanics, but Schrödinger argues that to deprive a physicist of the possibility of making space-time models of subatomic structures and phenomena will inhibit further progress in the theory, for one cannot operate entirely with abstract ideas such as transition probabilities, energy levels, and the like. For instance, he cites the interaction of an electromagnetic field with an electron in an atom, which he considers to be 'an especially important question, perhaps the cardinal question in all atomic dynamics'.

To attack this question, he proposed a new interpretation of the ψ-function: the assumption that the space density of electric charge is given by the real part of $\psi\, \partial\psi/\partial t$. He can consider the wave mechanical representation of the hydrogen atom to be a sum over its discrete eigenfunctions:

$$\psi = \sum_k c_k u_k \exp(2\pi i E_k t/h)$$

and the charge density would then become

$$2\pi \sum c_k c_m \frac{E_k - E_m}{h} u_k u_m \sin \frac{2\pi t}{h} (E_m - E_k)$$

The frequencies of the vibrating charge density occur only as term differences, and the radiation absorbed or emitted by the atom can be ascribed simply to the atomic dipole vibrating with these frequencies.

The third paper

The third paper in the series, Q3, was received by *Annalen* on May 10.[30] It is an extensive account (53 pages) of perturbation theory and its application to the Stark effect on the Balmer lines. In many instances an exact (analytic) solution of a differential equation cannot be obtained, but it is possible to obtain an approximate solution by applying a small disturbance (or perturbation) to the condition of a system for which an exact solution is available. Such methods were often used in classical theories, for instance, the treatment by Rayleigh of vibrations in a string with small inhomogeneities. In atomic theory there are many interesting cases in which more than one eigenfunction for the unperturbed system has the same eigenvalue (so-called *degenerate* eigenvalues) and a perturbation causes a splitting of these eigenvalues. In the Stark effect, the lines in an atomic spectrum are split by an applied electric field, and Schrödinger undertook the calculation of this effect for the hydrogen spectrum, by calculating the effect of the electric field on the eigenvalues of the hydrogen-atom problem.

He first outlined the general theory of the perturbation method for a differential equation of the Sturm–Liouville type,

$$L(y) + E\varrho y = 0$$

where

$$L(y) = py'' + p'y' - qy$$

and p, p', q and ϱ are continuous functions of the independent variable x, p' denoting dp/dx, etc. It is supposed that the eigenfunctions $u_i(x)$ and eigenvalues E_i of this equation are known. A small perturbation, $-\lambda r(x)y$, is added to $L(y)$, so that the equation becomes

$$L(y) + (E\varrho - \lambda r)y = 0$$

where r is an arbitrary continuous function of x. The new eigenvalues are then $E_k^* = E_k + \lambda \epsilon_k$, and the new eigenfunction $u_k^* = u_k(x) + \lambda v(x)$.

To a first approximation

$$\epsilon_k = \int r u_k^2 dx = r_{kk}$$

(assuming u_k to be normalized). Thus the effect on the eigenvalues is calculated simply by averaging the perturbation function over the unperturbed eigenfunctions. The new eigenfunctions are obtained as an expansion in terms of the old ones,

$$u_k^*(x) = \sum_{i=1}^{\infty}{}' \gamma_{ki} u_i(x), \qquad \gamma_{ki} = \frac{\int r u_k u_i dx}{E_k - E_i} \qquad (i \neq k) \qquad (1)$$

where Σ' denotes the sum with term $i = k$ excluded. The next approximation in perturbation theory uses these new u_k^* to calculate ϵ_k^* again. These methods are now a standard part of quantum mechanics and many important problems have been solved by their use.

In the case of partial differential equations, the perturbation theory is the same, but one must now consider the possibility of degenerate eigenvalues. If two distinct eigenfunctions u_1 and u_2 have the same eigenvalue, the new eigenfunctions cannot be expanded in terms of the original set, for the eigenfunctions corresponding to a degenerate eigenvalue are not in general orthogonal to one another. One must first take suitable linear combinations of the u_k to form an orthogonal set, the correct zero-order eigenfunctions, to which the perturbed eigenfunctions converge as $\lambda \to 0$. [We can see that the formula of (1) fails when $E_k = E_i$, or even when E_k is very close to E_i.]

Schrödinger applied the theory to the Stark effect on the Balmer lines of the hydrogen spectrum, which rise from transitions between levels with principal quantum number $n = 2$ and higher values. For example, the red H line is from the transition $n = 2$ to $n = 3$. The $n = 2$ level is fourfold degenerate, and $n = 3$ is ninefold degenerate. The first-order Stark effect splits $n = 2$ into three levels, and $n = 3$ into five. Schrödinger's wave mechanical theory yielded the famous Epstein formulas for the Stark displacements, and the calculated intensities and polarizations were in fairly good agreement with Stark's experimental values. Better agreement can be achieved by calculating the second-order perturbations. Schrödinger considered these calculations to be the first major application of his new theory.

Correspondence with Lorentz

Schrödinger sent reprints of Q1, 2, 3 and the paper showing the relation between wave and matrix mechanics to Hendrik Lorentz in Leiden, now seventy-three years old, the grey eminence of theoretical physics, whose words commanded universal attention and respect.

On May 27 Lorentz wrote a long letter with some cogent criticisms. In his opinion there were many difficulties yet to be overcome in the theory.[31]

Lorentz was particularly impressed by the way in which matrix mechanics could be derived from wave mechanics by equating the operators q and $(h/2\pi i)\,\partial/\partial q$ with coordinate and momentum, respectively. 'If I had to choose between wave mechanics and matrix mechanics, I would give preference to the former because of its greater *Anschaulichkeit*, so long as one is concerned only with the coordinates x, y, z. With a greater number of degrees of freedom, however, I cannot interpret physically the waves and vibrations in q-space and I must decide for matrix mechanics.'

Lorentz objected to the general representation of a particle as a wave packet, since such a packet does not remain compact with time but gradually spreads out. In the field of the hydrogen atom, this attenuation would occur rapidly, since a wave packet can persist for an appreciable time only if its dimensions are large compared to the wavelength, a condition not fulfilled in the Bohr orbits of the atom. 'Your surmise that the transformation which our mechanics must undergo is similar to the transition from ray optics to wave optics sounds enticing, but I still have doubts about it.'

Next, he raised the question of why the field of the electron itself is omitted from the wave function. He admitted that to include it would cause untold difficulties and spoil the calculation of the energy eigenvalues. Schrödinger's view might, however, solve this difficulty, for 'if the electron as such is no longer there, then one can be more readily satisfied that only the term for the nuclear charge appears'.

Finally he pointed out that it will be difficult to explain the Bohr frequency condition in terms of beats between different internal vibration frequencies, as was suggested in Q1, since only a nonlinear system would be able to respond to such beats.

Schrödinger replied to Lorentz on June 6. 'You have done me the extraordinary honor, in eleven closely written pages, to submit the ideas of my last works to a deeply searching analysis and critique. I can find no words to thank you sufficiently for the valuable gift that you have thereby presented to me.' He then responded to some of the points raised.

He admitted the difficulty of interpreting waves in q-space in terms of projections in ordinary three dimensional space, and referred to his interpretation of $\psi\bar{\psi}$ as a density of charge. If ψ is a function of $3N$ variables, R_1, R_2, . . . , R_n he would let R_1 be identified with the real space and integrate $\psi\bar{\psi}$ over R_2, . . . , R_n. He would then repeat the process with R_2, etc., and add all the results to obtain the charge density. 'What is unpleasant here', he said, 'and indeed directly to be objected to, is the use of complex numbers. ψ is surely fundamentally a real function . . . '. At this time he had not realized the fundamental, and indeed epoch-marking, significance of the introduction of complex numbers into the theory, as carriers of the unobservable phase information of the ψ waves.

Concerning the spreading of wave packets, he sent a copy of his paper on the harmonic oscillator for which a wave packet can be constructed that does not spread out with time. He still thinks the same sort of thing can eventually be constructed for other cases, although the problem now appears 'hopeless'. In the event, this optimism was not justified, since the harmonic oscillator is a special case arising from its equidistant energy levels.

In regard to the beat frequencies, he admitted his error in not distinguishing between beats and combination tones. 'I was at first so exceedingly happy to have achieved a picture that actually had anything at all to do with the frequencies observed in emitted light that I threw myself panting like a hunted fugitive on this thing in the form in which it directly made its appearance.'

Lorentz replied on June 19. He supposed that some of his difficulties were due to the fact that he was still involved in the thought patterns of the old quantum theory. He was pleased to get the harmonic oscillator paper and at first he thought that the wave-packet idea might be satisfactory after all, but his joy was soon extinguished, for in the case of the hydrogen atom, one does not have available the short wavelength vibrations that would permit a wave packet to be constructed. He sent a detailed 12-page calculation to demonstrate this fact. Schrödinger soon de-emphasized the wave-packet picture, which was not an integral part of his theory, although it helped to support his idea that waves constitute the basic reality of the subatomic world. Lorentz, however, thought that the demise of wave packets also meant the end of the analogy between wave mechanics and wave optics. Certainly the analogy cannot be extended to electrons in atoms, but it is still valid for electron trajectories in space.

This exchange of letters with Lorentz was very helpful to Schrödinger, since it provided a profound yet sympathetic criticism of some of his physical ideas. It is surprising that it did not lead him to examine more deeply the classical roots of his own interpretations. Lorentz belonged to an older generation of physicists, and Schrödinger might have drawn from their discussions the conclusion that his new discoveries cannot be fitted into a classical framework at all. From June, 1926, his original conviction of the primacy of wave motion as the source of physical reality began to waver.

Part of his difficulty was the conflict between his philosophical beliefs and his scientific methodology. A persuasive case can be made that at this time Erwin was a disciple of Mach in his epistemology, of Vedanta in his ontology, but a follower of Boltzmann in his scientific methodology.[32] Thus he demanded that a scientific theory provide pictures of reality in space and time, but he did not believe that the pictures portrayed a real world. He had a much more complex and subtle view of the world than the simplistic positivism of Göttingen and Copenhagen, but in 1925–26 he was creating new science at such an unbelievable pace that he had no time to consider the philosophical implications of his own creations. This is not to say that he was unaware of them, but for the time being at least, they would have to remain latent.

The fourth paper

The fourth and final paper (Q4) of the marvelous series was received by *Annalen* on June 23.[33] So far Schrödinger had treated the wave mechanics of stationary systems only. Now he turned to problems in which the system is changing with time. These include scattering problems, in particular the scattering of radiation by atoms and molecules, and also the absorption and emission of radiation. In principle also, the general theory comprises all of chemical kinetics, although in practice the calculations are feasible only for simple gas reactions.

Q4 was the culmination of six months of research activity that is without an equal in the history of science, both in the intensity of its creativity and in the importance of its results for subsequent progress in physics and chemistry. The results were achieved by one man, working alone, except for occasional discussions with Hermann Weyl about mathematical questions, and with Erwin Fues as a young audience for the new ideas and occasional helper with numerical calculations.

Up till now Schrödinger had been convinced that ψ must be a real function. He was troubled by the $\sqrt{-1}$ factors that occurred in the theory, but he thought that they could be avoided by simply taking the real part of complex terms, so that the use of $i = \sqrt{-1}$ would be merely a convenient device for calculations, such as that used in electrical-circuit analysis. Now he saw that a wavefunction periodic in time, represented by $\Psi = \psi \exp(-2\pi i Et/h)$, implies an operator for the energy $E = -(h/2\pi i)\,\partial/\partial t$, and hence the time-dependent wave equation is

$$\nabla^2 \Psi - (8\pi^2/h^2)\,V\Psi = (4\pi i/h)\,\partial\Psi/\partial t$$

In terms of the Hamiltonian operator H, and writing $h/2\pi = \hbar$,

$$i\hbar\dot{\Psi} = H\Psi$$

This is the equation that appears on the first-day postmark of the Austrian stamp commemorating Schrödinger's 100th anniversary.

Schrödinger arrived at the conviction that Ψ is complex sometime between June 11 and June 21.[34] Heisenberg had introduced $\sqrt{-1}$ in his commutation relation, $pq - qp = -ih/2\pi$, but without any interpretation. From Schrödinger's Q4, it can be seen that the wavefunction is a product of an amplitude factor and an imaginary phase factor. As Dirac said later, the phase quantity was very well hidden in nature and it is because it was so well hidden that people had not thought of quantum mechanics much earlier. The real genius of Heisenberg and Schrödinger was to discover it. It is the source of all interference phenomena, but, even in 1972, Dirac thought that 'its physical significance is obscure'.[35]

The general solution of the wave equation for time-dependent potentials is usually very difficult, but fortunately Schrödinger was able to attack the problem of radiation interaction by applying his perturbation theory to the known solution of the stationary equation, i.e., by considering the time-dependent perturbation $V(t)$ as small compared to the V_0 of the unperturbed system. The theory that he gave can be called 'semiclassical', since the atom is treated as a wave-mechanical system whereas the radiation field is treated classically. [In 1927 Paul Dirac introduced the method of 'second quantization' in which the radiation field is also treated by quantum-mechanical theory, similar to a many-particle Schrödinger equation. This theory effectively answered the question that had been so troubling for Lorentz and Schrödinger, how absorption and emission of radiation was restricted to energy quanta exactly equal to differences between two energy eigenvalues of the system.]

The electric field of the incident radiation ($F \cos 2\pi vt$) induces an oscillating moment in the irradiated atom, which is calculated by perturbation theory as

$$M_y(t) = 2F \cos 2\pi vt \sum_{n=1}^{\infty} \frac{(E_n - E_k)a_{kn}b_{kn}}{(E_k - E_n)^2 - (hv)^2}$$

This gives the intensity of scattered radiation when $(E_k - E_n) \neq hv$. It is a *dispersion formula* which relates the polarization of an atom by an oscillating electric field to the sum of spectral transition probabilities for all the possible atomic spectral lines. Schrödinger identified $a_{kn} = \int M_z u_k u_n d\tau$ and $b_{kn} = \int M_y u_k u_n d\tau$, where $M_z = \Sigma e_i z_i$ and

$M_y = \Sigma e_i y_i$ are the dipole-moment operators as functions of the configuration of the system of point charges, and the integrations over $d\tau$ refer to all the configuration coordinates.

He also made what he called a 'heuristic hypothesis' that for a single electron $\psi\bar\psi$ represents the density of electric charge as a function of space coordinates, like the hypothesis that led to the successful treatment of the Stark effect in the hydrogen atom (Q3). In case there is more than one electron, he would extend the hypothesis as outlined in his letter to Lorentz.

His final scattering formula is similar to that obtained by Kramers from a correspondence-theory argument, work which in turn harked back to the original idea of Slater that an atom provides a set of virtual oscillators of all frequencies which interact with the radiation field, an idea that would be preserved by Dirac in his quantum field theory.

Schrödinger also discussed briefly the 'two-state' (second order) processes in which scattering may occur with a change in frequency, $[\nu \pm (E_k - E_n)/h]$. Such processes had been previously predicted by Kramers and Heisenberg (1925) and by Smekal (1923), and were experimentally observed by Raman in 1928.

Only a brief nonquantitative discussion of the resonance case, $\nu = \nu_{nm}$ was given, with the explanation: 'I do not wish to attempt to work over the calculation of the resonance case fully here, because the result would be of little value so long as the reaction of the emitted radiation on the emitting system is not taken into account.' This 'back coupling' requires the addition of 'damping terms' in the denominators of expressions such as that for $M_y(t)$ to prevent their going to infinity as the resonance condition is approached.

In the final section of the paper, he returned to a discussion of the physical significance of the wavefunction ψ, which he called the 'field scalar'. His discussion is modest and tentative; although he was later to be accused of pushing forward an untenable interpretation, such an accusation does not seem to be borne out by his actual words.

$\psi\bar\psi$ is a sort of *weight function* in the configuration space of the system. The *wave mechanical* configuration of the system is a *superposition* of many, strictly speaking *all*, the kinematically possible point-mechanical configurations. Thereby every point-mechanical configuration contributes to the true wave mechanical configuration with a certain weight, which is given precisely by $\psi\bar\psi$.

If one likes paradoxes, one can say that the system is found simultaneously in all conceivable kinematic locations but not in all of them 'in equal strength' . . . That the $\psi\bar\psi$ itself in general cannot be given a direct interpretation in three-dimensional space, as in the one-electron problem, because it is a function in configuration space, not in real space, has been stated repeatedly.

This statement is remarkably similar to the statistical interpretation of the wavefunction as given by Max Born, who submitted his first paper on this subject just a few days after Schrödinger submitted Q4.[36] This similarity makes all the more surprising Schrödinger's rejection of the statistical interpretation as given by Born.[37]

In the last paragraph of Q4, he stated 'that a certain difficulty no doubt still lies in the use of a *complex* wavefunction. If it was fundamentally unavoidable, and not a mere convenience for calculation, it would mean that actually *two* wavefunctions exist, which only together give conclusions about the state of the system.' Apparently, he did not yet fully appreciate that the complex function contains the hidden phase information essential for a complete specification of the ψ waves, but the light of this truly remarkable fact would soon dawn upon him.

The physicists

The last of the six major papers was sent to the journal on June 21, and with it Schrödinger's greatest scientific achievements were concluded. Such is the natural history of theoretical physicists. Schrödinger was unusual only in the relatively advanced age at which his great work was achieved. Heisenberg reached his peak at 27, Dirac at 28, Bohr at 29, Pauli at 31, and Einstein at 36. It should not be thought, however, that these scientists no longer made important contributions to their subject. After the research breakthroughs came the time for teaching, writing, philosophical discussions, and the inspiration of the next generation.

No sooner had Schrödinger laid down his pen at the completion of Q4 than he had an opportunity to present his ideas to an international audience. Debye and Scherrer had arranged a 'Magnetic Week' in Zürich from June 21 to 26, with physicists coming from all over the world, notably Sommerfeld, Pauli, Langevin, Pierre Weiss and Otto Stern. Besides the scientific meetings, there was a delightful excursion on the lake, with opportunities for shipboard confidences enlivened by cold white wine. Charles Mendenhall from Wisconsin was there and he invited Schrödinger to visit Madison early in 1927 for a series of lectures.

The general reaction to wave mechanics was enthusiastic, but the younger physicists especially were sceptical about any attempts to restore classical concepts of the continuum on the basis of the new theory. During the past year Schrödinger had been concentrating so intensely on new results that he scarcely had time to think deeply about all the implications of his wavefunction. Thus his interpretations relied more upon his background as a Vienna physicist and his general philosophical viewpoint. His first idea was that the ψ-function represents a real disturbance, a matter wave in space (or even perhaps in some novel kind of aether), analogous to the electromagnetic waves of Maxwell theory. The 'vacuum' was becoming an extraordinarily prolific and responsive medium and physicists were no longer reluctant to endow it with new properties. Schrödinger soon abandoned these matter waves, however, because he saw no way to make them interact with electromagnetic waves in processes of emission and absorption of radiation. He thus turned to an 'electromagnetic interpretation' in which $\psi\bar{\psi}$ is a measure of the density of electric charge. For the hydrogen atom, for example, this picture implies that the electron is not a point particle but a smeared-out distribution of negative charge. When the atom is subjected to an oscillating electromagnetic field (as in a light ray) the charge distributions corresponding to all its eigenfunctions are simultaneously excited to different extents. He was not able to treat quantitatively the 'resonance case' in which light absorption occurs at a particular frequency ν_{nm}, but he suggested that these frequencies would be combinations of the eigenfrequencies of the atom. The process of light absorption or emission would be smooth and continuous without the instantaneous quantum jumps of the Bohr theory.

Although there were no personal hard feelings between Schrödinger and Heisenberg, they were unsparing in their criticisms of each other's interpretation of quantum mechanics. In printed papers, these criticisms were diplomatic, but in personal letters their true feelings were not concealed. Schrödinger felt that the lack of any pictorial model in matrix mechanics made its application to new problems practically impossible. He wrote Lorentz that 'the frequency discrepancy in the Bohr model seems to be . . . something *monstrous* and I should like to characterize the excitation of light in such a way as really almost *inconceivable*.'[38] On June 8, Heisenberg wrote to Pauli: 'The more I think of the physical part of the Schrödinger theory, the more abominable I find it. What Schrödinger writes about *Anschaulichkeit* makes scarcely any sense, in other words I think it is bullshit [*Mist*]. The greatest result of his theory is the calculation of the matrix elements.'[39]

Pauli tended to side with Heisenberg and he once referred to the *Züricher Lokalaberglauben* [local Zürich superstitions]. Erwin was upset when he heard this phrase, but Pauli wrote a soothing explanation: 'Don't take it as a personal unfriendliness to you but look on the expression as my objective conviction that quantum phenomena naturally display aspects that cannot be expressed by the concepts of continuum physics (field physics). But don't think that this conviction makes life easy for me. I have already tormented myself because of it and will have to do so even more.'[40]

Erwin was pacified and wrote: '. . . we are all nice people, and are interested only in the facts and not in whether it finally comes out the way one's self or the other fellow supposed. If outsiders, all the same, find us capricious, we know that such capriciousness serves science better than uniformity.'[41]

Erwin had received several invitations to lecture in Germany, and on July 11 he traveled with Fues to Stuttgart, where he stayed with the Regeners for several days, before proceeding to Berlin, where he stayed with the Plancks. His lecture before the German Physical Society on July 16 was entitled 'Foundations of Atomism Based on the Theory of Waves', and on the 17th he gave a more specialized talk to the university physics colloquium, which was followed by a party that evening at the Planck house. The older generation of Berlin physicists, Einstein, Planck, Laue, and Nernst, were enthusiastic about both his mathematics and his semiclassical interpretations. Planck began to consider seriously his plans to bring Schrödinger to Berlin as his successor when he retired in the following year.

Erwin's next stop was in Jena, the university where he had been a lowly assistant just five years previously. His old friends there gave him an enthusiastic welcome mixed with admiration for his accomplishments. Nothing makes a scientist happier than to lecture as a great man at the scene of his apprenticeship.

On the 21st, in Munich, he talked at Sommerfeld's 'Wednesday Colloquium' and on the 23rd repeated his Berlin lecture for the Bavarian branch of the Physical Society. Heisenberg happened to be visiting Munich and attended this lecture.[42] In the question time, he asked how Schrödinger ever hoped to explain quantized processes such as the photoelectric effect and black-body radiation on the basis of his continuum model. Before the speaker had a chance to reply, Willy Wien angrily broke in and, as Heisenberg reported to Pauli, 'almost threw me out of the room', saying 'Young man, Professor Schrödinger will certainly take care of all these questions in due time. You must understand that we are now finished with all that nonsense about quantum jumps.' Schrödinger was less dogmatic in his answer, but he was confident that all the problems eventually could be cleared up by his methods. Sommerfeld, however, was not so sanguine. He wrote to Pauli a few days afterwards: 'My general impression is that "wave mechanics" is indeed a marvelous micromechanics, but that the basic quantum riddles have in no way been solved by it. I cease to believe Schrödinger the moment he starts to calculate with his c_k (the amplitudes of the simultaneously excited eigenvibrations).'[43] The old master of Munich was the wisest of his generation.

Heisenberg was quite upset by the Munich seminar and the fact that he could make no impression with his arguments. He wrote immedi-

The Junger twins, Withi (left) and Ithi, 1926.

ately to Bohr to report on the situation, and Bohr wrote to Schrödinger to invite him to Copenhagen for some serious discussions.

When Erwin returned to Zürich for a belated summer vacation, he had earned the professional and personal esteem of the mandarins of German physics. He found that Anny had arranged for him an unexpected but tempting diversion, twin nymphets awaiting his instruction in high-school mathematics.

The Salzburg twins[44]

The nonidentical twin sisters, Itha and Roswitha Junger, known to their friends as Ithi and Withi, were fourteen-year-old schoolgirls from Salzburg. Their grandfather Georg Junger (1831–1908) had been a famous citizen of that city, who founded in 1858 a firm of wholesale merchants in the *Altermarkt*, which brought considerable riches to him and his two sons Hans and Carl who continued the business. Anny Bertel's mother is said to have been the illegitimate daughter of Georg Junger. The two families remained close, and Hans Junger's wife, Josefina Kohler, was Anny's godmother. The twins were born in August, 1912.

During the academic year 1925–26, the twins had been in the fourth form at St Ursula's convent school near Salzburg, and unfortunately

Ithi failed her mathematics course. They were both exiled to the convent at Meinzigen, not far from Zürich, to try to make up the failed material. Here Anny met their mother, and is said to have suggested that since Erwin was competent in mathematics, he would give them some special tutoring. It was arranged that they would come once a week to the Schrödinger house in Zürich for a lesson, with most of the time devoted to Ithi whose need was greater. Erwin did not know what maths was prescribed for the fourth form, but he consulted Hermann Weyl, who outlined the material for the lessons. As Erwin recalled in a poem sent to Ithi in the next year,

> Auf Herrn Professor Schnitzer's Spuren
> Mit Algebra und Dreiecksfiguren
> Das Ithilein zu Tode pflegt –
> Das arme Kind war ganz verjägt.
> Manch' anderes wäre zu berichten
> Von Zürich – ich will darauf verzichten.*

* On Herr Professor Schnitzer's traces / With algebra and three cornered races / He ran Ithy-bitty almost to death – / The poor little kid was quite out of breath. / Of Zürich there is much more could be told / But about such things I won't be so bold.

The 'such things' included a fair amount of petting and cuddling, but despite these distractions, the tutorials were a success, Ithi caught up with her class and entered the fifth form at St Ursula's. The tutor, however, had fallen in love with his pupil, and he began a patient campaign to bring about her surrender. He told her about his great discovery: 'I did not write everything down at once, but kept changing here and there until finally I got the equation. When I got it, I knew I had the Nobel prize.' He talked to her about religion: 'I believe more in God the Father with the white beard than I believe in Nothing.'

During the Christmas holidays the following year, Erwin joined the twins and their mother at Kitzbühel. Ithi was a better skier than Erwin, and the excursion was not a complete success, since he sprained his ankle. Their rooms were icy cold, even the wash water froze. The stove smoked and a great spark burned a hole in a new dress. Spirits were restored by champagne for New Year's Eve.

Erwin sent Ithi a poetic birthday wish:

> Als man Euch die ersten Windeln gebreitet
> Hat an der Wiege die Schelle geläutet.
> Hat der Narrenkönig sein Zepter geschwenkt
> Und Euch Fröhlichkeit ins Leben geschenkt.*

* When they unfolded your diapers the very first time / The bells on your cradle did joyfully chime. / The king of the fools gave his scepter a shake / And bid you in life every happiness take.

This seems like an invitation to Freudian analysis, but as Ithi recalls, Erwin was very proud of his verses, although Stefan Zweig once told him 'I hope your physics is better than your poetry.'

Erwin did not try too seriously to get Ithi to bed until she was sixteen. Then, one time in Salzburg, he came into her room in the middle of the night and sat on her bed and told her how much he loved and needed her. He also promised that he would take all precautions to ensure that she did not become pregnant. All to no avail at that moment, but not long after her seventeenth birthday, they became lovers.

Against Born and Bohr

After returning from his German travels, Schrödinger concentrated on the problem of how to reconcile his interpretation of continuous electron waves with phenomena such as the photoelectric effect, which seemed to demand particulate electrons and discrete light quanta. On August 25, he wrote to Willy Wien to confess his failure.[45]

That the photoelectric effect . . . offers the greatest conceptual difficulty for the achievement of a classical theory is gradually becoming ever more evident to me. Unfortunately I can find so far . . . no solution at all to the problem, I mean I see no concrete idea or calculation that could bring one closer to understanding it. And to phantasize about it, as could perhaps be done, is in my view as easy as it is worthless . . .

I have the feeling – to express it quite generally – that we have not yet sufficiently understood the identity between *energy* and *frequency* in microscopic processes . . . What we call the energy of an individual electron is its frequency. Basically it does not move with a certain speed because it has received a certain 'shove', but because a dispersion law holds for the waves of which it consists, as a consequence of which a wave packet of this frequency has exactly this speed of propagation. What we call the energy content of a stream of electrons depends for a given frequency more upon the number of electrons and this determines how often those processes that are permitted by its frequency occur in the electron stream.

But today I no longer like to assume with Born that an individual process of this kind is 'absolutely random', i.e., completely undetermined. I no longer believe today that this conception (which I championed so enthusiastically four years ago) accomplishes much. From an offprint of Born's last work in the *Zeitsch. f. Phys.* I know more or less how he thinks of things: the *waves* must be strictly causally determined through field laws, the *wavefunctions* on the other hand have only the meaning of probabilities for the *actual* motions of light- or material-particles. I believe that Born thereby overlooks that . . . it would depend on the taste of the observer which he now wishes to regard as real, the particle or the guiding field. There is certainly no criterion for reality if one does not want to say: the *real* is only the complex of sense impressions, all the rest are only pictures.

Bohr's standpoint, that a space-time description is impossible, I reject *a limine*. Physics does not consist only of atomic research, science does not consist only of physics, and life does not consist only of science. The aim of atomic research is to fit our empirical knowledge concerning it into our other thinking. All of this other thinking, so far as it concerns the outer world, is active in space and time. If it cannot be fitted into space and time, then it fails in its whole aim and one does not know what purpose it really serves.

Reading the first of these last two remarkable paragraphs, one must ask why Schrödinger has abandoned the idea that the most fundamental laws of physics are statistical, so that pure chance and chaos lie at the heart of nature. Of course, Born's concept of 'probability' is not the same as that used by Maxwell and Boltzmann in the kinetic theory of gases, and it would soon be shown to have some strange properties indeed. Nevertheless, one might have expected Schrödinger to welcome the Born interpretation as a victory for the ideas of Franz Exner, especially since he had himself suggested that $\psi\bar{\psi}$ may be a sort of 'weighting function'. The reason for his apostasy from the doctrine of pure chance may have been his own discovery of wave mechanics, which appeared to offer a renewal of confidence in the kind of causal, determinate laws found in Maxwellian electrodynamics and Einsteinian gravitation. It was ironic that by choosing to be a conservative in physics after the revolution of 1926, Schrödinger joined the radical minority who dared to dissent from an orthodoxy known as the Copenhagen Interpretation.

Copenhagen

At the end of August, Erwin and Anny went for a well earned holiday, to Fontane Fredde, in the mountains of the Alto Adige in South Tirol. They stayed about three weeks in this region that they always loved. Erwin then stayed with the Wiens for a few days at their summer place at Mittenwald, while Anny visited her family in Salzburg. Erwin then traveled to Copenhagen to see Bohr, arriving there towards the end of September. On October 4, he presented a formal lecture on his theory.

The most detailed account of his visit is given by Heisenberg in his remarkable book *Der Teil und das Ganze* [*The Part and the Whole*].[46]

The discussion between Bohr and Schrödinger began at the railway station in Copenhagen and was carried on every day from early morning till late at night. Schrödinger lived at Bohr's house so that even external circumstances allowed scarcely any interruptions of the talks. And although Bohr as a rule was especially kind and considerate in relations with people, he appeared to me now like a relentless fanatic, who was not prepared to concede a single point to his interlocutor or to allow him the slightest lack of precision. It will scarcely be possible to reproduce how passionately the discussion was carried on from both sides.

SCHRÖDINGER: You surely must understand, Bohr, that the whole idea of quantum jumps necessarily leads to nonsense. It is claimed that the electron in a stationary state of an atom first revolves periodically in some sort of an orbit without radiating. There's no explanation given of why it should not radiate; according to Maxwell theory, it must radiate. Then the electron jumps from this orbit to another one and thereby radiates. Does this transition occur gradually or suddenly? If it occurs gradually, then the electron must gradually change its rotation frequency and its energy. It's not comprehensible how this can give sharp frequencies for spectral lines. If the transition occurs suddenly, in a jump so to speak, then indeed one can get from Einstein's formulation of light quanta the correct vibration frequency of the light, but then one must ask how the electron moves in the jump. Why doesn't it emit a continuous spectrum, as electromagnetic theory would require? And what laws determine its motion in the jump? Well, the whole idea of quantum jumps must simply be nonsense.

BOHR: Yes, in what you say, you are completely right. But that doesn't prove that there are no quantum jumps. It only proves that we can't visualize them, that means, that the pictorial concepts we use to describe the events of everyday life and the experiments of the old physics do not suffice to represent also the process of a quantum jump. That is not so surprising when one considers that the processes with which we are concerned here cannot be the subject of direct experience . . . and our concepts do not apply to them.

SCHRÖDINGER: I don't want to get into a philosophical discussion with you about the formation of concepts . . . but I should simply like to know what happens in an atom. It's all the same to me in what language you talk about it. If there are electrons in atoms, which are particles, as we have so far supposed, they must also move about in some way. At the moment, it's not important to me to describe this motion exactly; but it must at least be possible to bring out how they behave in a stationary state or in a transition from one state to another. But one sees from the mathematical formalism of wave or quantum mechanics that it gives no rational answer to these questions. As soon, however, as we are ready to change the picture, so as to say that there are no electrons as particles but rather electron waves or matter waves, everything looks different. We no longer wonder about the sharp frequencies. The radiation of light becomes as easy to understand as the emission of radio waves by an antenna, and the former unsolvable contradictions disappear.

BOHR: No, unfortunately that is not true. The contradictions do not disappear, they are simply shifted to another place . . . Think of the Planck radiation law. For the derivation of this law, it is essential that the energy of the atom have discrete values and change discontinuously . . . You can't seriously wish to question the entire foundations of quantum theory.

SCHRÖDINGER: Naturally I do not maintain that all these relations are already completely understood . . . but I think the application of thermodynamics to the theory of matter waves may eventually lead to a good explanation of Planck's formula . . .

BOHR: No, one cannot hope for that. For we have known for 25 years what the

Planck formula means. And also we see the discontinuities, the jumps, quite directly in atomic phenomena, perhaps on a scintillation screen or in a cloud chamber . . . You can't simply wave away these discontinuous phenomena as though they didn't exist.

SCHRÖDINGER: If we are still going to have to put up with these damn quantum jumps, I am sorry that I ever had anything to do with quantum theory.

BOHR: But the rest of us are very thankful for it – that you have – and your wave mechanics in its mathematical clarity and simplicity is a gigantic progress over the previous form of quantum mechanics.

The discussion went on in this way day and night, without reaching any agreement. After a few days, Erwin became ill with a feverish cold, perhaps as a result of the enormous strain. Mrs Bohr took care of him and brought tea and cake to his bedside. But Niels sat on the edge of the bed and continued the argument, 'But surely Schrödinger, you must see . . .' But Erwin did not see, and indeed never did see, why it was necessary or how it was possible to destroy the space-time description of atomic processes. The conversations, however, deeply affected both men. Schrödinger recognized the necessity of admitting both waves and particles, but he never devised a comprehensive interpretation of quantum phenomena to rival the Copenhagen ortho-doxy. He was content to remain a critical unbeliever. Heisenberg began the analysis that soon led to his principle of indeterminacy, and with this as a foundation, Bohr ventured more deeply into philosophical waters and emerged with his concept of complementarity.

After his return to Zürich, Schrödinger reported in a letter to Wien:[47]

Bohr's . . . approach to atomic problems . . . is really remarkable. He is completely convinced that any understanding in the usual sense of the word is impossible. Therefore the conversation is almost immediately driven into philosophical questions, and soon you no longer know whether you really take the position he is attacking, or whether you really must attack the position that he is defending.

A basic reason for the failure of Bohr and Schrödinger to communicate more effectively was that their minds belonged to two different categories. In the analysis of human personalities devised by Francis Galton, Schrödinger was a 'visualizer' and Bohr was a 'nonvisualizer', one thought in terms of images and the other in terms of abstractions, and it is virtually impossible for such twain to agree in any kind of discussion.[48]

Nevertheless, Erwin was deeply impressed by the spirit of Niels Bohr, and wrote to him a rather ornate letter of appreciation.[49]

The lovely, sunny, hospitable home with its kindly people, which received a stranger like me as an old friend, surrounding me with every comfort, was an experience that the heart can never forget. But also this city, this home, this family – they are those of the great Niels Bohr; he is the one I have to thank for all this kindness through which I could speak with him for hours about those things that are so close to my heart, and hear in his own words about the positions he takes toward the many attempts to build one stage further upon the broad substantial foundation that he has given to modern physics. That was for a physicist . . . a truly unforgettable experience.

Schrödinger then said that despite the firmness with which he had advanced his own views, the psychological effect of the arguments of Bohr had been serious indeed and he was now more aware of the unresolved problems in his attempts to provide a continuous picture of the interaction of radiation with atoms, but he still believed this process could be *visualized* in space and time. Bohr[50] did not choose to enter a detailed debate by correspondence, but he called Schrödinger's attention to recent work in Copenhagen by Oskar Klein:[51] 'If you are not able to kill off the ghosts completely in ordinary space and time, perhaps you can achieve a settlement in five dimensions.'

7 Berlin

Schrödinger had accepted an invitation from the University of Wisconsin to give a course of lectures early in 1927; they would pay $2500, which included an allowance for travel costs.[1] He sent a paper to *Physical Review* which was published in the December, 1926, issue, so that the American physicists would have an excellent summary of his work on wave mechanics, and he would not need to review all the fundamentals when he lectured at American universities. Shortly before he left for America, he heard that he was a leading candidate for the succession to Max Planck at the University of Berlin.

American voyage

On December 16, Erwin and Anny attended a Christmas and Farewell Party held in the Physics Institute of the University. Edgar Meyer was master of ceremonies, and he recited one of the long doggerel poems for which he was famous. It included the verses:[2]

> Schon Galileo hat es uns gezeigt
> Das jeder Körper in Ruhe bleibt
> Zwingt ihn nicht eine auss're Kraft
> Zu ändern die Bewegungseigenschaft.

> Und so auch hier; denn glaubt Ihr lieben Leute
> Wir könnten Abschied feiern heute?
> Hatt' nicht die Anny zart getrieben,
> Der Erwin war zu Haus geblieben.*

* Galileo showed us long ago / That every body stays just so / Unless it's pushed by some outer force / To change its predetermined course.

  And so dear friends are you impressed / That we gather here for a farewell fest? / For if Anny had not so tenderly nudged / Erwin from home would never have budged.

Since Edgar Meyer always had some basis for his verses, it is likely that considerable urging was necessary to get Erwin to leave Zürich and thereby sacrifice his Christmas vacation.

They set out on December 18 for the new world, by train to Basel and Paris, and then the boat train to Le Havre where they boarded the French liner *de Grasse* for a ten-day voyage to New York. Erwin grumbled from the beginning. He found his fellow passengers distinctly unattractive 'examples of the modern "society" that I usually manage to keep at arm's distance.' At dinner he was seated between two painted and powdered ladies 'beyond the canonical age' and he found the 'hard, ruthless expressions' of their consorts equally repulsive. Their French manners, for him, made the bad company even worse. His disposition was not helped by the fact that he was cooped up in a small cabin with Anny, who was frightfully seasick throughout the voyage.[3]

The sight of the Statue of Liberty did nothing to alleviate his spleen, and he thought this monument to be 'grotesque, between the comic and the ghastly', and needing only a giant wristwatch on the upraised arm to complete the picture. They were met at the dock by Karl Herzfeld, who had come especially from Baltimore to welcome them to New York. Unfortunately, they had been booked into a poor hotel, the Martinique on Broadway, at $9 a night. Erwin was so distressed by the noise and dirt of New York that he threatened to take the next ship home, but Anny persuaded him to stay. He was then so frustrated by his attempt to arrange train tickets to Madison, Wisconsin, that he became nearly hysterical until Anny took over, but the comfortable drawing room and the attentions of the cheerful black porter began to restore his equilibrium. He had spent only twenty-six hours in New York, but he found the experience of the city 'shattering'. Chicago was even worse as there was the added fear of being shot by 'bandits who spring with loaded guns from speeding autos'. Anny was favorably impressed by ubiquitous drinking fountains, built-in wardrobes, and large railway stations.[4]

Madison was a great relief. Here at last was 'a real city' in the European sense. They received a warm welcome from the physicists and Sunday night attended a reception and buffet dinner for eighty persons at the Mendenhall home. Erwin was particularly pleased to meet young John van Vleck; he found that the depth of his ideas surpassed his ability to express them, 'just the opposite of the situation with most women'.

When Erwin decided to buy Anny a fur coat, he was amazed by the informality of the salesman, who treated him like an old friend. Unfortunately, they did not find a fit for Anny, and the salesman apologized with 'Well you see you are a pretty big girl.'

At Madison, Schrödinger's lectures were, as usual, an intellectual banquet, and he was offered a permanent professorship. He was not at all tempted by an American position, and he declined on the basis of a possible commitment to Berlin. While in the Midwest, he lectured also at the Universities of Chicago, Iowa, and Minnesota. On February 14, the Schrödingers set out for Pasadena on the California Limited. They saw the Grand Canyon and visited some Indian reservations, where Anny was dismayed by the poverty and ill health of the Indians, while Erwin was fascinated by their ceramics and textiles.

Anny found Pasadena to be 'unbelievably beautiful, like a great garden'. Even Erwin felt cheerful. He wished only that California was inhabited by Italians and Spaniards, instead of Americans. On their first evening, there was a party at the Millikans, and everyone was so kind and welcoming that Erwin had to admit that Americans were considerate of others to a degree quite unknown among Germans, especially northern ones.

The grand old man of theoretical physics, Hendrik Lorentz, was also in Pasadena at this time. He lectured at 3:45 p.m., after which there was a cup of tea, and Schrödinger followed at 5 p.m. Lorentz was not happy about the present state of physical theory. Once he said, 'I have lost the conviction that my work has led to an objective truth, and I don't know why I have lived. I only wish I had died five years ago, when everything seemed clear to me.'[5] The two Nobel prize winners, Lorentz and Millikan, sat in the front row. The California Institute of Technology was then one of the major centers of American physics, and there was a knowledgeable audience of about sixty, including all the professors of physics and physical chemistry.

Among the professors was Paul Epstein from Zürich, whom Erwin called 'one of the kindest, most unassuming, most worthy companions I have ever known'. Paul took them to visit the beach at Santa Monica, the movie studios in Hollywood, and later on a drive up the old Mount Wilson Road. Erwin was terrified by the precipitous roadsides, but Anny enjoyed the climb, and the famous view of the lights of Pasadena and Los Angeles from the summit.

They were both sorry to leave Pasadena, and Erwin wished they had stayed six weeks in warm California and only two in the cold midwest. Their return trip took them through Salt Lake City, where he listened with interest to the story of the Mormons and their devotion to polygamy. There was a stop for a lecture at Ann Arbor and a visit with Otto Laporte, a former Sommerfeld student. They lived there in the University Club, where women were admitted only by the back door. A week in Boston included three lectures at M.I.T. and three at Harvard, and a pleasant party given by the Bridgmans at a country inn.

Thence to Baltimore where Schrödinger was happy to meet Robert Wood, the great spectroscopist. He was much impressed by Wood's researches on fluorescence, and in October, 1927, nominated him for a

Nobel prize in physics (which, however, he never received). They saw Washington bedecked in cherry blossoms, and stayed one night in the villa of the millionaire Edward Loomis, a great friend and benefactor of physics at Johns Hopkins University. Through the efforts of Herzfeld, Schrödinger was offered a permanent professorship at Hopkins at the excellent salary of $10 000 a year, but he did not consider it seriously as an alternative to the Berlin chair.[6]

An important factor in Erwin's intense dislike of the American way of life was 'the great experiment'. An occasional glass of good beer or bottle of fine wine would doubtless have made everything seem more bearable. With a plate of succulent Chesapeake Bay oysters he was offered a choice between sweet ginger ale and chlorinated icewater. 'To the devil with prohibition', he exclaimed.

After a final lecture at Columbia, they embarked for home on the S.S. *Hamburg*. It had been a strenuous journey, including over fifty lectures in three months. They arrived back in Zürich on April 10, and almost immediately were faced with a decision about a move to Berlin.

The Berlin chair

The foundation of the University of Berlin was due to the efforts of two distinguished native sons, Wilhelm Humboldt, lawyer and diplomat, and his younger brother Alexander, the famous naturalist and explorer. They were both graduates of Göttingen, at that time the most eminent Prussian university. Alexander was active in the early planning but it was Wilhelm who drew up the final proposal to the king in July, 1809, and on the August 16 Friedrich Wilhelm III promulgated the order that established the university, which was duly named after him. The king gave the university a substantial endowment and also provided as a home the vacant palace of the late Prince Heinrich located in the heart of the city on the broad avenue Unter den Linden. This impressive building was to prove something of a tribulation for the growing university – even after extensive remodeling it was never well suited for academic purposes. In 1817, Friedrich Hegel from Heidelberg became the foundation professor of philosophy, and he declared that 'the German nation has the calling to care for and to propagate the sacred fire of philosophy'. After World War I, the university cultivated a nostalgia for the past. Its war memorial bore an inscription devised by a professor of theology, '*Invictis Victi Victuri* – To the Unconquered the Conquered who will Conquer'.[7]

In May, 1926, the Prussian Minister for Science, Art and Education had asked the Philosophical Faculty of the Friedrich Wilhelms Universität to consider the future of the chair of theoretical physics, which would become vacant with the statutory retirement of Max Planck.

The faculty appointed a committee to consider the matter, and after careful deliberations its report was sent to the minister by Dean Petersen early in December.[8]

The committee first considered the two distinguished theoretical physicsts already holding Berlin professorships, Laue and Einstein. They could see no reason to appoint Laue, since this would simply change one vacancy for another, and Einstein did not wish to accept a position that would require lecturing. It was thus necessary to 'cast a glance outside'. Their first choice would be Arnold Sommerfeld, holder of the chair at Munich. He was fifty-nine years old, probably the most proficient mathematician of all the physicists; he was not only a leader in research but also the best teacher of advanced students, graduates from his department filling many of the most important positions in the world of theoretical physics.

The other leading theoreticians, listed in order of seniority, were Max Born, Pieter Debye, Erwin Schrödinger, and Werner Heisenberg. Debye had directed an institute for experimental physics in Göttingen and he had a broad interest in both theory and experiment, but was not primarily concerned with mathematical analysis of problems at the frontiers of pure physics. Thus they thought he would be less suitable for the Planck chair and eliminated his name from the list. They recognized the genius of Heisenberg, 'at present the most outstanding theoretical physicist in Germany', but believed that a call to the Berlin chair at the age of twenty-three would not be in his own best interests.

Thus Born and Schrödinger remained to be considered. They found it difficult to rank these two, but Born had directed an Experimental Institute in Göttingen, and such an experience was not entirely in his favor. Regarding Schrödinger they wrote:

For some years already he has been favorably known through his versatile, vigorously powerful, and at the same time very profound style in seeking new physical problems that interested him and illuminating them through deep and original ideas, with the entire set of techniques which mathematical and physical methods presently provide. He has proven this method of working to be effective in the treatment of problems in statistical mechanics, analysis of optical interference, and physiological color theory. Recently he has succeeded in an especially daring design through his ingenious idea for the solution of the former particle mechanics by means of a wave mechanics in the differential equation he has set up for the wave function . . . Schrödinger himself has already been able to deduce many consequences from this fortunate discovery and the new ideas that he has inspired with it in many fields are even more numerous . . . It may be added that in lecturing as in discussions Schrödinger has a superb style, marked by simplicity and precision, the impressiveness of which is further emphasized by the charming temperament of a South German.

Schrödinger in Zürich days (c. 1927).

Thus the final recommendation to the ministry was (1) Sommerfeld, (2) Schrödinger, (3) Born. When Sommerfeld declined to leave Munich, Schrödinger became the leading candidate for the Planck chair.

From Zürich to Berlin

Soon after the beginning of the summer term in Zürich, Schrödinger received a formal invitation to accept the chair in Berlin. Despite the prestige of the position, it was not easy for him to leave Switzerland, for he had grown to love the stable and peaceful life and the nearness to the high mountains, and he was aware of the vicious forces lurking behind the democratic facade of Prussian society. As soon as he informed the dean of the Berlin offer, the University took up the matter with the Education Board and every effort was made to keep him in Zürich. The Berlin salary was much higher than the maximum in a Swiss university; as Anny put it, 'We got more and more and more – it was really very satisfactory.' The basic annual salary was RM 13 702 plus a cost-of-living allowance of RM 3000 and a guarantee of at least RM 10 000 for lecturing fees, making a total of RM 26 702 ($6676). The Swiss were able to arrange a joint professorship at the E.T.H., so that the combined salaries came close to the Berlin level, but the two professorships would have required almost twice the amount of teaching.

It was usual in Switzerland to publish the news of outside offers to professors, so that the students became aware of the possibility that Schrödinger might be leaving. The relatively small group of his students in physics organized a larger group of their friends into a torchlight parade that marched through the streets around the university to the courtyard of the Schrödinger house. Edgar Meyer paid for the torches. Such parades were traditional student demonstrations for favorite teachers, but they were quite rare, and Erwin was deeply moved. He accepted the petition offered by the leader of the students, and made a short speech expressing his love for Zürich and appreciation of their support.

He finally decided to accept the Berlin offer. An important reason for his decision was the personal persuasion of Max Planck. After he moved to Berlin, he wrote a heartfelt explanation in the form of a poem in Planck's guest book.[9]

> Die Ehre rief! – Doch was ist Ehre Dir,
> Des Gaukelspieles spöttischer Verächter!
> Ansehen – für drei Jahrzehnte oder vier!
> Kann die Physik der kommenden Geschlechter
> Mit Ruhm und Ehre Dir das Leid bezahlen,

Das, lebend eingesargt in Sand und Stein,
Dich bald Dein tolles Wagnis lässt bereuen,
Mit Ruhm und Ehre Deiner Sehnsucht qualen!

So dacht ich oft, und wenn ich dennoch kam,
War's nicht um Ruhm, das will ich wohl beschwören.
Und auch – ich sag es nur mit leiser Scham –
Auch Gold allein nicht könnte mich betören.
Den Ausschlag gab ein Wort – aus langen Reihen
Von briefen, von Gesprächen, bunt und kraus,
Verehrungswürdige Lippen sprachen's aus,
Nicht drängend zwar. Ganz kurz: Mich tät es freuen.*

*Honor called! – Yet what care you for honor, / Cynic despiser of the illusionist! / Prestige – for three decades or four! / Can physics in generations yet to come / With fame and honor pay you for the sadness, / Living entombed in sand and stone, / That makes you soon regret your senseless venture, / With fame and honor satisfy your longing!

So thought I often, and if for all that came, / 'Twas not for fame, which I will quite forswear, / And also – I say it only with faint shame – / Nor gold alone could now persuade me. / The scales were tipped with just one word – / From all the messages, various and intricate, / Respected lips expressed it thus, / Not pressing. Quite brief: 'It would make me happy.'

Before they moved to Berlin, Erwin remarked to Anny, 'You know, I have the feeling that we shall not be staying there for so very long.' This was an uncanny prediction, for at the time the Berlin chair was the apex of European physics, but he may have been thinking merely of an ultimate return to Vienna. They rented the spacious first floor of a substantial house in Grunewald (44 Cunostrasse), a western suburb adjacent to Dahlem where the Plancks lived. It was about twenty minutes by subway from the university.

The move to Berlin was completed in time for Erwin's fortieth birthday, August 12, 1927. He wrote for himself an 'Epitaph if I should die today':[10]

Er hatte mit seinen 40 Jahren
Von Leben weniger erfahren
Als manche von den Jüngern um ihn her.
Und dennoch wusste er erheblich mehr davon,
Wie dieses Weltgetrieb im Innersten zusammenblieb
Als er zu sagen sich erkühnte,
Obwohl er nicht den Namen 'Prud' verdiente.
Sein Wissen starb zum Glück mit ihm,
Jetzt teilt er's mit den Cherubim.
Ob denen Neues brachte sein Bericht,
Das wusste er zu Lebzeit selbst noch nicht.*

* In the 40 years that his course had run / He'd experienced less than many had done / Among the apostles around him here. / And yet he knew a great deal more / Of the ways the world is held together / To which he had never dared allude / Though he hardly deserved to be called a 'prude'. / It's lucky this knowledge died with him, / He can share it now with the Cherubim. / But whether his news would be a surprise / While still on earth he could never surmise.

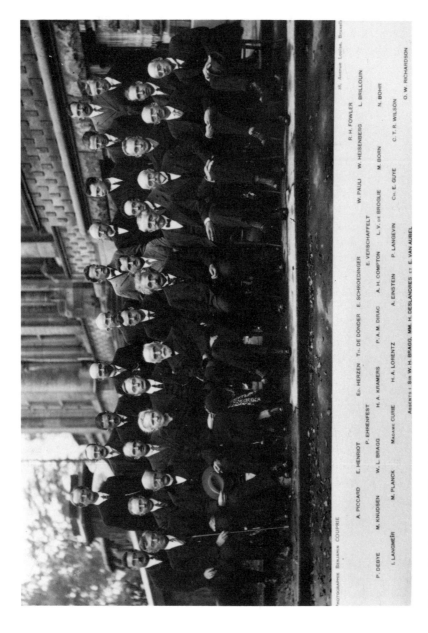

PHOTOGRAPHIE BENJAMIN COUPRIE   26, Avenue Louise, Bruxelles

R. H. FOWLER

A. PICCARD   E. HENRIOT   Ed. HERZEN   Th. DE DONDER   E. SCHROEDINGER   W. PAULI   W. HEISENBERG   L. BRILLOUIN

P. EHRENFEST   E. VERSCHAFFELT

P. DEBYE   M. KNUDSEN   W. L. BRAGG   H. A. KRAMERS   P. A. M. DIRAC   A. H. COMPTON   L. V. de BROGLIE   M. BORN   N. BOHR

I. LANGMUÏR   M. PLANCK   Madame CURIE   H. A. LORENTZ   A. EINSTEIN   P. LANGEVIN   Ch. E. GUYE   C. T. R. WILSON   O. W. RICHARDSON

ABSENTS : SIR W. H. BRAGG, MM. H. DESLANDRES et E. VAN AUBEL

Solvay Conference of Physics, 1927.

Before beginning his duties in Berlin, Schrödinger traveled to Brussels to participate in what would be the most historic of the Solvay conferences on physics.

Fifth Solvay Conference

The Fifth Solvay Conference met in Brussels from October 24 to 29, 1927, under the presidency of Hendrik Lorentz.[11] This was the last public appearance of this revered elder statesman of physics, for he died a few months later. Unlike his situation at the previous conference, Schrödinger was now recognized as a world leader in theoretical physics and he had been invited to give one of the 'thousand-dollar lectures'. Assured of his professional eminence, he felt more free to display his disdain for formal attire, and he shocked Paul Dirac by arriving at their luxurious hotel in one of his Tirolean costumes with a knapsack over his back; in the conference photograph, his jaunty sports jacket and bow-tie are conspicuous among the severe dark suits and high collars of his confreres.

The topic for the conference was 'Electrons and Photons' and it was intended to provide a searching discussion of all the implications of the revolution in physics associated with the recent great advances in quantum theory.

Schrödinger's lecture was on wave mechanics (*La Mécanique des Ondes*). It provides an excellent insight into his views at this time, after he had been subjected to all the subtle arguments of Bohr, the recently discovered uncertainty principle of Heisenberg, and the widely accepted statistical interpretation of Born. He began by pointing out that there were two different kinds of wave mechanics, the four-dimensional relativistic mechanics of Louis de Broglie and his own multidimensional nonrelativisitic mechanics. He refrained from mentioning that the de Broglie mechanics had not led to any practical results, and said he wished to consider the difficulties of the polydimensional conception, 'plus belle en elle-même'.

After an elegant deduction of his wave equation by a variation method, in which the equation appears as the Euler condition, he explained that if one wishes to deal with spectroscopic transitions, it is necessary to use the time-dependent wave equation,

$$\Delta\Psi - \frac{8\pi^2}{h^2} V\Psi - \frac{4\pi i}{h} \partial\Psi/\partial t = 0$$

The solution can be written

$$\Psi = \sum_{k=1}^{\infty} c_k \, \psi_k \, \exp(2\pi i \, \nu_k t).$$

The stationary states of Bohr then correspond to characteristic vibrations of the system.

To this point, Schrödinger's analysis appears to be incontrovertible. Now he asked, 'in reality what is the aspect in a space of three dimensions of the system described by this function Ψ?' He said that he will leave the defense of the probability interpretation to Born and Heisenberg, who were giving a joint paper at this meeting. His viewpoint, 'perhaps a little naive' was that:

the classical system of material points does not really exist, but something exists that fills all of space continuously, from which one may obtain an instantaneous 'photograph' . . . if one makes the classical system pass through *all* its configurations, letting the image in space sojourn in each element of volume $d\tau$ for a time proportional to the instantaneous value of $\Psi\overline{\Psi}$. In other words, the real system is a composite image of the classical system in all its possible states, obtained by using $\Psi\overline{\Psi}$ as a weighting function.

This probability interpretation, which he had originally put forward in Q4, differs from Born's probability interpretation in that it pictures a continuous cloud of 'something' (presumably charge and mass) whereas the latter assumes point particles (electrons) within the atom. An argument against the Schrödinger picture is that the wave equation includes a potential-energy term for electron–electron interactions, so that the continuous charge cloud must have contributions assignable to individual electrons. He mentioned this difficulty when he said that 'the pure field theory does not suffice . . . it must be completed by a kind of *individualization* of the electric densities coming from the various point charges of the classical theory, but . . . each individual can be distributed through all space in such a way that the individuals interpenetrate.' It is thus not so easy to dispose of the particle picture. Surprisingly, in his zeal to obtain a three-dimensional picture from his multidimensional Ψ, Schrödinger did not mention the most powerful property of Ψ, its ability to treat correctly problems in atomic and molecular structure that involve more than one electron. In the final part of his lecture, he outlined the general theory of the several-electron problem, but he did not apply it to any special cases. As usual, his interest was in the basic theory and not in computational applications.

Schrödinger's paper aroused considerable debate, and Born and Heisenberg attacked it quite vehemently. The latter remarked 'I see nothing in the calculations of M. Schrödinger that justifies his hope that it will be possible to explain or understand in three dimensions the results from polydimensions.' Schrödinger admitted that his analysis was not yet satisfactory but he thought the Hartree-type calculations showed how a three-dimensional picture could be approached. He said it was nonsense to talk about the trajectory of an electron inside an atom, but Born said 'It is not nonsense', and there this matter rested.

Einstein took no part in this particular discussion, but he did present an extended critique of the Copenhagen–Göttingen interpretation of quantum mechanics, and his debate with Bohr continued inside and outside the conference rooms. It provided the greatest excitement of the meeting and was an historic occasion, a battle of titans over the epistemological foundations of physics and over the way in which scientists should understand their world.[12] Einstein saw immediately that the Born–Heisenberg–Bohr interpretation would ultimately conflict with special relativity by requiring that the reduction of the wave packet at one point must instantaneously bring about effects at an arbitrarily distant point. When the meeting ended, however, most of the physicists departed with the belief that the positivist Copenhagen view had prevailed, a belief nourished by the anti-realist philosophical tradition of central Europe. But Einstein, de Broglie and Schrödinger were not convinced, and they left Brussels with at least a tacit resolve to fight another day.

University of Berlin

Schrödinger's appointment as Ordinarius in Physics and Director of the Institute for Theoretical Physics dated from October 1, 1927, and the lectures for the winter semester of 1927/28 began on November 1, continuing till the end of February. The summer semester began in May and continued through July. Schrödinger lectured four times a week, two-hour sessions on two afternoons. He also was responsible for a one-hour problem session (usually taken by an assistant), a proseminar for undergraduates, and a two-hour meeting each week on Recent Advances in Theoretical Physics.

In principle, his lectures in theoretical physics were supposed to cover the whole subject systematically in a three-year sequence and then to begin again. In the six years that he lectured in Berlin, however, no course appears to have been repeated exactly. The titles of his courses were as follows: (1) Mechanics (2) Mechanics of Deformable Bodies (3) Statistical Theory of Matter and Radiation

(4) Electron Theory I & II (Electromagnetic theory) (5) General Theory and Principles of Physics (6) Introduction to Quantum Theory (7) Field Physics I & II (8) Particle Physics I & II (9) Quantum Theory.[13]

During this time, the physics faculty of the University of Berlin comprised an extraordinary group of renowned scientists and also some of the most brilliant of the younger European theoreticians. Nernst lectured on experimental physics, Laue on various topics from thermodynamics to optics, Lise Meitner on nuclear physics and radioactivity, Fritz London on applications of quantum mechanics including chemical bonding, Ladenburg and Paschen on spectroscopic subjects. Although Max Planck was Emeritus, he gave a full course in the winter of 1928/29 on Theory of Electricity and Magnetism. From the winter of 1931, the Dozent Friedrich Möglich also gave full lecture courses in theoretical topics – the students naturally called him 'Unmöglich'. Since students were expected also to take courses in mathematics and in philosophy (from Hans Reichenbach usually), they had full schedules, typically thirty-seven hours of lectures and problem sessions a week, with classes scheduled from 7:30 in the morning till 8:00 at night. Saturday afternoons were free.

Schrödinger's lectures were the best, delivered in his beautifully expressive language, they have been described as 'always a pleasure and a delight'. Thus the students were pleased with him as a teacher. He did not, however, encourage them to work under his direction on their research problems. He once said, 'I am very busy, and so many research students want to come and study with me, and they ask for advice what to do. I'll tell you what I say to these students: "first year do nothing but mathematics, second year nothing but mathematics, in the third year you can come and talk with me."' He was acknowledged to be 'a grand mathematician'.[14]

Erwin introduced a style of informality into his lectures that was unlike anything ever seen before at the University of Berlin, where class lines were rigidly drawn. In Zürich the machinists had been treated as colleagues, in Berlin they were merely servants. Professors were expected to appear in a formal suit with a white collar and tie, but Erwin usually wore a sweater and bow-tie in winter and an open short-sleeve shirt on hot days in summer. On one occasion the guard at the university gate would not let him in until a student came and testified that he was indeed their professor and they were waiting for him to appear. Many professors practically read their lectures, he spoke without notes. Once when a zeppelin cruised over the university, he broke off his lecture, rushed to the window, shouting 'come, look at this!'.

For Schrödinger, however, as for most of the physicists in Berlin, the greatest intellectual delight in the university was the weekly colloquium, which met on Wednesday afternoons to discuss new discoveries and theories. Einstein took a leading role, as with careful questioning and explanations he sought to reach the heart of every problem presented. Schrödinger was a less persistent questioner, often preferring only to listen until he could make an important contribution, that is, he was less concerned than Einstein to make sure that all the members understood everything. The discussion after a paper was usually longer than the paper itself, and after an hour of this, the meeting was transferred to a local tavern where a special room was reserved. The Berlin physics colloquium was a happy recompense for the spartan facilities of the university buildings. Its only problem may have been too great a uniformity in the philosophical views represented – too many conservative thinkers and too few revolutionaries willing to shake the foundations of classical physics. This conservatism was also enshrined in the powerful Prussian Academy of Sciences.

The Prussian Academy[15]

In 1700 Gottfried Wilhelm Leibniz, under the auspices of the Elector (Kurfürst) of Brandenburg founded the *Kurfürstlich-Brandenburgischen Sozietät der Wissenschaften*, and became its first president. A year later, the kingdom of Prussia was established under Friedrich I and the society became 'The Prussian Academy of Sciences'. Since that time its members have included many of the most eminent scholars in Europe. The membership was divided into a Physical-Mathematical Class, which included all the sciences, and a Philosophical-Historical Class, which included also classical studies and philology.

When Schrödinger came to Berlin, as for many years previously, the membership was restricted to thirty-five Ordinary Members in each class; in addition there were provisions for a hundred corresponding and foreign members and a few honorary members. Membership in the Academy was considered to be a great honor, and many distinguished professors in Berlin never gained admittance.

The half-dozen or so academicians in physics exerted an inordinate influence on the development of their field in Germany, since the Academy was at the apex of a system of interlocking directorates of such institutions as the *Kaiser Wilhelm Gesellschaft zur Forderung der Wissenschaften* [KWG], the *Physikalische-technische Reichanstalt* [PTR], and the Astrophysical Observatory. In 1920, the Academy had helped to found the *Notgemeinschaft der Deutschen Wissenschaft* to obtain funds for the support of research. Thus the physicists of the Academy, Planck, Nernst, Paschen, Warburg and Laue, were in positions of unusual power and responsibility. Einstein was also a member, and he

Max Planck welcomes Erwin Schrödinger to membership in the Prussian Academy of Sciences, 4 July 1929.

held a special Academy professorship, but he took little part in the discussions of science policy. He once told Philipp Frank that he had experienced many strange things in his travels but 'nothing so strange as the sessions of the Prussian Academy of Sciences'.

At its meeting in February, 1929, the Academy elected Schrödinger to membership.[16] The nomination had been made by Planck and seconded by Laue, Nernst, Paschen, and Warburg (Einstein was ill at the time):

From the very beginning of his career, Schrödinger attracted favorable attention through the acuity and independence of his ideas and through the multiplicity of his interests. One after another, in different fields of physics, apparently without any definite preliminary plan, he would take up a problem that interested him, either with an experimental, or later usually with a theoretical approach, and come to grips with it in such a fundamental and skilful way as to bring to light many original and significant conceptions.'

After mentioning the fields of Schrödinger's earlier works, Planck continued:

Somewhat more than 3 years ago, he thus turned his attention to the idea of Louis de Broglie, that an electron moving with a certain velocity behaves in many respects like a wave of definite period and transmission velocity; and he succeeded in giving this idea a very general quantitative formulation through the statement of the differential equation now named after him, through which the hitherto somewhat mysterious wave mechanics in one stroke was placed upon a firm foundation.

The vote in favor of Schrödinger was unanimous in both the class and the plenum and his inauguration as an Academician took place at a formal (white tie) ceremony on July 4, 1929. At forty-two, he became the youngest member of the august society. In his short address, he paid tribute to Boltzmann and to his teacher Hasenöhrl. Of the former he said:

His ideas played for me the role of a scientific young love, no other has ever impressed me so much, no other will ever do so again. Only very slowly did I approach modern atomic theory. Its inner contradictions sounded like shrill dissonances compared with the pure, inexorably clear line of thought of Boltzmann. For a long time, inspired by Franz Exner and K.W.F. Kohlrausch, I took flight and saved myself in color theory . . . A certain enlightenment was first provided by de Broglie's idea of electron waves, which I built into an undulatory mechanics. But we are still apparently far from a real comprehension of the understanding of nature provided by wave mechanics . . . and by Heisenberg's quantum mechanics.

He summarized the goal of classical physics as the construction of general laws from which one can derive the multitude of experimental facts, the individual characteristics of each experiment being specified in the initial conditions of the particular system. For many years classical mechanics had been the prototype of such a scientific method, and there was hope that it could be extended to all of natural science. But now it is seen that classical mechanics is simply a first approximation to a more fundamental theory. 'One of the most burning questions . . . is whether we must give up, as well as classical mechanics, its basic concept that the individual case is uniquely determined by firm laws combined with initial conditions. It is a question of the suitability of an infallible postulate of causality.'

He pointed out that in classical mechanics, although the position and velocity of every molecule in an ensemble can in principle be specified, in practice this is not possible because of the vast number of molecules. He remembered listening as a young man to the inaugural lecture of Hasenöhrl, who explained that it would not be *impossible* for a block of wood resting on the lecture table to fly up into the air – it would only be extremely *improbable*. The problem raised by quantum mechanics, however, is more drastic, for now the initial conditions (e.g., position and velocity of a molecule) can not be observed even in *principle*, and 'for the scientist as a scientist what cannot be observed does not exist'. He then recalled his teacher Exner, who was the first to put forth a basically acausal interpretation of nature. He concluded, however, that there is scarcely any way imaginable to decide experimentally whether natural laws in reality are absolutely determined or whether they are partly indeterminate – at least we can expect a long wait before any such decision appears, or even before we can conclude which conception is more convenient for scientific work.

In his reply, Max Planck, the Secretary of the Academy, said that he thought that Schrödinger himself, despite the neutral stance just taken, had helped to restore a more classical foundation for physics. He would like to champion the cause of strictly causal physics. He would agree completely that it is a question of convenience – every theory is merely a scaffolding, which the mind of the researcher must knock into shape as well as he can, in order to fulfil his aim, the construction of as true a picture of nature as possible. But the scaffolding requires in every case a firm foundation, if it is not merely to stand in the air, and if the postulate of strict causality is not to serve this purpose, one must ask what an acausal physics would put in its place. Without some postulate, no physical theory can be developed, unless one wishes to consider the mere recording of individual observational facts as a theory. His strongest argument for causality is the progress that has already been achieved through its application, including the results of Schrödinger's own work. 'May you, my honored colleague, in forging ahead further along the road that you have pioneered, achieve many beautiful successes. That is the sincere and optimistic good wish with which I welcome you today in the name of the Academy.'

The Academy met every Thursday afternoon, alternate Thursdays being devoted to general meetings and those of one of the Classes. Schrödinger took quite an active role in academy affairs. He served on a committee that drew up plans for an Institute of Theoretical Physics to be headed by Laue, but owing to the economic recession this project was never realized. He was also one of the leaders in raising funds for a Planck Medal; the first recipient was Planck himself, afterwards the Academy members usually voted it to one another. The Academy was noted for the speed of publication of its *Sitzungsberichte*; two days after a manuscript was received, the proofs were brought to the author by a messenger boy who waited until they were corrected, and one week later reprints were available. When the Nazis came to power, Schrödinger and Einstein were the only scientists whose names were completely expunged from the membership records of the Academy. Even Franck and Born and the other Jewish physicists did not share this distinction.[17]

Travels and lectures

During his six years at the University of Berlin, Schrödinger was able to travel rather extensively in Europe, either in response to invitations to lecture at university colloquia and meetings of physical or philosophical societies, or sometimes, usually with Anny, during his holidays, which he held to be sacred times for rest and recuperation, never to be violated by professional duties.

In March of 1928, he traveled to London to give a series of lectures at the Royal Institution, which were published later that year by Blackie & Son under the title *Four Lectures on Wave Mechanics*.[18] At this time William Bragg (1862–1942), who in 1915 had shared a Nobel prize with his son Lawrence for their work on X-ray diffraction by crystals, was director of the Royal Institution. He had arranged the invitation to Schrödinger and he gave an introductory lecture to provide some background for the more difficult material to follow. It is said that, attracted by the title, at least one enthusiastic yachtsman turned up for the first lecture.

During the Christmas holidays of 1928, Erwin and Anny made a tour of the Dalmatian coast as far south as Dubrovnik. They were delighted with the warm weather, the lovely seascapes, and the remains of classical civilizations. It is surprising that they did not continue to Greece, in view of his love of ancient Greek history and literature. During the spring holiday in April, 1929, they went to Sicily, where the Greek influence had been so predominant. On the way back from this trip, they stopped in Zürich, where they enjoyed some happy times with the Meyers, the Bärs, and other old friends, and Erwin and Pauli thrashed out some of their differences over the interpretation of quantum mechanics. Pauli wrote to Bohr that Schrödinger 'now thinks much more mildly about the statistical meaning of wave mechanics'.[19]

It is true that during these Berlin years Schrödinger in some of his lectures seemed close to acceptance of the prevailing positivist philosophy of science. For example, in December 1928, he concluded a lecture in Frankfurt on 'The Epistemological Value of Physical Models' with the words: 'We must not forget that pictures and models finally have no other purpose than to serve as a framework for all the observations that are in principle possible.'[20]

The golden twenties

The Schrödingers moved to Berlin towards the end of the 'golden twenties'. The German economy, financed to a large extent by American loans, was prospering and the burdens of the vindictive Versailles treaty were less oppressive. In 1925, Field Marshall Hindenburg had been elected president, with the approval of the exiled Kaiser and even the support of the socialist party. He was a slow-witted man devoid of political ideas and distrustful of democracy, but he personified the past glory of 'an army that had never been vanquished' and provided an authoritarian anchor amid the shifting currents of unstable political coalitions. As he became more senile, he was easily swayed by unscrupulous politicians like Franz Papen, who ultimately used him to transfer power to the real dictator for whom so many Germans yearned.

Planck's remark to Schrödinger that in Berlin one could do anything without being troubled by censorious neighbors was indeed true. During the twenties it was the most licentious city in Europe. Politicians and financiers set the tone, and every night they could be observed in the cabarets along the Kurfürstendamm ogling the naked women or making drunken love to young sailors. The bitter cartoons of Georg Grosz were drawn from life, although he was once tried for obscenity for a picture of Christ in the gas works.

Yet, along with much hard-core pornography, there were great artistic achievements in the theater and cinema. Max Reinhardt mounted productions in two theaters, *Das Grosse Schauspielhaus*, a vast circus arena transformed into a theater in the form of a huge cave, and the more intimate *Kammerspiele*, where Marlene Dietrich made her debut in Somerset Maugham's *The Circle*. In 1930, as Lola-Lola in *Der Blaue Engel*, she portrayed the sexual ambiguity of the Prussian spirit, with its mixture of sensuality and sadism, which were to find full expression a few years later in Hitler's private army, the SA of Ernst Röhm. The leading actress in the Berlin theater, in both classical and modern roles, was Elisabeth Bergner, one of the brilliant German Jews who inspired the artistic and musical life of the city. Viennese theater was represented by Schnitzler's *Reigen*, which enjoyed an enormous success during its extended prosecution as an affront to public decency. The 'proletarian theater' also offered some famous productions, especially *Die Dreigroschen Oper* of Bert Brecht and Kurt Weill, with Weill's wife Lotte Lenya as Jenny Diver, which opened at the *Theater am Schiffbauerdamm* at the end of August, 1928. Brecht had joined the communist party (K.P.D.) in 1927 but he became a millionaire with this commercial success. His refrain, 'Erst kommt das Fressen, dann kommt die Moral' [First comes feeding, then comes morality'], expressed the feelings of all levels of Berlin society.[21]

Erwin was a lover of theater in Berlin as he had been in Zürich and Vienna. He did not make friends easily, especially outside the circle of his professional colleagues. Except for his fellow student Fränzel, he never in all his life had a close personal male friend, and he never had the experience of brothers or sisters or of children whom he could call his own, nor even of students for whom he could be an intellectual father and mentor. Thus perhaps he loved the theater because it provided vicarious experience of a gamut of personal relationships missing from his own life. In his love affairs he would act out romantic scenarios that sometimes seemed theatrical, and which probably did not always convince even their author, although it must be admitted that his heroines were not idealized with imaginary virtues, but loved with a certain scientific objectivity.

At this time, Ithi Junger was his principal romantic heroine, and he would meet her in his travels away from Berlin as she did not come to visit him there until 1932. There were other loves of whom little is known. For example, his affair with Erica Boldt was a serious episode. All that can be said of it is that this lady did not have to be pursued, she was as much the huntress as the quarry.

Late in 1929, Erwin visited Innsbruck to give a lecture and he stayed with Arthur March, who had married in July. On his return to Berlin, he told Anny how much he was impressed by the beauty and charm of Hildegunde, Arthur's young wife. Hilde was a tall, slim brunette, from an unpretentious family. She was devoted to skiing, even though she walked with a slight limp due to a bad fall on the slopes. She was enthusiastic about amateur theatricals, and her generally optimistic temperament was similar to that of Erwin. Like the majority of her sex at the time, she was unencumbered with much formal education. She was to become one of the great loves of Erwin's life, but it was not love at first sight, rather more like love held in reserve, for Hilde and Arthur were presumably still in a state of connubial bliss, and the bride would have no immediate romantic interest in the distinguished Berlin professor.

Schrödinger and Einstein became good personal friends in Berlin. Both of them were averse to social pretensions and they shared a dislike for the formality and stiffness of the Prussian professors. Erwin loved to visit Einstein at his summer house in Caputh on the Schwielow Lake near Potsdam and the two friends spent many hours together walking in the woods or sailing on the reaches of the Havelgewassern.[22] Erwin had learned to love sailing on Lake Zürich, and he and Anny kept a small sailboat on the Wannsee.

The social life of Anny and Erwin was paradoxical. They would have happy parties for their many friends, and became quite famous for their *Wiener Würstelabende* [Viennese sausage nights] which spanned the generations from distinguished elders like the Plancks and Laues to brilliant youngsters like Max Delbrück, Viktor Weisskopf and Eugene Wigner. When the guests were not there, however, the house was not so cheerful, as Anny fretted about Erwin's love affairs. After a lecture one afternoon in the winter of 1930, Walter Elsasser asked Schrödinger if he could discuss some points with him. 'Come along to my house', Erwin said, 'We can talk better there, I don't like my office here.' They traveled out to Cunostrasse. Every room in the flat was brightly lighted, but Mrs Schrödinger was not there. She came in later and disappeared almost immediately into her room at the back. After they had talked intensively for about an hour, Erwin said 'Are you getting hungry?' Then they went to the kitchen and found some bread and cheese and beer, over which they continued their discussions till quite late at night. Such visits were repeated several times in the course of the next few months, always in the same way. Elsasser concluded that the Schrödinger marriage was not a happy one.[23]

The revolution in physical concepts

In May, 1930, Schrödinger traveled to Munich to visit his old friend Sommerfeld and to deliver a lecture at the *Deutsches Museum*. His subject was 'The Transformation of the Physical Concept of the World'. Although it was not published until many years later (1962) it is an important statement of his ideas in these early years of quantum mechanics, displaying a view that is surprisingly favorable to the Copenhagen–Göttingen interpretations, more so than most of his subsequent statements.[24] As he once admitted to Pauli, the strenuous effort of Bohr to convert him had almost succeeded, even though he later relapsed.

He began his Munich lecture with the claim that 'the change in scientific thinking, which has been achieved by us, appears to be more profound than any other in historical memory'. The change occurred in two diametrically opposed forms: one claimed that the old picture of the world was too continuous and the other that it was too fragmented. One form demanded discontinuous quanta of energy and instantaneous jumps between energy states, the other contended that matter did not consist of individual mobile mass points but of continuous waves filling all space. 'The most remarkable thing was that both the opposed directions of thinking finally coincided in a formally mathematical way.' Thus both sides were led to admit that there must be something in the opposing view, 'but they were both not quite honest about the compromise' for a really existing thing, Schrödinger concluded, cannot be at the same time a mass point and a wave train. [This conclusion may not be justified, since reality may inhere in the entity and not in its properties.]

Even in principle, the deterministic picture of classical physics has failed.

Our mind, by virtue of a certain finite, limited capability, is by no means capable of putting a question to Nature that permits a continuous series of answers. The observations, the individual results of measurements, are the answers of Nature to our discontinuous questionings. Therefore, perhaps in a very important way, they concern not the *object* alone, but rather the relation between *subject* and *object*. For the philosophers that is an old truism, but it now gains a heightened significance. It is thus no longer so obvious that repetition of observations must lead . . . in the limit to an exact knowledge of the *object*. When we interpolate the actual measurements by the best possible means, they are imbedded in continua . . . that do not represent the natural object in itself, but rather the relation between subject and object.

The different wave forms, the old long-familiar electromagnetic waves as well as the new so-called matter waves, are not to be considered as purely objective descriptions of reality . . . The wave functions do not describe Nature in itself, but the *knowledge* that we possess at any given time of the observations actually carried out. They allow us to predict the results of future observations not with certainty and precision but with just that degree of unsharpness and probability with which observations actually made on the object permit predictions about it. The wave description that is presently accepted . . . is based on the fact that observations mutually disturb one another – a circumstance that in one respect increases our knowledge of the object, in another respect decreases it.

Most of us today feel that this necessary abandonment of a purely objective description of Nature is a profound change in the physical concept of the world. We feel it as a painful limitation of our right to truth and clarity, that our symbols and formulas and the pictures connected with them do not represent an object independent of the observer but only the relation of subject to object. But is this relation not basically the one true reality that we know? Is it not sufficient that it finds a solid, clear, unequivocal expression, wherein in fact all truth exists? Why must we exclude ourselves completely? Has not God himself through the mouth of our poet let it be said that we ourselves are that which must bring order out of chaotic nature? [Goethe, *Faust*]

> Das Werdende, das ewig wirkt und lebt
> Umfasst Euch mit der Liebe holden Schranken
> Und was in schwankenden Erscheinung schwebt
> Befestiget mit dauernden Gedanken.*

* The Becoming that eternally acts and lives / Encloses you with love in gentle bonds / And what hovers in wavering appearance / It makes secure with thoughts that last forever.

In this summary of his philosophy of science, Schrödinger seems to have met Bohr and Heisenberg more than half way. The important difference is that he still maintains the reality of the objects of physical investigation. Thus his philosophy is close to that called representational realism. It also accepts at least tacitly the duality of mind and body, subject and object, but the ontological status of the 'object' is somewhat changed in that the concept of object can no longer be employed without consideration of the subject who has knowledge of the object.

Schrödinger accepted this dualistic ontology only as a sort of working method for scientific research. At the deepest level, he never departed from his belief in the intuitive knowledge incorporated in

Vedanta. The English science writer J.W.N. Sullivan included Schrödinger in a series of 'Interviews with Great Scientists', published in *The Observer* in 1931.[25] His reply to a question about the meaning of life in the universe was:

Life here is confined to a very small space in the universe. It may also be confined to a very small time . . . I am no more frightened by time than I am by space . . . If this life is the only life, then the whole meaning of the universe, throughout its extent and throughout its history, is to be found here. Although I think that life may be the result of an accident, I do not think that of consciousness. Consciousness cannot be accounted for in physical terms. For consciousness is absolutely fundamental. It cannot be accounted for in terms of anything else.

One of Erwin's most interesting short essays was 'Science and Ethics', published in the Christmas, 1930, number of a Berlin newspaper *Die Vossische Zeitung*. He was not trying to derive ethics from science, but rather the converse, to show that science is based upon ethics. The first law of ethics is the golden rule, which in one form or another has been stated in all civilizations, both ancient and modern. It is a Kantian 'categorical imperative' – it cannot be proven, but is the principle from which practical rules of behavior are derived. Erwin calls it the 'personality principle'; it states that one's fellow human is a being like oneself. This principle makes science possible, because no person in isolation can create scientific works, and any person who embarks upon such works must know that he or she can use the contributions of other persons.

Yet the other person is not exactly like oneself. Erwin cites the 'scientific report on a crucial experiment' given to him by a three-and-a-half year old child: 'When you pinch yourself, it doesn't matter to me, but when I pinch myself, it hurts.' 'Take note', he says, 'of the wonderful "it", that recalls the famous "it" of Lichtenberg, with which he corrected the Cartesian "I think, therefore I am" (which should say "it thinks, therefore it is").' Erwin argues that science must accept the personality principle. Otherwise 'it must render all its own activities meaningless'.

Therefore science cannot be given a scientific foundation. The naive statement of the child leaves science in a painful dilemma, pushing it towards an acceptance of a multitude of egos which must have some incomprehensible, completely unexplained relation to the individual human bodies. 'Yet, the most wonderful and most sublime of all teachings, the Brahman doctrine that the all equals the unity of consciousness, culminates in a mystical victory over this dilemma, the words so obscure to the understanding, so close to the intuition: *Tat twam asi*, That Thou Art. One may see in the Brahman doctrine merely a facile play of ideas. But it would be a vast error to believe that science knows any better or clearer answer concerning these things.'

Ithi in Berlin

Everyone may judge for himself or herself whether or not the ethical principles that Erwin derived from Vedanta were rigorously applied in his relations with women.

During these Berlin years he would meet Ithi Junger whenever possible, in Salzburg or on skiing holidays. His wooing of Ithi led him to some spectroscopic reflections on sexual psychology:

Comparable in some way to the end of the spectrum, which in its deepest violet shows a tendency towards purple and red, it seems to be the usual thing that men of strong, genuine intellectuality are immensely attracted only by women who, forming the very beginning of the intellectual series, are as nearly connected to the preferred springs of nature as they themselves. Nothing intermediate will do, since no woman will ever approach nearer to genius by intellectual education than some unintellectual do by birth so to speak. It has often been said that no woman really has genius. The fact is that they all have. Genius is nothing outstanding in them, it is the rule, but it is usually too weak to withstand the contamination by culture and civilization. I am fairly convinced that the only sensible thing in this moment is to make [Ithi] give herself to me completely, and to make this not by any means, but by the means of real genuine love only. If I don't succeed I will take it to be the right thing that I don't, and I will not be blamed as a shy lover, but I am fairly sure that I will succeed.

By early August, 1928, success was imminent. They could hardly endure five minutes without each other, and her smile to him was like the very air he breathed. He thought these last days before Ithi surrendered were the loveliest and most fascinating experience of his life.

By the middle of August, she was his mistress, and Erwin was becoming more and more convinced that Ithi possessed all the qualities of a dear mistress but none of a good wife. He thought, however, that this was quite the natural thing with a girl aged seventeen, that the qualities of a married wife develop slowly in the course of time, long after sex maturity has been provided by nature. Yet in the case of Ithi, he did not doubt that his conclusion was justified.

Erwin never explained what he believed would be the ideal qualities of a 'married wife', but unswerving devotion to himself would have been one prerequisite. Ithi was kind and loving and her spirit was sensuous rather than sensual. She was unique among Erwin's women in the pervasive gentleness of her character. It was inevitable that she

Itha Junger (c. 1930).

would eventually be hurt by a lover so worldly and self-centered, so much her opposite in temperament.

Before they became lovers, Erwin thought that he might be prepared to quit Anny to marry Ithi, but afterwards he was almost horrified by such a thought, although he loved Ithi as dearly as ever, and felt only friendship and pity towards Anny. He thought that

> carelessness towards the welfare and wishes of the mate is now a standing feature in [Ithi] – she says herself that she is selfish and she is. What she is longing for in a marital union is her joy. This is an excellent thing if the union comprises nothing but the sex relationship. Sex relations in all their full splendor, and with all the beauty of the world that follows from them . . . [Ithi] loves to be adored, but her care and tenderness towards the man who adores her will diminish with any certainty of possessing him.

Even with due allowance for the male supremacist culture of Erwin and his friends, this was an egocentric analysis of the love affair. Ithi proved to be a constant mistress and during four years managed to meet Erwin whenever possible. Her most dramatic appearance was at a ski lodge in the Alps, where Erwin and Anny were staying. She had just fallen down a crevasse and was battered, bandaged and blood-stained, when she burst in and threw herself weeping into Erwin's arms.

One of their best times together was in the summer of 1932 in Berlin. Ithi came to stay at the Schrödinger house. Anny was conveniently away, and only the maid was there. They went walking, swimming, and rowing together. She met Planck and Einstein.

Their carefree happiness ended when Ithi became pregnant. Erwin tried to persuade her to have the child; he said he would take care of it, but he did not offer to divorce Anny. By now, he was thinking more of Hilde March, and already falling out of love with Ithi, and with him even love and the prospect of a family did not suffice for marriage. His continuing marriage to Anny provided a secure defense against importunate mistresses. She regarded him as a great man in all respects, who was above criticism on any ground. She was willing to tolerate his every *Seitensprung* [extramarital affair], and acted as an insurance against complications whenever he wished to end one. Even Erwin's desire for a son was not sufficient to cause a break with Anny. In desperation, Ithi arranged for an abortion. Although it was performed by a qualified medical practitioner, it seems likely that damage to the uterus occurred – she later married an Englishman and suffered a series of miscarriages and was never able to bring a pregnancy to term. When Ithi left Berlin in 1932, she was broken hearted, yet she never lost her loving regard for Erwin.

Some months later he recorded a strange dream. He and Ithi were going to bed, but there was another person present in the room who kept calling his mother by her first name, 'Georgie, Georgie' but the name was always strangely slurred so that it sounded like 'Zhorgie, Zhorgie'. Apart from obvious oedipal interpretations this may have been a voice calling his mother to punish a naughty boy, or it may have been the spirit of his unborn son. Erwin and Ithi met briefly once again in London in 1934, but their old love was not revived.

A trembling electron

In October 1930 and February 1931 Schrödinger presented before the Academy two of his most interesting works from the Berlin years. They were based on a detailed examination of certain properties of the electron according to the relativistic quantum mechanical theory that was published by Paul Dirac in 1928. We may recall that Schrödinger's first wave equation was relativistic but was not applicable to the electron because of its neglect of spin.

Dirac's approach to theoretical physics was quite different from that of Schrödinger; he was not interested in the construction of *anschauliche* physical models, but was content to let mathematics be his guide, confident that if the mathematical analysis was reasonable, the physical picture would in most instances emerge eventually. Thus his approach to the problem of the electron was highly original and it is difficult to see how anyone else would have thought of it for a long time.[26]

The classical Hamiltonian for a particle, taking into account the variation of its mass with velocity in accord with special relativity is

$$H = c(m^2 c^2 + p_x^2 + p_y^2 + p_z^2)^{1/2} \qquad (1)$$

If he wrote a wave equation $H\psi = E\psi$, where E here denotes $i\hbar\, \partial/\partial t$, it would become

$$[E/c - (m^2 c^2 + p_x^2 + p_y^2 + p_z^2)^{1/2}]\, \psi = 0 \qquad (2)$$

where the p's are the usual operators. This equation, however, is unsatisfactory because it is very unsymmetrical between E and the p's. He therefore multiplied it on the left by the operator

$$[E/c + (m^2 c^2 + p_x^2 + p_y^2 + p_z^2)^{1/2}] \qquad (3)$$

to obtain

$$(E^2/c^2 - m^2 c^2 - p_x^2 - p_y^2 - p_z^2)\, \psi = 0 \qquad (4)$$

This equation is relativistically invariant but is not in accord with the general laws of quantum mechanics because it is quadratic rather than linear in E. He then wrote an equation that is linear in E and 'roughly equivalent' to (4):

$$(E/c + \alpha_x p_x + \alpha_y p_y + \alpha_z p_z + \beta)\, \psi = 0 \qquad (5)$$

where the α's and β, which denote some quite new dynamical variables since they are independent of x, y, z, or t, are chosen so as to make (5) imply (4). These new variables turn out to describe the spin of the electron. The Dirac equation also allows solutions for both positive and negative values of the energy; at first Dirac thought that this was a defect, but soon afterwards came the discovery of the positive electron (positron) by Carl Anderson at Caltech, and the apparent defect became the crowning triumph of the new theory.

Schrödinger was not interested in simply making straightforward applications of well established theories. He was inspired by the problems and difficulties of the theories themselves and the ways in which they could be further developed or modified. He would rather work on a difficult conceptual problem even at the risk of failure.

Thus he asked himself what are the properties of a Dirac electron and he found an answer that is curious indeed to this question.[27] His starting point was a relation first derived from Dirac's theory by Breit, $dx_k/dt = c\alpha_k$ ($k = 1, 2, 3$). Since $\alpha_k^2 = 1$, $(dx_k/dt)^2 = c^2$, and the velocity components can take instantaneous values of $\pm c$. The measured (expectation) values of the electron velocity are smaller, and are related to the values of momentum in the usual relativistic way, and thus one has the picture of an electron as undergoing a rapidly oscillating *Zitterbewegung* [trembling motion] which is superimposed on the motion of the center of the wave packet representing the free electron. For an electron whose center of mass is at rest, he found that the amplitude of the *Zitterbewegung* $\bar{\xi}_k \approx h/4\pi mc \approx 10^{-11}$ cm; but h/mc is the well known Compton wavelength below which one cannot, in accord with the Heisenberg uncertainty relation, reduce the size of the wave packet without exceeding a momentum mc. There is also an angular momentum associated with the *Zitterbewegung*, so that the electron can be imagined to move through space while describing a tight spiral. It is tempting to associate the electron spin with this spiral motion (we know that a simple picture of a rotating sphere of charge is incorrect) but Schrödinger's attempt to derive the spin angular momentum on this basis failed by a factor of two.

The *Zitterbewegung* originates in the occurrence of both electron and positron states, so that both positive and negative frequencies occur in the Fourier spectrum of the particle. The electron and the positron are thus considered not as distinct elementary particles but as two possible states of the same particle. One criticism of the model of the trembling electron has been that in reality, if the expression may be excused, an electron can never exist in an isolated state in a field-free space.

In his second paper, Schrödinger discussed the 'quantum dynamics' of the Dirac electron in a coulombic field, as in the hydrogen atom. He showed how the introduction of the *Zitterbewegung* displacement into the potential $-e^2/r$ makes it possible to separate the Hamiltonian operator into even and odd terms, and thereby to avoid a problem with the Dirac theory of the hydrogen atom, the occurrence of both positive and negative eigenfunctions.

A message from the future?

In March 1931, Schrödinger presented to the Academy a remarkable paper, 'On the Reversal of Natural Laws', which opened up the possibility of a new interpretation of wave mechanics, thus revealing again the originality of his thinking, and his ability to discover new depths in older fields of mathematical physics.[28] In this case his starting point was the resemblance between his wave equation and a diffusion equation with an imaginary diffusion coefficient. One inspiration must have been the recollection of his work on the Fokker–Planck equation during the cold and hungry days in postwar Vienna.

Suppose, he said, that one has an initial probability distribution $w(x, t_0) = w_0(x)$; for $t > t_0$, the distribution is given by the solution of the diffusion equation. $D\, \partial^2 w/\partial x^2 = \partial w/\partial t$. Now suppose that the distribution is specified not only for t_0, but also for some later time t_1, so that $w(x, t_1) = w_1(x)$. What is the distribution at any intermediate time $t_0 \leqslant t \leqslant t_1$? The answer is not a solution of the diffusion equation, nor of the equation for diffusion in the reverse direction, $D\, \partial^2 w/\partial x^2 = -\partial w/\partial t$, but it must be a product of solutions for both these equations.

The general solution of this problem presents considerable mathematical difficulties. Schrödinger expressed it as a pair of integral equations, which he made no attempt to solve, merely commending them to the attention of mathematicians. Twenty-one years later he wrote on his reprint of the article that 'this is still a challenge to mathematicians to solve a certain boundary-value problem'. He was then arguing with Max Born about the origin of irreversible processes from reversible mechanics.

In one case, however, the solution is obvious, i.e., when $w_1(x)$ is exactly the value that would be reached by the forward diffusion process. The same conclusion is also true for the reverse diffusion process: if the final state is $w_0(x)$, it must arise by a natural diffusion process from the initial distribution $w_1(x)$. This conclusion may appear remarkable: if one does find a highly improbable distribution, it will have arisen from the more probable distribution (or in fact from a uniform equilibrium distribution) by a perfectly normal diffusion process. Schrödinger would later use this demonstration to controvert Born's idea (in his famous *Natural Philosophy of Cause and Chance*) that irreversibility arises from reversible mechanics as a consequence of our ignorance of the precise initial conditions for the locations and velocities of particles in an ensemble.

As Schrödinger says, 'the most interesting thing about our result is certainly the striking formal analogy to quantum mechanics.' One discrepancy, however, is the fact that the diffusion equation gives the probability density $w(x, t)$ directly, whereas the wave equation gives a probability amplitude that must be multiplied by its complex conjugate to give the probability density $\bar{\psi}\psi$. Also the occurrence of the $\sqrt{-1}$ factor in the wave equation gives it a hyperbolic, or physically speaking, a reversible character, in contrast to the parabolic, irreversible character of the Fokker–Planck equation. Schrödinger emphasizes that the kind of situation treated in this paper is much more closely related to that in quantum mechanics. The probability density is the product of two solutions of a diffusion equation and is thus bilinear as in quantum mechanics, and since the two solutions differ only in the direction of the time t, the product has no specified time direction.

Unfortunately, Schrödinger seems to have abandoned this fascinating analysis at this point, with the statement that 'I cannot foresee whether the analogy will prove useful for the explanation of quantum mechanical concepts.' He concluded with a few lines from Eddington's *Nature of the Physical World*, 'unclear though they may be':' The whole interpretation is very obscure, but appears to depend upon whether it is a probability after one knows what has happened, or a probability in the sense of a prediction. $\bar{\psi}\psi$ is obtained if one introduces two symmetric systems of waves, which are traveling in opposite directions; one of them presumably has something to do with the known (or supposed to be known) state of the system at a later point in time.'

Many years later, in 1986, John Cramer of the University of Washington, probably quite unaware of this Schrödinger paper (whose title does not suggest any relevance to quantum theory) devised a 'transactional interpretation of quantum mechanics'.[29] Cramer showed that 'the complex character of the wavefunction is a manifestation of its time structure'. Its real part is even under time reversal and its imaginary part is odd under time reversal. The time-reversal operator of Wigner is the operation of complex conjugation. Thus the Born interpretation of $\bar{\psi}\psi$ implicitly tells us that the probability of a particular observation is obtained by taking the product of a component of ψ with its time reverse.

Cramer then advances his transactional interpretation, in which the basic element is an emitter-absorber *transaction* through the exchange of advanced and retarded waves as first described by Wheeler and Feynman. The ordinary Schrödinger equation, $i\hbar\,\partial\Psi/\partial t = -(\hbar^2/2m)\,\nabla^2\Psi$ is first order in the time variable and thus does not have advanced solutions, but Cramer points out that the relativistic wave equation in the nonrelativistic limit ($v/c\ll1$) yields two Schrödinger type equations, one of which is the time reverse of the usual equation. 'The state vector [Ψ] of the quantum-mechanical formalism is a real physical wave with spatial extent and is identical with the initial "offer wave" of the transaction. The particle (photon, electron, etc.) and the collapsed state vector are identical with the completed transaction.' Or, as Lawrence Bragg once said, everything in the future is a wave, everything in the past is a particle.

One cannot yet say whether this new interpretation of quantum mechanics will supplant the Copenhagen interpretation, but it seems to avoid the sterile 'positivism' of the latter, which would deny any reality to the constructs of theoretical physics by focusing entirely upon relations between variables. As Cramer remarks, 'the positivist curtain is no longer needed to hide the nonlocal backstage machinery'. I think that Schrödinger would have liked the transactional interpretation.

Decline and fall of the Republic[30]

Early in 1929, the German postwar economic recovery began to falter. It was a bitterly cold winter in Berlin and schools had to be closed for a week in February. Unemployment rose to 450 000 in the city and on May Day there was a huge demonstration at Alexanderplatz; the police used tanks to disperse the unemployed with thirty killed and many arrests. The Wall Street panic on Black Friday, October 25, signaled a worldwide economic collapse. In the elections for the Berlin city council in November, the Socialists held 64 seats, the Communists (K.P.D.) 56, the German Nationals 40, and the National Socialist German Workers Party (N.S.D.A.P. or Nazis) 13 (including one for Goebbels). Early in the new year 1930, there were bloody street battles between communists and Nazis, which neither the city nor the national government seemed able or willing to prevent. The Nazi army was better organized and more disciplined and usually gained the upper hand in battles for control of the German cities. On March 1, the SA leader Horst Wessel was killed in a communist attack on his house and the song in honor of this 'martyr of the battle for Berlin' became a rousing Nazi anthem. The government banned the SA but Goebbels merely told them to change their brown shirts for white ones and they continued much as before.

The republican parties appeared to fear the communists more than they did the Nazis and there was a growing feeling that the defense of any capitalist system might require the Nazi methods. In the national elections of September, the N.S.D.A.P. emerged as the second largest party, after the socialists, gaining 18.3% of votes nationally and 12.8% in Berlin. Most of the Nazi strength came at the expense of the German national parties, which represented Protestant middle-class interests, but the street fighters of the SA were mainly recruited from embittered unemployed war veterans. The power of the Nazis to control the streets of Berlin was demonstrated in December when they forced the closing of the movie *All Quiet on the Western Front* on the grounds that its antimilitaristic theme was an affront to the national honor. By 1931 Hitler's private army the SA numbered 100 000, and in the next two years its enrolment tripled.

As the depression deepened, 25% of the population of Berlin was on some kind of relief, and the city finances were tottering. Less than 5% of university students came from working-class families, but there was nevertheless considerable privation and even near starvation among the students. As the Nazis gained more adherents, there were riots on the university grounds and sometimes it was necessary to suspend classes. The students agitated especially for a quota limitation on enrolment of Jews. Viktor Weisskopf had an office from which he could survey the student fighting in the courtyard, and sometimes he would bring the victims of Nazi assaults into the building and help to clean their bloody heads. The police were not allowed to enter the university grounds so that no help could be expected from them.[31]

The university teachers, in their pleasant suburban homes, were considerably insulated from the effects of disorder in the central city, and also from effects of the economic depression, since as prices fell their government salaries became more valuable. They were not usually political extremists; more than half of them supported republican parties, but many adhered to the German nationalists who were readily converted to the so-called 'national socialism'.

In 1931, an old friend of Anny's, Hansi Bauer, came to Berlin to pursue her art studies at the Charlottenburg Academy. She lived with a potter, a niece of Michael Polanyi, who was then a member of the Kaiser Wilhelm Institute for Physical Chemistry. Hansi joined the Schrödingers on skiing excursions to the Riesengebirge, southwest of Berlin, and later took driving lessons with Anny. She was a raven-haired beauty, both rich and talented, and she quickly made many friends in avant-garde circles, enjoying in particular a close relationship with Arthur Köstler, who had come from Vienna by way of Paris to seek his fortune. Convinced that the socialists were 'doddering and spineless', he had recently joined the K.P.D., but was working as science editor for the conservative Ullmann newspaper, *Vossische Zeitung [Tante Voss]*.[32] Hansi brought Arthur to a party at the Schrödingers, and Erwin at first disliked him intensely: 'Why did you ever introduce me to that shit?' Later, however, as Köstler cultivated the acquaintanceship, Erwin began to appreciate his journalistic ability and his interest in science, and he used to send him occasional material for the Ullmann papers.

Shrovetide, [*Faschingszeit*] the period just before Lent, was an occasion for carnivals and sometimes quite wild parties. A memorable fancy-dress ball was organized at Harnack-Haus in Dahlem, the residential center of the Kaiser Wilhelm Gesellschaft.[33] Erwin and Anny appeared as Amenophis and Nofretete. The next year the Fasching party was held at the Schrödinger house, which was turned into the $\Psi\Psi^*$ Hotel, with Erwin as Innkeeper. Max Delbrück collected guests at the station in the uniform of a bellboy with a cap emblazoned with $\Psi\Psi^*$. Erika Cremer, then a young student with Otto Hahn and Lise Meitner, came as a 'dairymaid from the Resonance Mountains' with a pitchfork shaped like a Ψ. The Plancks, Polanyi, the Defants, Grotrian, and many young people were there. It was to be the last of their happy Fasching balls.[34]

Although their personal living standards were scarcely affected by the depression, the leading physicists did suffer from severe cutbacks in their research funds, so that there was considerable hardship among the younger untenured staff members. Hendrik Casimir, who was visiting Berlin from Leiden in the summer of 1932, recalls attending with Lise Meitner a lecture by Schrödinger at which an admission fee was charged to raise funds for student aid. Schrödinger tried to show that the tendencies of modern science reflected those in contemporary society. Casimir remarked to Meitner that at least it was a witty talk, but she shrugged and replied 'Ganz wittig', her tone implying that the subject was really too serious for humor.[35]

At that time Schrödinger made no secret of his dislike for the Nazis and their fascist allies, but he never sought in any way to oppose them actively or to join organizations dedicated to such opposition. In the words of his East German biographer, Dieter Hoffmann, 'His political position did not transcend the boundaries of an abstract humanism as defined by his bourgeois-humanistic basic outlook. Such an appraisal naturally in no way diminishes the personal fortitude and strength of character that led to his decision to leave Berlin and his secure position there.'[36]

The summer of 1932 marked the beginning of the end for the

Weimar Republic. As soon as Papen became prime minister, he lifted the ban on the SA, and in the next six weeks there were almost daily street fights, with 72 killed and 500 severely wounded in Prussia. Student rioting forced the closure of the university. The state government was overthrown by a rightist putsch, and in the national elections on July 31, the N.S.D.A.P. became the largest party in the Reichstag, but still far short of a majority. In November, there was another election in which Nazi strength fell markedly from 37 to 33% nationally. They had passed the peak of their electoral popularity, but Papen conspired with the Army generals, the Junkers, and some leading industrialists to force the now senile Hindenburg to call Hitler to power.

On January 30, 1933, Adolf Hitler was appointed Chancellor of Germany. The transfer of power was accomplished with an observance of legal forms, and the Germans quietly accepted their new political master. Most German scientists considered the public expression of any political allegiance to be inconsistent with the dignity of their profession. Einstein was exceptional in his open avowal of socialism and pacifism, as were Lenard and Stark in their vociferous devotion to Hitler. The German universities, however, both staff and students, had been the strongest supporters of antisemitism, and the Weimar Republic had few friends among the professors.[37] As early as 1930, the University of Jena appointed the virulently antisemitic Hans Gunther as Professor of Racial Science [*Rassenkunde*] and his inaugural address was wildly cheered by the students.

The hierarchy of professors was not only antisemitic, it was also antisocialist, and in Prussia, anticatholic. These prejudices played an important role in the selection of new professors and the promotion of associate professors to personal chairs. When Göttingen appointed several distinguished Jewish mathematicians and physicists (Courant, Born, Franck) it was derisively called 'the Jewish university'. It should be remembered, however, that academic antisemitism was not a peculiarly German institution. During the 1930s prestigious American universities such as Harvard strove to limit appointments of Jewish professors and admission of Jewish students.

Scientists and Nazis

Despite the political violence and economic deterioration, Schrödinger carried on with his teaching, but he accomplished little research during the dismal academic year of 1932–33. On February 8, he gave a popular public lecture at the Prussian Academy, 'Why are Atoms so Small?'[38] Small in relation to what, he asked? In relation to the sizes of highly organized living structures, that have complex functions, such as cogitating about atoms. Thus the question can be rephrased as 'Why are there so many atoms in living organisms?' The answer is then that the laws of physics are statistical laws, and they become precise only for systems containing a very large number of particles. If living organisms contained many fewer atoms, they would be subject to random fluctuations that would make any determinate behavior impossible. In view of this general rule, however, the permanence of the genetic elements in living cells after billions of cell divisions is truly remarkable. The explanation must be that the genes are composed of complex molecules in which the atoms are maintained in unchanging structures as a result of the laws of quantum theory. This lecture contained in outline many of the ideas that he was to present ten years later in his Dublin lectures and book *What is Life?*

On February 16, Schrödinger gave a more specialized paper before the Academy of Sciences on 'The Second Law of Thermodynamics'. Only the abstract was printed.[39]

If one expresses the Second Law as the law of increase of entropy, then one certainly cannot establish it through the statistics of reversible systems. If one wishes to do that, then one must give the law a formulation that contains no reference to past or present. One has to see its important assertion to be that in all the energetically isolated partial systems, branching off transiently from the same 'mother system', the entropy changes monotonically in the same sense. This statement can be formulated as a minimum principle, which is satisfied by the sum of the absolute amounts of all entropy changes occurring in isolated systems.

It is interesting to see that in the midst of political turmoil, Schrödinger was pondering the fundamentals of classical thermodynamics, an ivory-tower subject indeed. Unfortunately the complete paper was never printed in the *Sitzungsberichte* of the Academy, since meanwhile Schrödinger had left Berlin and his ordinary membership had been converted to a corresponding membership.

Many Germans thought that having achieved power, Hitler would moderate some of the more extreme Nazi policies, but they were soon to be disillusioned. On February 27, the Reichstag Building was set afire, probably by the Storm Troopers (SA) as part of the campaign for new elections on March 6, in which, after excluding the communists, the Nazis gained a majority. On March 24, the new Reichstag passed the 'Enabling Act' which voted all power to Hitler and thus, still with some pretence of legality, Germany became a dictatorship. While politicians in France and England merely wrung their hands, the Nazis revived the old program of the Prussian military machine and began to prepare systematically for World War II.

March 31 was declared to be a day of 'National Boycott' of the Jews. Mobs of hoodlums led by storm troopers with swastika armbands roamed through Berlin and other cities preventing anyone from entering shops owned by Jews, and beating up anyone they took to be Jewish. The police stood by laughing and joking with the storm troopers. Anny has related[40] that Erwin happened to be downtown in front of Wertheim's, the largest Jewish department store. He was infuriated by the scene and went up to one of the storm troopers and berated him. The troopers turned to attack him, and he would have been seriously mauled had it not been that one of the young physicists was also there, wearing Nazi insignia. This was Friedrich Möglich, who was an assistant to Max Laue; after the war he became a professor at the Humboldt University in East Berlin. Möglich was able to get Schrödinger away safely. Whether or not this story is strictly factual, there is no doubt that Erwin had nothing but contempt for the Nazis; however, he never expressed any public criticism of the regime.

Einstein was in America when Hitler came to power; he declared that he would never return to a Germany ruled by the Nazis. They seized his property and offered a reward for his capture. On the day his ship docked, he wrote to resign his membership in the Prussian Academy. His resignation was accepted, but Bernhard Rust, the Nazi minister responsible for the Academy, demanded that the academicians formally denounce Einstein.[41]

On April 1, without consulting any of the physicists, the Academy published a statement on Einstein, essentially saying that they were glad to get rid of him. Max Laue then summoned an extraordinary general meeting of the Academy and tried to have this statement withdrawn. Only 14 members attended, and only two supported Laue. Schrödinger did not attend this meeting and took no part in the commotion surrounding the Einstein resignation; in fact, he ceased going to the Academy at about this time, perhaps as a tacit protest.

On April 7, the Reichstag passed the 'Law for Restoration of the Career Civil Service', the purpose of which was to eliminate Jews and socialists from all government positions, including teaching staffs of universities and technical institutes. During the first year, 1700 faculty members were dismissed.

Several efforts were made to persuade Max Planck to make a public protest against the Nazi dismissals of Jewish and socialist professors. Otto Hahn asked him to support a statement by a dozen eminent physicists, but Planck easily persuaded him that it would be useless and might even make things worse, and Hahn made no further efforts in that direction.[42] Marie-Elisabeth Lüders, a former member of the Reichstag and old family friend, 'with mixed feelings of hope and doubt', twice visited Planck in January, 1933, and asked him to support a protest in which scientists would withdraw from all teaching and research.

I hoped that this general strike of the intellect would tear from the eyes of millions half-blinded the veil behind which the Nazis concealed the dangerous abyss towards which they were leading Germany. I left Planck with the dreadful knowledge that there was no way to stop the coming downfall of German science, and no remedy against the disgraceful readiness of so many to sell out teaching and research. Planck himself felt that what was to come, must come. After a long silence, he went to his grand piano and played a Bach chorale, then he offered me his hand, and I left.[43]

Planck's personal situation was so prestigious that he could have spoken against some of the excesses of the regime without fear of reprisals, but he chose to remain silent. Later Einstein told a friend who was going to visit Germany, 'Give my regards to Laue'; 'and Planck?', the friend asked. Einstein shook his head sadly.

There is no known instance in which a professor of physics or chemistry without any Jewish family ever made an open protest against Nazi activities. Even among the German intellectual elite, the scientists were conspicuously unanimous in this respect, since a few protestors can be found among scholars in other fields. It is true that after 1934 open opposition would have been dangerous, with the establishment of the dreaded concentration camps and the revival of what Goering called 'the good old German custom of the axe'. In the early years of Nazi power, however, opposition was not yet suicidal, and it was during 1933 and 1934 that the scientific establishment, led by Max Planck and Walther Nernst, washed its hands of the growing terror and concentrated on defending its own special privileges. On the 25th anniversary of the Kaiser Wilhelm Gesellschaft, in 1936, Planck was able to send a telegram to Hitler thanking him for his 'benevolent protection of German science'.[44]

The only notable German scientist who was conspicuous in his disapproval of the Nazis was Max Laue, and even his actions were taken within the physics establishment and not in open criticism of the regime. Laue tried without success to persuade Heisenberg to help resist the worst Nazi excesses, and called his attention to the biblical injunction against casting pearls before swine.[45]

This unbroken record of collaboration does not mean that every nonjewish scientist was a convinced Nazi. Many were merely opportunists who welcomed the chance to advance their careers by taking the positions of dismissed Jews, while the more eminent were either mildly anti-Nazi like Planck and Hahn, or moderately pro-Nazi like Heisenberg. On June 2, 1933, Heisenberg wrote to Max Born urging him not to leave Germany but to take advantage of certain provisions that might exempt him from the general 'cleansing' of the civil service.

'Since only the very least are affected by the law – you and Franck certainly not, nor Courant – the political revolution could take place without any damage to Göttingen physics . . . In spite of [some dismissals], I know that among those in charge in the new political situation, there are men for whose sake it is worth sticking it out. Certainly in the course of time the splendid things will separate from the hateful.'[46] The 'very least' referred to the associate professors, Dozents, and others, who had not achieved international eminence with Nobel prizes and the like. They indeed had the most difficult time, since foreign universities were not anxious to make places for them, and often foreign governments would not even accept them as refugees. Born rejected this advice from Heisenberg with polite contempt.

In fairness, one should mention the 'splendid things' that caused Heisenberg and most young Germans of his generation to accept national socialism. As summarized by Ernst Krieck,[47] the Nazi rector of the University of Frankfurt, the movement exalted feeling above abstract thought, honor above profit, bonding and fellowship above freedom, and the nation above the individual. The Nazi state promised to eliminate the corrupt political parties that were merely tools of special interests. It called for a new kind of man, who would no longer be a cog in the bourgeois capitalist machine, but one of a heroic band of blood brothers ready to march through heaven and hell to accomplish the will of a charismatic leader. The idealistic students failed to notice that while the Nazis rhetorically denounced the philosophy of liberal capitalism, in practice they left its economic system untouched, so that liberals and Nazis could join forces against their common enemy, which was marxist socialism.

'The Prof' to the rescue

Scientists all over the world anxiously watched these events in Germany but only one of them resolved immediately to do something. This was the eccentric professor of physics at Oxford University, Frederick Alexander Lindemann, known to his closest friends as 'Peach' but to everyone else as 'the Prof'. His mother was an American and his father an Alsatian who had made a great fortune in the construction business. Lindemann was a bachelor, a teetotaler, a vegetarian, and played an excellent game of tennis. He had studied with Nernst and Planck in Berlin and showed early promise in research. During the first war he worked at the Royal Aircraft Factory at Farnborough, and is reputed to have figured out how to prevent tailspin in small airplanes and then to have risked his life to test his theory personally. In 1919 he was appointed professor of physics at the Clarendon Laboratory in Oxford, as the successor to Ralph Clifton, who had served forty years as a resolute opponent of all research. The Prof himself did little research, but he was good at raising money and finding talented young staff. By Oxford standards, he was wealthy, having three men servants to look after his needs and to pilot his Rolls Royce through the narrow streets. In the way of cars, however, he could not rival his brother Sepi who had two Rolls, a white one driven by a negro and a black one driven by an albino, but these were on the Riviera.[48]

Even his friends did not like the Prof, owing to his malicious and sarcastic remarks; he could not mention the name of a lady without suggesting some intrigue. He acted as if he believed himself to be gifted with a most powerful intellect, and did not hide his scorn for those less well endowed. Actually he had an inferiority complex: when he sent his book on quantum theory to Max Born, the Prof suggested that it would be useful only as toilet paper. Among the rich, however, he was capable of a certain charm. Now this may seem a most unlikely man to have set about a single-handed rescue of Jewish physicists from the Nazis. He was not himself Jewish, even mildly antisemitic at times, and at first his idea was simply to get one or two outstanding theoretical physicists for Oxford University.

Lindemann soon realized that more than one or two scientists would require help and he discussed the problem with the Academic Assistance Council, which had been set up in London under the chairmanship of Ernest Rutherford. The Council had limited funds, about £13000 which had been raised by public subscription. The economy was in the depths of worldwide depression; many young British scientists could not find jobs of any kind. The idea of importing a considerable number of scientists from Germany was unacceptable to the British Ministry of Labour.

Now Lindemann's friends in high places became helpful, and he approached Harry McGowan, chairman of Imperial Chemical Industries (I.C.I.) for financial support. The positions would be additional to any existent ones, and thus no British scientists would be displaced. Only scientists of established reputation would be offered places. With these understandings, the Ministry of Labour withdrew its opposition. Refugee scientists would be accepted only from academic institutions, thus avoiding any conflict between I.C.I. and German companies such as I.G. Farben, with which it had cartel agreements.[49]

The Prof had already been to Germany to survey the situation, arriving in Berlin about April 16. He interviewed many physicists and several chemists, and drew up a tentative list of those who might be forced to leave Germany. He came to afternoon tea at Schrödinger's home, and Erwin expressed freely his disgust with the Nazi policies.

Lindemann talked with him about his assistant Fritz London, who was a Privatdozent at the University. It was not yet certain that the racial laws would be extended to such positions, and London was still in some doubt as to whether he should leave. The Prof said that he had offered a position to London. 'Naturally he accepted', said Schrödinger. The Prof replied that London had asked for time to think it over. 'That I cannot understand', said Schrödinger, and added on the spur of the moment, 'Offer it to me, if he does not go, I'll take the position'. The Prof was astonished by this unexpected news, since he knew that Schrödinger was not Jewish and the Nazis had nothing against him. He was aware that Erwin hated the Nazis, but he had not realized that he was willing to give up his prestigious position in Berlin to face the uncertainties of emigration to England. They discussed the possibilities in some detail and Lindemann promised to try to arrange a suitable position at Oxford. They devised a code based on the atomic numbers of the elements with which Lindemann could telegraph the financial level of the position. Erwin then asked whether a temporary fellowship could also be obtained for his old friend Arthur March, who was an associate professor at Innsbruck. March had recently published a good book on atomic structure, and Schrödinger said that they would like to work together on another book. Lindemann promised to approach I.C.I. on the basis that Schrödinger would need an assistant at Oxford.[50]

At this time Erwin, much in love with Hilde March, was reading Thornton Wilder's *The Bridge of San Luis Rey*, and he confided to his *Ephemeridae* the following quotation: 'Now he discovered that secret from which one never quite recovers, that even in the most perfect love one person loves less profoundly than the other. There may be two equally good, equally gifted, equally beautiful, but there may never be two that love one another equally well.'

Before he had visited Germany, Lindemann thought that 'the Nazi madness' would be over in a couple of years, so that the displaced scientists could then return home, but he now began to realize the extent of Hitler's hold on the country. On May 4, he wrote to Einstein: 'I was in Berlin 4 or 5 days at Easter and saw a great many of your colleagues. The general feeling was much against the action taken by the Academy . . . Everybody sent you their kind regards, more especially Schrödinger . . . It seems that the Nazis have got their hands on the machine and they will probably be there for a long time.' It was a good time to get one or two Jewish theoretical physicists to Oxford. Sommerfeld had suggested London and Bethe. 'I have the impression that anyone trained by Sommerfeld is the sort of man who can work out a problem and get an answer, which is what we really need at Oxford rather than the more abstract type who would spend his time disputing with the philosophers.' Einstein replied at once, offering to devote one-third of his salary for this year to help establish positions for young displaced scientists.

On May 6, the Nazis announced that the racial law would apply to all instructors in institutions of higher learning, and Fritz London was dismissed. He would join Schrödinger at Oxford. Max Born and his family had already left Göttingen and were staying in the Italian Tirol near Bolzano. He wrote to Anny to invite the Schrödingers to visit them there. Northern Italy was to be a sort of assembly place for many of the refugee scientists during the summer of 1933.

On May 10, the ceremony of the 'Burning of the Books' was held in most of the university towns of Germany. In Berlin about 40 000 people assembled in the great square between the university and the opera house. A torch-bearing parade of 5000 students, carrying Nazi banners and singing the 'Horst Wessel Lied', accompanied the trucks and cars bearing the books to be burned. Each student carried an armful of books to the huge bonfire as the name of the offending author was read out. Finally Dr Goebbels appeared, limped to the rostrum, and gave his speech: 'Jewish intellectualism is dead. The German national soul can again express itself.' The students responded with cheers and more songs, and the burnings continued for most of the night.

On May 27, the Nazis imposed a visa fee of 1000 marks for Germans wishing to visit Austria. This was an act of economic warfare by Hitler against the clerical fascist government of Dollfuss in Vienna. As a government employee, Schrödinger legally was considered to be a German citizen, and he was infuriated by this regulation, which prevented him from visiting his native land. He made an appeal to the Ministry for a special exemption so that he and Anny could attend her mother's seventieth birthday celebration, but it was rejected. Any lingering thoughts he may have had about staying in Germany were removed and he quietly arranged to give up their house and send their furnishings to England. He sent two large trunks to Switzerland. At this time it was still possible for those leaving Germany to take with them household goods, and even personal jewelry, although the export of money in excess of 1000 marks was prohibited. Later the Nazis confiscated all the property of anyone who managed to escape.

Anny Schrödinger was an enthusiastic motorist but an inexperienced driver. About the middle of May, she saw a bright new grey cabriolet in a B.M.W. show window in Berlin and fell in love with it. The Schrödingers bought the car, and made plans to travel to the South Tirol.[51]

Meanwhile Lindemann had returned to Oxford and obtained formal approval from I.C.I. for the necessary grants. The company agreed to provide £15 000 for thirteen two-year grants to Jewish refugees. The initial list of grantees included the physicists Heinrich Kuhn, Franz Simon, and Fritz London, who went to Oxford. Kurt Mendelssohn, assistant to Simon at Breslau, was already there on a previously arranged assistantship. The appointments of Schrödinger and March were to be made separately, since they were not Jewish refugees. Towards the end of July, both Einstein and Schrödinger called Lindemann's attention to the plight of Leo Szilard, a brilliant but temperamental Hungarian physicist whom they had known in Berlin, and the Prof arranged for him a temporary position at Oxford.

Other organizations began to follow the lead given by Lindemann. The munificently endowed Rockefeller Foundation helped to place almost 200 refugee scholars in American and European institutions in the period 1933–1939, at a cost of $743 000. The Guggenheim Foundation, however, did not respond to the emergency situation in any way.

In Oxford, the Prof surveyed the possibilities of a college fellowship for Schrödinger and found that Magdalen College might be able to elect him. On July 21, George Gordon, who was Professor of Poetry and the President of Magdalen, wrote to the Prof that the election of Schrödinger had been set down for the next college meeting, on October 3. It would be a supernumerary nonstipendary fellowship for a period of five years, with an annual grant of £200 to £250 a year. The post would be like that of a senior fellowship and the holder would be a member of the governing board of the college. Magdalen is one of the oldest, most famous, and most beautiful of the Oxford colleges. It was founded in 1458 by William Waynflete, Bishop of Winchester. The great tower, completed in 1509, has become a symbol for Oxford itself. The college buildings are surrounded by spacious gardens, riverside walks, and a large deerpark. Election to a fellowship at Magdalen is a signal honor, and it is pleasant to note that Erwin's election came before and not after he received the Nobel prize.

Tirolean adventures

Schrödinger did not formally resign his professorship in Berlin, but he wrote to the Minister of Education to request a 'study leave'. This letter was not answered at the time. He also sent a postcard to the porter in charge of notices in the physics department to inform him that his lectures would not be given in the fall semester. On September 1, payment of his salary was stopped. His manner of leaving his important Berlin position was like a slap in the face to the regime; even though he made no overt political statement, the Nazis recognized him as an enemy. Yet whenever he was asked why he had left, he gave a standard answer: 'I could not endure being bothered by politics.'

Schrödinger was exceptional – very few nonjewish professors refused to knuckle under to the Nazis. Among the distinguished resisters were Karl Jaspers, E.I. Gumbel, Theodor Litt, Julius Ebbinhaus, Hermann Bruck the astronomer, and perhaps a dozen others less famous. Most emigrated to America. Professor Theodor Lessing escaped to Czechoslovakia, but was tracked down and murdered by Nazi thugs. In the autumn of 1933, 960 professors published a vow to support Hitler. They were led by Heidegger, the existentialist philosopher, Sauerbruch, the surgeon, Pinder, the art historian, and the Rector of Berlin University, Eugen Fischer. Heidegger was appointed as Rector at Freiburg University and announced that 'The much praised academic freedom will be rooted out of the German university.'

Erwin had been ardently pursuing Hilde March for some months. He had even offered to divorce Anny, but Hilde did not want to leave Arthur. In May, Erwin wrote in his diary (in French, which must have seemed the appropriate language for amorous intrigue): 'Now for the first time I must make a little plot against her. To make her sleep with me. To hold her embraced – even if it is only for one night. It has never happened that a woman has slept with me and did not wish, in consequence, to live with me for all her life. I swear in the name of the good God that it will be the same thing with her.'

Hilde spent most of the month of June in Berlin. Erwin gave her his diary to read at leisure. This must have convinced her that he really loved and needed her, but she left to rejoin Arthur in Brixen with his 'little plot' still in abeyance. On July 6, however, he wrote, 'I think indeed she is mine.' It is worthwhile to note that when Erwin arranged the fellowship for Arthur March at Oxford, he and Hilde were not yet lovers, hopeful as he may have been.

About a week later, at the end of the summer semester at the University, Erwin was in a somber mood as the day of his final departure from Berlin approached. He was almost forty-six years old, and he reflected 'might it not be the case that I have already learnt enough of *this* world. And that I am prepared . . .'. Anny, however, was in good spirits as they left Berlin in the little B.M.W. with a chauffeur, who provided driving instruction as he drove them through Leipzig, Bayreuth and Nürnberg to the Swiss border at Singen, where he left them on their own.

They proceeded to Zürich where Anny met Wolfgang Pauli on July 21. She told him that they intended to visit the Borns at Selva Gardena (near Bolzano). Hermann Weyl had resigned his professorship at Göttingen, and also planned to be there. Pauli wrote to Heisenberg to try to persuade him to meet them all in Italy for some mountain climbing. It seems that he also had in mind the possibility of persuading him to leave Germany, but Heisenberg was not to be drawn away. Pauli discussed with Erwin the possibility that he might return to Berlin to try to oppose the Nazis. Erwin said only 'Ich habe der Nase voll'. [I've had a nosefull – I want to get out.']⁵² Pauli's suggestion supports the idea that opposition to Hitler was still feasible in 1933, but it would have been hopeless in view of the support of the Nazis by the big industrialists and landowners and the military establishment. Schrödinger was careful never to make an open denunciation of the Nazi regime. He maintained a formally correct relation to the University of Berlin, and later even received a form letter signed by Adolf Hitler thanking him for his services. Nevertheless he was entered in the Nazi records as 'politically unreliable' and marked for future retribution.

From Zürich, Erwin and Anny proceeded via Lenzerheide, the Engadin and the precarious Ofen Pass to Merano and Bolzano. This was a glorious journey through high mountain country, and if Anny was not an expert driver when they started, she would be able to manage almost any road by the time they reached Italy. She did most of the driving, since Erwin at the wheel would sometimes fall into a meditation, and this could prove fatal on an Alpine road.

By early August, the Schrödingers were installed on the Via Croce, Bressanone, amid the beauties of their much loved Tirol. Bressanone is the Italian name of Brixen, which was the birthplace of Arthur March, when it still belonged to Austria. The Marches were of course also there, while the Borns were with their children at Selva, nestled in mountains surrounding the Val Gardena, only about 40 km away. Max had worked at Cambridge as a postgraduate student with J.J. Thomson, and he received an invitation to return there. Anny visited the Borns with her beloved Peter Weyl.

Hilde's resistance was at an end, and before long, she and Erwin took off together on a bicycling tour of some of the nearby beauty spots. It is difficult to say to what extent their love affair was founded on mutual passion and to what extent it was based on the more reasoned estimation that they would make highly suitable parents. In any event, when the loving couple returned from their bicycle tour in the South Tirol, Hilde was pregnant. Hilde and Arthur had been married for four years and had no children, while Erwin and Anny had been married for thirteen years without children. The latter situation was not of Anny's choosing, for she loved children, but since early in their marriage, she rarely had sexual relations with her husband. 'I like Anny as a friend, but I detest her sexually', confided Erwin to his diary. Anny's love affair with Weyl was also childless.

Hildegunde March.

This was apparently the only instance in which Erwin had a love affair with the wife of a close friend and professional colleague. Arthur was certainly aware of the situation, and he gave it his blessing, at least after the pregnancy. Probably his admiration for Erwin was so great that he considered it an honor to share a wife with him. When they were all together later that summer at Lake Garda, however, Arthur's discomfiture was revealed by attempts at light-hearted humor, which he said were as necessary for his equilibrium as a balancing rod for a tightrope walker, but which tended to make Erwin rather nervous.[53] Anny had long since given up any objections to Erwin's love affairs, especially when Weyl was nearby, and she was always a good friend to Hilde as well as to Erwin. Conventional standards of sexual morality were irrelevant. Love and friendship were important and they could co-exist in many permutations without engendering anger or jealousy. Tensions inevitably occurred, but they were repressed, sometimes causing latent psychological traumas.

On August 12, Schrödinger wrote to Einstein from Solda, a mountain resort in the Alto Adige, where he was staying with Hilde. Strangely, this letter uses the formal 'Sie' instead of the intimate 'Du' of previous letters, perhaps because he was not sure how Einstein felt about him after the events in Berlin.

After the long time since I last saw you in beautiful Caputh before your American journey, I need to give you at least a sign of life. I hope and believe that I need not fear that the dear and beautiful friendship which you have always bestowed upon me has become dimmed in your thoughts, although unfortunately you indeed have had official cause enough. Spare me from speaking of these things, and let me hope that you know me well enough to make that speaking superfluous . . . I very much hope to see you in Brussels in October . . . Unfortunately (like most of us) I have not had enough nervous peace in recent months to work seriously at anything.[54]

On August 23, Erwin wrote to Lindemann. 'I intend to stay at Solda till 8 or 10 September, then go to Malcesine on Lago de Garda for about 14 days. You are sure to find me at Solda or there. I'd like to know of your visit 3 or 4 days ahead so as to advertise it to the Marches and my wife, who will be at that time about the Alto-Adige and would like to join us if possible. Arthur March, who spends the summer at Bressanone, intends to travel to Innsbruck to arrange about his leave.' Anny was visiting the Borns. Hansi Bauer, who was on her honeymoon in the Tirol, also visited Malcesine. She was already somewhat disappointed in her bridegroom, and Erwin advised her to put love above anything else. He met her once in the grocery shop and a 'spark of enlightenment' arced between them, which was destined to kindle later a more enduring flame.

Lindemann reached Italy early in September and conferred with

Schrödinger at Lake Garda and Born at Selva. He brought details of the Oxford appointments and news about prospects for the Magdalen fellowship. The I.C.I. appointment would run for two years from the date of Schrödinger's arrival at Oxford; the exact stipend was not yet arranged but it was hoped that his total salary would equal that of a professor. Among the conditions were that the appointee would normally reside in Oxford and carry out research there, that he would not disclose to others results of research of interest to I.C.I. without its consent, and any patent applications would be made through the company. Formally the appointment was like a consultancy agreement, but there were no formal duties required.

Autumn had now come to the Tirol. There were snowfalls on the mountains and chilly nights in the valleys. The long lazy summer was over and the refugees began to prepare for departures to their temporary positions, their uncertain futures. On September 27, Max Born came to Malesina with his small son Gustav to say farewell to the Schrödingers. By October 17, the Borns had moved into their 'little home' in Cambridge. They were happy there, and Max was attached to St John's college, where Paul Dirac was a fellow. Unlike the Clarendon Laboratory at Oxford, the Cavendish had an array of first-class physicists: Rutherford, Wilson, Aston, Chadwick, Fowler, Oliphant, and Cockroft. It was one of the great physics centers of the world.[55]

On October 3, the Prof wrote to Erwin: 'Cher ami, I am glad to tell you that you were elected a fellow of Magdalen [today]. Don't refer at all to the financial aspect in your acceptance. Simon, Kuhn, Mendelssohn, London, all are here. There is no worry about the entry permit.' Gordon wrote to the Prof that Schrödinger's total income would be at least £600. Funds were limited and demands on them were many. He thought, however, that 'If S is so brilliant and agreeable as I believe him to be, there will probably be a move . . . to transfer him to a more permanent post.'

Anny and Erwin loaded the B.M.W. and set out via the St Gotthard Pass for a meeting of the French Society for Physical Chemistry in Paris, where they arrived the third week in October. Anny bravely drove in the Paris traffic, where Erwin did not dare to take the wheel at all.

The Nazi control of the press had already become so effective that Schrödinger's departure from Berlin was noted in only one German newspaper, the Berlin *Deutsche Zeitung* of October 24, 1933. Under the heading 'Loss for German Science', it reported:

The German world of learning has suffered a severe loss: Professor Erwin Schrödinger, the successor of Planck as Ordinarius in Theoretical Physics at the University of Berlin, has received a call to the University of Oxford.

Schrödinger had already scheduled lectures for the winter in Berlin; hence the call to the English university, which will entail no teaching duties, apparently occurred not without his agreement, and must lead to the conclusion that the distinguished scholar, who has worked in Berlin since 1927, is leaving Germany. This is all the more to be regretted, as only a short time ago Professor Hermann Weyl, Ordinarius for Mathematics at the University of Göttingen, also in the field of theoretical physics, accepted the call to the American university of Princeton.

8   Exile in Oxford

President Gordon had hoped that Schrödinger would arrive in Oxford by October 21, so that he could matriculate and receive the M.A. degree that had been arranged for him (a requirement for the fellowship), but the VIIth Solvay Conference was to be held in Brussels, October 22 to 29, and he was obliged to attend this. The Schrödingers stayed at the elegant Hotel Gallia et Brittanique, and as usual Erwin arrived looking more like a Tirolean mountaineer than a distinguished scientist. Hilde March joined them there, and she wrote to Lindemann that she would be coming to Oxford with the Schrödingers, while her husband would be delayed. He replied that they were looking forward to seeing her and he hoped that she would 'persuade Professor March to come over as soon as possible'.

Seventh Solvay Conference[1]

The VIIth Council of Physics of the Solvay Institute, devoted to 'Structure and Properties of Atomic Nuclei', was opened by its President, Paul Langevin, on October 23 at the Université Libre de Bruxelles. Thirty-five of the most eminent workers in the field comprised the members of the scientific committee and the invited members of conference. Six members were from Cambridge. The German universities were represented by Walter Bothe (Heidelberg), Werner Heisenberg, Pieter Debye (Leipzig), and Lise Meitner who was still holding out at the Hahn Institute in Berlin. Marie Curie and Frederic Joliot, and both Maurice and Louis de Broglie came from Paris. The only American member was Ernest Lawrence from Berkeley. This was the first conference devoted to nuclear structure and nuclear reactions; with the discoveries of the neutron and positron, dramatic progress had been achieved in this field during the past few years and the results presented at the conference were important and exciting.

The discussions after the formal papers were extensive and lively, but Schrödinger took little part in them. The published reports contain

With Lindemann at Oxford (photograph by Charlotte Simon).

no comments either by him or by Louis de Broglie. Niels Bohr, however, contributed an essay on 'The Correspondence Method in the Theory of the Electron', as an appendix to Dirac's fascinating lecture on the 'Theory of the Positron'. Schrödinger had evidently not thought much about nuclear theory, and thus he remained silent and listened to the debates between the active workers in the field.

A.F. Joffe from Leningrad had a long discussion with the Schrödingers about the Hitler regime. He reported that 'they expressed openly antifascist convictions and showed themselves to be devoted to socialism'.[2]

Arrival in Oxford

The Schrödingers and Hilde March arrived in Oxford on November 4. Anny had mastered the art of driving on the left side of the road but had been careful to detour around London. They stayed at the Isis Hotel before moving into a furnished house, 12 Northmoor Road, and early in 1934, they moved to a lovely house with a large garden, at 24 Northmoor Road in one of the best residential areas of Oxford. The lease on this house (£150 per annum) was taken by I.C.I. and Schrödinger reimbursed the company on a monthly basis.[3] It was just around the corner from the house of Franz and Charlotte Simon. The

Marches rented an attractive smaller house, not far away at 86 Victoria Road.

Erwin found at Magdalen a letter of October 31 from Luther Eisenhart, Dean of the Graduate School of Princeton University, inviting him to lecture for 1, 2, or 3 months at $1000 a month plus $500 for travel expenses.

Erwin's formal welcome as a fellow of Magdalen College was accompanied by some exciting news. According to Armin Hermann:[4] 'At the conclusion of the ceremony, there was the usual "shake hands" with the other fellows and the almost ritually spoken words "I wish you joy, I wish you joy", and the fellows took their places at the high table for the festive dinner. But at exactly this time, *The Times* of London called Schrödinger's hotel to inform him that the Nobel prize in physics for 1933 had been awarded to Paul Adrien Maurice Dirac and Erwin Schrödinger. The prize for 1932 went to Werner Heisenberg. When they could not reach Schrödinger at the hotel, they called the college.' As Erwin recalled later in a letter to Max Born. 'On 9th November, 1933, dear George Gordon, the President of Magdalen College, called me to his office to tell me that *The Times* had said I would be among that year's prize winners. And in his chevalieresque and witty manner, he added, "I think you may believe it. *The Times* do not say such a thing unless they really know. As for me, I was truly astonished, for I thought you had the prize."'[5]

Although some years were to pass before James Watson destroyed the last vestiges of the myth that scientists are immune to worldly incentives, a Nobel prize was a reward far beyond its monetary value. Henceforth every word of the laureate, not about physics only, but also in fields of philosophy, education, and world affairs, would be brought to a wide and generally respectful public. 'Fame is the spur that the clear spirit doth raise (That last infirmity of noble mind) To scorn delights and live laborious days.' Erwin certainly had lived some laborious days, but he had not scorned all delights, and had even found time 'To sport with Amaryllis in the shade, or with the tangles of Neaera's hair.'

Congratulations poured in from all over the world. Lindemann wrote immediately to Rintoul at I.C.I. 'I was amazed to see this a.m. that Schrödinger has been awarded the Nobel prize . . . He should be given the maximum salary possible . . . £1000.' Owen Richardson, who had received the prize five years previously, wrote from King's College, London, with practical advice about the procedures in Stockholm. He warned Erwin that they would put him up in an expensive suite at the most expensive hotel, the bill for which would be presented just as he was leaving. He would then have a cheque for the prize, but it would be difficult to cash. Richardson kindly offered to

lend him some money for the trip if he was short of funds (as he well
might have been because of the German restrictions).

Nobel prize

On December 10, 1896, Alfred Nobel at the age of sixty-three died of a
stroke at his villa in San Remo. After accumulating vast wealth
through his inventions of dynamite and smokeless powder, he had
become a merchant of death who hated war, and a lonely prey to
sadistic fantasies who yearned to do something for humanity. The
bulk of his fortune of thirty million crowns (8×10^6 at the then current
exchange rate) was left to establish annual prizes in physics, chemistry, physiology and medicine, literature, and peace. Each prize was
worth about 200 000 crowns ($54 000). This was thirty times the annual
salary of a professor and two-hundred times that of a skilled worker,
so that the Nobel prizes were the richest ever known; by placing a high
monetary value on intellectual discoveries, they greatly enhanced the
respect of middle-class citizens for such achievements.[6]

Selection protocols are similar for the different prizes. In the case of
physics, nominations are solicited from a select group of scientists,
consisting of former prize winners, other noted scientists chosen on an
ad hoc basis, members of the Swedish Academy of Sciences, professors
of physics at the Swedish universities, at Copenhagen, Helsinki and
Christiana [Oslo], and at six other universities which vary from year to
year. The nominations are considered by a committee of five: the
president of the Nobel Institute for Physics and four members elected
by the physics section [*Klass*] of the Academy. Committee members are
elected for a term of four years, but in most cases appear to have been
re-elected for life. The committee report on the nominations, usually
with a recommendation for the prize, is considered by a meeting of the
Klass, which transmits its nomination to a plenary meeting of the
Academy. Members of the Academy receive a gold medal each time
they attend a meeting to vote, and members of the committee are paid
a generous salary. No records are kept of the deliberations of the
Academy, and members are admonished not to reveal any details. The
final decision must be made before November 15. On December 10,
the anniversary of the death of the founder, the prizes are awarded in
formal ceremonies in Stockholm and Oslo (for the peace prize). The
prizewinner is expected to give a public lecture on his work.

The will of Alfred Nobel had specified that the prize should be given
to the person whose 'discovery or invention' shall have 'conferred the
greatest benefit on mankind'. Over the years, this provision gave rise
to considerable controversy, since the members of the Academy were
almost all engaged in research in basic or 'pure' science, whereas

Nobel had been an inventor concerned with practical applications. The problem was solved by a gradual acceptance of the theory that pure research leads to an expansion of man's knowledge of the universe and therefore must ultimately confer 'the greatest benefit on mankind'. In the case of Fritz Haber, who received the chemistry prize in 1919, his development of poison-gas warfare did not delay the award.

The first physics prize was given in 1901 to Wilhelm Röntgen for 'the extraordinary services he has rendered by the discovery of the rays subsequently named after him'. Until 1915, there was a strong emphasis on the recognition of experimental work; of the twenty prizewinners, only Lorentz, who shared the 1902 prize with Zeeman, was a theoretician. The prevalent feeling was that physics is an experimental science, in which significant advances can be expected only from new discoveries in the laboratory.[7],[8]

The absence of theoretical eminence among the Swedish scientists led them to underestimate the importance of the revolution in twentieth-century theoretical physics. Einstein, who had published great discoveries in 1905, and again in 1915, did not receive the prize till 1921, and then for his work on the photoelectric effect. The Swedish Academy had hardly come to terms with the old quantum theory with an award to Bohr in 1922, arising from his great discovery in 1913, when it had to face a new barrage of revolutionary theories, which began with the discovery of wave-particle duality by Louis de Broglie in 1923.

At this time the members of the physics committee were S. Arrhenius (Stockholm), V. Carlheim-Gyllenskold (Stockholm), A. Gullstrand (Upsala), Manne Siegbahn (Upsala), and Carl Oseen (Upsala). When Arrhenius died in 1927, H. Pleijel (Stockholm) was elected, and in 1929, E. Hulthen (Stockholm) replaced Gullstrand. The task of evaluating nominations of theoreticians devolved mainly upon Oseen. He was a competent but conservative mathematical physicist, a specialist in hydrodynamics, who approached new ideas with extreme caution.[9]

The first nomination of Erwin Schrödinger for a Nobel prize in physics was received in 1927 from an unlikely source, David Starr Jordan, chancellor emeritus of Stanford University and a famous expert on American fishes. He had been invited in October to make nominations for the prizes in physics and chemistry, and after consulting David Webster of the Stanford physics department, he nominated A.H. Compton, E. Schrödinger, and P. Debye, in that order, for the physics prize. Jordan was pleased to learn that Compton had just been awarded the 1927 prize, shared with C.T.R. Wilson, inventor of the cloud chamber.

In his review of the 1927 nominations, 'Newer Trends within Atomic Theory', Oseen discussed the 'fall of the Bohr theory'. After a cursory account of matrix mechanics, he turned to the work of Louis de Broglie and objected that it was presented by the author in 'an awkward and dilettantish way'. He was favorably impressed by Schrödinger's solution of the hydrogen-atom problem, but believed that 'the task that lay before Schrödinger after this great success was to formulate a physical theory from the mathematical results achieved'. Oseen described Schrödinger's theory of the Stark effect, and his proof of the equivalence of wave and matrix mechanics. He mentioned the need to develop a relativistic theory and the difficulties in Schrödinger's attempt to interpret $\psi\bar\psi$ as an electron density. His conclusion was that during the past eighteen months, a revolution had indeed occurred in atomic theory.

This competent report shows that as early as 1927 Oseen, and hence the other members of the Nobel committee, were aware of the 'revolutionary' importance of the work of Schrödinger. Yet Oseen at this time was primarily concerned with the way in which theoretical physics might elucidate the relative importance of different experimental 'discoveries'. He was reluctant to consider a new theory to be itself a 'discovery' worthy of a Nobel prize, and this narrow interpretation of 'discovery' would become the basis of his opposition to a prize for the quantum theoreticians, which became increasingly obvious in subsequent years, and finally led to a critical situation as the nominations of Schrödinger and Heisenberg became more numerous and more forceful. It is curious that Oseen, who was himself a theoretician, adopted such a limited view of physical theory, but his own field was essentially a branch of applied mathematics.

In 1928, Edgar Meyer, Schrödinger's former colleague at Zürich, nominated him for a prize to be shared with Heisenberg. Another nomination that year came from Julius Wagner-Jauregg in Vienna, who had received the 1927 prize in medicine for the use of artificial fever in the treatment of tertiary syphilis.

In its report to the *Klass*, the committee noted the nominations of the 'pioneers of atomic physics', Heisenberg and Schrödinger, but, following the advice of Oseen, it expressed reservations in view of the highly mathematical character of their work. 'Heisenberg has, from the beginning, abandoned any attempt to create a theory possible to understand with physical thought processes and has been satisfied with mathematical processes in the matrix calculations . . . The same holds for Schrödinger's wave mechanics as can easily be seen from the fact that the space in which his waves propagate is not the usual three-dimensional space.' Furthermore, although Schrödinger uses more easily accessible mathematical methods, his results can be translated into the Heisenberg language. 'Work on the interpretation of the theory has not led to an experimental discovery of basic importance, nor has it clarified the logical basis of the theories.' Thus the committee could not suggest that de Broglie, Heisenberg or Schrödinger be awarded a prize either in whole or in part. The 1928 prize was reserved, but it was awarded the following year to Owen Richardson for his experimental studies in thermionic emission, which were basic to the rapidly developing electronics industry.

The number of nominations received by a candidate does not appear to have strongly influenced the selection committee, although eventually, as in the case of Planck, the 'nomination pressure' might become too insistent to be disregarded. Most nominations were one-page letters with concise mention of the work concerned, or even a remark that it was too well known to require comment; only occasionally were detailed supporting statements provided.

In 1929, Vienna was one of the universities invited to submit nominations. Hans Thirring, now the professor of theoretical physics, nominated Heisenberg and Schrödinger, as did Friedrich Kottler. The other Viennese physicists (G. Jäger, S. Meyer, K. Prizbram, and E. Haschek) nominated Otto Stern and Walter Gerlach for their measurements of magnetic moments by atomic-beam techniques, and their demonstration of space quantization in a magnetic field. In case theoretical candidates were desired, they suggested Sommerfeld, de Broglie and Schrödinger.

Jean Perrin, from Paris, who had received the prize in 1926, nominated Louis de Broglie, 'who was the first to propose that light and matter have an essentially analogous structure (particles guided by trains of waves) and who predicted quantitatively the frequency of the waves necessarily associated with a moving electron'. He included an electron-diffraction picture to emphasize the experimental confirmation of de Broglie's idea. He also mentioned Schrödinger, who, 'without being a creator of the same rank [*titre*] as M. Louis de Broglie, has contributed to the new theory developments of capital importance'. Eugene Bloch, in Paris, recommended that the prize be divided between de Broglie and Schrödinger.

Enrico Fermi from Rome nominated Heisenberg, with no mention of either de Broglie or Schrödinger: particularly valuable was his work on the 'conceptual content of quantum mechanics. This work has established an interpretation of the formalism of q.m. whereby the final elucidation of the most important atomic questions appears to be only a matter of time and the overcoming of mathematical difficulties.'

Schrödinger received seven nominations in the 1929 prize, with one exception jointly with de Broglie or Heisenberg or both. Altogether twenty-eight different persons were nominated by forty-eight nominators, and seventeen of the nominees were in the field categorized as 'atomic physics'.

Once again, Carl Oseen had the responsibility of guiding the physics committee through the unfamiliar pathways of quantum mechanics. He recalled his previous remarks about the theories, but now he believed that abundant experimental proof of wave-particle duality had been obtained, not only in the work of Davisson and Germer and that of G.P. Thomson and A. Reid, but also in the electron diffraction studies of Rupp and of Kikuchi. He would like to recommend splitting the prize between de Broglie and Davisson, Germer and Thomson, but this cannot be done since Thomson was not nominated. One cannot give the prize to Schrödinger without honoring Heisenberg. Thus, 'if the discovery of the wave nature of electrons is to be honored this year, the prize should be given undivided to L. de Broglie'. This recommendation was adopted by the committee, supported by the *Klass*, and voted by the plenary meeting of the Academy.

For the 1930 prize a new set of nominations was received. The Svedberg of Upsala nominated Heisenberg with reference to his paper on nuclear-spin isomerism in ortho-para hydrogen. He thought that this might be the important experimental discovery based on quantum mechanics which had hitherto been lacking. As a member of the physics *Klass*, he would be aware of Oseen's emphasis on this point.

Schrödinger received eight nominations for the 1930 prize, out of a total of thirty-seven received for twenty-one candidates, most of whom were in the field of atomic physics. Among his nominators were Frenkel (Leningrad), Planck and Laue (Berlin), and Nagaoka (Tokyo).

In his report, Oseen first considered an award to Born and Heisenberg, which he decided would be impossible since if Heisenberg shared a prize, Jordan should also be included, but he had not been nominated. He rejected the proposal of Raman and Heisenberg, because the first theoretical prediction of the effect was due to Smekal. Next he took up the case for Schrödinger and Heisenberg, and devoted a long analysis to the suggestion of Svedberg that ortho-para hydrogen might be the 'experimental discovery of great importance' that would justify a Nobel prize for quantum mechanics. In accord with tradition, however, the properties of atoms and molecules are the concern of the Nobel chemistry committee. [It was this division of territory that had led to a chemistry prize for Rutherford in 1908, and his amazement at his overnight transmutation from a physicist to a chemist.] Thus Oseen was dubious whether 'the allotropy of hydrogen' could be considered as an important discovery in *physics*. He concluded, however, that it did support the work of both Heisenberg and Schrödinger, and if a prize were to be given in theory, it should be shared between the two. In the event, the committee chose the experimental option and recommended that the 1930 prize be given to Raman.

Only a few new nominations were received for 1931. Louis de Broglie nominated Schrödinger; Niels Bohr, Heisenberg and Schrödinger. The committee had now reached an impasse. Oseen had become adamant in his opposition to an award for quantum mechanics, and most of the other advances in contemporary physics depended in one way or another on that theory. After long and inconclusive discussions, it was decided to reserve the 1931 award till the following year.

For the 1932 prize, Pauli nominated Heisenberg with the remark: 'If I place Heisenberg above Schrödinger for the Nobel prize, it is for two reasons, (1) Heisenberg's matrix mechanics preceded Schrödinger's wave mechanics, (2) Heisenberg's creation must be considered as more original, since Schrödinger could rely for the idea to a consider-able extent upon de Broglie.' In an interview in 1963, Dirac was asked 'Where do you place Schrödinger?' His reply was 'I'd put him close behind Heisenberg, although in some ways Schrödinger was a greater brain power than Heisenberg, because Heisenberg was helped very much by experimental evidence and Schrödinger just did it all out of his head.'[10]

The Swedish theoretician David Enskog wrote that the most impor-tant works in recent physics were matrix mechanics and wave mechanics. While Heisenberg's work had been developed with the collaboration of Born and Jordan, Schrödinger had worked single-handed. He outlined the vast range of problems that were solved by the new mechanics, including the most important of all chemical problems, the nature of the covalent bond, which was elucidated by Heitler and London.

An important nomination of Schrödinger and Heisenberg came from Einstein.

The contributions of the two men are independent of each other and so significant that it would not be appropriate to divide a prize between them. The question of which should receive the prize is difficult to decide. I personally value Schrödinger's contribution higher, because I have the impression that the concepts created by him will extend further than those of Heisenberg. If I had the decision, I would give Schrödinger the prize first.

In a handwritten footnote to this letter, he added: 'This, however, is only my opinion, which may be wrong.'

A Swedish member once remarked in a mixed simile that 'sitting on a Nobel committee is like sitting on a quagmire, one doesn't have a firm foundation under one's feet.' Until now the committee had been meekly following the advice of Carl Oseen that the award of a prize to the quantum mechanicians would be premature. As more nominations were received from many of the most eminent physicists in Europe, it was becoming obvious even to the experimentalists that quantum mechanics was probably the most important theoretical discovery of the twentieth century. A whole series of major advances that depended on quantum mechanics could hardly be considered while the basic discoveries remained uncrowned with the Nobel laurels. Yet the opposition of Oseen became even more determined, and in 1932 he prepared a report that finally made evident his intransigence.

I believe I should develop somewhat more closely the reasons that have made it impossible for me in previous years to give the prize to either Heisenberg or Schrödinger. The obstacles that have stood in the way relate both to the by-laws of the Nobel Foundation and also to inner difficulties in the theory. According to the will of Nobel, the prize in physics should be awarded to the person who has made the most important discovery or invention, words which used in the ordinary sense mean either an advance in our knowledge of factual reality or, secondly, a useful application of such knowledge. It is apparent to me that neither of these theoreticians has been able to make such a discovery or invention. The condition for an award must then be that their work has led to some discovery or invention of such importance as to deserve a Nobel prize. In my view, I do not believe that such a condition has been satisfied.

Now that Oseen had taken such a negative stand, the committee thought that an additional report from a less rigid viewpoint would be desirable, and they asked E. Hulthen to prepare a 'special submission'. He viewed the situation as follows: 'Heisenberg and Schrödinger have by different routes set up a symbolic quantum mechanics which gives a formally unobjectionable correlation of an area of empirical fact the extent of which is without parallel in earlier physics.' This statement would please the ghost of Ernst Mach, but it is hardly more positivistic than the interpretation of quantum mechanics that was then popular among philosophers of science and many scientists, including even Bohr and Heisenberg. Hulthen, however, almost immediately went beyond the Machian framework when he wrote, 'on the basis of the indistinguishability of electrons and of nuclei, Heisenberg formulated a new theory for resonance phenomena which revealed new aspects of the atomic world'. He recommends that the prize for 1931 be divided between Heisenberg and Schrödinger, and that for 1932 be reserved till the following year.

Despite this dissenting report, the influence of Oseen was still too powerful. The committee members were anxious not to allow any

The Nobel party at the Stockholm Station, Anny, Mother Dirac, Paul, Werner, Erwin.

open discussion in the meeting of the *Klass*, which would have taken the decision out of their hands. After a long and bitter debate, they persuaded even Hulthen to join in a unanimous recommendation that no prize should be given for 1931 and that the 1932 prize should be reserved till the following year.

Thus for two years no Nobel prize in physics had been awarded, even though eminent scientists such as Einstein, Bohr, Planck, and de Broglie had testified to the epoch-making importance of quantum mechanics. It is difficult to appreciate the reasoning of the committee, but none of them except Siegbahn was working in an area close to atomic physics, and he could be relied upon to support his Upsala colleague Oseen. When a prize was not awarded, the money could revert to the Nobel Institute for Physics.

In 1933, many new nominations were received. Schrödinger received nine nominations, including those of five Nobel laureates, and Heisenberg received eight. In the event, a key to the impasse in the committee was provided by the suggestion of Lawrence Bragg that the prize be divided between Heisenberg, Schrödinger, and Dirac. This was the first time Dirac had been nominated. The capricious effect of nominations is illustrated by the fact that over the years Heisenberg and Schrödinger had each received twenty-five nominations.

Oseen prepared a long report which reviewed all the work of Dirac from 1926 to 1933. He concluded that 'it is indeed an impressive body of work'. He believed that Dirac's first researches were mainly based upon the ideas of Heisenberg, and only after this influence waned, did Dirac make highly original contributions. Such 'creations' were the theory of the one-electron problem (1928), the theory of magnetic poles (1931), and the attempt to provide a new basis for quantum mechanics (1932). After all these years, Oseen was prepared to call a theory a 'creation' rather than a mere 'correlation'.

With reports on Dirac and Pauli available, and the previous reports on Heisenberg and Schrödinger brought up to date, the committee now roused itself to come to a decision.

The Committee is of the opinion that the point in time has now arisen at which the question of the founders of the new atomic theory should be decided . . . A great discovery from recent times, the discovery of the positive electron, has totally changed one of the most difficult objections against the new theory into a support for it . . . Under these circumstances, a sufficient cause is now at hand to propose an award to the founders of the new atomic physics.

The Committee proposes that the Nobel prize for 1932 be given to Professor Heisenberg (Leipzig) for the presentation of quantum mechanics and applications of it, particularly the discovery of the allotropic forms of hydrogen, and that the Nobel prize in physics for 1933 be shared between Professor Schrödinger (Berlin) and Professor P.A.M. Dirac (Cambridge) for the discovery of new forms of atomic theory and applications of them.

Stockholm

Erwin and Anny arrived in Stockholm on December 8, two days before the formal presentation of the prizes. This took place on Sunday evening at 5:00 p.m. in the grand ballroom of the Concert Hall, which was decorated with an array of Swedish flags and a profusion of flowers. A large bust of Alfred Nobel occupied a place of honor. Almost all the places in the Hall were filled with invited guests, including the prime minister, members of the cabinet, and the ambassadors of Britain and the United States. The prize in physiology and medicine had been awarded to Thomas Hunt Morgan, but he was not able to be present owing to important departmental business at Caltech. The laureate for literature was Ivan Bunin, a Russian author living in the south of France. The prizes for chemistry and for peace were not awarded in 1933.[11]

Exactly at five o'clock, King Gustavus arrived, accompanied by members of the royal family. The royal anthem was played by the orchestra and sung by the standing audience. Then the laureates entered, escorted by Swedish prizewinners of past years, Siegbahn,

Svedberg, von Euler, Barany and Dalen. Students paraded on either side of the platform with German, British and American flags.

A short opening discourse by M. Hammarskjold was followed by a detailed account of the life and work of Nobel, this year being the 100th anniversary of his birth. The orchestra played the *Polonaise Festivale* of Alfven, and then Pleijel, president of the physics committee, presented the prizes to Heisenberg, Schrödinger, and Dirac. His citations were brief but scientifically precise. Each laureate received a diploma, a gold Nobel medal, and a medal from the Academy of Sciences. The obverse of the Nobel medal was the excellent portrait of the founder by Erik Lindberg, and the reverse bore the name of the laureate and an inscription from the *Aeneid*, 'Inventas vitam juvat excoluisse per artes' [It is delightful to discover life embellished by the arts]. The Academy medal portrayed Nature as the goddess Isis emerging from clouds and holding in her arms a cornucopia while the veil which covers her face is lifted by the Genius of Science.

After the *Egmont Overture*, the physiology prize was awarded, and following Grieg's *Danse Symphonique* the prize in literature. The ceremonies were concluded with the Swedish National Anthem.

The next event in the festivities was the ceremonial banquet in the Winter Garden of the Grand Hotel. The Prince and Princess Royal and all the princes and princesses were there but not the king and queen. Heisenberg and Dirac were accompanied by their mothers. After formal toasts to the King and to the memory of Nobel, toasts were offered to the prizewinners, who replied at different lengths. Dirac's words were somewhat incongruous. He said that the world was in the depths of economic misery, but anything to do with numbers should be capable of theoretical solution, and that the cause of the great depression was that people much preferred to collect interest 'through all eternity' instead of taking a single payment for their goods or services, and hence there was a great shortage of buyers for the world's goods. Anny thought that his talk was 'a communistic propaganda tirade'.[12] Bohr once wrote that Dirac had the 'purest soul of all the physicists'. Heisenberg spoke very briefly to thank everyone for their hospitality.

Schrödinger, perhaps with some inspiration from the champagne, gave the most spirited response, full of 'enthusiastic ardor' in the words of the reporter.

There are things in life that one cannot learn from experience – one must get it right the first time . . . You may say perhaps that I had time yesterday and today to prepare myself . . . But that is easier said than done. Put yourself please in the place of a man who for the first time wakes up early in the morning with the bright sunshine to set foot in this wonderful city with its expanses of water, splendid buildings, its proud castle, its stones and spires, the happy, kindly people, who accept you with friendship; this city, which is so modern, so different from any other in the world, so that it has you immediately under its spell, and you want only to walk about and look and climb here and there, in order to drink in its picture as fully and completely as possible. Yes, one can do nothing else, one does not hear the call of duty, but thinks that somehow or other tomorrow will go well – people who live in such beauty will surely not be such stern critics.

He thanked everyone in glowing terms – from his Majesty and the royal family to the committee members who so carefully consider the prizes. And then he came to his 'most difficult part' – the ending.

Let me close with an egoistical wish. I hope that I may come again soon . . . It will not be to a celebration in halls bedecked with flags and not with so many formal clothes in my luggage, but with two long skis over my shoulder and a knapsack on my back. And I hope then to learn to know this country that has shown me so much generosity and affection and – if that be possible at all – to learn to love it even more deeply than today.[11]

These words were received with thunderous applause. The party continued with an unceasing flow of champagne for another two hours before the exhausted but happy participants retired for some much needed sleep.

The next evening there was a smaller dinner given by the royal family at their palace. There were 100 guests and 40 servants to wait upon them. The king and the crown prince were served by a special servant wearing a high hat with feathers. Last night the menu had been consommé, sole, chicken, and ice cream, but tonight they dined on caviar, salmon, pheasant, and venison. After the dinner there was a reception for 1200 persons at the National Gallery. The Nobel festivities are the high point of the Stockholm social season, and the entire city takes part in them.

Schrödinger's formal Nobel lecture was given on December 12, on 'The Fundamental Idea of Wave Mechanics'. The lecture was intended for an audience interested in science but without specialized knowledge of mathematical physics. It was essentially a more popular account of the way in which he had developed the analogy between optics and mechanics in the second of his great Zürich papers of 1926. It was a modest lecture in that it did not give any emphasis to his own contributions to the subject and the revolution they wrought in physics and chemistry. The lecture would have been quite appropriate as an account of the earlier work of Louis de Broglie. The felicity of style and the careful explanations are nevertheless typical expressions of the genius of Erwin Schrödinger.

When Erwin and Anny returned to Oxford, they found a Christmas present addressed to Erwin from Peter Weyl. This was a fine edition of

Shakespeare's *Venus and Adonis*, with beautifully erotic illustrations. Peter had written an inscription: 'The sea has bounds but deep desire has none.'

Princeton[13]

At the beginning of the fall term of 1933, the physics department at Princeton University was considering how to fill the Thomas D. Jones Professorship of Mathematical Physics. The holder of the chair was expected to devote himself to research; there were no formal teaching duties but if he so wished he could give courses for advanced students. This was one of Princeton's most distinguished professorships. In October a meeting of the physics professors decided, after much discussion, that their first choice was Heisenberg and their second, Schrödinger, but they would like to invite the latter for a visiting lectureship. Accordingly, Eisenhart had invited Schrödinger to visit Princeton for one to three months. He accepted and left for the U.S.A. on March 8 on the *President Harding*, returning to England on April 13. A special working desk was installed in his cabin, so that he might waste no time on shipboard.

While in Princeton, Erwin lived at the Graduate College, a structure in collegiate Gothic style in imitation of an Oxford college, with tower, common room, a large refectory with stained glass windows, and accommodations for students and fellows. The English atmosphere was not recreated perfectly, however, since the plumbing was modern, no alcoholic drinks were allowed in the public rooms of the college, and the servants were not old British characters with traditional lower-class accents, but young and breezy Cypriots with an erratic command of American slang. Since Erwin did not care for the English original, it is doubtful that he was much impressed by the American imitation.

Schrödinger's lectures, as usual, were models of scientific exposition, and, early in April, he was offered the Jones Professorship. He wrote immediately to Lindemann to advise him of the situation:

Other things have come up here, on which we shall have to have *ausführliche* [extensive] talks after my return. There is not much use discussing them in a letter, yet since – in spite of all confidentiality – rumours spread out almost in contradiction with the principle of light, I should like to avoid your hearing them in this way. Yesterday a permanent chair at this university was offered to me. Strange situation! It is only in three weeks from now that my wife will be busy to move us *at last* into definite and permanent quarters in Oxford. And precisely during these weeks I am forced to take into preliminary consideration a new aspect, under which the state reached in Oxford would not be so very 'permanent'. I am not frightfully happy at that, yet I am aware that I have to consider things seriously. For after all and in spite of all so-called scientific reputation, I am actually without what a man of my age and métier considers a *Lebensstellung* [permanent position]. And if, e.g., I were drowned on the passage, I am afraid that my wife could neither live upon the German pension nor on the '*Schrödingergleichung*' [Schrödinger equation].

After his return to Oxford, Schrödinger delayed any decision on the Princeton offer. Since Ladenburg was going from Princeton to a meeting in Oxford at the end of June, he was asked to discuss the matter with him. Schrödinger finally decided to decline the Princeton professorship. He stated some of his reasons in a letter to Ladenburg, written in October, after he had seen him at the Oxford meeting:[13]

Thus the situation is this. The conditions of appointment with respect to salary and pensions are not sufficiently agreeable to me, especially when I compare them with those provided in several cases (I am told: normally) at the Institute. I have simply spoken [to them] about the pension in terms of its provisions, since this seems to me to be unsuitable regardless of any comparison, and because I wished to avoid any express reference to the Institute, which perhaps would cause annoyance.

I . . . do not believe that the university can bake such extra sausages for me as to make up the difference. It is very embarrassing to still want something when one has been told: This is about the best there is, nobody has any more, and nobody has a better old-age pension.

Now, listen, dear Herr Ladenburg, intrinsically in spirit and in work, fortunately no dividing line exists between University and Institute. Those of my friends who have wanted me to be in Princeton would not care whether I came to one or the other and it's also basically all the same to me . . . I think I recall that there are even 'amphibians'.

Whether Herr Flexner possibly is thinking about inviting me, I do not know. I hear he is now strongly inclined to an expansion in the direction of theoretical political economics. But this I do know: if he should [want to] invite me, the thought that he might thereby give the university a knock in the head . . . might perhaps prevent him – for there are certainly enough others, without wanting to trouble good relations because of one individual.

I hope you don't take it as a very serious calamity that Flexner 'devalues the prizes'. We, dear friend, you and I and our friends, we are indeed merchandise and not merchants . . .

Considering the above mentioned alternative plans of Flexner (which I learned from himself) the probability is not on the whole very great that I should still perhaps come to Princeton in this way. But it would be effectively nil if Flexner cannot be completely at ease that no misunderstandings are to be feared with the university authorities.

And now 'bon voyage' [Glück auf dem Weg]! Please give my best regards to your dear wife and children. And 'remember me to all my friends' [English] who are so many that I do not want to count them, but greet Veblen especially, for I am really *fond* of him and his wife.

The salary of the Jones professorship was $10 000 a year, about as good as any American professorship at the time. When Einstein was thinking of coming to Princeton, he asked Flexner for a salary of $3000 a year ('Could I live on less?'); his salary for 1933 was set at $15 000 a year. This was probably the figure that Schrödinger mistakenly regarded as 'normal' at the Institute; it was probably unique in the U.S.A. at the time. The top salary of $10 000 at Princeton was about twice the £1000 he was receiving at Oxford, and this was regarded as excellent for an English professor at the time. The Princeton University pension scheme of 1934 was indeed unsatisfactory. It took no account of possible inflation, and in the event of a professor's death, the widow would receive no more than $2000 a year. Erwin had an almost neurotic fear of an impoverished old age, not so much for himself as for Anny. This doubtless had its origin in the experience of his family in the postwar inflation in Vienna, when his parents were reduced from affluence to penury, and he himself had to live on Anny's earnings as a secretary, while his university salary became virtually worthless. Thus old-age pensions became something of a fetish for him; his worry about future disasters prevented a reasonable view of present opportunities. He still had his Nobel prize money safe in a Swedish bank. He was one of those fairly well-to-do people who always feel afraid of poverty.

The reasons given by Erwin for not accepting the prestigious American professorship may not tell the whole story. There have been persistent rumors at Princeton that he discussed with President Hibben his plans for an unusual family arrangement, in which he would live with Anny as an official wife and Hilde as a second unofficial wife, both women sharing the care of his expected child. He wondered if this extended family would be acceptable to the Princeton community, and he even worried about a New Jersey law against bigamy. President Hibben was a conservative man, and he would have expressed to Erwin his doubts about the desirability of such a menage. Despite the fact that everyone in Princeton believes this story, it seems improbable that Erwin would have gone out of his way to outline this family problem to the President. On the other hand, he did write in his journal that it would be sad to go away and 'leave the mother and the child'. Since he had been served ice-water with his oysters in 1927, Erwin had never overcome his aversion from the American way of life and, after Vienna, Zürich and Berlin, Princeton must have seemed somewhat as it did to Einstein, 'a quaint ceremonious village of puny demi-gods on stilts'. He would probably have accepted a position at the Institute with salary equal to Einstein's, but despite considerable pressure, Flexner was not persuaded to make this available.[14]

Life in Oxford

During term time at Oxford, Schrödinger was scheduled to give one lecture a week, on Elementary Wave Mechanics. This was held at the Clarendon Laboratory at noon on Saturdays. The lecture hall was icy cold in winter, warmed only by the steamy breaths of the students. In the academic year 1933/34, there were 59 men and 4 women enrolled in all the physics courses at the university, 8 in the first year, 23 in the second, 21 in the third, 8 in the fourth, and 3 in the fifth. Most of these students would have been poorly prepared to follow Schrödinger's lectures and C.H. Collie was asked to give a class in theoretical physics leading up to quantum mechanics. This met during Trinity term from 5 to 7 p.m. at the Mathematical Institute. Unlike the German system in which all of physics was covered systematically in comprehensive lecture courses, the English universities relied upon a few schematic lectures which were intended to suggest to the students lines of independent study. Often even the few lectures provided were not especially good. Lindemann lectured twice a week on different topics each term, but owing to a slight speech impediment, his lectures were somewhat unintelligible.

Erwin thought that these teaching arrangements compared unfavorably with those in Germany, and he often complained to Anny that he was being paid for doing nothing or that he was being treated like a charity case. His lectures, though few, were highly appreciated and were said to be the best physics lectures ever heard in Oxford.

While Erwin was away at Princeton, Anny had visited the Borns in Cambridge. Max intended to visit Oxford early in May, but had to postpone his visit because Erwin was recovering from a cold that had settled, as usual, into his chest. Then on May 23 he came down with a case of chicken-pox [varicella], never having had an opportunity to catch it at a primary school.

Max Laue visited Oxford from May 29 to June 2 and stayed with the Schrödingers. He also visited Fritz London, and wrote:[15]

London has a charming small house in the Thames valley but he is always complaining about the lack of furnishings and facilities, which in England I have only a slight feeling for. Schrödinger also feels himself to be not at all happy, although he is the only one of the German scholars who has had the honor to be made a fellow of a college. He speaks of leaving and alternates in his mind between the most distant lands of the earth.

It was only to be expected that these refugees, with nostalgia for their homelands, should shed tears among the alien corn. Erwin did not mention to Max that he was on the point of becoming a father, a fact that may have contributed to his distraught air.

It was late Spring in Oxford when Erwin's first child, a daughter, was born. Hilde March had entered the Maternity Home attached to the Acland Hospital and the baby was born on May 30, 1934. She was christened in the Church of England as Ruth Georgie Erica. The middle name was in memory of Erwin's mother Georgine who was always called Georgie. Erica Boldt was a recent mistress of Erwin's and Hilde also had a dear friend called Erica, so that this name had fond connotations for both parents. The birth was registered by Hilde on July 2, and of course Arthur March, professor of physics, was named as the father.

It is said that Hilde suffered a post-partum depression and tended to reject the baby. Anny, however, was very supportive and for several months took over much of the care of the infant, although it is not certain whether she kept it temporarily in the Schrödinger house. Once while watching a nurse change the two-week-old infant, Erwin remarked, 'Goodness, what a lot of things you need to know to treat such a little mite correctly.' Matter-of-factly she replied, 'Oh, I suppose so, but it takes quite a lot to kill it.' He remembered this remark twenty-five years later.[16]

Before the birth of the baby, Erwin and Hilde used to go everywhere together in Oxford, and he made no attempt to conceal their special relationship. He did not regard her as a mistress, but rather as a second wife who happened to be married also to another man. Conventional sexual morality was simply not worth bothering about, so long as those directly concerned accepted the unusual situation. His understanding of personal feelings was often naive. More than one of his friends have even called it 'childish'. He seemed to carry over into his mature life the patterns and attitudes that he had learned as an only child surrounded by three doting women with little to do but cater to his whims. Yet Musil cautions us: 'How narrow-minded it is to ascribe to a person by way of a character a tendency to repetition that he has acquired involuntarily, and then to blame his character for the repetitions.'

Erwin loved and appreciated women, but his attitude towards them was essentially that of a male supremacist. Oxford society, on the other hand, was based on a sort of official misogyny. The kind of thing that he found uncongenial was the traditional Christmas party at Magdalen College. The main party was held in Hall – that is where the fellows and their male guests and the choir boys gathered around a big Christmas tree; wives and other 'female appendages' were confined to the gallery of the Hall. The first part of the evening's entertainment consisted of the choir's singing part of the Messiah. Then there was a break during which the men and boys ate and drank and the boys received their presents from the Christmas tree. The ladies during that

Anny Schrödinger in Oxford.

interval went into the Senior Common Room and had their refreshments there, after which they came back to the gallery for the second part of the entertainment, the singing of Christmas carols. Just before midnight, there would be silence until the College clock sounded midnight and at the last stroke the choir would break into an anthem by Pergolesi. The bell ringers would start pealing the bells. Oxford fellows recollect that 'it was very fine', but Erwin could never agree that choir boys were preferable to ladies.

Alfred J. Ayer who was a lecturer in philosophy at Christ Church College, where Lindemann lived, has described Oxford society at this time in his autobiography.

It was still almost wholly masculine . . . The men had spent most of their time at boarding schools in a homosexual atmosphere. They were ill at ease with women. Such sexual experience as they had was almost always homosexual. The active homosexuals were not a majority of Oxford men, but they were very much in evidence. The tone was set by a number of celebrated 'Queens', whose flamboyant appearance was joined to a studied formality of manner. Many of those who paid court to these queens were not radically homosexual, but simply continuing their schoolboy practice of using boys as a substitute for girls . . . Life in the colleges was organized on monastic lines. The good food and drink were for men only. If fellows had wives, they were not admitted. Many of the permanent staff were homosexual.[17]

Schrödinger regarded a society without women as detestable and barbaric. He complained to Max Born, 'These colleges are academies of homosexuality. What queer types of men they produce.' He did not enjoy the college dinners. 'You never know who your neighbor might be. You talk to him in your natural manner, and then it turns out that he is an archbishop or a general – huh!'[18]

It is not surprising that Erwin found it difficult to adapt to Oxford society. When Lindemann, who was certainly a misogynist if not a homosexual, discovered Erwin's liaison with Hilde March, he became furious, with words to the effect that: 'He said he wanted to bring the husband here as his assistant, but he was really after the wife. We ought to get rid of the bounder.' It was bad enough to have one wife at Oxford – to have two was unspeakable. Relations between the Prof and Schrödinger became rather strained. Lindemann was trying to get I.C.I. to renew the grants for refugee scientists beyond the two years originally promised. I.C.I. was reluctant to spend more money, but finally agreed to an extension through 1936. One director said they already had been extremely generous since they had paid not only for the scientists but in some cases for their mistresses.

Spain

In the summer of 1934, while Anny went to Switzerland to be with Weyl, Erwin visited Spain, where he had been invited to give a course of lectures at the Universidad Internacional de Verano in Santander, which was directed by the philosopher José Ortega y Gasset, and also to lecture in Madrid. Ortega had recently gained international recognition outside the field of philosophy by his book *La rebelion de las masas* (*The Revolt of the Masses*) which expressed in rather flamboyant terms his aristocratic view that democracy and mass society posed a growing threat to authentic personal existence.

The theme of the Santander 'summer university' was 'El Siglio XX'. Schrödinger's course of six lectures was given in August in the Castiglio de Magdalena in Santander. Because of the great heat he dressed in tennis shorts and sports shirt. The lectures were translated by Xavier Zubiri and published under the title *La Nueva Mecanica Ondulatoria*.[19] They were intended to explain the fundamental ideas of wave mechanics to a nonmathematical but somewhat philosophical audience. They are written in an unusually informal style, and one can almost hear the author explaining the deepest questions with a charm and lucidity that no other physicist could have equaled. His lectures on the vectorial representation of the wave functions (in Hilbert space) and on the equivalence of wave and matrix mechanics are masterpieces, made even more luminous by the beauty of the Spanish language. A lecture on 'Causality and the Concepts of Classical Physics in the New Mechanics', despite the elegance of its presentation, did not go deeply into the concept of causality, which he defined as the principle that the state of a physical system is determined unequivocally for all future times once its initial state is specified. He pointed out that, even classically, the inevitable lack of experimental precision in the definition of the initial state may, after a sufficient lapse of time, lead to large uncertainties in later states, but having thus disposed of his own definition of 'causality', he devoted his lecture to a discussion of the microscopic indeterminacies of the Heisenberg type, especially in the 'two-slit experiment' with electrons and the emission of light by atoms. His final lecture brought forth the incompatibility of quantum theory and relativity, and the serious problems caused by the privileged situation of the time variable.

Erwin was greatly impressed by both the personality and the ideas of Ortega, who despite his aristocratic background was a loyal citizen of the republican Spain (later he escaped into exile when the fascist forces captured Madrid). Ortega thought that politics and government are inevitably degrading and the sensitive intellectual must avoid such activities and devote himself to 'rational and responsible' personal relations like love and friendship. It is doubtful, however, that Erwin would have gone so far as to accept the view of Ortega that 'philosophy, science, and mathematics are pure exact fantasy, games played according to strict but arbitrary rules, by a minority who seek to escape the tedium, vulgarity, and deadly seriousness of the world of beliefs.'

He liked Spain so much that he decided to return as soon as possible. The following Spring, he set out with Anny in the little B.M.W. to make a wider tour of the peninsula. They covered a great figure eight, with the midpoint in Madrid, south through Valencia, Gibraltar and Cadiz, north through Salamanca, Altamira and Roncevalles, altogether about 8000 km. They were delighted with everything, especially the hot sunshine. Erwin was astonished by the prehistoric paintings in the caves at Altamira, which he thought 'were not talented children's drawings, but a free flowering of powerful forms'.

In Madrid, Schrödinger gave a series of lectures which were received with enthusiasm, so that Blas Cabrera and the other Madrid physicists began to explore the possibility of attracting him to a permanent professorship. As might be expected he found time in Madrid to meet some beautiful Spanish ladies, including one Susanna Mocaroca, but he did not have enough time to fall seriously in love with any of them.

While in Madrid he wrote to the authorities in Berlin to request a formal release from his professorship there and the grant of emeritus status. On March 31, his resignation was accepted; on June 20 Hitler sent a letter of thanks for his services and in July he was given the status of professor emeritus. These amicable arrangements would have given him no cause to believe that he was *persona non grata* to the Nazi regime.

In the June, 1935, number of *Anales de la Sociedad Espanola de Fisica y Quimica* Schrödinger published a short paper in Spanish, 'Are the True Equations of the Electromagnetic Field Linear?'. He developed an analogy between the pressures in sound and light waves, showing that a linear approximation to the wave equation for sound gives the correct pressure, as do the linear Maxwell equations for light. But, since we know that the linear theory is only an approximation for the true nonlinear equation for sound waves, he argued that the exact equations for electromagnetic waves may also be nonlinear. He wrote to Einstein about these ideas, but the latter was not impressed by the analogy between light and sound, even though he favored a nonlinear theory for light.

A talk on freedom

In May, 1935, Erwin went to London to deliver a twenty-minute talk on 'Equality and Relativity of Freedom' in the B.B.C. National Programme.[20] It was the ninth of ten talks in a series that had led to much discussion and correspondence in *The Listener*. He was the only foreigner invited to talk in the series. The final word was to be 'Faith and Freedom' by William Temple, the Archbishop of York.

It was unprecedented for Schrödinger to speak publicly on a subject that was quite unrelated to science or philosophy, and his short talk provides more of an insight into the inner man than he may have intended. The series was being given against a background marked by the rise of the dictatorships in Germany and Italy, and most of the previous speakers had said something in defense of political freedom. At the beginning of his talk, however, Schrödinger said, 'I beg to state that my acquaintance with politics in this or any other country is as feeble as my interest in them.'

He next complimented the British on the way in which they tend to protect the individual against

objectionable intrusion into his privacy . . . You esteem the merits of this custom if you have ever entered a circle where there is an unwritten law to start fathoming every newcomer to the depths of his soul with respect to right or left wing, fighting or supporting the government, pro-Jews or contra-Jews, in favour or not of free love, religious or anti-religious, etc. etc. It is true that among the possible answers to these questions there are some, which in most circles of this country are 'impossible'. You are tacitly assumed to abhor adultery (even by those who commit it), not to be in favour of free love and even, as a rule, not to be anti-religious.

Considering the speaker's own life, the choice of examples was rather poignant.

He next spoke approvingly of the independence of *justice* in Britain. Tyranny in making laws, he thought, is not so bad as unfairness or undue influence in enforcing them. Had he forgotten already his experiences in Berlin?

He then proceeded to the principal point of his talk: that freedom may be relative and thus depend on a balance of demands by different members of society. His examples were trivial but served to make his logical point: the freedom to spit vs hygiene, animal rights vs bull fighting or fox hunting, blowing a horn vs the need for sleep. 'One man will call a capricious whim what to the other appears an indispensable necessity. You can, of course, let the majority decide what to *do*. But whether it brings you nearer to freedom or further away from it, this theoretical question cannot be decided by a majority.'

Not all such situations are trivial. Under the influence of tradition or prejudice, atrociousness can claim to be a sacred right. He cited as an example the vendetta or murderous revenge. Then he mentioned an example of atrociousness closer to home. 'The individual, this time a female, in order to avoid contempt and rejection by all the "respectable people", is compelled to commit an action, which is threatened by the law of most countries with penal servitude. Most of you will know what I mean.' Erwin was remembering here the bitter experience of Ithi Junger. At that time an illegitimate child was a mortal affront to middle-class morality, and abortion was often accepted as the only alternative.

After starting his talk in a detached, almost offhand manner, Erwin had gradually progressed to major confrontations. The worst threats to freedom arise when a real claim is confronted by a fictitious claim that is strongly supported by deeply rooted beliefs. 'You all know that there have been communities who felt aggravated at persons who did not *think* and *believe* the same as everybody was supposed to think and to believe. You know that gallows and stake, sword and cannons have served to free respectable people of such annoyance, usually in the name of almighty God.'

The general rule suggests itself, that whenever a real claim competes with a fictitious one, the latter should be dismissed, giving way to the former in the name of freedom. This rule left unanswered the questions of what is the criterion to distinguish reality from fiction, and who is to draw the distinction. The talk revealed in the speaker a contempt for politics, a distrust for democracy, a liking for privacy, and a disdain for bourgeois hypocrisy. This was a fair yield for twenty minutes. The talk is also revealing for what he did not say – there was no word of criticism of the suppression of all freedom in Nazi Germany.

Early in 1935, Schrödinger's book *Science and the Human Temperament* appeared in London under the imprint of George Allen and Unwin. It comprised a selection of his more popular German lectures and essays. They had been translated by James Murphy, a defrocked Irish priest, who had enjoyed a picaresque career including a translation of *Mein Kampf*. Erwin met Murphy in Berlin in 1929, and in 1932 plans for the book were made at a café on the Unter den Linden. Despite some arguments between author and translator, the final result displays an urbane style and provides a cross-section of Schrödinger's ideas in easily readable form.

Spooky action at a distance

Einstein had attended the 1933 Solvay Congress and listened to Bohr's paper without raising any objections. All was peaceful after the dramatic confrontations of the 1927 and 1930 Congresses. Einstein did not like the Copenhagen–Göttingen interpretation of quantum mechanics but he seemed to have become resigned to it.

Then, in the May 15, 1935, issue of *Physical Review* he launched a brilliant and subtle new attack, one which was to have consequences of major importance for the world view of physics. The paper, 'Can Quantum-Mechanical Description of Physical Reality be Considered Complete?', written with two young co-authors, Boris Podolsky and Nathan Rosen, has come to be called the *EPR paper*.[21] They began: 'In attempting to judge the success of a physical theory, we may ask ourselves two questions: (1) Is the theory correct? and (2) Is the description given by the theory complete?'

EPR did not question the correctness of quantum mechanics but its completeness. 'Whatever the meaning assigned to the term *complete*, the following requirement for a complete theory seems to be a necessary one: *every element of the physical reality must have a counterpart in the physical theory* . . . The second question is thus easily answered, as soon as we are able to decide what are the elements of the physical reality.'

EPR then proposed a sufficient condition for an element of physical reality: *if, without in any way disturbing a system, we can predict with certainty . . . the value of a physical quantity, then there exists an element of physical reality corresponding to this physical quantity.* This criterion rests upon the assumption that the physical world can be correctly analyzed in terms of distinct and separately existing elements of reality. It has been called the assumption of *local realism*.

The rest of the EPR argument will be presented in a form different from that of the original paper, merely as a simplification. Electrons and other elementary particles have an intrinsic angular momentum or spin, so that in a magnetic field they act as little magnets that can take only either one of two orientations to the field direction, which may be called $+$ or $-$. [Actually the experiment cannot be done with free electrons but it can be done with neutrons.] It is the experiment of subjecting the particle to a magnetic field that causes it to declare itself as $+$ or $-$. A basic tenet of quantum mechanics is that the particle itself does not possess the property of being $+$ or $-$; before the experiment it has a probability 0.5 of becoming $+$ and an equal probability 0.5 of becoming $-$. If two particles, designated a and b, are brought together so that they interact, they will usually go into a state of lowest energy, called the *singlet state*, in which if one particle is $+$, the other is $-$. There is no way to say which is which, and the wavefunction to express this situation is written as $(1/\sqrt{2})[a(+)b(-) - a(-)b(+)]$.

Now suppose, say EPR, that after their interaction the particles are separated to such a great distance that there is no possibility of any further interaction between them, and an experiment is then made to measure the spin orientation of *a*. If it is found to be *a*(+), it is absolutely certain that *b* must be *b*(−). EPR now apply their criterion of physical reality: since the value *b*(−) was predicted for the spin of *b* without in any way disturbing *b*, it must correspond to an existent element of physical reality. Yet this conclusion contradicts a fundamental postulate of quantum mechanics, according to which the sign of the spin is not an intrinsic property of the particle, but is evoked only by the process of measurement. Therefore, EPR concludes that quantum mechanics must be incomplete. For example there may be 'hidden variables' not yet discovered, which determine the spins as intrinsic properties of the particles.

On June 7, Schrödinger dashed off a letter to Einstein: 'I was very happy that in the paper just published in *P.R.* you have evidently caught dogmatic q.m. by the coat-tails.' After a detailed analysis of some points in the paper, he concluded: 'My interpretation is that we do not have a q.m. that is consistent with relativity theory, i.e., with a finite transmission speed of all influences. We have only the analogy of the old absolute mechanics . . . The separation process is not at all encompassed by the orthodox scheme.' Schrödinger had recognized the essential point of the EPR paradox – the separation process and hence the assumption of local reality.

Einstein immediately replied:[22]

The actual difficulty lies in the fact that physics is a kind of metaphysics; physics describes reality; we know it only through its physical description.

All physics is a description of reality; but this description can be 'complete' or 'incomplete'. To begin with, the sense of this expression is even a problem itself. I will explain with the following analogy:

In front of me stand two boxes, with lids that can be opened, and into which I can look when they are open. This looking is called 'making an observation'. In addition there is a ball, which can be found in one or the other of the two boxes when an observation is made.

Now I describe the situation thus: the probability that the ball is in the first box is ½. Is this a complete description?

(1) NO. A complete description is: the ball is in the first box (or is not). This is the way to express the characterization of the state by a complete description.

(2) YES. Before I open the box the ball is not in *one* of the two boxes. This existence in a definite box first occurs when I open one of the boxes. In this way arises the statistical character of the world of experience or its empirical system of laws [*Gesetzlichkeit*]. The state *before* the box is opened is completely described by the number ½.

The Talmudic philosopher doesn't give a straw for 'reality', a bogy of naiveté, and explains both statements as only different ways of expression.

I bring in the *separation principle*. The second box is independent of anything that happens to the first box. If one holds fast to the separation principle, only the Born description is possible, but now it is incomplete.

It is the violation of Einstein's *separation principle* that introduces the 'spooky action at a distance' into the EPR experiment.

In a later letter (August 8), Einstein wrote:

> You are the only person with whom I am actually willing to come to terms. Almost all the other fellows do not look from the facts to the theory but from the theory to the facts; they cannot extricate themselves from a once accepted conceptual net, but only flop about in it in a grotesque way . . . My solution:
>
> The ψ-function describes not the state of an individual system but (statistically) of an ensemble of systems. Compared to one ψ, a linear combination $c_1 \psi_A + c_2 \psi_B$ signifies an enlargement of the totality of systems. The change which the system consisting of two parts undergoes when I make an observation on A signifies, conversely, the separation of a partial totality from the whole ensemble. The separation occurs differently according to the choice of quantities that I measure on A. The result is an ensemble for B which also depends on this choice.
>
> Naturally this interpretation of quantum mechanics displays especially clearly that, through its restriction to statistical statements, it has necessarily accepted the feasibility of only an incomplete representation of real states and processes.
>
> You, however, see something quite different as the reason for the inner difficulties. You see in ψ the representation of reality and would like to change its connection with the concepts of ordinary mechanics or do away with them altogether. Only in this way could the theory be made to stand on its own two legs.

Einstein did not believe that Schrödinger's approach would overcome the difficulties. He asked him to consider the case of a mass of gunpowder that would probably explode spontaneously in the course of a year. During this time the ψ-function would describe a sort of superposition of exploded and unexploded gunpowder. 'There is no interpretation by which such a function can be considered to be an adequate description of reality.' Einstein's gunpowder would soon reappear in the form of Schrödinger's cat.

Leon Rosenfeld was in Copenhagen when the EPR paper arrived. He recalls that 'the onslaught came down upon us as a bolt from the blue. Its effect on Bohr was remarkable.' We were in the midst of important theoretical work.

> A new worry could not come at a less propitious time. Yet, as soon as Bohr had heard my report of Einstein's argument, everything else was abandoned: we have to clear up such a misunderstanding at once. We should reply by taking up the same example and showing the right way to speak about it. In great excitement, Bohr immediately started dictating to me the outline of such a reply. Very soon, however, he became hesitant. 'No, this won't do, we must

try all over again . . . we must make it quite clear'. So it went on for a while, with growing wonder at the unexpected subtlety of the argument . . . 'We must sleep on it'. The next day he was calmer. He put aside everything else and worked on his reply for six weeks.[23]

This reply was published in *Physical Review* (Oct. 15, 1935).[24] Bohr rejected the criterion for reality as stated by EPR. 'There is . . . no question of a mechanical disturbance of the system under investigation during the last critical stage of the measuring procedure. But even at this stage there is essentially a question of *an influence on the very conditions which define the possible types of predictions regarding the future behavior of the system.*'

As the EPR paper arrived in other laboratories, it also left consternation in its wake. Pauli in Zürich was infuriated. Early in July, Schrödinger wrote to him from Oxford:[25]

Now 'state' is a word that everyone uses, even the holy PAM [Dirac], but that does not make it any richer in content . . . I say: different psi-functions correspond definitely to different factual conditions, or states. I don't consider that to be an illegitimate dragging-in of the reality dogma. But I should like to know . . . whether you really think that the Einstein case . . . gives absolutely nothing to think about, but is quite clear, and simple, and self evident. This is how everyone I have spoken to about it thinks at first, because they have learned well their Copenhagen Credo in Unum Sanctum. Three days later they usually come back with: what I recently said was naturally quite wrong . . . Or (like Szilard) I must first think over what I must forbid you. But I've not yet received a clear explanation of why everything is so clear and simple . . . So hearty greetings, dear friend, from your old Schrödinger.

Pauli replied virtually by return mail.[26] He thought that 'old gentlemen like Einstein and Laue' considered quantum mechanics to be something like statistical gas theory, in which the hidden variables are in the motions of the individual molecules. He believed, however, that you cannot introduce such variables into quantum mechanics without destroying its effectiveness. This view is similar to one set forth by John Neumann in *Mathematical Foundations of Quantum Mechanics*, published in 1932. It was shown later by David Bohm, however, that at least nonrelativistic quantum mechanics can be made compatible with hidden variables.[27] The hidden-variable theories, however, have not led to any new results: it is a matter of the interpretation rather than the application of the theory.

Schrödinger's cat

Motivated by the EPR paper, Schrödinger published in 1935 a three-part essay in *Die Naturwissenschaften* on 'The Present Situation in

Quantum Mechanics'.[28] He said he did not know whether to call it a 'report' or a 'general confession'. It is written in a sardonic style, which suggests that he found the 'present situation' to be less than satisfactory. It would be his definitive statement about the theory that he and Heisenberg had discovered.

He first explained in detail how physics, on the basis of experimental data, creates *models*, which are representations of natural objects idealized or simplified so that mathematical analysis can be applied to them. The deductions from such analysis are then tested by experiments, the results of which may lead to refinement or even drastic alteration of the model. The model can be described in terms of certain specifications [*Bestimmungsstücken*]. For example, the Rutherford model of the hydrogen atom consists of two mass points, and the specifications might be the two times three coordinates of these mass points, and their two times three components of momentum. Such specifications are often called *variables*. In addition there are *constants* of the model, for example, the masses (m, M) and the charges ($+ e$, $- e$). In classical physics one can define a *state* of the model by giving the values of the twelve specification variables. In quantum mechanics, however, not all the variables can be simultaneously specified. If one measures exact values for the position coordinates, one can determine nothing about the values of the six momentum components. This situation is a result of the Heisenberg uncertainty relation, which is derived directly from the fact that the operators for position q and momentum p do not commute. It is, however, possible to measure values of q and p that fall within certain ranges in accord with the uncertainty relation, so that one can speak of the specification variables of the model as being washed out or blurred [*verwaschen*]. ·

Nevertheless the wavefunction ψ defines the *state* of the model unequivocally. It constitutes a complete catalog of the probabilities of finding any specified result for a measurement made upon the physical system for which the model was designed. Schrödinger points out that in the classical era scarcely any physicist would have believed that the determining parts of a model were directly measurable on the natural object.

As an example of the strangeness of this way of thinking, he considers the angular momentum of a material point M about a reference point O, which is defined in classical mechanics as $\mathbf{L} = \mathbf{r} \times \mathbf{p}$, the vector product of the position and linear-momentum vectors. In quantum mechanics, one must say that the probability is zero for any value of the magnitude of L that is not equal to $\sqrt{n(n + 1)}h/2\pi$ where n is an integer. Whereas r and p may each have a continuum of values, L is limited to discrete values. Since L does not commute with r or p, a measurement of L leaves r and p indeterminate. We can move the reference point by an arbitrarily small amount and still we must get one of the discrete values for L. There seems to be no relation between measured values of r and p and measured values of L. 'Does one not get the feeling', Schrödinger asks, 'that the essential content of what is being said can only with some difficulty be forced into the Spanish boot of a prediction of a probability of finding this or that result for a measurement of a classical variable?' Would it not be more reasonable to consider the 'angular momentum' of quantum mechanics to belong to a new class of properties, having only its name in common with the classical property?

In Section 5 of his paper, Schrödinger asks 'are the variables really blurred?'. He points out that the classical description with its sharp values for the variables can be replaced by the ψ-function as long as the blurring is restricted to atomic dimensions which escape our direct control. But when the uncertainty includes visible and tangible things, the expression 'blurring' becomes simply wrong.

One can even construct quite burlesque cases. A cat is shut up in a steel chamber, together with the following diabolical apparatus (which one must keep out of the direct clutches of the cat): in a Geiger tube there is a tiny mass of radioactive substance, so little that in the course of an hour *perhaps* one atom of it disintegrates, but also with equal probability not even one; if it does happen, the counter responds and through a relay activates a hammer that shatters a little flask of prussic acid. If one has left this entire system to itself for an hour, then one will say to himself that the cat is still living, if in that time no atom has disintegrated. The first atomic disintegration would have poisoned it. The ψ-function of the entire system would express this situation by having the living and the dead cat mixed or smeared out (pardon the expression) in equal parts.

It is typical of such cases that an uncertainty originally restricted to the atomic domain has become transformed into a macroscopic uncertainty, which can then be resolved through direct observation. This inhibits us from accepting in a naive way a 'blurred model' as an image of reality . . . There is a difference between a shaky or not sharply focused photograph and a photograph of clouds and fogbanks.

This conclusion has been called 'the principle of state distinction': states of a macroscopic system which could be told apart by a macroscopic observation are distinct from each other whether observed or not.

Only a few commentators on the cat paradox, the most notable being Eugene Wigner and John Neumann,[29] have defended the uncompromising idealist position that the cat is neither alive nor dead until a human observer has looked into the box and recorded the fact in a human consciousness. It might be, of course, that the cat itself has a consciousness quite adequate to complete the experiment and resolve the probabilities by passing from a superposition of two states to a single state. Even without an animal consciousness, the experiment would be decided as soon as the atomic disintegration activated the Geiger counter. In the state $\psi_A + \psi_B$, the waves ψ_A and ψ_B must represent solutions of the time-dependent Schrödinger equation for the macroscopic system, including the cat. The system cannot be effectively isolated from perturbations by the rest of the universe. Thus formulation of the quantum mechanical problem becomes impossible, and it is meaningless to talk about a superposition of states, ψ (live) + ψ (dead). The cat paradox, however, was useful as an antidote to the view that the wavefunction refers not to a physical model but to human knowledge about an object, and perhaps even more importantly, it served to emphasize the principle of state distinction.

A delayed choice experiment that is meaningful can, however, in principle be carried out with neutrons, in which two alternate paths, A and B, are possible for a neutron passing through an interferometer.[30] The superposition of states $\psi_A + \psi_B$ can be demonstrated by the interference pattern produced. At the end of the experiment, the determination of which path was taken corresponds to opening the box and looking at the cat. Such a determination, of course, prevents the interference effect. It is interesting to note that a neutron of energy 0.02 eV is actually a macroscopic 'object', being represented by a wave packet about the size of a postage stamp.

Schrödinger's discussion of EPR

Between two measurements, the ψ-function, the 'catalog of expectations', is in many ways similar to the *model* of classical physics; it changes with time in a well-behaved way in accord with a differential equation [the time-dependent Schrödinger equation] that is first-order in the time. When a measurement is made, however, ψ changes abruptly in a way that cannot be predicted. It is at this point that interpretations such as the Copenhagen one break with realism, 'because from the point of view of realism, an experimental observation is a natural process like any other and cannot in itself cause a break in the orderly flow of natural events.'

Schrödinger then examines the logical consequences of what he sees as a 'rejection of realism'. It implies that:

a variable in general has no definite value before I measure it: to measure it does not mean to determine a value that it *has*. Then what does it mean? . . . Just playing around with an indicating instrument in the vicinity of another body whereby at some time or other one gets a reading can certainly not be called a measurement on that body. Well, it is pretty clear; if reality does not determine the measured value, at least the measured value determines reality:

after the measurement it must be really present in the sense that it alone will be recognized again.

He now considers the situation of the EPR experiment: If one has a ψ-function for each of two completely separated bodies, one has maximal knowledge for each of them and therefore for the two taken together.

but the converse is not true. *Maximal knowledge of a total system does not necessarily imply maximal knowledge of all its parts, not even when these are completely separated from one another, and at the time cannot influence one another at all.*

In the EPR example, we know that if measurement gives $a(+)$, then a subsequent measurement must give $b(-)$. When such conditional statements occur in the combined 'expectation catalog' it cannot possibly be maximal in regard to the individual systems. Thus 'the whole is in a definite state, the parts, taken individually, are not.' One has a ψ-function for the whole system, but one does not have ψ-functions for the separate individual parts. *'The inadequacy of the ψ-function as a substitute for a model rests exclusively on the fact that one does not always have it.'* The 'entanglement' of the subsystems must arise as a result of an interaction between them sometime in the past. If they had never interacted, there would be no such entanglement and hence no conditional statements in regard to the subsystems. All these formulations place the two subsystems in a common three-dimensional space in which a physical distance between them can be defined. In general their wavefunctions span a multidimensional space, in which the subsystems cannot be said to be far from each other.

He next considers 'Untying the Entanglement [Aufhebung der Verschränkung] – the Result Dependent on the Intention of the Experimenter'. Any measurement on either of the separated parts a or b of the entangled system has the effect of breaking the entanglement and restoring the independence of a and b. By a suitable program of measurements on b, one can infer the state of a without at any time disturbing a. Schrödinger shows that the state which is thus inferred for a may depend upon the program applied to b. This result may be called the theorem of *non-invariance of inferred state description*. He has correctly isolated the crux of the problem, but he is unwilling to reject Einstein's principle of local realism, which he restates as follows: 'Measurements on separated systems cannot directly influence each other – that would be magic.'

Schrödinger believes that the paradoxes may be due to an inadequate consideration of the role of time in the theory of the measurement process. All the predictions have referred to values of variables at some identical point in time. It may not in fact be possible to predict the values of variables for two different subsystems at the same sharp point in time. The thought of simultaneity in time led him naturally to the fact that the quantum mechanics that is being used is not in accord with the principle of special relativity. [The Schrödinger equation is not invariant under a Lorentz transformation.]

Schrödinger also wrote at about this time two important papers that discussed in more mathematical terms the problems raised by Einstein's 'separation principle' and the EPR paradox.[31] These were submitted to the Cambridge Philosophical Society, the first through Max Born and the second through Paul Dirac, but there is no record that Schrödinger actually read either paper before the Society. He believes that the 'entanglement' of the wavefunctions of two separated systems that have previously interacted is 'the characteristic trait' of quantum mechanics. Disentanglement requires a measurement to re-establish the ψ-function of one of the systems, whereupon the other one can be inferred simultaneously. This procedure has a 'sinister importance', since it is the basis of the quantum theory of measurement. The fact that all conceivable decompositions of one system can be realized by applying all possible measuring programs to the other must be repugnant to some physicists, 'including the author'.

In his second paper, Schrödinger derived an even more striking result: 'In general a sophisticated experimenter can, by a suitable device which does *not* involve measuring noncommuting variables, produce a nonvanishing probability of driving the system into any state he chooses.' He reiterated the thought that such a situation was intolerable, but might be resolved by a revised treatment of time in quantum mechanics.

The subsequent history of the EPR paradox would fill several volumes. In 1935, no physicist would have thought it worthwhile actually to perform an EPR-type experiment, because they were all convinced that the result would in no way be able to decide between the ideas of Einstein and those of Bohr. Almost all the theoreticians were under the spell of John Neumann's book, in which abstruse mathematics was used to 'prove' that hidden variables cannot be added to quantum mechanics without destroying its verified results. Nevertheless David Bohm, Louis de Broglie and others produced hidden-variable theories that were capable of yielding at least some of the results of quantum mechanics. Then, in 1954, John Bell, an Irish physicist working in Switzerland, derived an important theorem which shows how an *experimental decision* can be made between the pure quantum mechanical predictions about entangled systems and the predictions of local hidden-variable (deterministic) theories.[32] Some beautiful experiments were made by John Clauser,[33] Alain Aspect,[34] and others, and the results showed that measurements on

correlated pairs of particles follow the quantum-mechanical predictions. This result has been found even when the particles are
separated by a distance such that even an influence propagated at the
speed of light could not affect the result. These results appear to have
left three options for theoretical physics: (1) to deny the validity of
inductive logic in physics, so that one could not predict the results of
future experiments from those already done; (2) to deny that the
microscopic entities of physics have any objective reality; they would
exist only in the mind of the observer; and (3) to assert that an
influence can be propagated from one part of an entangled system to
the other at faster than the speed of light. Such instantaneous
correlation is equivalent to a denial of Einstein's postulate of *local
realism*.[35]

Most physicists, with varying degrees of reluctance, have opted for
the third alternative. It means that the reality behind the appearances
of physics is an unbreakable whole. If the physical world is real, it is
holistic – it is not merely the sum of separate parts.

Schrödinger never drew such a conclusion from quantum
mechanics. He believed that the problem was caused by the extension
of nonrelativistic quantum mechanics beyond its legitimate range of
application, which should be to distances small enough to permit
neglect of the time light takes to travel across the system. Many
present-day interpretations of the EPR experiments do not agree that
the 'spooky action at a distance' can be ascribed to the neglect of
special relativity. Rather it is believed to be an ineluctable part of the
quantum world. Besides the physical paradox of quantum mechanics,
we are faced with psychological paradoxes in the mind of its creator.
His religion of Vedanta taught the unity of the world, he was to devote
the latter part of his life to research on the unification of physical field
theories, he rejected the duality of wave and particle in favor of a
world of waves alone, yet he was unwilling to accept the indivisible
nature of the quantum world.

Teasing Niels Bohr

On October 13, Schrödinger wrote a long somewhat teasing letter to
Bohr about his *Physical Review* article on 'the avoidance of the Einstein
paradox', finding it in the final analysis to be unsatisfactory.[36] In
particular, he wished to question Bohr about his often expressed
conviction that measurement must be described through 'the indispensable use of classical concepts'. Would Bohr not admit that future
developments in physics might show that what is now called an
indispensable necessity may yet become an inadequate approximation
which 'one might sometime succeed in escaping?'

There must be quite definite and clear grounds, why you repeatedly declare that one *must* interpret observations classically, which lie absolutely in their essence . . . It must belong to your deepest conviction – and I cannot understand on what you base it.

Schrödinger pointed out that when we consider the history of science we find many instances in which it was difficult to overcome old ways of thinking, so that now we can hardly refrain from thinking 'Incredible, that people till then were so limited?' The fact that our thought has not yet been adapted to the new theory

cannot possibly provide a basis for the idea that experiments must always be described classically, with neglect of the important characteristics of the new theory. It may be a childish example . . . but after the elastic theory of light was replaced by the electromagnetic, one did not say that the experimental results must be expressed afterwards as before in terms of the elasticity and density of the aether.

He asks 'Forgive me that I have been so long winded' and explains that he wishes only that Bohr would explain clearly why he insists again and again that at the basis of measurement lies the principle that it can be interpreted only classically, and above all, whether that is a temporary self-limitation or is something from which we can never escape.

I should like very much to see and talk with you again, but the times are now little suited for pleasure trips, and soon there arises in me the philistine wish once again to be somewhere permanently, that is, to know with considerable probability what one is to do for the next 5 or 10 years. It is fearfully philistine and not at all passing with time, presumably a reflex from the present environment, which indeed still rolls along richly undisturbed by the course of the world.

Bohr answered briefly that

my emphasis on the unavoidability of the classical description of the experiment refers in the end to nothing more than the apparently obvious fact that the description of every measuring apparatus basically must contain the arrangement of the apparatus in space and its function in time, if we are to be able to say anything at all about the phenomena . . . The argument is thus above all that the measuring instruments , if they are to serve as such, cannot be included in the actual range of applicability of quantum mechanics.

[If the measuring instrument were described quantum mechanically, its interaction with the object would merely extend the catalog of expectations [*Erwartungskatalog*] without leading to a definite result (eigenvalue) for the observation. The assumption that the measuring instrument behaves classically is the way in which the Copenhagen interpretation escapes this regress.]

In March, 1936, Schrödinger met Bohr in London, and reported to Einstein:[37]

Recently in London spent a few hours with Niels Bohr, who in his kind, courteous way repeatedly said that he found it 'appalling', even found it 'high treason' that people like Laue and I, but in particular someone like you, should want to strike a blow against quantum mechanics with the known paradoxical situation, which is so necessarily contained in the way of things, so supported by experiment. It is as if we are trying to force nature to accept our preconceived conception of 'reality'. He speaks with the deep inner conviction of an extraordinarily intelligent man, so that it is difficult for one to remain unmoved in one's position.

I found it good that they strive in such a friendly way to bring one over to the Bohr–Heisenberg point of view . . . I told Bohr that I'd be happy if he could convince me that everything is in order, and I'd be much more peaceful.

An uncertain future

The I.C.I. grants were due to expire at the end of 1935, and in October the company notified all the grantees that they would not be renewed. A special exception was made for Schrödinger, and he was awarded a two-year extension. Simon wrote to Born that one should not blame I.C.I. alone, since the company had always stated that the grants were meant merely to provide a buffer against an emergency situation, and they expected that the refugees would be absorbed into more permanent positions. Unfortunately this had not happened in many cases and that was the principal problem. 'Schrödinger is . . . not very active in such matters, and especially as one of those concerned cannot bring himself to do anything.'[38]

In March, 1936, the Academic Assistance Council was reorganized into a new body, The Society for the Protection of Science and Learning. The Council had assisted in the placement of 363 out of the 700 refugee scholars who came to Britain. On March 25, a letter from Albert Einstein, E. Schrödinger and V. Tschernavin was published in the London *Times*, with a prominent headline on the editorial page, 'THE FREEDOM OF LEARNING, Help for Displaced Scholars, Gratitude to England'. The letter concluded: 'The Academic Assistance Council is coming to an end in its emergency form, but we and our friends will endeavour to make it unforgotten. May we hope that the continuation of our scientific work – helped in no small part by its activities – will be an expression of our gratitude.'

Arthur, Hilde, and Ruth March had returned to Innsbruck, where Hilde made a stay of several months in a sanitorium, until her physical and mental equilibrium was restored. She was an unsophisticated woman, with no bohemian pretensions but with a wonderful zest for

Hansi Bauer-Bohm (c. 1933).

life, and the stress of living with two men, with their often conflicting demands upon her, had temporarily exhausted her psychological reserves.

Meanwhile, Hansi Bauer-Bohm had escaped from Berlin with her husband and was living in London. In addition to her drawing, she had now reached an almost professional standard in photography. She became a frequent visitor in Oxford. Anny had taken a small flat in London, so as to take some special courses, and also to give Erwin more freedom to be with Hilde or other friends. Erwin had always been attracted to Hansi, who was different in many ways from the other women he had known. Since his youthful love for Felicie he had not had a close relationship with a woman whose social class, or at least financial status, was definitely superior to his own. Hansi combined the cultural background of a wealthy Viennese Jewish family interested in music and art, with the more daring sophistication of avant-garde Berlin before it was crushed by the storm troopers. Although Hansi found the Schrödingers conventionally middle class in their artistic tastes, she could not escape the fascination of Erwin himself, his brilliance of intellect, his insights into feminine psychology, his boyish elfin charm. The spark of recognition in the Tirolean grocery shop now kindled a flame of passion. It was inevitable that they should become lovers, and after Hilde left Oxford in 1935, Hansi and Erwin went on a short holiday together to the Channel Islands.

In his previous conquests, Erwin had been forced to undertake a prolonged siege to win his lady's affections, but with Hansi he was as much the pursued as the pursuer. He was delighted with her both intellectually and amorously – a combination of satisfactions that he had not previously experienced.

Although his personal life had its compensations, he was becoming increasingly dissatisfied with his professional situation. A fellowship at an Oxford college and a grant from I.C.I. hardly equalled in status the professorship at an important university which his achievements deserved. In view of his inability to control or predict his own future, it is interesting to consider the short philosophical letter that Erwin published at this time in *Nature* under the heading 'Indeterminism and Free Will'.[39]

It has become the orthodox view of physicists to-day, that the momentary state of a physical system does not determine its movement or development or behaviour to follow; Nature is supposed to be such that a knowledge of state, sufficiently accurate for sharp prediction of the future, is not only unobtainable but also unthinkable. All that can be predicted refers to a large number of identical experiments, and consists in definite statistics among all the possible developments to follow. The relative margin of indeterminacy (the 'spread' of the statistics) is large for a small system, for example, an atom; but for large systems the margin is usually, though not necessarily, small, which makes it possible to account for the *apparent* determinacy of inanimate Nature.

Many eminent scientific workers, especially physicists, have tried to play with the idea that the *apparent indeterminacy* of animate Nature, that is, of living matter, might be connected with the theoretical indeterminacy of modern physics. What makes this play so fascinating and thrilling is evidently the hope . . . of extracting from the *new* physical dogma a *model of free-will*, which the *old one* would refuse to yield. I consider this hope an illusion, for the following general reasons.

After this exciting introduction, the analysis that Schrödinger gives of the problem is disappointing. He seems to have forgotten his famous cat, which showed how large-scale consequences can follow from an amplification of microscopic uncertainty. His conclusion is that 'free-will' itself is an illusion because 'the plurality of possible events, in the case of an action under free-will, is a self-deception'. Thus all human actions are essentially deterministic and in accord with the requirements of classical physics. 'There is no ψ-function in life.'

The centuries-old argument between free will and determinism is simply one aspect of the mind-body problem. Schrödinger's view in this note seems to be strictly materialist, dismissing even the possibility of a nonphysical 'mind'. Elsewhere, and with much greater depth of analysis, he espouses quite a different view. This note can be regarded as a casual comment on a recent article by Frederick Donnan, professor of physical chemistry at University College, London.

It must have seemed to Erwin at this time that his own future was no longer subject to the dictates of free will. No permanent position that he regarded as suitable had yet appeared. He had been offered a professorship at Madrid, and he was considering this seriously, when the Spanish civil war erupted on July 18, putting an end to that particular future.

In May, he had written to Einstein about his interest in a professorship that would soon become available at the University of Graz. He said that he would be inclined to accept this chair provided certain financial provisions were made, such as freedom to keep his securities in Sweden and exemption from tax on travel allowances from abroad. [His Swedish assets amounted to $12 000.] He may have been politically naive, but he had become financially sophisticated.

It is not that I can't stand it in one place for long. Up till now I've generally been contented wherever I was except in N[azi] Germany. Also it is not they haven't been very nice and friendly to me here. But nonetheless the feeling grows stronger of having no employment and living on the generosity of others. When I came here I thought I would be able to do something for the teaching, but no value is placed on that here. And if I think more about it, really I must say to myself: . . . I am sitting here waiting for the demise or the complete decrepitude of a very dear old gentleman (Love) and the possibility that they might make me his successor. I don't want to be hypocritical that this hurts my feelings, but it does hurt my self-esteem.[22]

In December, 1935, Erwin visited in Austria. He stayed in Graz with his good friend Kohlrausch and they discussed the possibility of a professorship there. On December 18 and 20, he lectured in Vienna, and consulted the Ministry of Education about appointments in Vienna and Graz. The Christmas holidays were spent skiing at Obergürgl, and he had an opportunity to be with Hilde and Ruth again. It also seems likely that he made a brief visit with Max Laue in Berlin, for Laue's son Theo, who had then just obtained his driving license, remembers bringing him from the station at about that time. He returned to Oxford via Brussels.

Edinburgh 1936[40]

Towards the end of 1935 an important professorship of physics became vacant at the University of Edinburgh when Charles Darwin, the grandson of the great biologist, resigned from the Tait Chair of Natural Philosophy to become Master of Christ's College, Cambridge. The professorship had been established in 1922 to perpetuate the memory of Peter Guthrie Tait, a famous Scottish mathematical physicist who, among other noteworthy accomplishments, had worked out the dynamics of flight of the golf ball. In February, 1936, a committee on the future of the position was convened under the chairmanship of Edmund Whittaker and in May it reported to the University Court that 'they had considered a number of names, and were unanimously of the opinion that of those available Dr Erwin Schrödinger was much the most distinguished, and unless it was considered undesirable to appoint a person of other than British nationality, they recommended that the Chair be offered to him conditionally on the Court being satisfied that he would be permitted to continue residence in this country.' The Court approved this report and authorized the Secretary to communicate the conditional offer to Dr Schrödinger.

Some weeks before this formal offer, Erwin arrived in Edinburgh to 'spy out the land'. He was wearing one of his Alpine costumes as protection against the Scotch weather; Whittaker met him at the railway station and was rather shocked by the informality of his attire. Darwin showed him around the university and explained the heavy administrative and teaching duties expected of a professor in Scotland. He met Robin Schlapp, a young lecturer in the department, who actually did much of the routine work. Erwin must have expressed considerable interest in the post, since the formal offer was sent to him. He told Hansi that he would go to Edinburgh if she would come with him, but this was a completely impractical suggestion, since she was expecting her first child and was planning to return to Vienna.

Perhaps Schrödinger would have accepted the Edinburgh post, even without Hansi, had bureaucratic incompetence not intervened. The university Secretary wrote to the Home Office to secure permission for Schrödinger's permanent residence. For some reason 'an accidental delay' occurred. As time passed with no decision from London, Erwin became less attracted by the offer. The salary of £1200 was much less than that of the Princeton professorship he had declined the year before; however, the retirement age was seventy and the pension provisions were better. Meanwhile an offer had come from the University of Graz with an honorary professorship at Vienna, through the efforts of his friend Hans Thirring, and he decided to accept this.[41] He told Darwin that the call of the mountains of his native Austria was too strong to resist. Also the Austrian pension provided full salary after retirement. Hilde and Ruth were back in Austria, and Hansi would soon be there, so that, everything considered, the opportunity to return home was irresistible.

The Tait Chair was then offered to Max Born, who accepted it gladly. This time the Principal of the University wrote a personal letter to the head of the Home Office to say that they had lost Schrödinger as a consequence of delays, and they hoped that this time the Office would show proper diligence. Approval for Born's residence came practically by return mail.

9   Graz

Niels Bohr was fond of quoting 'an old Danish proverb' to the effect that 'prediction is always difficult, especially of the future'. Schrödinger later described his decision to return to Austria in 1936 as a miscalculation of the political situation that was 'an unprecedented stupidity', but it would not have seemed so at the time. He was not satisfied with the quality of life in either Britain or America, and old friends like Fritz Kohlrausch in Graz and Hans Thirring in Vienna were urging him to return to his native land. Hans even offered to relinquish his Vienna post if Erwin preferred that one.

Nevertheless, Erwin knew that in going to Graz, he was stepping into a cesspool of Nazi activity. The university was a center for the Styrian Nazi party, whose leader was Armand Dadieu, professor of physical chemistry. More than half the students were active Nazis and they dominated the campus. The principal Graz newspaper, *Der Tagespost*, followed the Nazi line and the provincial *Heimwehr* was controlled by the party.

The Austrian situation

During the fifteen years that Schrödinger had lived abroad, the history of Austria had been marked by political strife between two major forces, the Christian Socialists or 'blacks' and the Social Democrats or 'reds'. Antisemitism in Austria, with its Catholic culture, was more prevalent but less extreme than in Germany. The clerical party was traditionally antisemitic; for example, in 1933, its chairman, with the good German name of Czermak, announced that 'the religious German must decisively reject baptism as an "entrance ticket" for Jews'.[1] There were about 190 000 Jews in Austria, with 176 000 in Vienna (9% of population of city) where they were prominent in learned professions, the press, music and the arts. As in Germany, the universities were citadels of antisemitism and it was difficult for even distinguished Jewish scholars to obtain professorships.

In 1932, the clericals selected Engelbert Dollfuss as chancellor. This diminutive politician (*Millimetternich*) was motivated by a colossal vanity and a paranoid hatred of the Social Democrats who comprised almost fifty per cent of his compatriots. At the urging of Mussolini, he suspended parliament on March 4, 1933, and established a fully fascist one-party state, committed to the fight against two internal enemies, the 'reds' and the Austrian Nazis. Britain and France urged him not to oppose Hitler, and his only support came from Italy. In 1934, Mussolini asked Dollfuss to crush the socialists and on February 12, Major Emil Fey led army and police in an attack on the workers' suburbs in Vienna, in which about 1000 were killed, including many women and children. As soon as the socialists were destroyed, the Nazis stepped up their terrorism, and on July 25 they assassinated Dollfuss, but the putsch then failed, as Hitler did not intervene as expected, having been warned off by Mussolini.[2]

Kurt Schuschnigg now became leader of Austria, a position he was to hold for almost four years, until he handed the country over to Hitler. As long as Mussolini was willing to support him, his position remained fairly secure, but as the Italian dictator became embroiled in Abyssinia, and as Hitler's military strength grew, Nazi pressure on Austria became more insistent. This was the situation in Austria as Schrödinger was negotiating the conditions for the professorship at Graz. Although he had lived for many years abroad, he had frequently visited Austria, and maintained close connections with family, friends, and lovers there, thus he was well aware of the instability of the country. He had no sympathy for the black dictatorship, and its excesses and incompetence confirmed his contempt for politics in general.

Before leaving England he had to make final arrangements with the tax commissioner. His total income for the tax year 1935/36 amounted to £1358, on which he paid tax of £189. The salary at Graz was finally set at 19 490 schillings and 90 groschen. (£734 or $3684 at the then current rate of exchange). This was far less than the Princeton salary that he had deemed inadequate, but his Vienna appointment would yield a fifty percent supplement, and the retirement provisions were excellent.

The Schrödingers found a pleasant house at 20 Merangasse, which they rented for AS 260 a month. Hilde and Ruth were in Innsbruck with Arthur, living in separate parts of the same house in a rather tense atmosphere. Erwin arranged for the third floor of the Graz house to be made into a separate apartment, where the mother and child would come to live early in the new year 1937. It would be a strange household. Anny spent most of the time in Vienna with her mother. Erwin confided to his journal some years later: If you should ever want to write an epitaph or a memorial about me and, as they say, my fate as a man, but you will not want to, then you can say: Firstly, that a woman who never loved me, on account of external circumstances was in a situation to bear me a child, which is not so remarkable because it happens every day. Secondly, that another woman loved it more than if it were her own, it seems to me for the reason that it was mine, which is not remarkable because under such circumstances (I mean especially with the almost uninterrupted understanding of the child's mother) it would occur roughly once in a thousand years.

Schrödinger began his duties as *Ordinarius* in physics at the Karl Franzen University of Graz on October 1, 1936. His inaugural lecture was a replay of his Nobel lecture; it was a lackluster performance that set forth his ideas essentially as they had been ten years previously.

Within a few weeks Schrödinger received a new honor, selection as one of the foundation members of the Pontifical Academy of Sciences.[3] In January, 1936, Pope Pius XI had announced the formation of the new academy – its purpose was to be the service of Truth. The Pope declared that 'God is Truth and science expresses one of the most beautiful harmonies, one of the most admirable splendors that can possibly be imagined, equalled and rivalled only by charity and goodness.' Besides being scientists of world renown, the members of the Academy were to be 'men of irreproachable civic and moral conduct who had always assumed a respectful attitude to religion, without allowing a humanistic evaluation of strictly scientific results to lead them to conclusions opposed to the faith'. The Vatican Secretariat of State, headed by Cardinal Eugenio Pacelli, undertook to provide precise evaluations of the moral qualities of those nominated for membership, and some illustrious scientists were eliminated on the basis of this scrutiny. The list of seventy members was published in October, 1936, including thirty Italians, eight Protestants, and two Jews. The physicists were Bohr, Debye, Keesom, Millikan, Planck, Rutherford, Schrödinger, and Zeeman.

The inauguration of the Academy was held in Rome on June 1, 1937. The Holy Father could not attend owing to his serious illness, but Secretary of State Pacelli welcomed the academicians, and conferred upon each of them the gold chain of membership. Erwin's letter of thanks concluded: 'Never will I forget this visit to the Eternal City, a visit that was the first for me and for that reason all the more impressive, and never will I forget my profound obligation to demonstrate as much as I am able my enthusiastic veneration toward a divine institution which for all of us represents the most powerful support against the frightful perils that are menacing human culture.'

Sylvester Abend

On New Year's Eve, 1936, Anny was in Ober-Gürgl for a skiing holiday. This was her fortieth birthday and she received a friendly greeting from Erwin, which she answered in a long letter:[4]

I shall now use the rest of this Sylvester-Abend to send my thoughts to the two men who have played the greatest roles in my life . . . I can understand so well when believers speak of a state that they call 'contemplation' [*Sammlung*], or as philosophers aptly call it, 'meditation'. Your birthday letter to me seems to have arisen from a similar feeling. It made me happy as it is the first personal emotion spoken from you to me for a long time. For the first time I again had the feeling of a mutual understanding and I believe, as you do, that this is the first step towards a change for the better in the relation between you and me. I believe you have understood very rightly wherein – among other things – a principal failure lies. I understood it much earlier but apparently I could not express it well, as you can so skilfully. You remember perhaps that I said many times: you cannot make any happiness while out of it you make me so unhappy . . . I would only be mistaken if I thought that for you I belong to the same category of person into which you place yourself, although I neverthe-less know for certain that for all other persons I definitely belong to another category. Where is the logic there? Erwin, believe me, I don't see fog and clouds at all, happily and confidently I see a future anyway. Even if the love between Peter and me should sometime come to an end, I would always be blessed that it had formerly existed, as I know that fate has given me the greatest happiness that a person can ever be given. And believe me, Erwin, I would not want to exchange this form of happiness for a marriage that allowed me always to be together with Peter. You yourself were the one who instilled in me that bourgeois marriage is definitely the finish of the deepest feeling of love because everyday living destroys the magic. What I experienced in Selva was the greatest happiness that I can imagine, and that it has endured to the present day, for that I cannot be thankful enough and all my wishes concentrate on this: that it may survive Peter and me. An episode such as you have experienced this year would for me be inconceivable. But I know that women are much truer by nature than men: from this arises no end of sorrows.
   Word for word I wish to accept the impression of your birthday letter. Gladly do I let my birthday today be a divide between the distressing past 'and a happier life that insensibly passes through the shadows and is prepared for the bright daylight'. This past ten years of my life was so overflowing with experiences, so overflowing with pain and happiness, as can rarely happen. I believe that I myself have drastically changed. When I think back on ten years ago when we landed in New York and you carried on so – were so unhappy – at that time I was still very young and inexper-ienced. The best thing that life has taught me is unquestionably self reliance. That was for me very hard to learn, for my innermost and earliest wish was to be a part of another person whom I loved. Today I know that you were tormented by this above all, but I believe you did not know how difficult it was for me to free myself of it.

Anny may have lacked formal education but she was not deficient in psychological insight. To a considerable extent, however, the emotional situation of the couple was a destiny beyond their control. Anny did not have children in whom she could satisfy, as in one typical course of psychosexual development, her need for self-sacrifice, while leaving Erwin free to pursue his adolescent ideals of romantic love and sexual conquest.

Turning to more practical matters, she reported that she had visited Innsbruck and found that Hilde was not making adequate preparations for the move to Graz, and she advised Erwin to light a fire under her if he wanted the move to occur on schedule. Erwin was to come to Innsbruck about a week later to pick up Hilde for their own skiing holiday.

During these months Schrödinger was carrying on a bizarre feud with I.C.I. about the disposal of the plumbing fixtures in the house he had rented in Oxford.[5] At the beginning of his tenancy, he had paid £31 for certain fixtures on the understanding that when he left the sum would be refunded less an allowance for depreciation. I.C.I. had taken a four-year lease on the house and after some delay they found a new tenant to whom they sold the fixtures for £20. They had a number of expenses in preparing the house for the new tenant, notably due to the neglected condition of the garden, and thus they applied the £20 to offset part of these costs. Schrödinger became furious at this 'insolence' and sent complaining letters to Lindemann, Gordon, the British tax authorities, and of course to I.C.I.. Lindemann hardly knew what to make of Schrödinger's behavior, in view of the generosity with which I.C.I. had provided for him at Oxford. Eventually Melchett, the chairman of I.C.I., sent him a cheque for £20 to conclude the matter, but it left a bad impression. Possibly Erwin was already finding the situation at Graz oppressive. He was being dunned for contributions to various official charities and political causes. He refused an invitation to lecture at Brunn in March, 1937, even though his old friend Fränzel lived there. He said that he was feeling out of sorts with aches and pains of unknown origin, but he also objected to the use of the catchword *völkisch* in the invitation.

Schrödinger and Eddington

For the next few years Schrödinger's research was inspired by the cosmological theories of Arthur Eddington. Eddington (1882–1944) was the most distinguished astrophysicist of his time.[6] His brilliant early work on stellar dynamics and on the structure and evolution of stars virtually created the discipline of theoretical astrophysics. During World War I, he received copies of Einstein's papers on the general theory of relativity from their mutual friend deSitter in the

Netherlands. Eddington was captivated by the 'magic of the theory', and soon mastered all its mathematical intricacies. The Astronomer Royal, Frank Dyson, pointed out that the eclipse of the sun on May 19, 1919, would afford an excellent opportunity to test a prediction of Einstein's theory, the bending of light rays from stars by the gravitational field of the sun. Despite wartime conditions, he organized two expeditions, one to Brasil and one to West Africa. Eddington went to Africa with E.T. Cottingham and they obtained the first confirmation of the Einstein theory. In 1923, he published his *Mathematical Theory of Relativity*, a masterly account that included several original advances.[7]

After the advent of quantum mechanics in 1926, Eddington began to consider the deep and difficult question of how to bring together relativity theory, so effective in the cosmic domain, and quantum theory, so effective in the atomic domain – how to reconcile the physics of the macrocosm and the physics of the microcosm. From his middle years, he also devoted much time to the philosophy of science, and this study influenced his approach to problems of theoretical physics. The later Eddington was a lonely and controversial genius who combined numerical legerdemain and idealist philosophy in a prose style of such eloquence that it charmed even those who could not follow his reasoning. He believed that scientific knowledge is derived not from an external world but from the abstract structure of human thought.

In 1927, Eddington gave the Gifford lectures at Edinburgh University, a series endowed to explore the relations between science and religion, and they were published the next year as *The Nature of the Physical World*.[8] The editor of *Die Naturwissenschaften*, Arnold Berliner, sent the book to Schrödinger for review.[9] He found it to be a

deep, original consideration of the world picture . . . He does not try to replace the mathematical structure by a creaking verbal scaffolding – as indeed it is not replaceable – but he views the form and appearance of the modern explanation of Nature striding forth free of the shackles of rigorous deduction – *vues à travers un temperament* it is true – but such eyeglasses are necessary when the more neutral ones of mathematics are forsworn . . . Like a Shakespearean fool, his laughing manner often encloses a deep truth in a bad joke.

The French quotation was from Emile Zola: *L'art c'est la nature vue au travers d'un tempérament*. In a lecture in Berlin, Erwin had contended that, to some extent, the same can be said of science.

Eddington's view of the world was similar to his own, perhaps somewhat more subjective, as in this passage.

If we were to try to put into words the essential truth revealed in the mystic experience, it would be that our minds are not apart from the world; and the feelings we have of gladness and melancholy and our yet deeper feelings are not of ourselves alone, but are glimpses of a reality transcending the limits of our own particular consciousness – that the harmony and beauty of the face of Nature is, at root, one with the gladness that transfigures the face of man.[8]

Schrödinger never reported a personal mystical experience but his feeling for the unity of nature and man was akin to that of Eddington.

While at Oxford, Schrödinger visited Eddington at Cambridge, where he was Plumian Professor of Astronomy and Experimental Philosophy. In 1936 he reviewed for *Nature* Eddington's *Relativity Theory of Protons and Electrons*.[10] He was impressed by the book but cautious: 'We have here before us a sketch of unusual grandeur, of which not the details alone need further development and, maybe much modification. I am convinced that for a long time to come the most important research in physical theory will follow closely the lines of thought inaugurated by Sir Arthur Eddington.'[11] The simple but profound idea of Eddington was that the atomic structure of matter – the very existence of particles such as electrons and protons – is a consequence of the curved finite space of the universe devised by Einstein in his theory of general relativity. It is the finiteness of space that leads to discrete quantized energy levels and hence to atomicity.

Eddington concluded that 'the curvature of space-time introduced in relativity theory and the waves of 'ψ' introduced in wave mechanics are equivalent. Both devices are used for the same purpose, to represent the distribution of mass and momentum in physical systems. Both are *devices*; it is not suggested that either the curvature or the waves exist in a literal objective sense.' When relativity is used, the gravitational constant occurs; wave mechanics uses Planck's constant h, and the masses m_p and m_e of proton and electron. Eddington's idea was to find a problem that can be treated by both methods and hence to obtain a relation between the constants. The problem he chose was the Einstein universe, a three-dimensional surface of radius R on a four-dimensional hypercylinder, idealized to contain a uniform distribution of protons and electrons in its ground state. The universe is treated as a giant atom, and indeed Eddington remarked in his paradoxical way that the universe is like an atom, only smaller. He did not calculate the energy levels exactly but took them to be similar to the translational levels of a particle in a box. 'In quantum theory, mass corresponds to the periodicity of the waves. The only direct connection between periodicity and curvature is the periodicity which arises from "going around the world", the corresponding wavelength being $2\pi R$.' The corresponding mass is $h/2\pi Rc$. The exclusion principle forces successive particles to have periodicities corresponding to higher harmonics. The rest mass of a particle is deduced from the topmost energy level when all the particles in the universe have been accommodated.

Schrödinger was enormously impressed by these ideas of Eddington. They extended his wave mechanics to the ends of the universe and gave his wave equation a cosmic significance. Other theoretical physicists were more critical. A long review of *Protons and Electrons* by Edmund Whittaker emphasized the speculative nature of the work and the fact that it led to no predictions of new experimental phenomena.[12]

A crucial result obtained by Eddington was his *fundamental relation,*

$$mc^2 = hc \ \sqrt{N}/R$$

where m is the rest mass of an average elementary particle, N is the total number of particles in an Einstein universe with radius of curvature R. To obtain this relation, Eddington compared the ordinary quantum-mechanical treatment of an isolated small system, such as a particle in a box, with an analysis of the same system considered as part of the entire Einstein universe. Since the space of this universe is not infinite, wavefunctions are not infinite plane waves but standing waves with discrete frequencies. Each proper vibration can be assigned to one particle. The state of the universe is close to equilibrium so that almost all the lowest energy levels are filled, as in the case of electrons in a metal at low temperature. The energy level at the top of the filled states (like the Fermi level in a metal) is identified with the rest energy mc^2 of a particle in the small isolated quantum-mechanical system. If the analogy given were complete, however, elementary considerations would show that the energy should be proportional to $N^{1/3}$ instead of $N^{1/2}$. The derivation of the $N^{1/2}$ factor, given on p. 271 of Eddington's book, is unusually opaque, and probably nobody except possibly the author has ever understood it.

In reply to a request from Schrödinger for elucidation, Eddington wrote to him on 17 June 1937:

I think I understand the reason for the discrepancy. Your equation gives possible states of equilibrium of one particle by itself. If R is given, the equation determines [for each spherical harmonic] the mass m which the particle would need to have in order to form a world of that radius. But if there are N particles, your equation does not apply; because we know from relativity theory that it is the total mass of all the particles which determines the radius R for which equilibrium is possible.

After further discussion, Eddington concluded:

I am delighted to hear that you like the main lines of the theory. I know that there is a lot that needs getting into better order and formulation; this can only be done by many people taking it up and worrying over points like these.

With Arthur Eddington (1942).

A meeting of the Societa Italiana di Fisica was to be held in Bologna, October 18 to 21, 1937, an international reunion to commemorate the 200th anniversary of the birth of Luigi Galvani, the discoverer of the electric current.[13] Most of the great European physicists would be there: Bohr, Debye, Fermi, Broglie, Heisenberg, Hevesy, Kramers, Perrin, Oliphant, Richardson, Siegbahn, Aston, and Sommerfeld. They were invited to give papers based on their recent research. Schrödinger was of course invited to present a paper, but he decided to talk not about his own work but about 'Eddington's Theory of the World'.

As he was preparing this paper he continued to worry about the $\sqrt{N}$ factor, and he wrote again to Eddington on September 14: 'Dear Sir Arthur, I feel like the woman of Kanaan in teasing you over and over again with the same argument. Forgive me and be merciful – I am helpless.'

The Bologna conference was officially opened by the puppet king of Italy in the Great Hall of the University. On October 20, news of the death of Ernest Rutherford was received, and Niels Bohr delivered a spontaneous eulogy of his old teacher – it marked the end of a great epoch in physics, during which the structure of the atom had been worked out in detail by theory and experiment.

Schrödinger's paper was given in French. The choice of language was a subtle political statement for an Austrian professor at that time.

He began with a theatrical metaphor: since the advent of general relativity, one can no longer regard space-time as the stage and matter as the actor; the properties of the one are so intimately related to those of the other that one can almost call them two different designations of the same thing. Thus when one applies quantum mechanics to an apparently isolated system, the mere fact of setting up the problem in space and time means that one must consider how the system reacts with the rest of the world. 'One could say that the two theories in question are concerned with two complementary aspects of the world, one its connection on a large scale, the other its connection on a small scale . . . What we are missing is evidently the union of the two.'

Schrödinger presented the basic ideas of the Eddington theory in clear and unequivocal terms. He did not try to criticize or amend the theory. Thus he bravely gave the 'fundamental relation' with the $\sqrt{N}$ term. There was little formal discussion after his paper. Pauli asked two questions that implied some scepticism about the whole procedure. To these, Schrödinger made only a 'hand waving' reply. In the corridors and over the delicious Italian meals and copious wines, the discussions were intense. Schrödinger braved the intellectual assaults of all his quantum-mechanical confreres. Since he was not presenting his own work, his defences were hardly adequate.

On October 23, he wrote to Eddington:

> I have just returned from Bologna, where I endeavoured to give a brief report of those parts of your theory which seem the most important to me. I met with an unvanquishable incredulity of the important group, Bohr, Heisenberg, Pauli, and their followers. I was in an extremely difficult position – spiritually I mean – because so many of your arguments are as ununderstandable to me as they are to them. They wished to know the way in which your famous $\sqrt{N}$ formula is derived. But on that way lies the terrible second half of page 271 . . .

Eddington replied immediately with a long letter enclosing another derivation of the $\sqrt{N}$ term. There was, however, no true meeting of the minds. They seemed to be talking about the same questions, but Schrödinger was too polite and Eddington was too introspective to allow a real thrashing-out of the area of misunderstanding.

Schrödinger was never satisfied with any of Eddington's several explanations of the $\sqrt{N}$ term. In a paper written in 1940, he confessed:

> I have tried to arrive at something like [Eddington's formula] in many more elaborate ways . . . in the faint hope of letting the *square root* of N slip in instead of the cube root. I have really tried hard, but . . . I always get the same result. Eddington arrives at the square root, which makes the formula agree with probable values of R ($= 10^{27}$ cm) and of N ($= 10^{79}$). His main derivation, which accepts the exclusion principle, is beyond my understanding.[14]

Despite his misgivings about some of Eddington's arguments, Schrödinger remained convinced that a quantum mechanical theory of the universe might be achieved. Thus, for several years, he devoted a major effort to problems derived from Eddington's cosmology. At the end of December 1937, he sent to the Pontifical Academy of Sciences a long memoir on 'Proper Vibrations of Spherical Space'.[15] In the introduction to this paper, he noted that Eddington had considered the proper vibrations of a homogeneous, isotropic space of constant curvature, but that no explicit derivation of these wave functions had been provided. It was to this lack that his memoir was directed, 'a desirable preliminary for a clear understanding and continuation of Eddington's train of thought'. He considered his paper to be primarily an abstract collection of mathematical results, which later he hoped to apply to the 'treasure of ideas' contained in Eddington's work.

He gave part of the problem to his assistant at Graz, Robert Muller, for a doctoral thesis.[16] In the Einstein universe, the line element is given by

$$ds^2 = - R^2[d\chi^2 + \sin^2\chi(d\theta^2 + \sin^2\theta d\varphi^2)] + c^2 dt^2 = - ds_0^2 + c^2 dt^2$$

To describe a location in space requires three angles, χ, θ, φ; R is the radius of curvature. In the vicinity of the origin θ, φ become the usual polar coordinates and $R\chi$ corresponds to the radius vector. Note that the Einstein world is 'cylindrical', space is curved (spherical) but time is not. In 1918, W. deSitter introduced a cosmology in which time is also 'spherical':

$$ds^2 = - ds_0^2 + c^2 \cos^2\chi dt^2$$

For a resting clock, $ds = c \cos \chi dt$, so that duration varies as $1/\cos \chi$. In such a world, a light beam requires infinite time to travel from the origin to the horizon. Muller solved the wave equation for both the Einstein and the deSitter universes.

Schrödinger's Pontifical Academy memoir contains much of interest in his detailed description of the wave equations and eigenfunctions on the surface of a hypersphere. He also included a discussion of the rotation group of this space, so that his interest in group theory must have deepened since deprecatory comments made to Wigner in Berlin. His most remarkable result was obtained when he introduced the Dirac equation and applied it to the test case of the hydrogen atom in the limit $R = \infty$ (flat space). He found 'unexpectedly' that the wavefunctions ψ were required to be two-valued. Eddington had already shown that the Lorentz-invariant Dirac equation might yield two-valued wavefunctions, but now it appeared that such functions *must* be used in certain cases.

In January 1938, he sent an expanded discussion of 'The Many Valuedness of the Wavefunction' for publication in the special issue of *Annalen der Physik* dedicated to Max Planck on his eightieth birthday.[17] He pointed out that: 'the wave function is not observable. Therefore one cannot place any requirement upon it directly, but only as a statement about the observable that must derive from it. The values of observable quantities, or more precisely their probabilities, must be clearly expressed. One cannot reject the conceptual possibility of dividing configuration space like a Riemannian surface and ascribing several branches to the wave function. Only the same assertions must be derived from each branch.' He found that for any particular problem one must have either all single-valued or all double-valued wavefunctions, and the two branches of the latter can differ only in sign. His acceptance at this time of a probability interpretation of the wave function may be noted.

Vienna visits

For the summer semester, 1937, Schrödinger's courses consisted of two 2-hour lectures a week on Field Physics, one hour a week on Introduction to Quantum Theory, and the Seminar in Theoretical Physics. In the winter semester, 1937/38, he lectured on Corpuscular Physics (4 h), Discussion of Physical Questions (1 h), in addition to the seminar and a proseminar. This was not a heavy teaching load, but he also traveled to Vienna once a week for lectures and seminars, and he kept an apartment there.

In 1932 Hermann Mark had moved to Vienna as professor of physical chemistry, and he often joined Erwin and Hans Thirring in various outings.[18] He was about eight years younger than the other two, but since he was a professor that made no difference. Mark had been a friend of Dollfuss dating from their wartime service together, so that he was known to be an enemy of the Nazis and was well aware of his dangerous situation in Austria. The three men often discussed politics and Mark thought that both Hans and Erwin were socialists. 'Thirring was a "red" as we would say at that time and Schrödinger disliked every politician and every government as a matter of principle. So they both disliked Dollfuss and hated Hitler . . . Neither Hans nor Erwin hated Schuschnigg, he was too colorless, he was a moderate, a philosopher, not a politician . . . He was no match for Hitler . . . The Nazis were very strong in Graz, in fact Schrödinger hated them there as much as he hated them in Berlin, and since they were much weaker in Vienna, he liked to come there to get away from them.'

Hans Thirring.

During the summer, Hermann, Hans, Erwin, Adolf Smekal, and a few others often went swimming in the Danube, changing into swimming trunks behind riverside trees. Sometimes the Thirring boys, Harald and Walter, would accompany them. Erwin was a strong swimmer and thought nothing of swimming across the river. They also would go in Mark's car to a big lake nearby, the Neusiedlersee.

Anny became a friend of Mimi Mark and Fanchi Thirring, and there were often lively parties in Vienna. Apparently Hilde stayed home in Graz on almost all these occasions. As Mark put it, 'Those were happy days – but most people did not know they were dancing on a volcano.'

Towards the end of June, Max Laue drove from Berlin to lecture in Graz and in Vienna. He took Erwin and some of the other physicists from Vienna on an excursion to the famous abbey at Melk. Unfortunately there is no record of what Max and Erwin had to say about the Nazi menace.

Mark had good contacts with the Swiss chemists Karrer and Ruzicka and they warned him in mid-1937 that Hitler was preparing to take over Austria. The Swiss papers were reporting that Mussolini had told Hitler that he would take no action if such a move was made. Mark told Hans and Erwin, 'I shall leave Austria.' They said, 'For God's sake, why do you want to leave Austria, you are a professor.' 'It's all going to go wrong.' They said, 'Impossible.'

Erwin also had another interest in Vienna, for Hansi Bohm had returned from England to the family mansion. Her father had suffered a heart attack in March 1937, and he died later that year. Thus Erwin now had the problem of correlating his schedules with those of three women. Hansi and Erwin used to meet in the Vienna woods, where tiny villages and rustic inns provided ideal refuges for those who wished to escape the tumult of the city. Erwin loved the *Heuriger* [new wines] for which the local taverns were famous and loved even more the delightful company of the young woman who understood him so well.

Anschluss

Early in 1938, portents of disaster appeared in Austria. An intense aurora lit up the sky so that in many districts fire alarms were sounded. The plague birds appeared on the streets of Vienna, albino sparrows with russet splotches like dried blood on their wings, and little girls were recruited by the Nazis to distribute pornographic broadsheets to passers-by.[19]

On February 12, Hitler summoned Schuschnigg to his mountain hideout near Berchtesgaden and bullied him into signing over control of the police and foreign affairs to the Nazis in return for a promise of a guarantee not to invade. When Ciano, Mussolini's son-in-law, heard about it the next day, he wrote in his diary, 'The Austrian chicken has fallen, or almost fallen, into the German soup pot.'

At the height of the crisis, Schrödinger overcame his contempt for politics and made a cautious but unmistakable political statement to a huge audience.[20] He was scheduled on Friday afternoon, February 18, to give a talk on 'World Structure in the Large and in the Small' in the Auditorium Maximum of the University of Vienna. The hall was filled to capacity. Hans Thirring introduced the speaker as a 'pathfinder of modern physics who has decisively accomplished the liberation of natural science from purely mechanical concepts . . . Although he has become world famous, he is a child of Vienna and has remained a true Austrian.' In his lecture, Schrödinger first outlined the structure of matter at the atomic level, and then showed that the large-scale structure of the world is related to this small-scale structure, so that the ancient dream of the mystical *Naturphilosophie* of the middle ages, the unity of macrocosm and microcosm, was now becoming a reality. The nuclei of all the elements in the universe are composed of tiny particles, protons and neutrons, but the universe is not infinite in extent. If one imagined a pearl necklace made of neutrons and protons which extended through all space and finally returned to its origin, the string would contain about 10^{40} particles. Since the whole universe contains about 10^{80} protons, Schrödinger triumphantly concluded that this shows how a property of the elementary particle determines the size of the universe. The spirit of Eddington was still strong within him.

At the end of the scientific part of his lecture, he expressed two closing thoughts: 'When one returns again from the kingdom of the stars to our world, one finds there a liking for a concept that wants to place one of the nations that live in this world over or under another one.' At these words thunderous applause broke out, and only after some time could the speaker continue. 'The science of inorganic nature compels a respect for a higher order that lies between pride and extreme modesty, an order which is over us and which always reveals itself to us in image and in likeness.'

After Schuschnigg returned to Vienna, his spirits revived, and he made a defiant speech to the parliament. This resulted in huge Nazi demonstrations. In Graz, a crowd of 20 000 gathered in the city square to hear Schuschnigg's address on the radio. The Nazis smashed the loudspeakers, tore down the Austrian flag, hoisted the swastika, and called for an uprising to take over the city. When Schuschnigg responded forcefully, sending an armored train to Graz and an overflight of light bombers, the Nazis canceled their rally. On March 1, Seyss-Inquart, the leader of the Austrian Nazis, who had recently been in Berlin to consult with Hitler, came to Graz, leading a parade of 5000 brownshirted storm troopers.

On March 6, Schuschnigg decided to hold a plebiscite on Austrian independence on March 13. Hitler was furious; he summoned his generals to prepare for an invasion on March 12. Schuschnigg was again cowed and called off his plebiscite. He asked the British government what to do. Ribbentrop, the German ambassador in London, invited Halifax, the foreign secretary, to tea, and explained that Hitler was only following a policy similar to that of Britain in regard to Ireland. Halifax advised Schuschnigg that he could expect no help.

On March 11, Schuschnigg resigned and broadcast a plea to the citizens and armed forces to offer no resistance. In fact there was not a single gesture of resistance anywhere in Austria as the local Nazi bands took over.

Hitler arrived as a 'tourist' to visit the grave of his mother in Linz, where he met Seyss-Inquart. He was greeted everywhere with delirious enthusiasm. Actually, up to this point, he had not decided to take over Austria completely, but Goering now advised *Anschluss* and Seyss said it was obviously the will of the people. The takeover was announced on March 13 in the *Wiener Gazette* and the next day Hitler arrived in Vienna, which welcomed him with unfeigned rejoicing.

The sadistic behavior of the victorious Austrian Nazis in Vienna was worse than anything seen in Germany up till that time.[21] Jewish shops and businesses were looted and anyone who resisted was beaten to the ground. Only the German military police kept the terror within any limits. For days a favorite pastime was to capture an old Jew and set him to work cleaning the sidewalks with an acid solution that severely burned the hands, while passers-by gathered to enjoy the spectacle. Jewish professionals were forced to clean the latrines of Nazi hangouts. Within a few days there were 76 000 arrests in Vienna alone, and 6000 dismissals from government offices and teaching posts. Hans Thirring was one of the first to be sacked; storm troopers appeared and seized his office but he was not arrested. Hermann Mark was arrested, but fortunately one of his former students was in charge of the jail and Mark was able to obtain his freedom by payment of a large bribe to a top Nazi named Indra. He converted the rest of his funds into platinum wire, fashioned it into coat hangers, and escaped to Switzerland. When the borders were closed, many could not escape and chose suicide. The Nazi control was enforced with many executions; no attempt was made to keep these secret and they were announced on large red posters throughout the cities.

Cardinal Innitzer sent greetings to Hitler and ordered all the Catholic churches to hoist swastikas and to ring their bells in jubilation. He visited Hitler at his hotel and promised him that the

Catholics 'would become the truest sons of the great Reich into whose arms they have been brought back on this momentous day'. The Führer was delighted with the cardinal. The Lutheran bishops ordered services of thanksgiving for the Anschluss to be held in all the Protestant churches, so that the Christian prelates agreed that Hitler's coming was a blessing from God.

There was no official objection in Britain or France to Hitler's takeover of Austria. On March 12, Cordell Hull, the American Secretary of State, published a calm statement approving an Anschluss. Two days later, however, Roosevelt intervened, and a denunciation of the German action was issued from Washington. Litvinov, the Soviet foreign minister, urged common action against the fascists by the League of Nations, but there was no response.

In Graz, there were fewer Jews to torture and Nazi control was already so complete that the Anschluss was less brutal than in Vienna.[22] The university was forced to close and the Rector, Josef Dobretsberger, was dismissed. David Herzog, professor of Semitic languages and Rabbi of Graz, was seized outside his synagogue and thrown into the River Mur. Victor Hess was a native of Styria and a graduate of the University of Graz. In 1936, he had shared the Nobel prize in physics with Carl Anderson of Caltech. He was dismissed because he was a strong supporter of Schuschnigg, but he was able to leave the country for a professorship at Fordham University in New York. About fifty Jews in the medical faculty were summarily dismissed and many were imprisoned. Otto Loewi had received the Nobel prize in medicine in 1936 (with Henry Dale) for his discovery of the chemical transmission of nerve impulses across synapses. He was sixty-five years old and had been a professor at Graz for twenty-nine years, choosing to stay there despite many offers from more prestigious universities. He and his wife and two sons were thrown into prison. After two months he was allowed to leave for England after he had paid a ransom by transferring his Nobel prize money to a Nazi bank.

Schrödinger had planned to visit Oxford in the fall term of 1938, but early in March, George Gordon received a letter saying that 'the great and important events we are living through in these days might make my presence necessary so that I could not get away'. William Bragg told Gordon that he had heard that Schrödinger was in a concentration camp. Gordon wrote to Lindemann: 'I propose to apply to Halifax to do something, since E.S. is still a fellow of the College.' Lindemann himself telephoned Halifax on April 6, and the foreign secretary promised to make inquiries through the British ambassador in Berlin, Neville Henderson, who was a great admirer of the Nazi regime.[23]

Appeasement in Graz

Although Schrödinger was kept under surveillance and his house was searched, he was not otherwise molested at this time, but he knew that if he wished to remain in Austria, he would have to try to appease its new rulers. He thought that he might have a situation similar to that of Laue and Heisenberg, and he did not understand that Laue was protected by the generals and Heisenberg by his family connections with Himmler. The new Nazi rector of the University of Graz, Hans Reichelt, had been assigned the task of compiling a list of the staff who were to be 'cleansed'. He advised Erwin to write a letter to the Senate setting forth his changed attitude. He did this and the letter was published in all the German and Austrian papers on March 30. The Graz *Tagespost* gave it a prominent headline 'Confession to the Führer':[24]

In the midst of the exultant joy which is pervading our country, there also stand today those who indeed partake fully of this joy, but not without deep shame, because until the end they had not understood the right course. Thankfully we hear the true German word of peace: the hand to everyone willing, you wish to gladly clasp the generously outstretched hand while you pledge that you will be very happy, if in true cooperation and in accord with the will of the Führer you may be allowed to support the decision of his now united people with all your strength.

It really goes without saying, that for an old Austrian who loves his homeland, no other standpoint can come into question; that – to express it quite crudely – every 'no' in the ballot box is equivalent to a national [*völkisch*] suicide.

There ought no longer – we ask all to agree – to be as before in this land victors and vanquished, but a united people [*Volk*], that puts forth its entire undivided strength for the common goal of all Germans.

Well meaning friends, who overestimate the importance of my person, consider it right that the repentant confession that I made to them should be made public: I also belong to those who grasp the outstretched hand of peace, because, at my writing desk, I had misjudged up to the last the true will and the true destiny of my country. I make this confession willingly and joyfully. I believe it is spoken from the hearts of many, and I hope thereby to serve my homeland.
E. Schrödinger

The paper commented that there were many scholars in ivory towers who had been misled by the Dollfuss–Schuschnigg system, but the scales had fallen from their eyes as a result of the storm of enthusiasm that had swept the country. 'The voice of blood calls also these men to their people [*Volk*] and thereby to find their way back to Adolf Hitler.' Although Schrödinger's letter repeated the key words *Volk* and *völkisch*, he did not conclude with the ritualistic *Heil Hitler*.

In future years, Schrödinger often regretted this letter, and the closest he came to an explanation was in a letter to Einstein (July 19, 1939):

I naturally knew there was a certain danger when I went back to Austria. But that the fortress would be surrendered without striking a blow, that I never reckoned until the end. Just a few days before, I was with a section chief in the Ministry and said to him: if you put a rifle in my hand I will be glad to defend myself, but don't let me remain as a hostage in nazified Graz. You can imagine with what feelings just a few weeks after the overthrow [*Umschmiss*] I read the signature of the same gentleman under the orders of the new Minister! I hope you have not seriously taken amiss my certainly quite cowardly statement afterwards. I *wanted* to remain free – and could not do so without great duplicity.

On April 3, Karl Renner, the socialist leader, asked his followers to approve the *Anschluss*. On April 10, the Austrians voted in a plebiscite 99.73% in favor of the Anschluss, 4 453 000 yes, 11 929 no. The votes were cast under the eyes of Nazi storm troopers, so that probably no more than 90% of the Austrians really approved what they later called 'Hitler's first conquest' or 'the rape of Austria'.

A brief notice of Schrödinger's attempt to appease the Nazis was published in *Nature* [May 21].[25] His friends assumed that his letter had been written under duress and expected the worst. Much to their surprise, however, it turned out that Anny and Erwin were enjoying a peaceful spring holiday, skiing in the Tirol. W.J.M. Mackenzie, a member of Magdalen College, reported to President Gordon that he had met them there, and Gordon asked him to send Lindemann an account of this meeting, which he did in a letter of April 21:[26]

We had a very long talk about things in general . . . The following points may be of interest:

(1) Schrödinger himself is in good health, has not at any time been under arrest, and has not been personally molested; Mrs. Schrödinger is still convalescent after an operation but seems to be getting on well. Their correspondence however is probably being tampered with . . .

(2) The University has been shut since the disturbances in February, and after the Nazis took over several of the professors were arrested . . . The University re-opens this week, and as far he knows S will proceed as usual with lectures in Graz and once weekly in Vienna.

(3) His personal position is one of complete uncertainty, for better or for worse; he summed it up by saying that they ought logically either to promote him to a better job or put him in a concentration camp, but he had no idea which it would be. He doesn't know who is responsible for deciding his fate; nor what his attitude to the regime is officially supposed to be; nor whether he is debarred completely from leaving the country. But for the present he

certainly cannot come to Oxford to lecture without specific authorisation from some high power, and doesn't even know whether he needs permission to travel to Berlin for some celebration there (I think in honour of Planck) . . .

(4) In regard to future plans; he has clearly no feeling that it is necessary for him to get out of the country at all costs as soon as possible, and would on the whole like to make his peace with the regime if they will let him. So that his policy is to establish a reputation as a non-political person, with reasonably patriotic principles; I don't think Schuschnigg's fate troubles him much, as a matter of principle – the 'Schuschniggers' are only getting as good as they gave – but dislike of the anti-Jewish policy is as strong in him as ever, and may perhaps cause some difficulty.

(5) There is practically nothing we can do to help him directly, unless things get much worse than they are; letters to him would be rather dangerous, but he said he would be very glad to see anyone from Oxford who was in Austria – apparently thought it would be quite safe . . . In general anything which stresses his international importance may be useful, and may filter through to the relevant department, provided always that it contains no hint *whatever* of *political* importance.

We had four or five hours of talk, and naturally there was a great deal more than this said, but I think these are the main points.

Schrödinger's remarks about 'promotion to a better job' could have referred only to the Vienna professorship from which Thirring had been dismissed. In his political naivety he did not realize that all such vacancies would be filled by eager and dedicated Nazis. He did begin to worry about his Swedish securities and undertook a delaying action to avoid their transfer to a German bank.

On April 14, Anny sent a postcard to Paul Ewald in Stuttgart:

We fled here for a week from dirty Graz and found (Tauernpass height 1738 in the Radstadter Tauer) still deep mountain winter. Besides a quite wonderful skiing district, perhaps we can meet here sometime. I have been indecently long without news of you. Apparently Erwin and I will travel next week to Berlin. On the other hand our previous plans to meet in England cannot take place. Otherwise things go well with us and we are happy to be at such a high altitude.

On April 23, a great international meeting of the German Physical Society was held at Harnack House in Berlin–Dahlem to celebrate the eightieth birthday of Max Planck. Max and Marga were delighted to see Erwin and Anny at the social events that accompanied the scientific sessions. By this time the German physicists had adapted themselves well to the Nazi regime, and the absence of Jewish colleagues would have been hardly noticed. They were probably in general pleased that Schrödinger, the one former exception to their compliance, had now rejoined the fold.

Schrödinger had ventured into the center of Nazi power and

returned uneventfully to Graz, but he found an unpleasant surprise awaiting him there. On April 23, he had been summarily dismissed from his honorary professorship at the University of Vienna, in a communication from the office of the Dean of the Faculty of Philosophy:[27]

On the basis of the decree of the Austrian Ministry of Education of 22 April 1938, No. 12474/I/1b, your authorization to participate in instruction in the Philosophical Faculty of the University of Vienna is withdrawn.

You are therefore to refrain from any professorial or other duties falling within the scope of your former appointment or especially assigned to you.

The Provisional Dean

His appointment at the University of Graz, however, was not affected at this time, since the rector had taken personal charge of the purging of the university staff. The Senate voted to change the name of the university to Adolf Hitler University, but the Senate was itself abolished before this change became effective. The University of Graz became a major center for Nazi education, with special courses in racial studies, war chemistry, and for training the SS medical corps who would operate the extermination facilities. After the defeat of Hitler, the Nazi staff fled, but within a few years most had returned, and the university archives were sealed so as to prevent any revelations concerning the Nazi period.[22]

Bad news travels fast, but in the case of Schrödinger's dismissal from Vienna, it surpassed itself: on the same day that the letter of dismissal was sent, Eamon de Valera wrote from the Office of the Taoiseach (Prime Minister) in Dublin to E.T. Whittaker in Edinburgh:

In an evening paper a couple of days ago I saw it noted that Professor Schrödinger had been dismissed from his post. I suppose that it has not been possible for you to get in touch with him? I am very anxious that we should secure his services in connection with the project we discussed when you were here. [Establishment of an Institute for Advanced Studies.] If you are able to communicate with him, will you please convey to him an invitation from me to come to Dublin. Whilst we are waiting to have the scheme worked out, some special financial arrangement can be made for him. The important thing is that we should not lose his services.[28]

Last days in Graz

The situation of the Schrödingers in Graz was more precarious than they realized. On May 9, Lindemann received a letter from Halifax:[26]

My dear Lindemann,
    With reference to your telephone message of the 6th April I am writing to let you know that our ambassador in Berlin has now had a reply from the German

minister for Foreign Affairs to the enquiries he made at my request about Professor Schroedinger.

I am sorry to have to tell you that Herr von Ribbentrop's reply is unfavourable. According to the German authorities Professor Schroedinger left Germany in 1933 for political reasons and, after a brief stay in Oxford, settled in Graz where he proceeded to busy himself as a fanatical opponent of the new Germany and of National-Socialism. A recent examination of his case had shown that Professor Schroedinger had, up till very lately, remained in constant communication with German emigrés abroad and it was therefore feared that permission for him to visit Oxford would merely offer him a further opportunity of resuming his anti-German activities. In these circumstances Herr von Ribbentrop fears that a further course of lectures by Professor Schroedinger at Oxford could only serve to harm Anglo-German relations and is therefore, to his regret, obliged to refuse my request.

I much regret that my intervention on behalf of Professor Schroedinger should not have been more successful but I fear that, in view of the attitude adopted by the German authorities, there is nothing more that I can do.

Yrs sincerely,
Halifax

After the Anschluss, Erwin felt it prudent to see less of Hansi, for the discovery of a liaison with a Jewish woman would have been fatal to his uneasy truce with the Nazis. He asked her to burn all his letters, and she complied, consigning them one by one to the flames while the tears trickled down her cheeks. She felt at this time that Erwin was not entirely immune from the virus of antisemitism that had infected most Austrians for so many years.

Richard Bär was an old friend of the Schrödingers from Zürich days. He had been using his family wealth generously to relieve the distress of refugees from Nazi Germany and elsewhere, or as Erwin rather preciously put it 'those uprooted by any kind of spiritual intolerance'. On June 10, Bär and his wife met Anny in Constanz. He reported to Franz Simon in Oxford: 'Erwin is no Nazi, despite the ominous letter that he freely wrote . . . He would now like to live in peace with the regime, like Laue and Heisenberg . . . He hopes to get leave for the fall semester to come to Oxford.' Anny had a very important reason for being in Constanz, but Bär did not mention it in his letter to Simon.

Simon replied to Bär with some political realism.[29]

If Schrödinger comes out, which appears to me doubtful, in my opinion he can never go back again. The situation is thus: The letter made an awful impression here. In order to restore his reputation – and in a certain sense that of all the emigrants – it has been spread abroad that he wrote this letter only under extreme pressure. When he now arrives here, he must either say that he has written it voluntarily, then he is in a really unpleasant situation, especially since people know very well what he thought before and what he just a little while previously expressed in writing to local colleagues. Or, he says that he was forced to do it, then he can no longer go back. (There are always informers here who make reports to Germany.) I wanted to make this clear to you, so that in case you have an opportunity you can pass it on to Schrödinger . . . I might mention that in this connection the famous saying of the King of Hanover is becoming cited here: Professors can be bought like whores.

Anny later told the story of the reason for her trip to Constanz as follows:[30]

Well, when the Nazis came to Austria, my husband got several invitations to foreign countries. He was not allowed to get the telegrams himself. They were brought to the university and he was called to the university and told 'Of course, you have to refuse. You can't go to Brussels and so on'. So it was really absolutely like a prison . . . De Valera knew that we were in danger and he let us know that there was a possibility for an institute for advanced studies in Dublin which he wanted to create if my husband said, in principle, 'Yes. I will come'. It was absolutely sure he could not write to my husband because everything was censored. So he asked [Whittaker] to ask Born; Born wrote to our friends in Zürich [Bär]. Our friends in Zürich told a Dutchman, who came to Vienna, about the possibilities. We were not in Vienna. He didn't come to Graz, he came to my mother and told her this important thing. She was afraid to take such a very important message without having anything written down, so she wrote just down in a few lines that de Valera wanted to create an institute for advanced study and whether he would come, in principle. This little piece of paper my mother sent to Graz. We saw it and we read it three times and then destroyed it, threw it into the fire, and told nobody about it at all. I went with my car as far as Munich. Thirring went with me and I went to . . . Constanz; I met our friends there and I told them, 'Yes, in principle, he will come. But nothing should be done that will let anybody know that we are going away' . . . My friends wrote to Born, and Born told Whittaker, and Whittaker told de Valera, and that was finished. He never spoke to de Valera [before all this]; he never knew de Valera – nothing at all.

Flight from Graz

One of the great advantages of the Graz position was its nearness to the mountains that Erwin loved so much. Winters were enjoyed for skiing and summers for hiking. In August, 1938, Erwin and Hilde were together in the Dolomites. They met the Plancks there. Their mountain tour must have recalled the Tirolean days five years ago when they first became lovers, and there was a renewal of their love, which was to have an important consequence for their lives during the coming war. Erwin let Richard Bär know that he would like to return to Oxford, and Bär relayed to Simon the information that Schrödinger would like a suitable offer of a permanent position.

Erwin was oblivious of the gathering storm of Nazi vengeance that was about to overtake him. Shortly after his return to Graz, he received the following curt notice from the Ministry of Education:[27]

Vienna, 26 August, 1938

Concerning: Measures on the basis of the ordinance for renovation of the Austrian professional civil service.

To: Herr Professor Dr Erwin Schrödinger

On the basis of Sect. 4, Paragraph 1 of the ordinance for renovation of the Austrian civil service of 31 May 1938, RGBl.I S.607, you are dismissed. The dismissal is effective as of the day of arrival of this notice. You have no right to any legal recourse against this dismissal.

The reason for dismissal specified in the ordinance was 'political unreliability'. After this action by the Austrian authorities, the Berlin University stripped Schrödinger of his title as Professor Emeritus and ordered that his name be stricken from all the records of the university.

In dismissing Schrödinger, the Nazis did not act in rational self-interest, since it would have been a great propaganda success to exhibit as a convert to their cause the only world-famous German scientist (without a Jewish family) who had spurned them. Evidently the contemptuous way in which he had left Berlin rankled Prussian self-esteem too deeply, for, as Plutarch remarked in his life of Timoleon, 'so true it is that men are usually more stung and galled by reproachful words than hostile actions: and they bear an insult with less patience than an injury.'

Schrödinger was dismayed by his unexpected dismissal and he went immediately to Vienna to talk with an official high in the government, whom Anny described as 'a very good man'. He was told not to worry because: 'It is easy for you with your name to get another job in industry, or somewhere.' His friend Hans Thirring, who had been dismissed earlier, managed to survive with the help of consulting jobs, but he had some friends in high places. Schrödinger said in effect, 'Well for a theoretical physicist that is not so easy, and look what happened to all the Jews, their papers became invalid or were misplaced, so they could not get work. I shall have to go to a foreign country to find my living again.' The official said, 'They won't let you go to a foreign country. Have you still got your passport?' This was apparently well meant; it was not a threat, but it gave Erwin a shock because he never thought he might be prevented from leaving the country.

Now he knew that there was no time to waste. He rushed back to Graz and in three days they were ready to leave. Everything they could take was packed into three suitcases. It was forbidden to take money or valuables out of the country, so they left almost all their possessions in Graz, including the gold Nobel medals and the chain of the Papal Academy. They bought two return tickets to Rome. They did not dare take a taxi to the station for fear their departure might be reported to the authorities. Anny brought the luggage to the station in her faithful grey B.M.W. and then brought the car back to the garage and asked them to wash it. They left Graz on September 14 with ten marks in their pockets.

As Anny continues the story, 'We didn't have the money to pay the porter in Rome, so we had the taximan pay the porter, and the hotel pay the taxi.' Fermi met them and lent them some money. He warned them, 'Don't write from Rome because it is already dangerous – it might be censored.' From the Papal Academy, which is beautifully situated in the Vatican Gardens, Erwin wrote three letters: one to Lindemann to tell him that they had left Graz, one to Bär in Zürich to ask him to send some money, and the third to de Valera who was the President of the League of Nations in Geneva. The letters were posted from Vatican City on Saturday.

'On Monday morning', Anny continued, 'we went again to the Academy and after half an hour's time there came a Diener [servant] who said that "his excellency" was wanted at the telephone. My husband didn't turn around – after having been thrown out of Graz. "Yes, yes", the man said, "it means you".' It was the Irish Embassy, with the news that de Valera had just telephoned from Geneva with instructions that everything should be done to get them there as soon as possible. That afternoon Erwin talked for the first time with de Valera, who said that he was very glad they they were out of Austria and they should come to Geneva to discuss a few things, but as soon as possible should go on to England or Ireland because there was such a great danger of war. It was near the time of the Munich conference and the betrayal of Czechoslovakia. It was forbidden to take any money out of Italy, and the Irish consul gave them each a pound and first class tickets to Geneva.

I was quite happy, I felt already safe but my husband didn't feel safe at all. At Domodossola, they looked at our passports, but the luggage hardly at all. The passports were all right. But before we came to Iselle – at one end of the Zipplertal [Simplon tunnel] – a carabiniere came into our compartment and we had to leave with all our luggage. He had a piece of paper with our name written on it. It was really the fright of my life. We had to take off everything [from the train]. [Erwin] was separated from me with the luggage and I was in another place. I couldn't speak Italian; there was a woman who looked through my things, saying, 'Put everything toward the X-rays'. They X-rayed every single bit of my things – my handbag, my teeth, everything. After about half an hour, or three-quarters of an hour – the train had to wait; it was an express train – we were allowed to enter the train again . . . They had looked at our passports and we had visas for all of Europe because de Valera had said that we can't go through France, we must go through Spain or Portugal or something. We had everything. We were asked if we had some money. We said that we had one pound. Then they thought that we had to be smuggling something, because one can't go through Europe on one pound.

De Valera was very, very pleased when we arrived. He was already in full dress because they had a banquet in the evening, but he received us and he was very kind, and we stayed for three days in his hotel and then went on.

Return to Oxford

During the month of September, 1938, war in Europe seemed inevitable as Hitler moved quickly to take over the northern borderlands of Czechoslovakia. With great resolve, the Londoners began to dig trenches in Hyde Park. The Czechs were at first prepared to fight, but the British government under Neville Chamberlain was determined to get Hitler what he wanted. Chamberlain and Daladier presented the Munich decisions to the Czechs and told them they must surrender. Chamberlain returned to 10 Downing Street to the cheers of his many supporters, to whom he brought 'peace in our time'. Churchill was one of the few sceptics: 'We have suffered a total unmitigated defeat.'

When Lindemann received Schrödinger's letter from Rome, he became quite angry: 'Schrödinger asks whether he can come again to Oxford. Is he mad? Doesn't he realize after this letter he has published what people think of him?'

Then suddenly, about the first of October, they appeared. They came to the Simons and it was an uncomfortable evening for everyone.[31] Franz Simon told Erwin, 'Well, we have tried very hard to tell people here that this letter was written under duress. We don't know the conditions, we said, probably somebody with a gun stood behind him and said "you sign this".' First he said 'What letter?' and then he became quite excited and said, 'What I have written, I have written. Nobody forced me to do anything. This is supposed to be a land of freedom and what I do is nobody's concern.' The Simons told him with more emotion than logic that he had made the situation more difficult for all the other refugees. They said that if war broke out tomorrow, he would be in serious difficulties. Erwin said he was grateful that they had made him aware of the matter, but it concerned only himself. Anny supported Erwin strongly – whatever he did was right. The next day he saw Lindemann who told him the same thing as Simon.

Schrödinger wrote a long letter to Born, which concluded. 'Ach, if only one could make it clear to these English how it really seemed over there! With them [the Nazis] you take care. You do not believe anyone.' Born commented to Simon, 'How are you supposed to believe a man who has published that pretty letter?'

On October 7, Simon wrote to Born, with more details about Schrödinger:[29] . . . 'He has in this matter a frightfully bad conscience, but does not have it in his heart to confess that what he has done is wrong (which does not surprise me in someone who so loves himself to the exclusion of all others) . . . One must unfortunately say that he behaves like a spoiled boy, besides this also with his immature sexual complex . . . These happenings have disturbed us greatly.'

Max Born does not appear to have replied to this letter. His friendship with Schrödinger survived the episode of the unfortunate letter. Max understood that the sources of political depravity lie deep in the human condition, so that no person can claim immunity. He was better able than Simon or Lindemann to appreciate the character of Schrödinger, who did not feel deeply about any political questions, and who honestly could see nothing wrong in writing an expedient letter to a bunch of contemptible gauleiters. Erwin had never claimed or aspired to be the conscience of the German physicists, and the idea of going to a concentration camp as a political protest had no attraction for him. In retrospect, harsh criticism of Schrödinger was hardly appropriate from those who were safely ensconced far away from the Gestapo. On October 20, Simon reported to Born: 'I met Schrödinger once in college. He is especially mad at Lindemann who is not willing to see him again. Yet Lindemann is certainly the Englishman who has done the most for him.'

The Schrödingers stayed two months in Oxford as 'paying guests' with John and Barbara Whitehead at their house, 22 Charlebury Road. John was a brilliant mathematician and fellow of Balliol College. He had taken his doctoral degree at Princeton with Oswald Veblen. His father was the Anglican Bishop of Madras and a brother of the philosopher Alfred North Whitehead.

In August, Hansi Bohm, by a mixture of bribery and good luck, had managed to escape from Austria to England via Switzerland. Sometimes Erwin lived with her in a small furnished flat in Hampstead. They wandered together through the neighborhood streets, stopping in the cheap teashops of prewar London, where awful sugar buns were displayed in flyspecked windows and the puddles of tea were seldom wiped from plastic tables. The conversation was always spirited as the lovers, by now completely at home with each other, lapsed into their native Viennese. Erwin would laugh at Hansi in the kitchen, saying she looked like a Burgtheater actress trying to act the part of a cook.

Belgian interlude

De Valera was actively preparing the legislation necessary to establish the Institute for Advanced Studies, but it would require at least a year to work its way through the parliamentary process. Meanwhile he arranged for a grant of £200 through University College in case Schrödinger wished to come to Dublin immediately. Erwin went over to Dublin on November 18 and stayed for several days, discussing future plans with de Valera and Arthur Conway, Professor of Mathematics at University College Dublin.

When he returned to Oxford, he was delighted to find a letter from Jean Willems, Director of the Francqui Foundation, which offered a visiting professorship at the University of Gent for the academic year 1938/39.[32]

The commission that you may kindly accept implies that in addition to the pursuit of your personal researches, you would consent to give several lectures and hold several colloquia, and to direct the researches of one or two young Belgian workers during the seven or eight months that separate us from the next academic vacation . . . The honoraria applicable to your position have been fixed at 75 000 Belgian francs [$10 000]. I should like you to consider the gesture of our Foundation as a hommage of particular esteem and admiration . . .

Erwin immediately accepted the appointment, and wrote, 'as regards the details of the letter, I am, of course, enchanted, since they meet my intense desire of being given again an opportunity to display some external efficiency and to do useful work.'

In early December Erwin and Anny visited Paul and Margit Dirac in Cambridge and made preparations for the move to Belgium. As Germany had completed its dismemberment of Czechoslovakia and was already stalking its next victim, Poland, Chamberlain's 'peace in our time' appeared less likely. Owing to the war clouds, there were difficulties in obtaining all the necessary residence permits, but eventually these were in hand and the Schrödingers arrived in Belgium about December 15.

From the beginning of the second term, Erwin gave a course of two lectures a week on Introduction to Wave Mechanics, and at fortnightly intervals a lecture series on Proper Vibrations of Spherical Space. An evening seminar on theoretical physics in the laboratory of Professor Verschaffelt was devoted to topics in molecular statistics. Visitors came from Brussels and other universities and the discussions sometimes continued late into the night, with 'indefatigable interest . . . in the advanced and subtle questions that were presented'.

During the next several months, Schrödinger lectured in Brussels, Louvain and Liège. In Louvain he came to know the famous cosmologist, the Abbé Georges Lemaitre, and they had deep and extensive discussions of possible models of the universe.[32]

Although his scientific life in Belgium was varied and satisfying, Erwin was continually worried about his future. As the parliamentary procedures in Dublin seemed to drag on and on, he began to consider alternatives, even including a poorly paid position that Raman had offered to provide in India. It must have been during this time that he wrote one of the most revealing documents that we have from his pen. It is an untitled sheaf of six handwritten pages attached to a few lines of Spanish verse suggested by Calderon, '*Gustas y disgustas no son mas que imaginares.*'

> No es consuelo de desdichas
> Es otra desdicha aparte
> Querer a quien las padece
> Persuadir que no son tales.*

* It is no solace of misfortune / It is another misfortune besides / To ask the one who endures them / To be persuaded they are not such.

The pages were in a folder marked 'Gent' that was with the papers he left in Dublin when he returned in Vienna in 1956.[28] Possibly they were used for a lecture at one of the Catholic universities in Belgium.

'Every second of our lives', he wrote, 'is saturated with the physical consequences of science or, as we could say, with excrements from the progress of research'. Schrödinger was not the first, nor the last, to suggest that technology has caused a deterioration in the quality of man's relation to the deeper sources of his being. Aldous Huxley, like Erwin a true believer in Vedanta, complained that: 'Modern man no longer regards Nature as being in any sense divine, and feels perfectly free to behave towards her as an overweening conqueror and tyrant.'

When Schrödinger performs scientific work, however, he has a full awareness of the divinity of Nature, so that his approach to research is in accord with his belief in Vedanta. He expressed it as follows:

Science is a game – but a game with reality, a game with sharpened knives . . . If a man cuts a picture carefully into 1000 pieces, you solve the puzzle when you reassemble the pieces into a picture; in the success or failure, both your intelligences compete. In the presentation of a scientific problem, the other player is the good Lord. He has not only set the problem but also has devised the rules of the game – but they are not completely known, half of them are left for you to discover or to deduce. The experiment is the tempered blade which you wield with success against the spirits of darkness – or which defeats you shamefully. The uncertainty is how many of the rules God himself has permanently ordained, and how many apparently are caused by your own mental inertia, while the solution generally becomes possible only through freedom from its limitations. This is perhaps the most exciting thing in the game. For here you strive against the imaginary boundary between yourself and the Godhead – a boundary that perhaps does not exist. You may indeed be given the freedom to unloose every bond, to make the will of Nature your own, not by breaking it or conquering it, but by willing it also.

The grave error in a technically directed cultural drive is that it sees its highest goal in the possibility of achieving an alteration of Nature. It hopes to set itself in the place of God, so that it may force upon the divine will some petty conventions of its dust-born mind, and it overlooks the possibility of reaching its goal in a way that expands the divine sparks to an indistinguishable image of the divine will or the divine action (which is one and the same).

The God-Nature wills everything that it does, scarcely everything that it wants, but otherwise nothing. This wonderful self-limitation is not really one, because outside of it nothing is possible which it has renounced, just as spherical space is finite without ever pushing against boundaries that separate it from anything else. This is what we worship in the Godhead. To become like it, we must accede to it, and not like naive children seek our own glory in a show of independence. Only if we follow this way, does everything maintain the character of the game, with that absence of gravity which is so characteristic of every higher spiritual activity.

Erwin thus found in the philosophy of Vedanta his understanding of the nature of scientific research. He may have been thinking of one of his best loved quotations from the *Bhagavhad Veda*:

> Who sees the Lord dwelling alike in all beings
> Perishing not as they perish
> He sees indeed. For, when he sees the Lord
> Dwelling in everything, he harms not self by self.
> This is the highest way.

Schrödinger's interest in cosmology was reinforced by the discussions with Abbé Lemaitre, who had shown, as had the Russian Alexander Friedman, that a static Einstein universe is not stable and must either expand or contract. At the end of August he sent to *Physica* a paper, 'The Proper Vibrations of the Expanding Universe', which was immediately published in the October issue.[33] This paper contains an important discovery: the first indication that an expanding universe under certain conditions may require the creation of matter.

The decomposition of an arbitrary wavefunction into proper vibrations is rigorous, as far as the functions of space (amplitude functions) are concerned, which, by the way, are exactly the same as in the static universe. But, it is known that with the latter, two frequencies, equal but of opposite sign, belong to every space function. These two proper vibrations cannot be rigorously separated in an expanding universe. That means to say, that if in a certain moment only one of them is present, the other can turn up in the course of time . . . Generally speaking, this is a phenomenon of outstanding importance. With particles it would mean the production or annihilation of matter, merely by expansion, whereas with light there would be a production of light traveling in the opposite direction, thus a sort of reflexion of light in homogeneous space. Alarmed by these properties, I have investigated the question in more detail.

He then found that if R is a linear function of time, i.e., if the universe is expanding at a constant rate, the alarming phenomena do not occur. However, with an accelerated expansion, as in the early stages of a universe after a 'big bang', these phenomena may become important. About ten years later, Hermann Bondi and Fred Hoyle developed a steady-state cosmology based on continuous creation of new matter in an expanding universe.

On July 31, while working on the expanding universe, Schrödinger sent to *Nature* a letter on 'The Nature of the Nebular Red Shift'.[34] In 1928, Edwin Hubble had reported his observations from the Mount Wilson Observatory, which showed that the wavelengths of spectral lines from distant galaxies are systematically shifted to higher values to an extent proportional to the estimated distance of the galaxy. Schrödinger pointed out that this red-shift can be explained as the consequence of the dilatation of an expanding universe during the time of travel of light from the distant source. It is not a Doppler effect and has nothing to do with dR/dt at the moment of emission or detection. He discussed the difficulty of proving that the shift is actually due to an expansion of the universe, but concluded that this is indeed the most likely explanation, a conclusion now accepted by almost all cosmologists.

On May 1, the rector of the University of Gent informed Erwin that the Academic Senate had recommended him for an honorary degree, the Doctor of Sciences. This was his first honorary degree. It was to be formally conferred in a ceremony on October 9, but the outbreak of war made this impossible.

In the late spring, the Schrödingers moved from Gent and settled in the seaside resort of La Panne, 7 Sentier des Lapins [Rabbit Lane]. Arthur March brought Hilde and Ruth to Belgium and returned to Innsbruck after a brief visit. Thus when war broke out on September 1, Erwin's unusual family was reunited.

He wrote to Professor Haesaert, the rector of Gent University: 'Permit me to place in your hands the offer of my loyal services, in any way in which it may be that your country may deign to make use of me. I can be sure that in this I find myself in complete accord with my loyalty towards my own country, Austria.'[35]

On September 29, he wrote again to the rector to say farewell and to thank him for:

the kind hospitality, that I have enjoyed at the University of Gent, and to express to you how deeply indebted I feel to your fatherland for the refuge and help that it unreservedly granted us in critical times. All this remains written in my heart; I should be happy if I could ever repay even a small part of it.

We travel next week for the time being to Ireland where I have for the autumn a temporary professorship. One is cured of further worries in a time that has made almost every individual fate in Europe dependent partly on the outcome of disturbances that shake it up completely, partly on any gust of wind that deflects a musketball by a handsbreadth.

There was a problem in obtaining a transit visa through England, since they were now technically enemy aliens. Erwin wrote to Frederick Donnan in London, who in turn conveyed the problem to Lindemann. Despite the ingratitude that Schrödinger has previously displayed, Lindemann took care of the visa problem. It is thought that de Valera also made representations. The Schrödinger party were granted a 24-hour transit visa, and they arrived safely in Dublin on October 6.

10 Wartime Dublin

The establishment of the Dublin Institute for Advanced Studies was due almost entirely to the efforts of one man, Eamon de Valera, the Taoiseach [prime minister] of Ireland. Dev, as he was called by friend and foe, was born in New York City in 1882, so that he was five years older than Schrödinger. His mother, Kate Coll, was an Irish servant girl, and his father, Vivion de Valera, a delicate Spanish artist. Eddie, their only child, was three years old when the father died, and he was brought to Ireland to be raised by his grandmother in a cottage at Bruree, County Limerick. As a schoolboy, mathematics was his best subject and he never lost his love for it.[1]

Dev was always a devout Catholic, in fact, a daily communicant. At that time it was not easy for a Catholic to get a university degree in Ireland, since Trinity College Dublin (T.C.D.) was almost exclusively Protestant and the Royal University of Ireland gave examinations but no courses. In 1904, however, Dev received the B.A. degree from the Royal University, and the following year had his first opportunity to attend some good lectures in mathematics, given by Arthur Conway at the newly established University College Dublin (U.C.D.). From 1906 to 1908 he followed a course in mathematical physics by Edmund Whittaker, who was professor of astronomy at T.C.D. and Royal Astronomer at the Dunsink Observatory. In 1908, Dev joined the Gaelic League and began an intensive study of the Irish language, which became an intellectual love second only to mathematics. In 1910, he married his Gaelic teacher, Sinead Flanagan, and they had seven children.

In 1913 Dev became convinced that Ireland would never obtain any measure of self government without a show of force. He joined the Irish Volunteers, an underground military organization, quickly rising to commandant of one of the Dublin battalions. In the Easter Rising of 1916, his assignment was to defend the southeastern approaches to Dublin, after the principal buildings in the city center had been seized by the Irish army and the Republic proclaimed by Patrick Pearse. The organization of the rising was bungled, but in any case it was doomed

352

to failure once British reinforcements arrived with artillery. The center of Dublin was destroyed by incendiary shells and the republican soldiers forced to surrender. The British commander ordered summary court martials and executions of all the leaders of the rebellion. The people of Ireland had not supported the rising, but the British revenge changed their feelings utterly. Dev had been sentenced to death a day later than the commandants in the city center, and when the British government halted the executions, he was sent to Dartmoor prison.

After the United States entered the war, Lloyd George released the Irish prisoners. Dev was elected to parliament as leader of the Sinn Fein [Ourselves Alone] party, dedicated to establishment of an Irish republic. He was again arrested and sent to Lincoln prison, but early in 1919 he effected a daring escape. In 1922, civil war broke out between the republicans and those willing to accept a sort of dominion status (the 'Staters'). Both sides fought without mercy; executions, reprisals, and assassination were the order of those days. The republicans were finally crushed and De Valera went to gaol again. To many he was still a hero, but others never forgave him for the bloodshed.

Released in 1924, he founded a new party, Fianna Fail [Band of Destiny] to pursue republican principles by electoral means, and in 1932 he achieved power in coalition with a small Labour contingent. In 1937, he won an absolute majority, scrapped the oath of allegiance to the British crown, and secured the adoption of a new constitution for Eire, the Irish Republic, without, however, completely severing ties with the Commonwealth.

Dublin Institute for Advanced Studies

Even in the midst of his political work, Dev never forgot his intellectual loves, mathematics and the Irish language. He noted the foundation in 1930 of the Institute for Advanced Study in Princeton, and the fact that Einstein and other great scientists had gone there as refugees from Nazi Germany. Abraham Flexner had described the ideal of such an institute: 'It should be a haven where scholars and scientists may regard the world and its phenomena as their laboratory, without being carried off in the maelstrom of the immediate; it should be simple, quiet, comfortable; quiet without being monastic or remote . . . Its scholars should enjoy complete intellectual liberty, and be absolutely free from administrative responsibilities or concerns.'[2]

Dev began to consider whether a similar institute might be established in Dublin. He thought that it could provide a common ground for scholars from the University of Dublin (T.C.D.) and the new National University of Ireland. His first idea was to acquire Dunsink

Observatory, 'for more or less sentimental reasons and because of its association with Sir William Rowan Hamilton'. In discussions with George Birkhoff of Harvard, Arthur Conway of U.C.D., and Edmund Whittaker now at Edinburgh, he was dissuaded from this idea. It was pointed out that, the Irish climate being as it is, only about forty nights out of a year would be suitable for visual astronomy at Dunsink.[3]

His final plan was to begin with two Schools in a Dublin Institute for Advanced Studies: Celtic Studies and Mathematical Physics. The 'humanist quality and national flavour' of the former would be balanced by the scientific precision and international character of the latter. He hoped to persuade Conway and Whittaker to become foundation professors in theoretical physics. They could not accept, but Whittaker stressed the importance of securing a physicist of world renown, and suggested that Schrödinger was under pressure since the Nazi seizure of Austria and might be persuaded to come to Dublin. De Valera, an expert in undercover activity, immediately arranged the secret contact with Schrödinger that led to their meeting in Geneva in September, 1938.

The bill to establish the Institute was introduced in the Dail on July 6, 1939. Since Fianna Fail had an absolute majority, there was no doubt about its eventual passage; the opposition, however, attacked de Valera fiercely, saying he was trying to satisfy his vanity with a pretence of scholarship at a time when war in Europe was imminent and Ireland faced military and economic perils.[4]

In introducing the bill, De Valera said:

The name of Hamilton is known wherever there is a mathematical physicist or theoretical physicist. This is the country of Hamilton, a country of great mathematicians. We have the opportunity of establishing now a school of theoretical physics which can be specialized as the school of Celtic studies can be specialized, and which I think will again enable us to achieve a reputation in that direction comparable to the reputation which Dublin and Ireland had in the middle of the last century.[5]

For Dev, William Rowan Hamilton was a hero of mathematics. Conway and A.J. McConnell were currently editing his works under the auspices of the Royal Irish Academy, over which he presided from 1837 to 1846. Hamilton was born in Dublin in 1805. He went to Trinity College in 1823, and his performance was so outstanding that in 1827 he was appointed Professor of Astronomy and Astronomer Royal, before he had even taken his degree. His most important works were the reformulation of analytic mechanics on the basis of what is now called the Hamiltonian function H, the demonstration of the analogy of mechanics with optics (Hamilton's principle and Fermat's principle), and various algebraic discoveries including his famous quaternions. His personal life was unhappy, disappointment in love leading to drink. Like Schrödinger, he wrote a considerable amount of poetry, which he valued highly. He died at Dunsink in 1865.

The Hamiltonian function,

$$H = T + V = \Sigma p_i^2/2m + \Sigma V_i,$$

is the starting point for all applications of quantum mechanics. Thus it was appropriate that Schrödinger, the discoverer of the new mechanics, should be the one to inaugurate a new era of Irish physics in the city of the illustrious Hamilton. When Erwin arrived in Dublin, however, the bill to establish the Institiuid Ard-Leighinn Bhaile Atha Cliath [Dublin Institute for Advanced Studies – D.I.A.S.] was still only in draft form. It was circulated to members of the Dail in March, 1940. It provided for two schools, Celtic Studies and Theoretical Physics. Their duties were specified to include research, training advanced students, provision of facilities for visiting scholars, organization of seminars, conferences and lectures, and the preparation of publications. Senior professors were to be appointed and removed by the president of the republic on the advice of the government.

Arrival in Dublin

Dublinn is an Irish word meaning 'a dark pool', but the official name of the city is Bhaile Atha Cliath, 'the town of the ford of the hurdles', after the site of a shallow crossing of Anna Livia, the River Liffey. Another etymology was given by an Englishman, John Head, in 1600: 'Many of its inhabitants call this city Divlin, quasi Divel's Inn, and very properly it is by them so termed; for there is hardly in the world a city that entertains such devil's imps as that doth.'[6] In the ensuing centuries, the city's reputation gradually improved, but when the Schrödingers arrived it was still known as 'dear dirty Dublin', for the living conditions of its poor were among the worst in Europe. It was, however, again the capital of a free country, had a population of half a million, three cathedrals and six theaters (not counting the popular movie palaces). The central area of O'Connell Street had not yet recovered completely from the devastation of Easter Week, 1916, and memories of the civil war had left a bitter residue in Irish politics.

On October 7, 1939, Erwin, Anny, Hilde, and Ruth arrived in Dublin, and put up at a hotel. One of the first friends they made was Albert J. McConnell, always called 'A.J.', who, as professor of mathematics at Trinity, was involved in planning for the Institute. They came around to his apartment on Pembroke Road and discussed the problem of finding a permanent place to live. Erwin displayed his best Viennese charm, but he made it clear in no uncertain words that

Mrs March was to be treated on exactly the same terms as Mrs Schrödinger.[7]

The family wanted to live near the sea, and they soon found a house in Clontarf, a fashionable suburb on the northern shore of the Bay of Dublin. It is famous as the site of the Battle of Clontarf on Good Friday, 1014, when Brian Boru, High King of Ireland, was killed by Brodir, a Viking from the Isle of Man. He had been warned by the family banshee not to fight on that day, but Brodir surprised him at prayer and provoked a fatal combat. The Schrödinger home was at 26 Kincora Road, a semidetached two-storey brick house in a row of similar dwellings on a pleasant tree-lined street about 300 m from the sea-shore, where the Wicklow mountains form a hazy blue background across the waters of the bay. They were to live here for seventeen years. Just down the street was Clontarf Castle, with its restored twelfth-century Norman square tower attached to a nineteenth-century house in Tudor style.

Their house was about 6 km from the Institute in Merrion Square. Erwin, an expert cyclist, had no difficulty in making the trip in all weathers. He often wore a professional-looking waterproof cycling suit and his jaunty beret, and he soon became a familiar figure among the many cyclists of wartime Dublin.

The house at 26 Kincora Road was not an elegant residence, as compared, for example, with the one they had rented in Oxford, but it had adequate room for the family, a small garage at the side, and a pleasant walled garden at the back. Erwin was not too pleased with the construction and the typically British plumbing and heating. After renting for several years, he bought the house for £1000 in 1943, selling it in 1956 for £2150.

The parish church of St John the Baptist was only a few blocks away and the school, conducted by the Sisters of the Holy Faith, which Ruth attended. The pastor, Father MacMahon, called at the Schrödinger home and met Hilde, who pretended not to understand either English or his attempts at German.

Erwin's unusual family did not cause much adverse comment in Dublin. Ireland was officially a puritanical society, under the strong Jansenist tradition of the Irish Church. At that time, the *Irish Catholic* was upset by women wearing slacks because 'their stylish cuts throw the feminine figure into undue prominence'.[8] Other ecclesiastics were troubled by schoolgirls who crossed their legs in public although 'they had presumably been educated by the nuns in Christian modesty'. There was a rigid censorship that tried to exclude not only most of contemporary European literature but even collections of Irish folk-tales that contained mild improprieties. At intervals, lists of forbidden books were printed in the *Irish Times*. The February, 1940, index

26 Kincora Road, Clontarf, where the Schrödingers lived for 17 years.

included such authors as Aldous Huxley, John Steinbeck, Stuart Cloete and Xavier Herbert, besides a collection of popular books of sexual advice such as *Is Birth Control A Sin?*. Actual pornography seems never to have come to the attention of the censors.

The official puritanism, however, was alleviated by an informal spirit of laissez faire. As Erwin once remarked, 'In Germany, if a thing was not allowed, it was forbidden. In England if a thing was not forbidden, it was allowed. In Austria and Ireland, whether it was allowed or forbidden, they did it if they wanted to.' Thus, in reality, Dublin was not a strait-laced city. For example, the inseparable actor-managers of the Gate Theatre, Micheal Macliammoir and Hilton Edwards, always called 'the boys', gathered only glances of admiration as they strolled about together painted and powdered, and Brendan Behan, on the run from the Garda, could find a refuge in many a respectable home.

The situation caused by the war in Europe was always officially called 'the Emergency'. The government declared a policy of strict neutrality and placed its small army on a war footing, expecting an invasion by England or Germany, or both. Nevertheless, more volunteers (per capita) went to fight with the British forces from the Republic than from Northern Ireland.[9]

In the first year or so of the war, there were few hardships. An endless series of letters in the *Irish Times* protested the persistence of daylight saving, and there was concern about the loss of stud fees for Irish stallions. Otherwise life went on very much as before. Gradually, however, as the U-boats disrupted shipping, conditions became more difficult. There was always plenty of food, although butter and sugar were eventually rationed. A visitor from England was astounded by the sumptuous helpings of roast beef at dinner at the Maynooth Seminary. The most grievous scarcity was of tea, sometimes cut to $\frac{1}{2}$ oz a week per person, but you could always get a cup or two in a tearoom. There was never any shortage of alcoholic drink. The bread became a uniform, wholesome, rather soggy brown loaf, made from local wheat mixed with barley and oats.

As the war went on, there were serious shortages of coal and oil. After 1942, there was no more petrol for private cars, even for doctors and priests. Taxis were still available and some horse-drawn buggies reappeared; buses stopped running at 10 p.m. Most people used bicycles. Eventually household gas supplies were strictly limited to a few cooking hours, and special police, called 'glimmer men' were empowered to enter any house to check the stoves for compliance. Remembering Vienna in 1919, Schrödinger found small reason to complain about wartime Dublin.

There were no tourists. In the words of John Ryan, 'the country was clean, uncluttered, and unhurried. All in all, it was not the worst "Emergency"' we ever had.'[9]

Dublin 1939–40

Soon after he arrived in Dublin, Erwin met Monsignor Patrick (Paddy) Browne [Padraig de Brun] who was then professor of mathematics at St Patrick's College, Maynooth, where most Irish priests received their training. He also had been actively involved in helping de Valera plan the Dublin Institute. Many people say that Paddy Browne became Erwin's best friend in Dublin, despite some serious differences they had from time to time. He was a Rabelaisian priest, in the literal sense that he knew much of Rabelais practically by heart, and also in the figurative sense that he had a great store of bawdy stories. He was also rather a gargantuan man, almost 2 m tall, with a leonine head and bushy hair, so that he stood out in any group, and really towered over

Erwin. Browne was an excellent teacher of mathematics but not a researcher in the subject; he was also a classical scholar and a Gaelic scholar; he had translated Aeschylus from Greek to Gaelic preserving the original metre. His elder brother Michael became head of the Dominican order, and was made a cardinal and member of the Curia, but Paddy was too broad a man to win preferment in the conservative Irish-Catholic church. Erwin's classical education was no match for Paddy's, and Paddy's mathematics was not in the same class with Erwin's, but they respected and loved the intellectual qualities they found in each other. They were also both poets by avocation.

Among other early friends of the Schrödingers in Dublin were Ludwig Hopf and his wife Alice. They were Jewish refugees from Germany, where Hopf had been professor of applied mathematics at Aachen. He had taken his doctorate in Munich and worked as an assistant to Einstein in Prague. Thus the Hopfs and Schrödingers had many friends in common, especially Arnold Sommerfeld. Hopf had been lecturing in mathematics at T.C.D. since July, but he became seriously ill in December and died on December 21. Anny and Erwin attended the funeral the day after Christmas; any celebration of their first Christmas in Dublin was overshadowed by this sudden loss of a friend. Anny did everything possible to help Alice, and Erwin agreed to continue Hopf's lecture course till the end of term so that she could receive the salary. Later Alice obtained a position teaching spoken German at T.C.D., actually one that Hilde March originally held, but resigned so that the widow would have enough income to support herself and her aged parents in Dublin.

Schrödinger's first public appearance in Dublin was on Wednesday afternoon, November 14, when he read a paper before the Dublin University Metaphysical Society on 'Some Thoughts on Causality'.[10] The talk was well publicized and attracted a large audience, by no means restricted to philosophers. He reviewed several themes that he had considered in earlier works: (1) the impossibility of prediction even in classical physics due to experimental uncertainties in the initial conditions, (2) the teaching of general relativity that matter is an inherent property of space and time and hence is not something predictable in space and time, (3) the uncertainty principle of quantum mechanics, which introduces indeterminism into the behavior of microscopic objects. He concluded that causality is not a necessity for logical thought, but he advanced this view tentatively, with a quotation from Miguel Unamuno to the effect that 'a man who succeeded in never contradicting himself was to be strongly suspected of never saying anything at all'.

In the ensuing discussion, Robert Ditchburn, the professor of experimental physics, granted the importance of chance but said, 'If I threw a brick out of the window just as Hitler was passing by, I would not plead with the Gestapo that I threw it on the chance that it would fly upwards.' The Reverend Dr A.A. Luce, who presided at the meeting, compared Schrödinger to a 'veritable Epicurus' who had restored the liberty of free will. He did not know that Erwin himself had rejected the idea that physics has anything to say about free will in a letter to *Nature* in 1936. By now, however, the meeting was ended in a round of applause, and no public debate on that question was possible.

This scientific dissection of causality did not please every philosopher in Dublin, since it tended to undermine the official Catholic philosophy based on Aristotle and Aquinas, which purports to trace an uninterrupted chain of cause and effect back to the first cause, which is God. Thus, strictly speaking, Schrödinger may have been guilty of heresy; since he spoke at Trinity College, this would not have been considered remarkable, but an interesting repercussion would occur a couple of years later.

In November, Schrödinger began an informal course of lectures on wave mechanics for undergraduates at U.C.D. The first meeting was so crowded that it was necessary to find a larger hall. His excellence in lecturing soon became famous. He used a blackboard with rare artistry – not for him the scribbled equation hastily erased or the words addressed to the board and imperfectly reflected at the audience. The written material appeared in logical order as if in accord with a mental map of the blackboard surface, and the words were spoken in his clear, light, precise English.

Erwin himself wrote a notice about these lectures 'for the press (if any)'[3]:

A course of lectures on the latest form of quantum theory intended for advanced students in physics was started yesterday in U.C.D. at 3 p.m. by the Viennese professor E. Schrödinger, Nobel prizewinner, honorary member of the Royal Irish Academy, honoured by the Nazi government with pensionless dismissal, without notice, from his academic chair in Austria. The first lecture will consider two discomforting features in the Planck–Bohr theory: (1) the quaint basic assumption about the discontinuity of states, (2) the frequency of spectral lines entirely different from the frequency of atomic oscillations, which the theory had to assume.

Thus Schrödinger would brook no delay in giving Dubliners his views on 'those damned quantum jumps'.

On December 11, he presented his first Dublin research paper to a meeting of the Royal Irish Academy on Dawson Street, and it was published in the *Proceedings* of 12 February, 1940.[11] He had been an honorary member of the Academy since 1931. Most of his papers were henceforth to be published in its *Proceedings*, where they were elegantly printed and appeared without delays. The volume of research from the Institute eventually became so great that the Academy was financially hard pressed to publish it at all, and the Institute was asked to pay part of the costs.

This first paper was 'A Method of Determining Quantum Mechanical Eigenvalues and Eigenfunctions', in which the second-order operator of the Schrödinger equation is factored into two mutually adjoint first-order operators. The method was applied to the one-dimensional harmonic oscillator and the radial part of the hydrogen-atom problem (Kepler motion). Then a new problem was attacked, the analogy of the Kepler motion on a four-dimensional hypersphere. The Coulomb potential becomes $\cot \chi$ instead of $1/R$, where χ is the radius vector divided by the radius of curvature. The result for the energy levels is:

$$E_n = B(- n^{-2} + (n - 1)(n + 1)a^2/R^2)$$

where B is the Rydberg constant, a the Bohr orbit, and R the radius of curvature. Instead of a continuous spectrum above a fixed dissociation energy as in the hydrogen-atom case, the energy levels become intensely crowded but still discrete. These results have no practical consequences, but they provide another interesting illustration of what happens when one does not allow a hydrogen atom to fill infinite space as it does in ordinary three-dimensional quantum mechanics.

About a year later, he presented a second paper on the factorization method, in which he gave more mathematical background and also showed how it could be extended through an approximate factorization to handle perturbation effects. He analyzed in this way the effect of an electric field on the hydrogenlike wavefunctions and energy levels (Stark effect). At the end of this paper he referred to an investigation published about ten years previously by Cornel Lanczos, which showed what happens in the Stark effect on very large orbits, a broadening of the energy levels due to the possibility of escape of the electron though a 'tunnel effect'. 'Indeed', he says, 'it is clear that the whole conception of *discrete* eigenvalues is merely a dodge – though a good one . . . Therefore the term spectrum is *really* continuous throughout, only with remarkably selective *peaks* of the weight function.'[12]

He sent copies of his papers on the factorization method to Max Born in Edinburgh, and pointed out its relation to Heisenberg's matrix mechanics. 'The latter was always advanced for its conceptional beauty and feared for its practical complicatedness. The present translation into wave mechanics shows, I would say, that conceptional beauty is not only a fancy-value. If you are hard-up, you can exchange it at the mont de piété against very substantial secular values.'[13] The factorization method has become an important technique for dealing with a variety of problems in mathematical physics.

The bill to establish D.I.A.S. was making its way through the Dail but Schrödinger as yet had no formal position in Dublin and thus no regular income. Therefore de Valera asked Conway, who was president of the Royal Irish Academy, to arrange a temporary professorship for him there. His appointment at a salary of £1000 a year, beginning April 1, 1940, was announced at a meeting on March 16, at which time also de Valera was admitted to membership. Schrödinger at once began a course of advanced lectures on quantum mechanics held in the Meeting Room of the Academy; almost all the mathematicians and physicists of Dublin crowded into these lectures.[14]

Irish holidays

In the first week of May, 1940, as springtime came to Europe, there was little activity on the battlefields of what was being called the 'phoney war'. The French were ensconced in the Maginot line and Neville Chamberlain was dithering in London. On May 7, the *Irish Times* published a brief account of a paper that had appeared in the U.S.A. in the February 15 issue of *Physical Review*. It confirmed the fission of the U-235 nucleus and showed that several neutrons are emitted in the process. Thus a chain reaction and probably an explosion were theoretically possible in a mass of fissionable material. On this subject, little more was published in the open literature for some time.

The long Whitsunday weekend was coming, and on May 10 Erwin took off for a short cycling holiday:[15]

It was my first spring in Ireland. I had got myself and my bike on to the train to Galway. The high sombre walls, which still today girdle the seats of the old country lords, recede and disappear. Get-at-able nature clambers up close to you; calm, silent, unstinted meadows, poplars, willows, grazing beasts. Not exciting romance, but so very soothing.

And so is that town at the other end – a tender old lady with a great past and all the charms of happy retirement after a well-spent life. When she had replenished my sidebags with drink and food, I embarked on the 50 miles to Clifden. The change, step by step, from the loveliest spring show, much like Italy or Dalmatia (you are almost surprised not to see vineyards) to the solemn barrenness of Inner Connemara, where the grand bleak loins of our mother rise aloft without drapery – that wonderful crescendo and decrescendo . . .

The mountain that he refers to as 'our mother' is shown on the maps as 'The Devil's Mother'. Erwin was almost never inadvertent – was this a subtle bit of gnostic theology?

After passing the first houses that announce the township I alighted to ask my way. 'To Clifden? You have just gone through it.' Good. So I inquired of the friendly Garda, knew he some private quarters or a lodging house (I hate hotels). That was splendid. A wee grocery with two or three guest rooms freshly decorated for the summer season. In the corners of the shop cower little unpaid messengers, gazing at the event. They take charge of signalling all important movements of mine to their mother. 'The man has come down. The man is going out. The man is coming back.'

The charge was five shillings 'Bed and Breakfast'. My reminder that I had had a good evening meal, as well, is greeted with a gentle smile: 'Well, if I thought it right I should make it six.'

It was Whit Sunday. I bought a newspaper and stuffed it unread into the cycle-bag. Off northward, not knowing how far it would take me. My pace was slowed by the beauty of the day and of the path, winding between the dark blue bays and blinding seas of gorse in full blossom. Unforgettable the view down on blue Ballynakill Harbour, the gentle lakes of Kylemore girded with the early green of young beech trees, the grave sombre fjord of Killary Harbour with the evening sun gilding the mountain ridges high above.

I landed late in Leenane and decided to crown the day by a bottle of claret. Waiting for my meal, I took out my morning paper – to learn that Germany had invaded the Low Countries. My return to Dublin was hasty, but imbued with gratitude for the fascinating beauty, whose spell had averted my spoiling the heavenly day by a preposterous grip into the fateful saddle-bag.

By June 3, the Nazis had overrun Belgium, the Netherlands, and defeated the French army. The British forces were being evacuated through Dunquerque.

Paddy Browne had a house at Dunquin, on the extreme southwestern tip of the Dingle Peninsula in County Kerry, just a few hundred metres from steep cliffs overlooking the Atlantic Ocean. It was a wild, Gaelic-speaking country of sea and mountains. Erwin and Anny were invited to visit there towards the end of their first summer in Ireland. The three children of Paddy's sister Margaret were there at the time, Maire (18), later Mrs Conor Cruise O'Brien, Seamus (16), and Barbara (12). The monsignor had practically raised the children since their father Sean MacEntee, who was now minister for Industry and Commerce in Dev's government, had been so occupied with Irish revolutionary politics. Paddy Browne had been a student at Heidelberg and he taught the children a few German folk songs. They became very fond of Anny who knew a number of unfamiliar ones – 'Phyllis ging wohl in den Garten', 'Horch was kommt von draussen 'rein'. Erwin was less popular since he wrote a poem for Barbara about neatness and good behavior. It began: 'You're growing a lady now, my dear. For instance, clean your nails.' Such an equation of personal grooming with virtue was incomprehensible to the MacEntee children, and it created 'an absolute cultural barrier'.[16]

Despite her dirty nails, Barbara was a beautiful child and Erwin became infatuated with her. She was the third instance of his 'Lolita complex', taking her place along with 'Weibi' Rella and 'Ithi' Junger. Perhaps it was the 'little boy' aspect of his nature that attracted him to nymphets, but to him these loves seemed like erotic arrows that pierced his soul unbidden and unexpected. The situation became so incongruous that someone, probably Paddy Browne, had a serious word with him, and muttering dark imprecations, Erwin desisted from further attentions to Barbara, although he listed her among the unrequited loves of his life.

On another visit to Kerry, the following year, Erwin and Anny went on a bicycle excursion from Dunquin, and had an adventure which was later related by Sean O'Kennedy who was on a hike with two teenage pals.[17]

It happened that three of us (all natives of the district) were bantering by the road-side when we observed the pair, who were then altogether unknown to us, cycling in our direction. Before passing us, we advised them to dismount the bikes as they were almost certain to have punctured tyres if they cycled over an extensive patch of jagged stones immediately ahead. They paid no attention, perhaps because we appeared to be loungers who were up to no good and to them the patch of stones posed a lesser danger. In the event, she got through without incident but his tyre was badly gashed, and however suspicious he might be of us, he was obviously glad when we offered to repair the damage.

In those days, it was almost impossible for teenagers to come by the cost of a packet of five Woodbine cigarettes (2d) not to mention that of a bottle of Guinness (6d). Hence we decided while conversing in our native Irish to charge one shilling for our labours. E.S. objected to the charge and offered to pay 6d. After considerable bargaining we agreed to 'split the difference' at 9d . . . We were enjoying the encounter with what to us was a most unique character, even had he not spoken with a strange accent and been dressed in an attire [*Lederhosen*] which we had never seen before. His mode of dismounting the bike was no less fascinating for, instead of throwing one foot back while the other rested on the pedals, he threw it forward over the handle bars so fast that the movement of his hands to make way for his foot was barely noticeable . . . We were more than grateful for the 9d with which we purchased the Guinness and the Woodbines . . . We did not know who he was until we saw his picture in the *Irish Press* about a week later . . . In the background was a blackboard with some figures, squiggles, and other mysterious symbols.

O'Kennedy later studied science in Dublin and became professor of physical chemistry at University College, Galway. After Schrödinger had returned to Austria, O'Kennedy once wrote to him about some problem suggested by *What is Life?*. By a remarkable coincidence he put on the envelope only a 3d stamp (the internal postage) instead of a 6d stamp (foreign postage). Erwin, who did not in any way connect him with the erstwhile tyre mender, wrote a rather sharp protest concerning the Austrian schilling he had to pay in postage due. As a little boy, he had observed the power of money when his parents were prosperous, as a young man he suffered from their ruin and humiliation in the postwar Austrian inflation. He would haggle with I.C.I. over a few pounds and with country boys over a few pence. He never achieved financial security until it was too late to enjoy it, but his insecurity was always more psychological than economic.

D.I.A.S. installed

On June 1, the bill to establish D.I.A.S. passed the Senate, and it was signed into law by President Douglas Hyde on June 19. Schrödinger began his tenure as a senior professor at a salary of £1200 a year, and it was a relief to be back on a payroll again. It was quite a good salary for Dublin, about what a successful solicitor would make. The Taoiseach himself was paid £3000, an effective upper limit on salaries. About a quarter of Erwin's salary would go to income tax; the salaries of politicians were tax exempt. Erwin was continually worried that the Institute made no provision for a widow's pension. The vicissitudes of his academic life and the loss of his retirement pay from Berlin and Graz meant that he did not have any appreciable financial reserves. He tried to save some of the Nobel prize as a safety net for Anny.

Each school at the Institute was administered by a Governing Board consisting of a Chairman, appointed by the president on the advice of the government, other members similarly appointed, and the Senior Professors of the School. The overall administration of the Institute was vested in a Council consisting of representatives from the Governing Boards of the constituent schools, the President of University College Dublin, the Provost of Trinity College Dublin, the President of the Royal Irish Academy, and a Chairman appointed by the President on advice of the government.

The Institute was located in two interconnected eighteenth century houses, Nos 64 and 65, on the south side of Merrion Square. This is one of the largest squares in Dublin, surrounding a beautiful park with a small lake and lovely rose gardens. The houses are of uniform Georgian style. To Brian O'Nolan the Georgian architecture of Dublin was 'a remote faded poignancy of elegant proportions, minute delicacy of architectural detail balanced against the rather charmingly squalid native persons who sort of provide a contrapuntal device in the aesthetic apprehension of the whole'.[18] Within ten minutes walk from Merrion Square are the grounds and buildings of Trinity College, in the heart of the city life of Dublin. The administrative offices of the National University of Ireland were on the east side of the Square and one of its constituent colleges, University College Dublin, had its medical school on the south of St Stephen's Green, adjacent to the Square. The Royal Irish Academy occupies Northland House which is nearby. All these learned institutions as well as the National Library and Art Gallery lie within an area not so large as the campus of many a university. At that time there were about 1400 students at Trinity and 1200 at U.C.D.

In July, both the University of Dublin (T.C.D.) and the National University conferred honorary degrees on Schrödinger, doctorates of science, D.Sc. The Trinity ceremony was on July 3, with the Vice Chancellor, Thomas Malony, and the Provost, W.E. Thrift, presiding. The Latin encomium for Schrödinger noted that : 'Among his merits, I make especial mention of the fact that using those functions (as they are called) by which the movements of waves are described, for the purpose of investigating the structure of matter, he shows that the atoms themselves (unless this is abhorrent to the literal meaning of the word) can all be composed of waves'. [The literal meaning of 'atom' is 'uncuttable'.][19]

On July 11, a similar investiture took place at the National University. Arthur Conway, the President, introduced Schrödinger. Francis Murnaghan from Johns Hopkins, a graduate of U.C.D., was also awarded a D.Sc. at this ceremony. An effort was made to attract him to the Institute as professor of mathematics, but influenced in part by the grim situation of the anti-Hitler forces in Europe, he declined.

The Institute buildings had been used for government offices and some remodeling was necessary before they were ready for occupancy. In October, 1940, the lecture room at Hamilton House, 65 Merrion Square, was finished. Erwin had already been occupying his office for some time. It was a spacious room, lined with bookshelves. There was a large desk with a comfortable chair, at the back of which was a Georgian fireplace in which the coals glowed on chilly days; even during the Emergency, the Institute managed to keep most of its fires burning. Erwin wore glasses and young visitors were at first intimidated by the 'ice-blue eyes glaring greatly magnified through the strong lenses' but they were soon reassured by his kindly manner. As he worked or talked, he usually smoked his pipe, enjoying the ritual of cleaning, filling, tamping and relighting. If a visitor was also a smoker, the room would sometimes become so filled with smoke that they would have to repair elsewhere to replenish their oxygen supplies.[20]

D.I.A.S. progresses

On October 5, the Council of D.I.A.S. was constituted under the chairmanship of Patrick Browne, who was uniquely qualified for the position, being both an Irish scholar and a mathematician. They held

First meeting of the Governing Board, School of Theoretical Physics, Dublin Institute for Advanced Studies (21 November, 1940). From left: Erwin Schrödinger, A.J. McConnell, Arthur Conway, D. McGrianna (Registrar), Eamon de Valera, William McCrea, Msgr Patrick Browne, Francis Hackett.

their first meeting on October 15 in the Council Chambers of the Government Buildings on Merrion Street. Dev was there to welcome them and to see his dream for the Institute enter the realm of reality.

The first meeting of the Governing Board of the School of Theoretical Physics was held in the Minister's Room at 1 Hume Street on November 21. The Taoiseach was there to welcome them on behalf of the President and the government. Patrick Browne attended as Chairman of the Council, and all members of the board were present except Whittaker, who sent apologies from Edinburgh: McCrea, Hackett, McConnell, Conway, and Schrödinger. The Registrar, Domhnall McGrianna was introduced; he was an old revolutionary and comrade-in-arms of de Valera. Conway proposed that Schrödinger be appointed Director of the School, which was carried unanimously.[21]

It was decided to offer a junior professorship to Walter Heitler at a salary of £400 per annum. He had worked in Schrödinger's department in Zürich, and was Privatdozent in Born's department at Göttingen when Hitler came to power. He left Germany for a research fellowship in Bristol, but had been interned on the Isle of Man for several months after the fall of France. He was now thirty-six years old and at the height of his theoretical powers. He was glad to accept the position arranged by Schrödinger in Dublin.

At the second meeting of the governing board, estimates of expenditure for 1941 were presented: director £1200, jr professor £600 (raised from £400), assistants £400, scholarships £600, library £500, visiting professor £250. The modest total was £4050, which did not include costs of nonacademic staff, maintenance and supplies.

Flexner's ideal of no administrative work for the research workers had not quite been realized, but it would have been impractical when so many decisions depended on scientific judgment. Erwin was happy with the situation. As he wrote to Born, 'The people in Dublin are good to talk to but overworked with lectures and exams particularly . . . [My position] is one of the most appropriate in the world. To be reinstated to absolute security (at least as regards yourself) at 53 by a foreign government, in my case fills you with – well, infinite gratitude towards *that* country.'[13]

Even amid wartime stresses and shortages, D.I.A.S. flourished, fulfilling the highest hopes of de Valera, and Dublin quickly became a world center for theoretical physics. Almost every year, special colloquia brought contingents of the best theoreticians from the U.K., including refugees from the continent of Europe. Special lectures were given by visitors at other times. Discussions were lively and the exchange of ideas stimulating. Graduate work in physics in the Dublin universities took on a new spirit and enthusiasm. In addition to the specialized lectures and seminars, the statutes of the Institute provided for an annual public lecture series to be given alternately at T.C.D. and U.C.D. These became one of the highlights of the intellectual life of Dublin.

Schrödinger recognized that D.I.A.S. was generously funded by the taxpayers of a relatively poor country with many demands on its resources. Therefore he always tried to give more than a full measure of service, especially during the war years when it was impossible to vacation abroad. He received many letters from the public, some of them long pseudoscientific documents about which his opinion was requested. Sometimes he passed them along to assistants, saying 'Now you see what I have to put up with', but every letter was answered.

Anny and Hilde were also kept busy, for visitors had to be entertained. Buffet lunches and evening parties at Clontarf were quite frequent, and Anny became famous for her Viennese cakes and strudels. The Schrödingers were always kind and helpful to younger scholars and staff members, having them to tea or supper, while Erwin provided detailed advice on how to ride a bike safely on the wintry Dublin streets. He abhorred any formality of dress or pretentiousness of style.

Erwin preferred the company of young people as he grew older, especially young women, although he also was attracted to young men who were handsome, lively and intelligent. The Austrian community in Dublin used to meet at the 'Old Vienna Club', located about a block from St Stephen's Green, and soon after he came to Dublin Erwin met a young biochemist there, Stephen Feric, and they became good friends despite the 25-year difference in their ages. Feric was full of youthful vigor and good cheer, and for a while he became practically a member of the Clontarf household. The housekeeping was divided on a rota system, with one week Anny, then one week Hilde in charge. It usually worked smoothly, but occasionally there were tempests, and Erwin would retire upstairs until all was calm again. Sometimes, however, he and the two women would get involved in frantic scenes over trivial domestic problems, and Stephen would take Ruth into the garden or for a walk by the seashore.[22]

Another young Austrian refugee encountered at the Old Vienna Club was Alfred Schulhof whose mother had gone to school with Hilde. He was practically adopted by the Clontarf family and they paid all the expenses for his study of electrical engineering at the university. When he thanked Erwin, the response was only, 'Well, I might have had a son.'[23]

Alfred Schulhof with Anny, 'I might have had a son.'

Dublin theater

By the time the Schrödingers arrived in Dublin, the great days of the
Abbey Theatre, of Willy Yeats, Augusta Gregory, Sean O'Casey, and
John Synge, were only memories. The old theater still stood, however,
at the corner of Marlborough and Lower Abbey Streets, and its ten to
twelve productions a year, except for those in the Irish language, often
played to packed houses.

The great rival of the Abbey, the Gate Theatre, was founded 24 years
later, in 1928, by Mícheál MacLiammoir and an Englishman, Hilton
Edwards. The 'boys' continued their partnership for almost fifty years,
as actors, directors, designers, and lovers, but not always everything
in every year. They did not neglect Irish drama, but their policy was to
stage in Dublin the best plays from all countries and all times. They
were enthusiastic about stagecraft and the Gate's sets were a relief
from the whitewashed cottage walls and dismal city flats of the more
realistic Abbey. As Micheal put it, 'At that time the theatrical fare of
Dublin consisted of an honest and supremely well cooked bacon and
cabbage at the Abbey, stewed prunes and custard at the Gaiety, and
kickshaws at the music halls.'[24]

Till 1936, the Gate was supported financially by Edward Longford,
but after a dispute about an Egyptian tour, Edward and his wife
Christine formed a separate group, Longford Productions. They took

turns with the Gate in its theater, but after 1940, many Gate productions were staged in the old Gaiety on King Street. The Longfords owned a large country estate (Pakenham Hall) in County Westmeath and an elegant town house on Fitzwilliam Street, where they gave fashionable and inebrious champagne parties.

Devoted to the theater since youth, Erwin soon came to know a lot of people in theatrical and artistic circles in Dublin, but he was never a close friend of 'the boys' or their leading ladies Meriel Moore and Coralie Carmichael. Sometimes, however, they would meet after the play for coffee and discussion. Charles Acton, who became drama critic of the *Irish Times*, recalls one night in Mitchell's Cafe when Erwin was trying to analyze the classic problem of whether tea cools faster when the milk is added before or after.[25] Hilde, who had a more fun-loving disposition than Anny, shared Erwin's devotion to the dramatic arts. She found a good friend in Peggy Manly who made theatrical costumes and knew everyone in the Dublin theater world.

Sometimes a group would gather at Madame Bannard (Toto) Coghley's place on Grafton Street. Patriotic Irish ladies despised the English 'Mrs' and found the Irish 'Beann' a bit cumbersome, and thus decided to be called 'Madame'. Another gathering place of the bohemian set, where Erwin sometimes appeared, without Anny or Hilde, was Des McNamara's flat on the top floor of the Monument Cafe on Grafton Street, where Mac ran a nonstop Fabian salon in a one-room studio. He was noted for his papier-maché theater masks. Erwin described himself to these friends as a 'naive physicist', whose hobby was weaving tapestries. John Ryan had a studio next door, where Brendan Behan and the poet Paddy Kavanagh used to take refuge when the worse for drink, and Erwin would certainly have met Paddy but probably not Brendan.[26]

Soon after their arrival in Dublin, Erwin, Anny, and Hilde were attending a play at the Abbey, and even Erwin had difficulty following the country brogue in which it was spoken. He turned to an attractive young man sitting next to him and asked, 'What did he say?' This was Ronald Anderson, a school teacher and part-time actor who knew everyone in the theater world. His wit and gaiety appealed at once to the Schrödingers and Ron became a good friend of both the ladies and also of Erwin.[27]

In the winter 1940/41, when the prospects of stopping Hitler seemed most bleak, the Irish censorship became increasingly concerned not to allow anything to upset the German government. Lennox Robinson wrote a play *Roly Poly* based on the Maupassant story *Boule de Suif*, which takes place during the occupation of France after its defeat in the Franco-Prussian War of 1870/71. A group of French refugees journey by stagecoach across occupied territory, a noble couple, a bourgeois couple, a shopkeeper and his wife, a republican, and a chubby kind-hearted whore. They are detained at an inn by a Prussian officer who demands that 'Roly-Poly' spend the night with him before they can be allowed to move on. She refuses on the grounds that she will never sleep with an enemy of her country. She resists for a couple of nights, but her fellow travelers finally force her to give in to the Prussian, the nobleman being the most persuasive. The next day, they travel on and she is treated contemptuously by all those whom she has saved. This is one of Maupassant's most caustic stories, for only the whore, and perhaps the innkeeper, display any decency at all. The play opened at the Gate on November 22, and the Schrödingers were there. Both the German and the French consuls protested and the government ordered that the play be withdrawn. Erwin was greatly interested by these events, and he used to relate with glee how an emissary from the French legation called on the playwright the morning after the opening to say that the French Minister wanted to fight a duel with him.[27] Another evidence of sensitivity to German feelings was censorship of a picture in the *Irish Times*; the National Bank building has on its facade a relief of a lion and unicorn; the censor ordered these to be erased from a picture of a parade that showed the Bank in the background. Following such events, Erwin became even more pessimistic about the prospects for an Allied victory.

There were three internment camps outside Dublin. The largest was for the I.R.A. – Brendan Behan spent a couple of years there. The one for Allied internees was lightly guarded and stray British airmen who had been rounded up could escape quietly after a decent interval. The Axis prisoners enjoyed parole, and often came in military dress to enliven some of the dances in Dublin. They also frequented the International Club, which Hilde and Peggy Manly sometimes visited.

Shortly before Christmas, 1940, David and Sheila (May) Greene returned from Glasgow, where he had been lecturing in Celtic studies. He was appointed assistant librarian at the National Library, a post he held until he joined the Dublin Institute for Advanced Studies in 1948. Sheila came from an artistic Donnybrook family, who owned a large Dublin music store; her brother Frederick was one of Ireland's best known composers. In 1936, she made her debut with the Gate Theatre in Elmer Rice's *Not For Children*, and after returning from a successful tour to Egypt, became engaged to David in August, 1938. Mícheál called her 'our new golden-haired ingenue'; she had rather unruly fair hair worn high off the forehead, large sympathetic blue-grey eyes set wide apart, and an air of serious animation. 'She was kind and thoughtful and passionate in all that she did.' Her good looks and charm caused 'the better half of Dublin to be in love with her', in the words of Prince Alexander Lieven.[28] Sheila's voice was not commanding enough for leading parts, so that she never became famous as an actress in major roles.

David Greene had a brilliant academic and social career as an undergraduate at Trinity College, taking double firsts in Celtic and Modern Languages, winning three gold medals, and captaining the water polo team. He was a big, handsome man, with an excellent brain but an irascible and self-centered disposition. Erwin was attracted not only by Sheila's beauty, but also by David's scholarship, and at that time he was making a serious, but ultimately unsuccessful effort to learn the Irish language. David considered that Erwin was 'a very Socrates, although the Irish Catholic city fathers are fortunately too blinkered to prescribe hemlock'.[29] Erwin thought that David was 'an amazingly charming rascal'.

In October, 1940, a writer from the *Irish Press* had visited Erwin in Clontarf for an interview called 'A Professor At Home'.[30] He was quoted as saying: 'I believe there is a deeper connection between us Austrians and the Celts. Names of places in the Austrian Alps are said to be of Celtic origin.' Erwin brought forth dozens of Celtic ornaments cut in wood and reproduced in plasticene or concrete. On his desk he had a copy of *Aids to Irish Composition* by the Irish Christian Brothers. He concluded the interview with the thought that: 'There is no wordly truth but mathematical truth. In politics, diplomacy, history, truth changes from day to day, and people get different concepts of right. But mathematics never lie.' Einstein in his Nobel address had been less optimistic: 'As far as the laws of mathematics refer to reality, they are not certain; and as far as they are certain, they do not refer to reality.'

Turn of the tide

Just before Christmas a few bombs fell on Sandycove near Dublin; there were no fatalities, but on January 3 three people were killed by 'stray' German bombs and the main Dublin synagogue was damaged. At this time the military power of Germany seemed irresistible. Thus the holiday season was not cheerful for everyone, although traditional pantomimes filled the theaters, and parties and dances continued as usual. In Britain and Northern Ireland, many German refugees were interned, and some were sent to Canada or Australia. The refugee ship sent to Australia was the infamous *S.S. Dunera*; in February, newspaper accounts appeared of what was called 'inhuman treatment' of the passengers. One of the internees sent out on this ship was Arnold Ewald, the younger son of Paul and Ella. Paul had obtained a professorship at the Queens University in Belfast, and Anny was delighted to have her best friends as close as this.

Fellowships were awarded to three new scholars in theoretical physics, Sheila Power and Hwan-Wu Peng, who had been with Born in Edinburgh, and Jim Hamilton, a native of Sligo who had studied at Belfast and Manchester. They shared an office, which was also the room in which tea was brewed.

Schrödinger usually arrived late at the Institute but he also stayed late. He would come in at about 11:30 while they were having tea, perch upon the table, and talk about all kinds of things, but mostly physics and philosophy, until it was time for lunch. As Sheila recalls,

Now this was wonderful but it wasn't specific, having nothing to do with any work anyone was doing, but one saw how his mind worked and that was absolutely fascinating – it never stopped . . . After the seminars we used to have something similar; tea was served and a few people, always the scholars and a few others, would wait on, but the core was Professor Schrödinger and Dr Paddy Browne. They matched each other. One set up a mode of thinking in the mind of the other, so it was a great to-and-fro-ing. There was nothing of a destructive kind of argument.[31]

Throughout the spring of 1941, the war continued to go badly for the Allies, as the Germans invaded and soon overran Yugoslavia and Greece. On April 15, a raid on Belfast left 500 dead. Erwin was worried and depressed, and he wrote to Weyl in Princeton:[32]

Dear Peter:
Thanks for your gifts to the embryonic library of the Institute and greetings between institutes. To create anything that would deserve this name is in the present moment almost impossible. Continental books are well-nigh inaccessible. My copies of Courant-Hilbert I & II, of Jahnke-Emde, which I was fortunate enough to procure in France, will have to be the only ones for some time.

. . . I should very much like to know how your people feel about the present situation . . . If you asked me my opinion, I should be at a loss to describe it properly. I have not a bit of hope. I have passed the stage of 'desesperance'. What happens is thus, that democracy has not been people's power, as the name indicates. It has been power of Eaton [*sic*] and Christ Church, and I know not what. Over here. Over there it has been Wall Street. May be all that has changed now. It was too late. Unless something spectacular comes forth from these democracies presently, I cannot see what should change the present state. I have believed long enough. I can no longer. It is all nothing. It is all old and weak and arteriosclerotic and inefficient.

I have found a good deal of consolation in a book of Somerset Maugham – *The Summing Up*[33] . . . He is a plain, simple, ingenuous man. You'll say that is not rare. Go to the workshops, to the docks, you'll find plenty of them. True. But to go along with high intellectual gifts you will not find it very often. Spinoza was, and Lichtenberg. Goethe came near to it, had he not been such a prig . . .

Maugham might not have been too surprised by this absurd assessment of his character. As he remarked in *The Summing Up*, 'the celebrated develop a technique to deal with persons they come across. They show the world a mask, often an impressive one, but take care to conceal their real selves. They play the part that is expected from them, and with practice learn to play it very well, but you are stupid if you think that this public performance of theirs corresponds with the man within.'

On the night of May 31, the most serious bombing of Dublin occurred, with thirty persons killed and many houses destroyed in the North Strand area. The German government apologized for the 'mistake', but most people regarded it as a warning. However, the course of the war was about to change.

Till now, Hitler had met no serious resistance in Europe, and he decided to complete his conquests by an attack on Russia, launching a massive invasion on June 22. That night Erwin noted in his diary: 'The most important thing is this: the banner is turned around, the Proletariat of the world now knows on which side it stands. The Proletariat of the world knows the reason for this war and will decide it. I don't like it very much. They are a miserable people – but one must put up with them. Considered quite soberly, it is really a great joy to see the two wretches [Hitler and Stalin] in battle against each other.'[34] For over two years the issue hung in the balance, but ultimately the Red Army beat back the Nazi onslaught. The German military realized that defeat was inevitable and began to hatch various ineffective plots against Hitler.

The war situation did not inhibit the progress of D.I.A.S. and during 1941 it became evident that the theoretical physics school was fulfilling all the hopes of de Valera. Many members of the two Dublin universities had feared that the Institute, through diversion of scarce resources, might weaken their own establishments, but by now it was evident that such fears were groundless, and the Institute was acting as a catalyst and unifying force for scholarly work in its fields, thus markedly strengthening the work at the universities.

From July 16 to 29, the first major seminar was held at the Institute, to coincide with the arrival of Heitler. Schrödinger gave a course of ten lectures on perturbation theory and the mathematical background for the quantum theory of the meson, and Heitler presented ten lectures on the meson theory of nuclear forces. Dublin had never before experienced such a deep and comprehensive treatment of an exciting topic in theoretical physics. Participants in the seminar came from all over Ireland: Hackett and Keating from U.C.D., Broderick and Rowe from Trinity, Power and McKenna from Galway, Atkins from Cork, Ewald from Belfast, and about a dozen of their junior colleagues. It was reported in the *Irish Times* as a conference on 'The Secret of the Mystifying Mesotron'.

Erwin wrote to Max Born to say how delighted he was to have Heitler at the Institute.[35] 'Scientifically he is the equal of his "milk brother" London [they were both Sommerfeld students] but as a man and a teacher he is much greater in every respect. Reading his book has led me back to your nonlinear field theory.' The book was *Quantum Theory of Radiation*, a pioneering exposition of this subject. The book marked the directions in which future great advances in quantum theory would be made and it is curious that even with Heitler as a colleague, Schrödinger was soon to turn away from this field entirely so as to pursue his researches in unified field theory.

Intellectuals at bay

On Wednesday evening, November 5, Erwin was a guest at the Inaugural Meeting of the College Historical Society.[36] This is the oldest surviving university debating society in the 'British Isles', having been founded at Trinity College by Edmund Burke in 1747. The Auditor [President] of the Society was now Alexander Lieven, scion of an old Baltic family, twenty-two years old and a student in history and languages. An ancestor of his had been governess to the children of Paul Romanoff, who became the mad Tsar of Russia (1796–1801), and ordained that all her descendants should be called 'Prince' or 'Princess'. Sheila Greene had introduced Erwin to Alex, with whom she was enjoying a casual love affair with David's mild disapproval. According to tradition, at the first meeting of the society each year the Auditor would read a paper and four distinguished guests would be invited to take part in the debate. The motion this year was 'Intellectuals are at Bay'.

The meeting was chaired by Gerald Fitzgibbon, a former judge of the Supreme Court; the guests were Cyril Connolly an essayist and literary critic who was then visiting Dublin, Stephen Spender a poet who was cultural attaché at the British High Commission, Jim Prendergast a leading Irish communist, and Erwin Schrödinger. It was quite a formal affair: the auditor and the judge wore white tie and tails, the communist wore a dark coat and sweater, and the other speakers wore black ties and dinner jackets. Erwin had resisted formal dress all his life, but he succumbed to the customs of Dublin undergraduates. Lieven based his opening talk on *La Trahison des Clercs* by Julien Benda, and deplored the decline of the influence of intellectuals. Connolly said there was a strong prejudice against intellectuals in England, but Ireland seemed actually to allow them into the government. Spender thought that intellectuals in Europe had acted ineffectively and now everything they stood for was threatened by tyranny. Prendergast said that Marxist principles provided the only guide for correct action by the intelligentsia.

In seconding the motion, Schrödinger said that he did not believe in the possibility of cooperation by intellectuals to improve the world's natural course. He had no confidence in such cooperation – firstly, because intellectuals were not likely to exert themselves very much, simply because they lacked the power to impose their wills. There seemed to be an absolute incompatibility between the sphere of action and the contemplative sphere. The major cause for his lack of confidence, however, was that he did not consider it a right view to split the phenomena of history into two parts, the natural course of development on one side, and on the other side, man's conscious and determined influence in directing events. Man was just as much a part of nature as anything else. There were instances of overwhelming influence of individuals, but he thought, nevertheless, that such persons were products of history rather than producers of it. The motion was carried.

Alex Lieven became a frequent visitor at Clontarf and Anny taught him how to make Apfelstrudel. Once Alex surprised Erwin watching a cricket match and teased him about it. He retorted with a quotation from Terence, 'Humani nil a me alienum puto.'

The Myles case

In 1942, after he had been in Dublin about two years, and while the debate over the founding of the Institute was still fresh in the public mind, Erwin had a sort of comic theological collision with one of Dublin's most famous literary figures. This was Brian O'Nolan, an author of great wit and style, whose novels *At Swim Two Birds* and *The Third Policeman* are considered by some critics to be in a class with those of Joyce and Beckett. O'Nolan was a hard-drinking civil servant who wrote under a bewildering variety of pseudonyms, which he pretended to believe were impenetrable. Dismayed by the rejection of *The Third Policeman* by Longmans, he hid it away in a cupboard and turned to journalism. He wrote a daily column for the *Irish Times* under the title, *Cruiskeen Lawn*, ['The Little Overflowing Jug'], which he signed 'Myles na Gopaleen' ['Myles of the Little Ponies'], after a character in a play by Boucicault. The local color and humor of his material often concealed a savage social commentary.

Jack White, the feature editor of the *Irish Times* was aware that Myles 'took an extravagant pride in his ability to circumvent the laws of libel . . . His copy was scrutinized for libel, scurrility, and double meanings, and any column that offended was chopped ruthlessly or thrown into the wastebacket.' His contract originally called for £1 a column, but so many were thrown out that he complained, and the paper agreed to pay the pound whether or not they were printed. It should be remembered that in the British tradition the laws of libel were designed to prevent the press from embarrassing the ruling classes, and material that would hardly raise an eyebrow in Boston could trigger legal salvos in Dublin.

Nevertheless, on April 10, 1942, *Cruiskeen Lawn* included the following paragraphs:[37]

That nothing but the Best is good enough for the Institute of Advanced Studies is another quip that must, in the name of reticence and that delicacy of manner which distinguishes the gentleman, remain unsaid. Here, at any rate. By all means pass it off as your own in the boozer to-night. The laugh that you will get will be as forced and as false as your own claim to be a wit.

*   *   *

Talking of this notorious Institute (Lord, what would I give for a chair in it with me thousand good-lookin' pounds a year for doing 'work' that most people regard as an interesting recreation), talking of it, anyway, a friend has drawn my attention to Professor O'Rahilly's recent address on 'Paladius and Patrick'. I understand also that Professor Schroedinger has been proving lately that you cannot establish a first cause. The first fruit of the Institute, therefore, has been an effort to show that there are two Saint Patricks and no God. The propagation of heresy and unbelief has nothing to do with polite learning, and unless we are careful this Institute of ours will make us the laughing stock of the world.

*   *   *

Before the seminar at the Institute that day, a retired Church of Ireland clergyman, Rev. J.R. Colthurst, brought a copy of the paper up to Schrödinger and showed him the column in question. He read it through and handed it back with a smile and dismissed it as a matter of no importance. Nevertheless, a real tempest of discussions, meetings, telephone calls, and the like raged for the next ten days, until the meeting of the Council of the Institute on April 23. Solicitors were consulted and the chief editor of the *Irish Times*, R.M. Smyllie, was accused of publishing defamatory material. Schrödinger, however, remained cool, at least publicly; after the polemics of Karl Kraus in Vienna, Myles seemed tame enough to be taken calmly. Thus he wrote to the Council the day before its meeting:

the draft of an apology by Mr Smiley [*sic*], Chief Editor, which I was shown yesterday by the Registrar and which was intended for publication in the *Irish Times*, includes the statement that I felt grieved by the article in question, and included an apology to me personally. From this I conclude that Mr Smiley must have understood some information he got to the effect that I felt grieved and asked for an apology. For, news-paper-men are not in the habit of apologizing without being asked to do so . . . I beg to decline emphatically the inclusion of any statement about my having been grieved by that article, or of any apology *to me* . . . or of anything that gives the wrong impression that I have asked for an apology.[38]

Erwin went on to suggest that it would be better not to have any public apology printed in the paper but simply to ask for a personal letter from the editor expressing his regrets and giving his assurance that such an incident would not happen again. This good advice was followed by the Council, but they went further and asked their solicitors to demand from Smyllie an assurance that Myles na Gopaleen would never mention the Institute again in his columns. This outrageous demand was actually accepted. The paper was also asked to pay £100 to a nominated charity and all legal costs. In the event they settled for £50 to the Red Cross and the costs.

We do not have a copy of Schrödinger's lecture on 'Causality', which aroused the ire of Myles, but a paragraph he wrote some years later on 'The Atheism of Natural Science' is relevant.

I shall quite briefly mention here the notorious atheism of science . . . The theists reproach it for this again and again. Unjustly. A personal God can not be encountered in a world picture that becomes accessible only at the price that everything personal is excluded from it. We know that whenever God is experienced, it is an experience exactly as real as a direct sense impression, as real as one's own personality. As such He must be missing from the space-time picture. 'I do not meet with God in space and time', so says the honest scientific thinker, and for that reason he is reproached by those in whose catechism it is nevertheless stated: God is Spirit.[39]

A full scale debate between Erwin and Myles would have been a memorable occasion, for they both were masters of language and serious devotees of drama. They remained on friendly terms and Myles consulted Erwin for advice about his adaptation of *The Insect Play* by the Čapek brothers, which was staged by the Gate Theatre early in 1943.

D.I.A.S. colloquium

During the spring of 1942, plans were made for a major event in the School of Theoretical Physics, the first colloquium with distinguished speakers from overseas. Dirac and Eddington agreed to come; they would each receive an honorarium of £50 for five lectures, first class travel to Dublin, and subsistence of a guinea a day while there.

In June, Erwin had his first serious disagreement with the Registrar of the School of Theoretical Physics. The Registrar was responsible for the nonacademic administration of the School. He had a rigid conception of his duties, a lack of sympathy for the requirements of intellectual work, a growing fondness for the bottle, and an unstable character which eventually led him to make off with some of the school funds. His first collision with Schrödinger was over management of the library, and it was necessary for the governing board to inform him that library policies were entirely the responsibility of the Director of the School.

The Colloquium met from July 17 to 26, with forty-five participants. Besides the two main lecture series by Dirac and Eddington, lectures were given by A.W. Conway, T.E. Nevin, C.O'Ceallaigh, and E.T.S. Walton. Ernest Walton was born near Waterford in 1903; he graduated from the University of Dublin in 1926 and took a Ph.D. at Cambridge in 1931. His great discovery of artificial nuclear disintegration was made at Cambridge with John Cockroft in 1934, by bombarding lithium with a beam of protons accelerated to 120 000 V: $\mathrm{Li}_3^7 + \mathrm{H}_1^1 = \mathrm{He}_2^4 + \mathrm{He}_2^4$. He had returned to Dublin as a Fellow of Trinity College in 1936, and fifteen years later received a Nobel prize for his early work.

Eddington's lectures were on 'The Combination of Relativity Theory and Quantum Theory'.[40] He began as follows: 'Twenty years ago relativity theory had unified molar physics – molar mechanics, molar electrodynamics, and the geometry of molar measurement – into a single formulation, rational and complete. By "complete" I mean that it was as complete as a theory which admittedly covered only one-half of physics had any right to be. Microscopic physics, as represented by quantum theory, was still in its infancy. Since then a highly developed microscopic theory has grown up almost independently of relativity theory, and the problem of unification arises.' In this first lecture, Eddington immediately took up the problem of the infamous $\sqrt{N}$ term, gave a new derivation based on an 'extraordinary fluctuation' of the number of particles N in a closed universe, and promised that 'we shall show that . . . space curvature is simply a way of taking the extraordinary fluctuation into account.' What makes Eddington's work so baffling is the way in which he unobtrusively slips magical transpositions into a clear and logical story. His Dublin lectures were masterpieces of exposition, using hardly any mathematics; everybody followed them without difficulty yet nobody understood them.

Dirac's lectures were on 'Quantum Electrodynamics'.[41] He began by saying that his lectures, like those of Eddington, were concerned with unifying relativity and quantum theory, but his approach was first to find a neat and beautiful mathematical scheme, and then to fit it to a physical interpretation. The corner-stone of electrodynamic theory is the simple harmonic oscillator. He shows that an assembly of harmonic oscillators is mathematically equivalent to an assembly of bosons, e.g., photons. He develops the classical relativistic (Lorentz invariant) treatment of such an assembly and translates it into quantum mechanics by deriving the Hamiltonian operator. As in his theory of the electron, negative as well as positive energy states occur, but now the parts of the wavefunction that refer to the existence of positive and negative-energy particles indicate emission and absorption of photons. He is thus able to calculate probability coefficients for all transition processes, 'which essentially is all that one needs for comparing the theory with experiment'.

Schrödinger reported to Born[42] that the colloquium was

a great success *externally*: the two dear men here for a fortnight – pleasure, contact, rejoicing on all sides, receptions, parties, climaxed by huge invitation cards to meet the President on Wednesday afternoon.

But the internal effect was negligible. Dirac gave a lucidly clear exposition of an intrinsically inconceivable business. Eddington entirely messes up a position which in its main points really is quite clear and tenable.

As a matter of fact, they are monomaniacs. Your idea of getting their opinion on Born's theory is pathetic. That is the thing beyond their linear thoughts. All is linear, linear, – linear in the n'th power I would say, if that was not a contradiction. Some great prophet may come.

'If everything were linear, nothing would influence nothing', said Einstein once to me. That is actually so. The champions of linearity must allow zero-order terms, like the right side of the Poisson equation, $\Delta V = -4\pi\varrho$. Einstein likes to call these zero-order terms 'asylum ignorantiae'.

The reception for the scientists was held at Arus an Uachtarain [President's mansion], with Douglas Hyde, Eamon de Valera, and other dignitaries. A more convivial party was a great banquet at Maynooth hosted by Paddy Browne. Smaller groups were entertained also by Erwin, Anny and Hilde at Clontarf. Hilde was disappointed, however, not to be invited to meet the President.

Myles na Gopaleen took note of Eddington's remark that less than a thousand people in the world can understand Einstein's theory and less than a hundred can discuss it intelligently. He proposed to make it compulsory in the schools and to have it taught in Irish. Then instead of the joke about 'being illiterate in two languages', Irish children could be 'illiterate in four dimensions'.

The lectures of Dirac and Eddington were published in editions of 500 copies as the first of the *Communications of the Dublin Institute for Advanced Studies*. Schrödinger made the decision that the Dirac work should be No 1 and the Eddington, No. 2, his reservations about the former evidently having less weight than his incomprehension of the latter. Eddington died in Cambridge in 1944 and most of the material from his Dublin lectures was incorporated verbatim into his posthumous *Fundamental Theory*.

As soon after the colloquium as possible, Erwin and Anny took off for a rest and holiday in Killarney, County Kerry. They stayed at the Swan Hotel, where they met a cheerful and friendly teenager, Lena Lean, whom they invited to come to Dublin to help take care of Ruth.

Nonlinear optics

Schrödinger's principal research during 1941–42 was devoted to a nonlinear classical electromagnetic theory that had been devised by Max Born in 1934. He continued to explore the consequences of this theory through 1943–44, and some of the methods were then taken over into the more general unified field theories that became his main scientific interest until he left Dublin in 1956.

The foundations of electromagnetic theory had been constructed by the Scottish physicist James Clerk Maxwell, while he was professor at King's College, London, from 1860–65. Starting from the experimental discoveries of Faraday, Ampère, and Gauss, he devised a consistent mathematical theory which not only comprised all the known facts of electromagnetism, but also predicted the existence of electromagnetic waves propagated through space and showed that light is an instance of such radiation. Maxwell's memoir of 1865, setting forth the definitive form of his theory, is generally considered to be the greatest paper of nineteenth-century physics. Maxwell disposed of the concept of action at a distance, which had dominated physics since the time of Newton, substituting the idea that space is permeated by *fields* which transmit effects from place to place at a finite velocity.

The Maxwell equations for fields in a vacuum are:

$$
\begin{aligned}
\nabla \cdot \mathbf{E} &= \varrho & &\text{(Coulomb Law)} \\
\nabla \cdot \mathbf{B} &= 0 & &\text{(Gauss Law)} \\
\nabla \times \mathbf{H} &= \mathbf{J} + \dot{\mathbf{E}} & &\text{(Ampère–Maxwell Law)} \\
\nabla \times \mathbf{E} &= -\dot{\mathbf{B}} & &\text{(Faraday–Lenz Law)}
\end{aligned}
$$

Here, $\mathbf{E}$ is the electric field strength, $\mathbf{H}$ magnetic field strength, $\mathbf{B}$ the magnetic induction, $\mathbf{J}$ the electric current, and ϱ the density of electric charge. A dot over a symbol indicates rate of change with time.

The Maxwell equations are all linear, i.e., the fields and their derivatives occur only in terms of the first degree. A fundamental property of such linear relations is that the fields cannot generate themselves, but certain sources must be introduced. An example of a source term is the ϱ in the first equation: one must assume a distribution of electric charge which acts as the source of the electric field $\mathbf{E}$. The connection of a field with its sources is always difficult to specify; in this case, for example, if one introduces a point charge such as an electron, the field goes to infinity at this point, which is physically unreasonable. This is the kind of difficulty that Schrödinger complained about to Born, in the letter cited above.

The idea of a nonlinear modification of the Maxwell equations had occurred to Born in 1933 'in the loneliness of the Dolomites'. After he was settled in Cambridge, he published a first account of a theory in which an electron of finite radius arises naturally out of the field equations, but still behaves in external fields like a classical point charge. Meanwhile Leopold Infeld had come to Cambridge from Poland on a Rockefeller Fellowship; he and Born published together a long paper in which the foundations of the new theory were set forth more deeply and carefully than before.[43] Born said that his theory was motivated by the fact that the Maxwell equations could not be made compatible with the requirements of quantum mechanics, but it turned out that his new equations were equally intractable in this respect, so that after a few years he lost interest in them. The development of quantum electrodynamics followed a quite different path, starting with Dirac's relativistic quantum theory of the electron and culminating in the work of Julian Schwinger and Richard Feynman from 1948 to 1953.[44]

Born and Infeld began their paper by pointing out that the relation of matter to the electromagnetic field can be interpreted from two opposite standpoints: (1) Unitarian: only the field exists and particles are singularities of the field; their masses are derived from the field energy. (2) Dualistic: particles are the sources of the field; they are acted upon by the field but they are not parts of it; they have the characteristic intrinsic property of inertia, measured by mass. Quantum mechanics is a dualistic theory and it has had great success, but it is worthwhile to explore in depth the unitarian approach.

They used a general method based on the minimization of a Lagrangian function **L** with respect to the field variables. A similar method had been used by Hilbert in his derivation of the equations of general relativity, and it will be described in more detail when Schrödinger's work on unified field theory is considered. The Maxwell equations can be derived in this way from $\mathbf{L} = \frac{1}{2}(\mathbf{H}^2 - \mathbf{E}^2)$. The Born Lagrangian is

$$\mathbf{L} = b^2 \left[\sqrt{1 + \frac{1}{b^2}(\mathbf{H}^2 - \mathbf{E}^2)} - 1 \right]$$

Born called his basic idea 'the principle of finiteness'. Just as there is an upper limit c to the magnitude of velocity, just as space itself is closed with a radius of curvature R in Einstein's cosmology, so Born decided to impose an upper limit on the strength of an electric field. This limit is controlled by the value of the parameter b; b is very large, of the order of 10^{16} esu cm^{-2}. The effective 'radius of the electron' becomes $r_0 = \sqrt{e/b} = 2.29 \times 10^{-13}$ cm. The ordinary Maxwell equations (and hence Coulomb's Law) are followed as long as $x = r/r_0 \gg 1$, but at shorter distances ('within the electron') the law is modified so that the potential does not go to infinity but to a limiting value $= 1.8541\ e/r_0$.

Schrödinger was greatly impressed by the Born–Infeld electrodynamics, and he proceeded to recast the Born equations into the form of complex six-vectors (instead of real three-vectors), thereby simplifying the mathematics, and he sent his version to the Royal Society (via Lindemann) in 1935.[45] The unitarian approach appealed to him for a number of reasons: his faith that the world is a universal consciousness, the idealistic elements in his view of science, his nostalgia for real waves as the ground of reality, and his growing estrangement from the advances in elementary-particle physics.

In his 1935 paper, Schrödinger showed that introducing $\mathbf{F} = \mathbf{B} - i\mathbf{D}$, and $\mathbf{G} = \mathbf{E} + i\mathbf{H}$, leads to a surprisingly simple Lagrangian, $\mathbf{L} = (\mathbf{F}^2 - \mathbf{G}^2)/\mathbf{FG}$. The concepts of mathematical simplicity, intellectual beauty, and scientific truth were so intimately combined in Erwin's mind, that the discovery of this Lagrangian lent the Born theory an irresistible fascination. Paul Dirac once said:

Of all the physicists I met, I think Schrödinger was the one that I felt to be most closely similar to myself. I found myself getting into an agreement with Schrödinger more readily than with anyone else. I believe the reason for this is that Schrödinger and I both had a very strong appreciation of mathematical beauty and this dominated all our work. It was a sort of act of faith with us that any equations which describe the fundamental laws of Nature must have great mathematical beauty in them. It was like a religion with us. It was a very profitable religion to hold and can be considered as the basis of much of our success.[46]

Whatever his motivation, Schrödinger devoted considerable effort in 1941–43 to working out the detailed consequences of the Born–Infeld electrodynamics. His work had great formal elegance and displayed his pre-eminent abilities as a mathematical physicist, but it lacked any new physical ideas, being like a virtuoso variation on a classical theme. Nevertheless, his conviction that nonlinear theories would be essential for future progress in physics has turned out to be abundantly justified.

The theory predicted a correction to the Planck Law for black radiation, but it amounted to only 1% at a temperature of 4×10^{10} K. Schrödinger recognized the aspect of unreality about this work, when he wrote that 'to many a reader it may seem lunatic to bother about corrections of that small order by classical methods', but he argued that classical understanding of electrodynamics must precede quantum-mechanical understanding.[47]

Dublin Colloquium, 1943, Paul Ewald, Max Born, Walter Heitler and Schrödinger.

He wrote frequently to Born about his progress,[48] expressing his great confidence in the theory. 'Please don't mind my machine-gunning you with reports . . . but it is such a pleasure to tell things.' He was particularly concerned by a pair of ordinary differential equations that had turned up in the treatment of diffraction of light by a point charge. They were 'a little more complicated than anything hitherto on the main road of theoretical physics'. After struggling with them for some months, he sent them to Weyl. 'There is a slight chance that there is something in the problem that arouses his interest sufficiently to do for us in this case what Gauss, Legendre, etc., did for us in previous simpler cases. I dare say he is of that same general calibre – and he has always been interested in problems coming from physics.' Apparently Weyl did not find time to attack these equations.

Unified field theory

From 1943 to 1951, Schrödinger's research work was dedicated almost exclusively to the search for a unified field theory that would encompass both gravitation and electromagnetism. He was inspired by a metaphysical belief in the unity of nature, a belief that had not essentially changed since he had written it down almost twenty years earlier in *Meine Weltansicht*. As he said then, only metaphysics can inspire the hard work of theoretical physics. The philosophy of Einstein in his later years was similar, leading to a feeling of wonder at the simplicity and beauty of the landscape of the universe as seen through the window of mathematical theory. Bruno Bertotti, who worked with Schrödinger from 1953 to 1955, called this view of the world 'rational mysticism'.[49] Roland Barthes, however, discerned in it the perennial gnostic themes: 'the unity of nature, the ideal possibility of a fundamental reduction of the world . . ., the age old struggle between a secret and its utterance, the ideal that total knowledge can only be discovered all at once, like a lock that suddenly opens after a thousand unsuccessful attempts.'[50] Erwin was more a gnostic than a mystic. He never displayed any inclination toward the life of asceticism and self-denial that mark the way of a religious mystic. He would attempt to reach the secret at the heart of the world through a labyrinth of mathematical symbols.

Einstein was thirty-six years old when he published his general theory of relativity; Schrödinger was thirty-eight when he discovered wave mechanics. By 1940, when he began to consider generalized field theory, he was fifty-three. Einstein, then sixty-one, had been working on the problem for twenty-five years without apparent success.

In the history of physics it is unusual to find anyone who has made a major theoretical discovery after the age of forty. Thus it might seem that Einstein and Schrödinger were facing an insurmountable psychological barrier. Revolutionaries in physics must be young men whose minds have not had time to become habituated to well worn pathways of thought. In fact, however, Einstein and Schrödinger were not seeking any revolution, they were simply trying to extend the range of a method that had already proved its worth in the theory of general relativity. The techniques they used were well established in the 1916 papers of Einstein and Hilbert, the 1918 book of Weyl, and the 1923 book of Eddington.

At any point in history there are different potential directions for the future of physics, and the choice of a particular direction may preclude exploration of any of the others. When the choice is made, the best minds, a few in each generation, dash eagerly along the new path, and physics follows them, while the other paths are blocked by an impassible psychic gate. Thus Schrödinger and Einstein may have had little chance of success in their efforts to discover a unified field theory in the 1940s by 1920 methods. One can see an example of such a blocking of pathways in Richard Feynman's 1962 Caltech lectures on gravitation:[51] 'None of these unified field theories has been successful . . . Most of them are mathematical games, invented by mathematically minded people who had very little knowledge of physics and most of them are not understandable.' Is it likely that any of his students would choose to work in such a direction? This closure of alternatives is consistent with Schrödinger's vedantic concept of the unity of mind. Even without overt telepathy, the unified mind of physics determines the history of the subject, and that mind is perpetually young.

It has often been suggested that Schrödinger and Einstein turned to unified field theory because of their disenchantment with the prevailing state of quantum mechanics. Einstein, however, certainly hoped that field theory would eventually include both the macroscopic systems of cosmology and the microscopic physics of elementary particles and quanta. Schrödinger was less optimistic than Einstein: 'At the back of our striving for a unitary field theory, the great problem awaits us of bringing it into line with quantum theory. This point is still covered with deep mist.'[52]

There is no convincing evidence that either scientist turned to field theory as an escape from the probabilities and uncertainties of quantum mechanics. Einstein was continuing to follow the way that had led to his greatest success. Schrödinger, influenced by Einstein, had always been interested in the geometrization of physics, and the concept of a unified field was an ideal counterpart of the unity of mind and nature in which he found the ultimate reality. 'The city of Brahman and the palace within it is the small lotus of the heart, heaven and earth are contained in it.'

Einstein's gravitation theory[53]

Einstein's special theory of relativity (S.R.) of 1905 was based upon two principles: (1) The laws of physics have the same simple mathematical form in all inertial frames. Inertial frames are physically defined coordinate systems moving uniformly (i.e., at constant velocity) relative to one another. (2) The speed of light *in vacuo*, c, is the same in all inertial frames. To some extent Einstein may have been influenced by the famous experiment of Michelson and Morley of 1887, which was interpreted to show that there is no absolute motion of light. In 1908, Herbert Minkowski, Einstein's former mathematics teacher at Zürich, showed that S.R. can conveniently be based on a four-dimensional geometry of space-time, in which the distance between two closely neighboring points, the interval ds is given by

$$ds^2 = - c^2 dt^2 + dx_1^2 + dx_2^2 + dx_3^2$$

Or, writing $-c^2 dt^2 = -dx_0^2$,

$$ds^2 = -dx_0^2 + dx_1^2 + dx_2^2 + dx_3^2$$

Einstein was not satisfied with S.R. because it arbitrarily selected a certain class of reference frames, *viz.* inertial frames, in which the laws of physics are invariant. He believed that physical laws should not depend at all upon the choice of reference frame. He was also strongly influenced by Ernst Mach's criticism of the foundations of mechanics, and the resulting Mach Principle that inertial forces have their origin in the total distribution of mass in the universe. For example, Newton had argued that the centrifugal force on a rotating mass was a demonstration of its absolute motion, but Mach ascribed the force to motion with respect to other masses in the universe, especially those of the fixed stars.

Einstein knew that it is possible to make gravity disappear in certain localized regions of space by a transformation of the reference frame. For example, freely falling sky divers (with acceleration $= g$) do not observe the force of gravity in their moving reference frame. If they release a tennis ball it does not drop away from them. Similarly, observers in an orbiting space shuttle are weightless, as the shuttle has a constant acceleration g directed towards the earth. These results are consequences of the equivalence of gravitational mass and inertial mass, but there is no single accelerating reference frame that can cause gravity to disappear *at all points in space.* Thus Einstein asked himself, 'Is it possible to find a geometry of space that accounts for all the gravitational effects of the masses in the universe, and yet in a small region reduces to an ordinary accelerated Euclidean frame of reference?'[54],[55]

Part of the mathematical answer to this question was already at hand, in the generalized geometry of curved spaces devised in 1854 by Georg Riemann. For example, the surface of a sphere is a two-dimensional curved space embedded in a Euclidean space of three dimensions. A very small area of the surface has essentially the properties of a flat Euclidean space. Einstein required a four-dimensional curved space-time. Such space-time can be embedded in a ten-dimensional Euclidean space, but it is much simpler to deal with the intrinsic curvature in four dimensions and to discard Euclidean geometry altogether.

In a curved space in two dimensions, x_1 and x_2, suppose that from a point P, infinitesimal vectors are drawn to a point Q. In the Riemannian geometry of curved space, the square of the distance between P and Q is a quadratic form in dx_1 and dx_2,

$$ds^2 = g_{11} dx_1^2 + 2g_{12} dx_1 dx_2 + g_{22} dx_2^2 \qquad (1)$$

This is the analog in Riemannian geometry of the Pythagorean theorem. For a space of any dimensions, the corresponding expression is

$$ds^2 = g_{ij}dx_i dx_j \tag{2}$$

According to conventional notation, summation is to be taken over all repeated indices, in the case of 4-space, over i, j from 1 to 4. The quadratic form in (2) is called the *metric form*, and the quantity g_{ij} is called the *metric tensor*. Note that the g_{ij} are functions of x_i and may vary from point to point. Physical laws must be expressed by tensor relations that are invariant for arbitrary continuous transformations of the x_i.

The metric tensor provides the *connection* between neighboring points in space, permitting the distance between points to be measured. Thus Riemannian geometry, like Euclidean, is a metric geometry, in contrast with the affine geometry that Schrödinger used in his color theory, and which he would use again in his unified field theory.

In Einstein's general theory of relativity (G.R.), the curvature of space-time is given by the *curvature tensor*, sometimes called the *Riemann–Christoffel tensor*, which is defined by

$$R^{\alpha}_{\beta\mu\nu} = \Gamma^{\alpha}_{\beta\nu, \mu} - \Gamma^{\alpha}_{\beta\mu, \nu} + \Gamma^{\alpha}_{\sigma\mu}\Gamma^{\sigma}_{\beta\nu} - \Gamma^{\alpha}_{\sigma\nu}\Gamma^{\sigma}_{\beta\mu} \tag{3}$$

The Γs, called *Christoffel symbols*, are related to partial derivatives of the components of the metric tensor.

$$\Gamma^{\nu}_{\beta\mu} = \tfrac{1}{2} g^{\alpha\nu}(g_{\alpha\beta, \mu} + g_{\alpha\mu, \beta} - g_{\beta\mu, \alpha}) \tag{4}$$

The notation $g_{,\mu}$ means the partial derivative of g with respect to x_μ, for example, $g_{\alpha\beta, \mu} = \partial g_{\alpha\beta}/\partial x_\mu$, etc.

The distribution of mass-energy in the world is represented by the energy-momentum tensor, $T_{\alpha\beta}$, defined as the flux of α component of momentum across a surface of constant x_β. Note that since x_0 is the time coordinate, the flux $T_{\alpha 0}$ is an energy flux.

Einstein was striving to relate the curvature of space-time to the distribution of mass-energy. He found that the only tensor related to space-time structure that follows the divergence theorem of the energy-momentum tensor is

$$G_{\mu\nu} = R_{\mu\nu} - \tfrac{1}{2}R g_{\mu\nu} \tag{5}$$

Here $G_{\mu v}$ is now called the *Einstein tensor*; $R_{\mu v}$ is the *Ricci tensor*, a contraction of the curvature tensor, and $R = g^{\mu v} R_{\mu v}$ is the curvature scalar, the trace of $R_{\mu v}$.

Thus, he set $G_{\mu v}$ proportional to $T_{\mu v}$,

$$R_{\mu v} - \tfrac{1}{2} R g_{\mu v} = G_{\mu v} = - \kappa T_{\mu v}. \tag{6}$$

To eliminate the possibility that a field could exist even when there was no mass-energy to produce it, he later added a term $\lambda g_{\mu v}$ where λ was called the *cosmological constant*, thus obtaining the final form of his equation as

$$R_{\mu v} - \tfrac{1}{2} R g_{\mu v} - \lambda g_{\mu v} = - \kappa T_{\mu v} \tag{7}$$

Einstein, after ten years of unremitting thought at the peak of his powers, did not in any sense 'derive' this equation from pre-existing principles. Nor was he guided by a need to explain particular sets of experimental data or observations. He was guided only by his quest for two abstract ideals: beauty and simplicity. The equivalence of gravitational and inertial mass and the need to make all the equations independent of any particular choice of coordinate were also basic requirements. The second condition is met automatically by the general covariant property of tensors: they do not depend on the choice of a frame of reference. The first condition was met by requiring that the dynamical paths of bodies be geodesics in a quasi-Riemannian space-time whose curvature is determined by the distribution of mass-energy.

First unified-field paper

Schrödinger's first paper on unified field theory, 'The General Unitary Theory of Physical Fields' was read at a meeting of the Academy in January, 1943, and published in the Proceedings for July.[56] The task he set for himself was to modify Einstein's general theory of relativity so that it would explain electromagnetism as well as gravitation.

He returned to an approach first given by Weyl in 1918, which was based on a closer look at the foundations of geometry. In the most general 4-manifold, the coordinates x_k are merely labels for the world points. There is no connection between the points. Thus there is no way to compare the magnitude and direction of a vector at point P with a vector at a nearby point Q, i.e., there is no *affine connection*, nor is there any way to measure the distance from P to Q, i.e., there is no *metric connection*. Weyl drew attention to the primordial nature of the affine connection, and tried without success to obtain both gravitation and electromagnetism as consequences of an affine geometry of space-time.

The affine connection is based on the operation of *parallel displacement* of a vector at P. In an article in *Nature*, Schrödinger explained the geometrical basis of his work with typical clarity.[52]

The *displacement vector*, dx_k, leading from a world point P with coordinates x_k ($k = 1, 2, 3, 4$) to a neighboring point Q with coordinates $x_k + dx_k$, is the prototype of a contravariant vector A^k at P. If you execute a general transformation of the frame, the A^k transform *by definition* as the dx_k, thus

$$dx_k = (\partial x_k / \partial x_j) dx_j, \qquad \text{and} \qquad A^k = (\partial x_k / \partial x_j) A^l \qquad (1')$$

This rule is obviously inherent in the world point P, because the partial derivatives which form the coefficients *vary* from point to point. As a consequence of this, you cannot directly compare a 'vector at P' with a similar 'vector at Q', even when Q is an infinitely neighboring point . . .

You have lost the connection between neighboring points. You can no longer . . . set up differential equations to control geometrical or physical *fields* . . . To recover this possibility, you must reinstate a connection . . . What you want to know, or rather, what you want to 'state by decree' in an invariant manner, is: which vector at the neighboring point Q ($x_k + dx_k$) is to be considered the *same* as a given vector A^k at P (x_k).

The components of that 'vector at Q' must differ infinitesimally from the A^k, and so we suitably term them $A^k + \delta A^k$. The simplest and most straightforward, and at the same time a fairly general way of making the required 'decree', is to demand that the δA^k shall be some bilinear functions of the dx_k and of the A^k, thus:

$$\delta A^k = - \Gamma^k_{mn} A^m dx_n \qquad (2')$$

the Γ^k_{mn} being 64 ($4 \times 4 \times 4$) coefficients, which can vary arbitrarily from world-point to world-point, that is, they are arbitrary continuous functions of the x_k (the minus sign is conventional).

If we adopt (2') without any further restrictions on the Γs, we say we impose (the most general form of) an *affine connection* or an 'affinity' on our continuum. The Γs are called the components of the affine connection. Formula (2') is said to determine the 'parallel displacement' according to this affine connection.

Equation (2') and the notions introduced with regard to it are common to all (generally invariant) field theories. But various theories differ widely as to the way in which the affine connection is introduced.

The components of the affine connection Γ are the Christoffel symbols. In G.R. they are symmetric. Einstein's first approach to a unified field theory was to discard the symmetry ($\Gamma^k_{mn} = \Gamma^k_{nm}$) of the affine connection, and then to try to derive the electromagnetic vector potential from the unsymmetric (skew symmetric) part, and the gravitational tensor from the symmetric part. Since the approach did not lead anywhere, he abandoned it in 1925.

In his first paper on unified field theory, Schrödinger followed a path suggested by Eddington: to make the affine connection primary and to make it symmetric. The number of Γs was thus reduced from 64 to 40. He derived the field equations by a method based on a variational principle similar to that used by Hilbert in his 1916 derivation of G.R. and by Born in his nonlinear electrodynamics.[57]

The variation principle originated in 1744, when Pierre Maupertuis read a paper before the Paris Academy on a general 'principle of least action', which he believed would explain all the laws of nature. 'Nature in the production of its effects always acts by the most simple means . . . When any change occurs in Nature, the quantity of action necessary for this change is the smallest that is possible.' His definition of 'action' lacked precision, but Maupertuis was certain that 'these so beautiful and so simple laws perhaps are the only ones that the Creator and Lawgiver of all things has established in matter to cause there all the phenomena of this visible world.' Beauty and simplicity were the magic words in the eighteenth as in the twentieth century. 'Cosmos' means 'universe' but it also means 'beauty'.

The principle of least action achieved mathematical precision in Hamilton's variational principle, which is the basis of a powerful general way to derive field equations. It can be used to obtain Newtonian mechanics, relativistic mechanics, and, as Schrödinger showed in 1926, quantum mechanics. Consider an actual path of a system from P to Q. Set up the Lagrangian function for the system, $L = T - V$, the difference of kinetic and potential energies. Then,

$$\delta \int_P^Q L\,dt = 0$$

This equation states that the integral of the Lagrangian along the actual path is stationary with respect to variations in the path between P and Q that are continuous with the actual path in space and time. The calculus of variations is used to show that the variational condition is equivalent to a set of differential equations, the Euler equations, for $L(x, \dot{x}, t)$. In Newtonian mechanics, these equations are simply the Lagrangian equations of motion.

The variational equation used by Schrödinger was $\delta \int \pounds d^4x = 0$, where $\pounds$ is a *Lagrangian density*, since the integrand must be invariant with respect to any transformation of coordinates. Thus $\pounds$ is the product of L and a functional determinant (Jacobian) arising from transformation of the product of differentials. If the field equations are derived from such a variation principle, the conservation laws are an automatic consequence of the general invariance. There is no way to derive the appropriate Lagrangian from first principles – it is matter of trying to match the mathematical form to the physical reality by a process that depends on contemplation, intuition, and considerable trial and error. One can assess the result only after the mathematics is worked through to a final set of equations and the symbols related to the physical field quantities.

Schrödinger decided to carry out the variation of the integral with respect to the affinities Γ. He calls this procedure 'imposing on the affine connection Γ the "field law" that a certain scalar density £, the Lagrange function, shall have vanishing Hamiltonian derivatives with respect to the Γ'. He assumes that £ is a function only of R_{kl}, the Ricci tensor, and writes

$$\delta \int £ d\tau = \int \frac{\partial £}{\partial R_{kl}} \delta R_{kl} d\tau$$

Carrying through the rather complex mathematical procedures, he obtains a set of general field laws of gravitation and electrodynamics. They result from the affine connection with no other specification of the Lagrangian save that it should depend only on the R_{kl}. At this point, he has delineated the geometric structure of space-time that can yield the field laws, but to establish the physical interpretation, he must introduce a special form of £ (R_{kl}). When he does this, he obtains a set of twenty coupled differential equations, which link the matter-tensor of the gravitational field with an electromagnetic energy tensor similar to that of the Born–Infeld theory.

Schrödinger then thought that with a second symmetric affine connection, he could also include the meson field for nuclear forces (in a classical form) within his grand scheme. When A.J. McConnell saw the paper, he objected to this awkward 'duplication' and pointed out that the same result could be achieved by using an antisymmetric connection in the first place, and this suggestion was adopted in subsequent papers.

It was by now apparent, from this work and similar studies by Einstein, that electromagnetism can be 'geometrized' in a bewildering variety of ways. The real problem is to establish some connection with the observable world, first by representation of the known physical laws, and then by prediction of hitherto unobserved relations. Einstein's G.R. had passed these tests perfectly, but so far all the generalized geometric theories had failed to do so.

Before Schrödinger addressed this problem, however, he temporarily put aside the unified-field theory to devote his mind to a problem that had fascinated him since student days – the mechanisms of heredity and the nature of life.

What is Life?

Schrödinger had agreed to give the statutory public lectures for 1943 and they were scheduled for February at Trinity College. He decided to prepare a semipopular lecture on a biological subject, the mutation rate caused by action of X-rays on the fruit fly, *Drosophila Melanogaster*. He knew that the absorption of one quantum of X-radiation can produce a mutation at one locus in a chromosome. Perhaps he recalled this result from discussions with Max Delbrück in 1933 in Berlin. Sometime towards the end of September, he realized that if he knew the absolute intensity of the absorbed X-rays and the mutation rate that they caused, he could calculate from the target area the effective size of the information center in the chromosome, the gene.

Either he or Anny must have mentioned this problem to Paul Ewald, for Paul sent him a copy of the paper from the 1935 *Göttinger Nachrichten* by Max Delbrück, Timofeef-Ressovsky, and Karl Zimmer, the famous 'three-man paper', also called 'the green pamphlet' from the color of its reprint covers.[58] This work had anticipated his estimation of the target area and suggested for the first time that a mutation is caused by a change at one place in a molecule. He began to think about the implications of this result for the physical and chemical mechanism of inheritance, and the ways in which this mechanism must be constrained by the laws of statistical thermodynamics, chemical kinetics, and quantum mechanics. There was far too much interesting material here for one lecture, and he decided to give a series of three lectures under the title 'What is Life?' The title was provocative and certain to attract a good audience.

Among all the great physicists, why was Schrödinger the one destined to make a major contribution to the history of biological thought? The answer may be found in his early life: his father, the spiritual guide of his formative years, was passionately interested in biology, and Erwin always referred to him as a botanist, even though he was actually an industrial chemist whose botany was an avocation pursued in the time that could be spared from business affairs. At the university, Erwin's intimate friendship with Franz Frimmel, the religious student of biology, reinforced his interest in this subject and led him into extensive reading in fields that were foreign to most physicists. Another great physicist very interested in biology was Niels Bohr, whose father was a distinguished professor of physiology. Bohr wrote several philosophical papers on biological themes, such as the one at the Galvani Congress in 1937 on 'Biology and Atomic Physics'. Max Delbrück, while he was a research fellow in Copenhagen, often discussed biology with Niels Bohr. Yet Bohr never went so deeply into biology as Schrödinger. Perhaps the subconscious reason for the difference was that Niels saw his father as a successful biologist, whereas Erwin sensed the unsatisfied life of his own father, and was stirred by a need to atone for it in some substantial way. Schrödinger was more than an amateur in biology; although he had not worked in genetics, he was still a world authority on the physiology and biophysics of color vision.

Erwin celebrated the New Year 1943 with Hilde and Ruth, while Anny was once again in Belfast visiting the Ewalds. He was feeling tired and out of sorts, but managed to prepare the 'What is Life?' lectures in his usual polished style. The first lecture took place on Friday evening, February 5, with two more on successive Fridays. De Valera was there with other notables from Church and State, cabinet ministers, diplomats, socialites, and artists. The number of people who tried to crowd into the Trinity lecture hall was so great that it was necessary to repeat the lectures on Mondays for those who were turned away. The total audience was estimated as more than 400, and the numbers did not dwindle between the first and last lecture.

Even *Time* magazine took notice of the excitement in Dublin and published a story in its issue of April 5. 'Schrödinger has a way with him', they wrote:

His soft, cheerful speech, his whimsical smile are engaging. And Dubliners are proud to have a Nobel prizewinner living among them. But what especially appeals to the Irish is Schrödinger's study of Gaelic, Irish music and Celtic design, his hobby of making tiny doll-house furniture with textiles woven on a midget Irish loom – and, above all, his preference for a professorship at the Dublin Advanced Studies Institute for one at Oxford.

Early in March, Erwin suffered a nasty attack of flu, and he had to drag himself from bed to give the final pair of lectures. By the time they were over, he was exhausted and longing for a holiday and some unattainable sunshine. As soon as the lectures were finished, however, he began to write them up for publication as 'a little booklet'.[59]

The question to be addressed in the lectures was clearly stated at the beginning: How can the events *in space and time* which take place within the boundary of a living organism be accounted for by physics and chemistry? At present, he said, these sciences cannot answer the question, but in the future they will be able to do so. The reason no answer can be given now is that the most essential part of the living cell, the chromosome fibre, is a piece of matter that differs entirely from any matter hitherto studied by the physicist. It may suitably be called an *aperiodic crystal*. By this is meant a regular array of repeating units in which the individual units are not all the same.

To show why such a structure is necessary, Schrödinger first tells how a 'naive physicist' would approach the study of living organisms by way of the statistical nature of ordinary physical laws. He admits that he is himself the naive physicist who followed this crooked path towards an unexpected destination. From his 1933 Berlin lecture he borrows the question 'why are atoms so small?' and shows, as he did then, that the real question is why do living organisms contain so many atoms? The physical laws that govern the behavior of macroscopic systems are based upon the statistical mechanics of the atoms and molecules that comprise the systems. Only as the number of component particles becomes very large, does the behavior of the system become reproducible and predictable to a high degree of accuracy. The probable scatter of results of observations on a system depends on $\sqrt{N}$, where N is the number of particles that it contains. Systems that contain only a few particles should have a random and unpredictable behavior. Thus, 'an organism must have a fairly gross structure in order to enjoy the benefit of fairly accurate laws, both for its internal life and for its interplay with the external world'.

He admits that this first conclusion of the 'naive physicist' might seem obvious, even trivial. It is, however, something far worse and far more important: it is a mistake, and a serious mistake can sometimes alert a scientist to a new line of thought. The physicist, no longer completely naive, now learns that 'incredibly small groups of atoms, much too small to display exact statistical laws, do play a dominating role in the very orderly and lawful events within a living organism'.

Schrödinger next gives an overview of the current understanding of the mechanism of heredity, in terms of chromosomes, genes, mitosis, meiosis, and crossing over. He introduces here what was to become one of the most fundamental concepts in the new science of molecular biology: *the chromosome is a message written in code.*

In calling the chromosome fibre a code-script we mean that the all-penetrating mind, once conceived by Laplace, to which every causal connection lay immediately open, could tell from their structures whether the egg would develop, under suitable conditions, into a black cock or into a speckled hen, into a fly or a maize plant, a rhododendron, a beetle, a mouse or a woman . . . But the term code-script is, of course, too narrow. The chromosome structures are at the same time instrumental in bringing about the development they foreshadow. They are law-code and executive power – or to use another simile, they are architect's plan and builder's craft – in one.

This was the birth of the concept of a 'genetic code'. A few earlier works had hinted at the idea of a code in the chromosomes, but

Schrödinger was the first to state the concept in clear physical terms.

The chromosome fibre is a linear code for the information that constitutes the genotype of a particular individual organism. The smallest locus in the chromosome coding for an individual difference in the genotype is called the *gene*. Schrödinger defines it as 'the hypothetical material carrier of a definite hereditary feature'. What is the maximum size of a gene?

Crossing-over data and microscopic examination of giant salivary-gland chromosomes from *Drosophila* roughly agreed that at most a gene can contain a few million atoms. 'That number is much too small (from the $\sqrt{N}$ point of view) to entail an orderly and lawful behavior according to statistical physics.' He thought that 'the gene is probably a large protein molecule, in which every atom, every radical, every heterocyclic ring, plays an individual role . . . This, at any rate, is the opinion of leading geneticists such as Haldane and Darlington.' Almost at the same time that these words were being spoken, however, Oswald Theodore Avery, then sixty-five years old and facing retirement from the Rockefeller Institute for Medical Research in New York, was writing up his most recent experiments which showed that the 'transforming principle' in *Pneumococcus* is not a protein but is in the DNA [deoxyribonucleic acid] fraction from the cell nuclei.

The permanence of the genetic information is almost absolute. 'The whole (four dimensional) pattern of the "phenotype", the visible and manifest nature of the individual . . ., is reproduced without appreciable change for generations.' When a change does appear it is discontinuous or 'jump-like', occurring in a few individuals out of a large population, and there are no intermediate forms between the few changed and the many unchanged. Such a discontinuous change is called a 'mutation'. 'It reminds a physicist of quantum theory – no intermediate energies occurring between two neighboring energy levels. He would be inclined to call de Vries's mutation theory, figuratively, the quantum theory of biology. We shall see later that this is much more than figurative. The mutations are actually due to quantum jumps in the gene molecule.'

In 1926, Hermann J. Muller, then at the University of Texas, irradiated a selection of fruit flies with X-rays and found an enormous increase in the number of mutations, by a factor of up to 150 over the spontaneous rate.[60] Schrödinger did not refer to Muller's pioneering work, but cited a later review by N.V. Timofeef-Ressovsky, a Russian emigré who used to run the genetics laboratory at the Kaiser Wilhelm Institute for Brain Research in Berlin. Soon after publication of Muller's paper, Timofeef began to study X-ray mutations, and in November, 1932, Muller came to Berlin and worked for a year in his laboratory. Timofeef was enthusiastic about the applications of physics to genetics. Delbrück had been attracted to biology by discussions with Niels Bohr, but he was then working as an assistant to Lise Meitner in Hahn's Physical-Radioactivity Institute at Berlin-Dahlem. Karl Zimmer was in the Radiation Section of the Cecilien Hospital. The three men met to discuss genetic mechanisms, often talking for ten hours at a stretch, and the result of these discussions was the 'green pamphlet' on which Schrödinger based his 'What is Life?'.

Delbrück, in the 1935 paper, had estimated the size of the target volume as about 1000 cubic Ångstroms, or less than ten average interatomic distances cubed. This is far less than indicated by the genetic data. Schrödinger identified this small target volume as 'the size of the gene'. Zimmer showed later that the mechanism of X-ray mutation is more complex than originally supposed. The beautiful linear dependence on ionization rate given in the 'green pamphlet' is actually an average over nonlinear curves for populations of flies with different sensitivies. Thus one of the foundation stones of the 'green pamphlet' was removed after it had served as a basis for the theories of Delbrück and Schrödinger. Delbrück later called it 'that silly paper'.[61] Nevertheless, the theoretical conclusion was abundantly confirmed by later work. The gene is a molecule and its mutation by X-rays is due to the chemical change they produce in that molecule. Schrödinger and Delbrück reached the correct conclusion that the stability of the genotype over hundreds of years can only be explained if the 'aperiodic crystal' which carries the genetic code is a stable molecule.

The existence of stable chemical bonds between atoms in molecules was explained only after the advent of quantum mechanics. Heitler and London, working in Schrödinger's department in Zürich, devised the first quantum-mechanical theory of the chemical bond in 1927. The development of ideas might thus be summarized as follows: discontinuity in physics → quanta of energy → quantum theory of chemical bond → stable molecules → genetic code → discontinuity of genetic mutations.

'Was it absolutely essential for the biological question to dig up the deepest roots and found the picture on quantum mechanics?', he asked. He said that although the conjecture that the gene is a molecule is now a commonplace, he wished to show that it is well founded in physics and that indeed no other explanation is even possible. The nature of the chemical bond is essentially the same in molecules and crystals; thus it is correct to call the gene molecule an aperiodic crystal, a long linear molecule in which the units are not identical but consist of a small number of moieties that can serve as the symbols in the code script, like the dots and dashes in the Morse code. As an example he supposes a code made up of twenty-five units consisting of five each of five different symbols. He calculates that there are 62×10^{12} possible combinations. If Schrödinger had been aware of the recent work on *Neurospora crassa* the bread mold, by George Beadle and Edward Tatum at Stanford University, and their one gene – one enzyme concept, he might have gone on to surmise that genes were coding for protein structures, but his knowledge of genetic research was not sufficiently up to date. Nevertheless, he has shown that 'the miniature code should be in one-to-one correspondence with a highly complicated and specified plan of development and should, somehow contain the means of putting it into operation.' How does it do it? He does not expect any detailed information on this question to come from physics in the near future, but he is confident that biochemistry guided by physiology and genetics will eventually provide an answer.

Schrödinger then made a statement that, when the lectures were published, inspired many young physicists to turn to research in biology. 'From Delbrück's general picture of the hereditary substance it emerges that living matter, while not eluding the "laws of physics" as established up to date, is likely to involve "other laws of physics" hitherto unknown, which, however, once they have been revealed, will form just as integral a part of this science as the former.'

Schrödinger believed that these other laws of physics would be related to the ability of a living organism to maintain itself in an ordered state and to evade the tendency towards equilibrium of all nonliving systems. 'It is by avoiding the rapid decay into the inert state of 'equilibrium', that an organism appears so enigmatic; so much so, that from the earliest times of human thought some special non-physical force (*vis viva*, or entelechy) was claimed to be operative in the organism, and in some quarters this is still claimed.' [Hermann Muller visited Copenhagen in 1933, and reported 'I was glad to meet the physicist Bohr, but I found that his ideas in Biology were hopelessly vitalistic.']

'How does the living organism avoid decay? The obvious answer is: By eating, drinking, breathing . . . What then is that precious something contained in [its] food which keeps [it] from death? . . . It can only keep . . . alive by continually drawing from its environment negative entropy.' Here Schrödinger fell into a partial error, which was later pointed out to him by Franz Simon and others. He seemed to identify the source of negative entropy in the orderliness of the molecules that are eaten: 'An organism maintains itself stationary at a fairly high level of orderliness (= fairly low level of entropy) [by] continually sucking orderliness from its environment.' This statement might be misleading since it emphasizes the inward flux and neglects the outward one. The organism is an open system; its entropy can decrease or remain constant provided the total entropy of organism plus environment always increases; thus it is not merely a matter of ingesting organized foodstuffs, rather the net overall flux of entropy from organism to environment must be positive.

The new kind of physical law that Schrödinger discovers in life is rather an anticlimax when he finally propounds it: the living organism is like a clock and life is like clockwork. 'Thus it would appear that the "new principle", the order-from-order principle, to which we have pointed with great solemnity as being the real clue to the understanding of life, is not at all new to physics . . . We seem to arrive at the ridiculous conclusion that the clue to the understanding of life is that it is based on a pure mechanism . . . The conclusion is not ridiculous and is, in my opinion, not entirely wrong, but it has to be taken with "a very big grain of salt".'

Actually true clock works are after all subject to the statistical law of entropy; they will gradually wind down owing to frictional effects. When does a physical system escape from the statistical laws and become truly dynamical and not thermodynamical? Quantum theory has a short answer to this question: 'at the absolute zero of temperature'. 'As zero temperature is approached the molecular disorder ceases to have any bearing on physical events.' This is one statement of the Third Law of Thermodynamics, discovered empirically by Walther Nernst. What temperature in any particular case is practically equivalent to absolute zero? For a clock, room temperature is already almost close enough. If you cool the clock, it will continue to run, but if you heat it, it will eventually melt. The cardinal point is that clocks are made of solids, which are kept in shape by the Heitler–London forces that make chemical bonds. The living organism depends on an aperiodic crystal, the chromosome fibre. One may call it a cog of the organic machine, but this cog 'is not of coarse human make, but is the finest masterpiece ever achieved along the lines of the Lord's quantum mechanics'.

On this pious note, Schrödinger ended the lectures at Trinity College, but when he prepared the written version for publication he added an epilogue, 'On Determinism and Free Will'.

Immediate experiences in themselves, however various and disparate they be, are logically incapable of contradicting each other. So let us see whether we cannot draw the correct, non-contradictory conclusion from the following two premises:
 (i) My body functions as a pure mechanism according to the Laws of Nature.
 (ii) Yet I know, by incontrovertible direct experience, that I am directing its motions, of which I forsee the effects . . .
  The only possible inference from these two facts is, I think, that I – . . . that is to say, every conscious mind that has ever said or felt 'I' – am the person, if any, who controls the 'motion of the atoms' according to the Laws of Nature.

. . . In Christian terminology to say: 'Hence I am God Almighty' sounds both blasphemous and lunatic. But please disregard these connotations for the moment and consider whether the above inference is not the closest a biologist can get to proving God and immortality in one stroke.

Schrödinger then relates this idea to the expression in the earliest Upanishads: Atman = Brahman, the personal self is identical with the all-comprehending universal self. Far from being blasphemous, this idea, the essence of Vedanta, is to him the grandest of all thoughts.

He continues in a more personal vein. In the West, such unity may at times be experienced by true lovers who, 'as they look into each other's eyes, become aware that their thoughts and their joy are *numerically* one – not merely similar or identical; but they, as a rule, are emotionally too busy to indulge in clear thinking, in which respect they very much resemble the mystic.' Erwin was not writing now on the basis of pure theory – he was falling in love again, at the beginning of what was to be the last major love affair of his life.

Much of what he says in the 'epilogue' repeats what he wrote in *Meine Weltansicht* eighteen years previously. He was a lifelong believer in Vedanta, and he never wavered in this belief. Here, however, he lashed out quite savagely at 'official western creeds', accusing them of 'gross superstition' in their belief in individual souls. Plurality of selves is merely an illusion produced by Maya, like a deception seen in a gallery of mirrors.

He concludes with a reference to his long exile.

You may come to a distant country, lose sight of all your friends, may all but forget them; you acquire new friends, you share life with them as intensely as you ever did with your old ones. Less and less important will become the fact that, while living your new life, you still recollect the old one . . . Yet there has been no intermediate break, no death. And even if a skilled hypnotist succeeded in blotting out entirely all your earlier reminiscences, you would still not find that he had killed *you*. In no case is there a loss of personal existence to deplore . . . Nor will there ever be.

What is Life? – the book

In view of his situation in Ireland, it is astonishing that Erwin at this time would publish a scornful denunciation of western religious beliefs, which was certain to dismay many of his colleagues, friends and neighbors. His tone would have been appropriate for a Voltaire or Diderot in eighteenth-century France, but it was strangely discordant with the spirit of tolerance that prevailed, at least in intellectual circles, in contemporary Dublin. He was then fifty-six years old, an age by which wisdom and prudence might have been expected to temper polemic with mildness and courtesy.

He made arrangements to publish *What is Life?*, including the controversial added chapter, with Cahill & Co, a respected Dublin publisher.[62] Schrödinger's negotiations with the company were handled by the Managing Director, John J. O'Leary. The typesetting, proofreading, and layout of final page proofs was supervised by the Publications Manager, Basil Clancy, who recalls that Erwin always arrived at the office of the company on his bicycle with his hair in windblown disarray. The book was in the final stages of publication, when someone raised a question with O'Leary about the controversial sections. Possibly this was Paddy Browne – he was helping Erwin to polish his English and when he came to the epilogue, he was furious, especially since his help was acknowledged in the preface. Thus not only did Schrödinger propose to publish derogatory remarks about the Church, he was even thanking a monsignor at Maynooth, the training center for Irish priests, for his cooperation.

The situation here recalls the letter that Schrödinger wrote in Graz at the time of the Anschluss, a letter which he refused to disown even when he knew that it would hurt the situation of his friends among the refugee scientists. 'What I have written, I have written', he said at that time in Oxford, and he probably used similar words in Dublin. The consequence was that O'Leary refused to publish the book and the type was dispersed. Erwin and Paddy Browne remained friends, but they never again achieved the warm and trusting relationship that had marked their earlier years at the Institute.

In October, Schrödinger sent a copy of the manuscript to a friend in London, the physical chemist Frederick Donnan, who recommended the Cambridge University Press as a suitable publisher. The book finally appeared late in 1944.

Hermann Muller gave *What is Life?* a favorable review in *The Journal of Heredity*.[63] He dealt gently with the fact that Schrödinger failed to mention his priority in the discovery of X-ray mutations: 'As a physicist, he was apt to attribute established conceptions to those authors in whose articles he happened to read about them.' Muller was appalled, however, by the mysticism of the final chapter: 'If the collaboration of the physicist in the attack on biological questions finally leads to his conclusion that "I am God Almighty", and that the ancient Hindus were on the right track after all, his help should become suspect.' Such a distrust of the mysticism of modern physics was to be echoed many times by biologists. With few exceptions, notably Charles Sherrington and his pupil John Eccles, they remained firmly attached to the world view of nineteenth-century physics, to mechanism, reductionism, and materialism. They experienced no metaphysical torments about ultimate reality, since they had found it in physics itself, an essentially classical, deterministic physics.

Schrödinger's book had an enormous influence, not only upon physicists who were persuaded that their methods might solve the problems of biology, but also among biologists who were encouraged to think more rigorously in terms of mathematically formulated and physically testable models. It was translated into seven languages and the total sales are estimated as over 100 000 copies. Erwin did not derive much profit from such sales; for example, a printing of 10 000 of the Japanese edition netted him £17.12s.8d.

In the Fall of 1946, the National Academy of Sciences sponsored a conference in Washington on 'Borderline Problems in Physics and Biology'. It was organized by George Gamow of George Washington University and Merle Tuve of the Carnegie Institution of Washington. Max Delbrück opened the discussion by saying that Schrödinger's book had caused them to come together for what would doubtless be the first of many meetings.

Twenty years later Gunther Stent, in a rather rueful essay called 'Waiting for the Paradox', wondered why *What is Life?* had such a great impact.[64] He thought that the ideas presented were neither particularly novel nor original, and could have had little or no influence on professional biologists. 'At their most charitable, they must have viewed the book with amused tolerance. And the title was a piece of colossal nerve.' He thought that the propaganda effect on physical scientists was great since their knowledge of biology was then limited to stale zoological and botanical lore. Many of these physical scientists were suffering from a general professional malaise in the immediate postwar period and they were eager for a new frontier. 'Schrödinger's book became a sort of *Uncle Tom's Cabin* of the revolution in biology,so that, when the dust was cleared, it left molecular biology as its legacy.' Stent underestimated the influence of the book on biologists, for in the early 1940s most of them were quite unaware of the nature of chemical bonds, the Second Law of Thermodynamics as applied to open systems, and the theory of reaction rates, subjects then to be found only in graduate courses in physical chemistry. Thus Schrödinger brought physics to the attention of biologists as well as biology to the attention of physicists.

What is Life? had a determining influence on the career of James Watson. He read the book in the spring of 1946, while he was an undergraduate at Chicago and undecided what to do and where to go for graduate work. 'From the moment I read Schrödinger's *What is Life?* I became polarized towards finding out the secret of the gene.' He applied for graduate work to Harvard, CalTech and Indiana. Harvard accepted him but offered no money, whereas Indiana offered $900. He recalls that 'Indiana had a great biology department because the eastern universities were antisemitic and people like Muller, Sonneborn and Luria could not get jobs in them . . . It was clear in those days that physicists were brighter than biologists.'[65]

Francis Crick thought that the book was 'peculiarly influential' and 'attracted people who might otherwise not have entered biology at all'. Maurice Wilkins reported that 'Schrödinger's book had a very positive effect on me and got me, for the first time, interested in biological problems'. Seymour Benzer was fascinated by Schrödinger's concept of the 'aperiodic crystal'. No doubt molecular biology would have developed without *What is Life?*, but it would have been at a slower pace, and without some of its brightest stars. There is no other instance in the history of science in which a short semipopular book catalyzed the future development of a great field of research. The influence of the book continues to be felt, and many people who know nothing else about Schrödinger will immediately associate his name with *What is Life?*.[66]

Sheila

On March 15, 1943, Sean O'Casey's new play *Red Roses for Me* opened at the Olympia Theatre, produced and directed by Shelagh Richards. This was the first new O'Casey play to be performed in Ireland since the Abbey refused *The Silver Tassie* in 1928. O'Casey was a Dublin labourer when he wrote his first play at the age of forty-two, and the contrast of the realism of his characters with the wild poetry of their language brought him instant fame in the western world. He was an outspoken enemy of capitalism and clericalism; the Irish recognized his genius but preferred to keep him at a safe distance from their church-dominated society. Richards was the only person in Dublin willing to take the risk of producing *Red Roses*. Sheila May was given the important role of Sheila Moorneen, the Catholic lass in love with Ayamonn Breydon, the young Protestant leader of the railway strikers who is shot down on the Vigil of Easter.

By this time Sheila had been married almost five years. There were no children and David at least was determined never to have any, thus she had decided to resume her career as an actress and to write a master's thesis in social work at Dublin University. The part in *Red Roses* was her best so far. As an active member of the Irish Labour Party, her political sympathies were all with O'Casey. The play received poor reviews in Dublin and closed after two weeks, but the London and New York productions had good runs.

Sheila May published a critique of *Red Roses* in the *Dublin Magazine*.[67] She liked the play but objected to the character she had played: 'Sheila is the latest addition to that dreary procession of Mary Boyles and Norah Clitheroes and Iris Ryans, weak and clinging, hugging security rather than her lover, with his dreams and aspirations, losing him to clammy death and realizing too late, etc. At least she wasn't "wronged" – we are spared that. Why is it that neither Sean O'Casey nor Paul Vincent Carroll can draw a decent, sensible, full-blooded young woman? There must be a few Pegeen Mikes still left in Ireland.'

Sheila was a political activist, who was not afraid to attack the social irresponsibility of the de Valera government. Since early 1942, Dublin had been stirred by a controversy over a plan by the City Council to provide a hot lunch for school children, instead of the customary bun and glass of milk. The wrath of the bishops was aroused and Sean Macentee, then Minister for Public Health and Local Government, declared that 'there are strong moral and social reasons for refraining from interfering with the normal family life of the people'. The Council replied that this statement was 'unctuous hypocrisy', but Macentee held the pursestrings and blocked the lunch program. The leaders of the Church, who watched every schoolgirl's knee with anxious eyes, had nothing to say about undernourishment and rickets.

As the controversy over the lunch program dragged on, Sheila attacked de Valera directly in an article in *The Bell*, Sean O'Faiolain's literary monthly.[68] Dev had said in a speech that 'there is nobody in this country who is not getting proper food', and later, 'every section of the community has had the careful regard of the government.' The Medical Officer of the Dublin Corporation [City Council], however, had found widespread malnutrition among children of the unemployed. Sheila wrote:

I have been puzzled for some time by these and other contradictory state-ments, which regularly appear in the daily press. They were all made by people of position and importance; at least one half of them must be either misinformed or deliberately lying. People are either starving or not starving; it should be comparatively simple to find out. I decided to investigate.

The first place I went to was an alley not far from Camden Street, one of the most ramshackle, insanitary slums it has ever been my misfortune to see . . . What do these people eat? I asked six of the housewives to make out a weekly budget for me. The menu seldom varies: milk and bread and butter for the children, tea and bread and butter for the adults . . . Not one of these people, children or adults, ever touches fresh fruit or vegetables. Nearly all the children in this alley suffer from one complaint or another – T.B., rickets, scabies and conjunctivitis are the most common. There is no place for them to play, except a passage in front of the houses about three feet wide, or the courtyard which houses the three lavatories for ninety people. The cold water tap, the sole vehicle for washing and drinking, lies plumb outside the doors. The last time I visited the place the sewer pipe under the front archway had

Sheila May Greene, from the painting by Barbara Robertson.

burst, and a harassed father was urging the children not to paddle in the sewage on the ground above it.

After visiting another slum even worse than the first one, Sheila pointed out that only one change had occurred since Ireland got its own government: the names of the alleys had been changed from English to Irish. She cursed the 'ignorance and inertia' which allowed such conditions. Her protests were disregarded, as de Valera did not dare to take any measures opposed by the bishops.

Meanwhile, for reasons that were not concerned with Erwin, Sheila and David began to have some serious quarrels. When angry, he could become quite violent, and she was rather afraid of him, but at the same time rather enjoyed the excitement. One time when they were visiting Dunquin, she hid herself in the bushes for fear he might throw her over the cliffs. Erwin and Sheila began to meet more often. They were not yet lovers, but his thoughts were turning to the composition of poetry as a change from tensor analysis.

In all his previous love affairs, except that with Hansi Bauer, Erwin had been the supreme male confident of his superiority over psychologically subservient women. 'Poor things,' he called them, 'they have provided for my life's happiness and their own distress. Such is life.' In Sheila May he found a woman who, except in science, was more nearly his equal. They were very different in their views of the world. In terms of Thomas Mann's categories, she was a politician and he was an aesthete.[69] She was committed to action, whereas he was devoted to works. As Schopenhauer had said, 'Actions are transitory while works remain. The most noble *action* still has only a temporary effect; the work of genius on the other hand lives and has beneficial and uplifting effect through all times.' Schrödinger was seldom inclined to risk noble action; in this restraint, as in so much else, he was a faithful disciple of the philosopher.

In the Springtime of 1944 Sheila and Erwin became lovers. He wrote in his journal: 'What Is Life?, I asked in 1943. In 1944, Sheila May told me. Glory be to God!' At first they met at her house, 10 Swan Place, and were involved in all the cryptic messages, stolen glances, anxious phone calls, and hasty assignations that mark the typical adulterous affair. On July 24, Erwin rented a little apartment in the center of the city, near to both her house and the Institute. Now they could enjoy their love in a habitation of their own, but still, as Erwin wrote, 'No word, no line, May reach the world, We quench our pride, And in ourselves abide.'

Most of the love poems that Erwin published in 1956 had been written to Sheila.[70]

Liebeslied

Niemand als du und ich
Wissen wie uns geschehn.
Keiner hat es gesehen
Wenn wir uns küssten inniglich.

Keiner, keiner weiss
dass uns der Himmel liebt
dass er uns alles gibt
was er zu geben weiss.

Und säh uns wer
er dacht es kaum
dass in weiten raum
sonst alles leer,

nur wir, nur wir
und unser glück
Nie nie zurück
als nur mit dir.*

* *Lovesong*

No one knows as you and I / How with us it came to be. / Not a one was there to see / When we kissed so fervently.

No one, no one knows / That heaven loves us so / That it gives us every thing / That how to give it knows.

And whoever might see us / Would scarcely think / That in wide space / Of all else void,

Are we alone, only we / And all our joy. / Never never do I return / Except with thee.

In his journal, Erwin recalled an old love to compare with the new. 'Sheila. Now let me tell you, of all I have told you, there is nothing quite real but Hansi Bauer. She was my great, great friend. She is the only one about whom I need not blush telling you of her. And she would approve. Knowing you for what you are, she would understand and thoroughly approve of my boundless love. She might not like it, most certainly not. But she has the same kind of clear, objective mind as you have.'

In this love, Erwin almost found the mystical union promised by the Hindu scriptures. In a passage from his journal addressed to Sheila: 'You might give me what you please in the future, and you probably will be pleased to give me quite a lot, but nothing will ever outweigh the moment when I saw the glory of God with all his angels, when your half-open lips, quivering as it were (or was that me?) . . . told me that you loved me.' This recalls Francesca and Paolo, 'questi, che mai da me non fia diviso, la boca mi bacio tutto tremante', or indeed Lancelot and Guinevere, and all the romantic lovers of the age of chivalry. Two months later, Erwin wrote that 'with all that – ostensibly the ultimate fulfilment of the supreme expression of love, yet there still remains a longing for something more, something more intense, a more complete union. Something like blood flowing into blood (she said) some more intimate unification.'

Unless he went to elaborate lengths to delude posterity, it is evident from his journals that Erwin was not, or not usually, a libertine. He speaks the authentic language of romantic love, seeking transcendence in the person of his beloved. In one respect, however, he is not a romantic: he does not idealize the person of the beloved, his highest praise is to consider her his equal. 'When you feel your own equal in the body of a beautiful woman, just as ready to forget the world for you as you for her – oh my good Lord – who can describe what happiness then. You can live it, now and again – you cannot speak of it.' Of course, he does speak of it, and almost always with religious imagery. Yet at this time he also wrote, 'By the way, I never realized that to be nonbelieving, to be an atheist, was a thing to be proud of. It went without saying as it were.' And in another place at about this same time: 'Our creed is indeed a queer creed. You others, Christians (and similar people), consider our ethics much inferior, indeed abominable. There is that little difference. We adhere to ours in practice, you don't.' Whatever problems they may have had in their love affair, the pangs of conscience were not among them.

Sheila was as much an unbeliever as Erwin, but in a less complex, more realistic way. She was never entirely convinced by his vedantic theology. She wrote:

The difficulty is to me that what we call consciousness, all the experiences that build us into the creatures that we are, each with his own individual memory – these experiences come always through the senses . . . I know Erwin through sight, touch, and sound. If I cannot see, or feel, or hear him, does he exist for me at all? If I had not got these senses, how could I be aware of his existence, or indeed of my own? Take away these things by death, and what is left? Is there anything else? Ordinary common sense tells me no. Of course I want to believe differently, but then we might as well embrace the Christian myth or any other. Is it possible that your intelligence being too great to accept the usual fairy stories is forced to invent a more rational explanation of its own. Please, darling, don't think I have the temerity to criticize a brain so far superior to mine. I am asking, asking questions all the time, because I want so much to know and I must know. I feel now as if I were standing before a half-open door, and I know that I can only go through it hand in hand with you. Perhaps I shall be left behind.

One thing I do know is that I love you in a way that will last as long as I last, and that when we have kissed and made love all we want to, this other stronger force will live on – at least for me . . . I have never felt like this before.

Now do be a good man and destroy this letter, or lock it away, but don't leave it in a telephone booth. I love you very much . . . Sheila.

When Erwin published a selection of his poems, the one he placed first was 'Geborgen' [Hidden] written in July 1944:[70]

Geborgen

Ueber eilenden silberbronnen
wo weisse vögel im blauen stehn
und weisse wolken in winde sonnen
und ferne berge in dunst vergehn,

im grünen halblicht von eichenzweigen
wo über das moos die sonnenflecken
kriechen und wachsen und höher steigen
und an den braunen stämmen lecken:

wo vergässest du sonnenflecken
wolken und berge leichter als hier
wenn hohe farne schützen und decken
vor neugier und neide – dich mit ihr.*

*Hidden

Beyond rushing silver streams / where white birds hover in the blue / and white clouds bask in the wind / and far hills fade in the haze

In a green halflight of oak bows / where over the moss sunbeams creep / and climbing higher stretch out / licking at the brown trunks:

Where do you forget sunbeams / clouds and hills more easily than here / When high ferns protect and shield / From envy and prying – you with her.

In the summer of 1944, Sheila began to join some of the Sunday bicycle excursions with Erwin and students and fellows from the Institute. For instance, they would pack a lunch and cycle down the Liffey valley to some old castle, returning through Phoenix Park in the long daylight evening.

Erwin sometimes would summarise his sayings to Sheila for his journal. Early in August:

Sheila, please do not think I am a complicated man for whom his so-called 'brain work' plays a prominent role in itself and is linked with his natural life by most involved, curved, branched channels. No. They are neighbours. Both are equally simple and straightforward, equally natural. I have never been able to understand, let alone explain, anything difficult or mysterious or involved. I hate it.

The simplest thing in the world is to go to bed. We have to do it every day. And everybody hates to do it alone. And you have given me more, more, a thousand times more than anybody else ever has: your clear, clean, simple, straightforward love. Not for one second has there ever been any petty play about it, nor will there ever be.

Yet, you are not the simple, plain, little fair girl. You're as finely and subtly organized – at least as finely and subtly as me. But love is to you the simple, great, elementary power as it is to me. The two of us accept with delight the verdict of the teasing, humourous God: ye love each other to the reach of your power. Nothing better do we wish than to love, love, love, where we are sure of unlimited return.

Arthur Schopenhauer saw 'falling in love' as the process whereby the noumenon, Kant's 'thing in itself', enters the world of phenomena.[71] He had more to say about sexual love than any previous philosopher, since 'the sexual relation in the world of mankind . . . is really the invisible central point of all action and conduct . . . The ultimate aim of all love affairs, whether played in sock or in buskin, is actually more important than all other aims in man's life; and therefore it is quite worthy of the profound seriousness with which everyone pursues it. What is decided by it is nothing less than the *composition of the next generation*.' The new individual who will arise from the love affair is like a new Platonic idea, and 'just as all the Ideas strive to enter into the phenomenal with the greatest vehemence, avidly seizing for this purpose the matter which the law of causality divides among them all, so does this particular Idea of a human individuality strive with the greatest eagerness and vehemence for its realization in the phenomenon. This eagerness and vehemence is identical with the passion for each other of the two future parents.'

It is not likely that Erwin was considering Schopenhauer's *The Metaphysics of Sexual Love* during his meetings with Sheila. He protested that he was not trying to regain his youth: 'I do not deplore anything I have lost, but quite often a thing I have kept. Considering my age, it is unlikely that anything more precious will come my way . . . Goethe, Böcklin, Legantini, the old Franks and Provencals, the Upanishads – and you.' And then 'Gib mir, gib mir Ihr Kind.'

A few days later, Sheila's pregnancy was confirmed. 'I am the happiest man in Dublin, probably in Ireland, probably in Europe. At least I cannot imagine greater happiness.'

As is inevitable in human affairs, the mystic union of sexual love did not endure for long – with Erwin it was never able to survive tidings of pregnancy, but Sheila had hoped for something more. In the middle of August, she wrote:

I looked into your eyes and found all life there, that spirit which you said was no more you or me, but us, one mind, one being, one loving. For two months that common soul existed. Today I saw the scales creep over your eyes and I watched it die. It slipped away without even a struggle. My mind went numb, there was nothing I could do, nor ever can do, to give us that again . . . You love me still, I know. I love my cat because he's soft and sweet and lets me play with him. You can love with tenderness, with devotion, you can love me all your life, but we are two now, not one. Why did you let it go? Wasn't it worth fighting for? My fault maybe in the beginning, for I am thoughtless and foolish, but surely age and learning bring some sort of wisdom to a man . . . Don't you know that anything could be achieved when you and I are together, that even though I am young and scatter-brained, when you open your mind to me, I can see with it and use it. But no, you talk of one lover placing the other on too high a pedestal. You talk of loving, but perhaps not approving. In a few brief sentences you kill the greatest thing I ever had, and then you ask me into bed, unless I would prefer to go out for a drink. Of course, in bed we're all right, we'll always do that well. But what is gone, will it ever, ever come again? I could stand deliberate cruelty from you and I wouldn't really mind, but the heartbreaking thing is that you didn't even know what you were doing. I can only pray that our child has been conceived a week ago, or two.[72]

In the middle of September, another cloud appeared on the horizon of personal relations. Anny was again in Belfast with the Ewalds and she reported a serious misunderstanding with Paul and Ella. It arose from a complex situation in which their daughter Linda, a medical student, was invited to stay with the Greenes, but Sheila went to meet Erwin in Cork leaving Linda without a chaperone. For the past five years Anny had lived as a loyal companion and housekeeper in the Schrödinger home, with only her visits to Belfast and her faithful dog Barney to provide any relief for her need for love and affection. It is doubtful that she and Paul were at any time lovers in a physical sense, but her love for him had become a psychological necessity. When the tension of this became too great for both of them, he tried to withdraw from the situation, and Anny was devastated by the threatened loss. When she returned to Dublin she was close to a breakdown. Erwin said that he could not help her in dealing with her sweetheart, but he was nevertheless affected by this emotional crisis.

The meeting of Erwin and Sheila in Cork was unsatisfactory for both of them. She was worried about the baby and about what to tell David, who would have little reason to think that it was his. Erwin was so concerned with the situation of Anny and Paul that he seemed uninterested in Sheila's problems, just at the time she most needed his love and attention. He told her that as soon as the three of them knew everything, the situation would become untenable. On October 7, however, he wrote to advise her that the most prudent thing would be to go to David and tell him that the affair was ended. In the middle of October, he wrote a curious poem, 'Die Jahre Gehen' [The Years Go By] in which the jingle of the rhyme is strangely discordant with the dismal theme; it reflected his confused state of mind at the time.

'Two months delight, and eight month's worries', he wrote in early November. But a few days later, he was more cheerful, writing a poem of the troubadours in both German and English, and then reflected the mixed emotions of his love with:

She muses and jubilates and weeps and laughs
In many a long winter night
And dreams in joy and many pains
Of my child under her heart.

In the middle of November, Sheila wrote to Erwin:

I think I told you that I believe that the love generated between two people is
far greater than the persons themselves. Sooner or later the strain of living up
beyond their powers will cause a fall and that is why so many marriages fail
and so many people are unhappy and disillusioned . . . Having no super-
natural God to worship I put you in his place . . . I want to finish my thesis . . .
I'd like a nice hole in the ground where I could live as a rabbit . . . P.S. I have a
lecture at 12 tomorrow so come at 10:30. I can send the maid out shopping.

As the old year came to an end, so did the brief but incandescent
love affair of Erwin and Sheila. David behaved very decently about the
whole thing. He knew that his wife wanted a child that he was unable
or unwilling to give her.

A few years later, Sheila and David separated, but David kept the
child and brought her up as his own. In 1959, he married the sculptress
Hilary Heron. She made a bronze nude of him in the style of Epstein's
Adam; it was kept in the foyer of their house and much admired by
visitors. He had a distinguished career as one of the foremost Celtic
scholars of his generation. Sheila became a political journalist, editing
the newspaper of the Irish Labour Party for a while. Late nights and
heavy drinking with male colleagues took their toll and she died in the
early 1970s.

Erwin's last poem to Sheila was 'Der Entäuschte' [The Disappointed
One]

Für kürze stunden glaubt ich zu entfliehn
Aus dem gefängnis dieser erde
zu dir zu dir mit fliehender gebärde,
du schienest mich hinanzuziehn.

Mit einmal wird es wieder nacht um mich
vielleicht die letzte nacht, der keine folgt.
Du hast die grosse hoffnung mir erdolcht.
Nur um die hoffnung wein ich bitterlich.

Du bist nicht schuld. Du gabst mir alles gern
worum ich bat. Gabst mir im überfluss.
Trägst mir ein kind. Doch schlimmerts den verdruss
dass du mir jetzt so tausend meilen fern.

Ist es nur alpdrück einer bangen nacht?
Erwachen wir und finden uns vereint?
Was soll der traum, – mir scheint
das ist die letzte nacht aus der man nie erwacht.*

* For a few short hours I thought to flee / From the prison of this earth / To you, to you with an imploring gesture / You seemed to draw me upwards.

All at once night falls again around me / Perhaps the last night, which no other follows. / You have stabbed to death my great hope / For that hope alone do I cry bitterly.

You are not to blame. You gladly gave me all / For which I asked. Gave me to overflowing. / Bore me a child. Yet that makes worse the discontent / That now you seem a thousand miles away.

Is it the nightmare of an anxious sleep? / Will we awake to find ourselves united? / What means the dream – , it seems to me / This is the final night from which one never wakes.

11 Postwar Dublin

After five years of World War II, the end of the conflict and the defeat of Germany and Japan could be foreshadowed, so that the Irish 'Emergency' was becoming less anxious, although shortages of supplies and restrictions on travel were scarcely improved. Schrödinger no longer confined his intellectual work to unified field theory and he began to take an interest in a variety of other problems, including some of a more philosophical nature. Physically, as he neared the age of sixty, he enjoyed good health, despite annual attacks of respiratory illness, aggravated by the dismal Irish climate and excessive smoking. When springtime came, Dublin broke out with great masses of yellow daffodils for sale at almost every corner and as he strolled down O'Connell Street Erwin exclaimed in joy at the sight of all the pretty girls who had emerged from winter coats to warm themselves in the occasional midday sunshine.

Statistics again

From January to March, 1944, Schrödinger had returned to one of his first loves in science in a course of lectures on Statistical Thermodynamics at D.I.A.S. They were published in a small hectographed edition and later (1946) by the Cambridge University Press.[1] In less than one hundred pages he covered the fundamentals of the subject with an insight and clarity that have never been equaled. The book is a distillation of his many years of creative work in the field, and one hears echoes of the passionate discussions of the twenties with Planck, Ehrenfest, and Einstein. He was pleased to learn that Max Born liked the little book on statistics, and wrote to him: 'For I have no higher aim than to work out the *beauty* of science. I put beauty before science. *Nitimur in vetitum* [Ovid: we strive for that which is forbidden]. We are always longing for our neighbour's housewife and for the perfection we are least likely to achieve.'[2]

During 1944 Schrödinger exchanged several letters with Lajos Janossy (1912–1978), a Hungarian physicist who had attended some of his courses in Berlin, and who was now working on cosmic rays with Patrick Blackett in Manchester.[3] He had become interested in a note by Janossy in *Nature* on the statistics of coincidences in Geiger counters – anything to do with statistical problems attracted his attention, and he would put aside even his general field theory to deal with it. Janossy was invited to lecture at a Dublin summer school on cosmic rays in 1945, and he stayed on to direct a new cosmic-ray section at the Institute. He was an intense young man with a shock of black hair that was continually falling into his eyes. A stepson of Gyorgy Lukacs, he was also himself an ardent Marxist, but he was a great admirer of Erwin and they got along well together.

The group of young researchers at the Institute now consisted of James McConnell, H.W. Peng from Peking, Jim Hamilton, and Sheila Power. Erwin was delighted with the progress of Peng; in a letter to Einstein he remarked that 'it is unbelievable how much the youngster has learned, how much he knows, and how quickly he understands everything.' Peng noted a basic reason why Schrödinger had so few students: while he was thinking about a problem and trying to obtain a clear picture in his mind, he did not like to communicate with anyone, and as soon as it became clear to him, the result was so polished that there was nothing left for a student to do. Once Schrödinger told him: 'Quantum mechanics was born in statistics and it will end in statistics.' This gnomic saying left open the question of whether the ending would be happy one.[4]

Peng recalls that Schrödinger once illustrated a lecture with a plaster model of a potential surface. Later, after a dinner at Clontarf, Peng was examining an artistic sculpture of Venus, and was amazed when Anny told him that Erwin had made it himself.

During 1945 Schrödinger had an extended feud with Registrar McGrianna over the supervision of the cleaning ladies. When his office was tidied, notes that he had thrown onto the floor sometimes disappeared into the trash bins. A special meeting of the governing board was called to consider this problem, and when it failed to affirm the absolute authority of the director, Schrödinger handed in his resignation from that office. He refused a plea to reconsider, and Heitler was appointed as director from the beginning of 1946.

Tests of unified theory

In 1944, Schrödinger published three papers on unitary field theory.[5] He enlisted Father McConnell in an attempt to find some experimental support for his predictions.[6] When electromagnetism is neglected, the equations of his 1943 paper yield Einstein's equations for the gravitational field, but when gravity is neglected, they yield not Maxwell electrodynamics but equations related to those of Proca and to the linear limits of the Born–Infeld equations:

$$\mathbf{H} = \text{curl } \mathbf{A}; \qquad \mathbf{E} = -\dot{\mathbf{A}} - \text{grad } V;$$
$$\text{curl } \mathbf{H} - \dot{\mathbf{E}} = -\mu^2 \mathbf{A}; \qquad \text{div } \mathbf{E} = -\mu^2 V$$

Thus the potential V itself acts as a source of negative charge and with a nonvanishing value of μ it might be possible to detect the resulting negative currents or the magnetic fields resulting from them around the earth or sun. McConnell made extensive analyses of the available geophysical and solar data, but the experimental uncertainties and effects due to variations in the magnetic fields were too large to permit any unequivocal demonstration of a nonvanishing μ factor. From the observed accuracy of Gauss's Law for the earth's magnetic field, a lower bound of $\mu^{-1} > 15\,000$ km could be inferred. This was the closest that Schrödinger ever came to any physical consequence of his unitary theory, and shortly thereafter he changed the geometrical basis of the theory by dropping the symmetry condition on the Γ.

Viewed in retrospect, the work on unified field theory does not appear to have led to major advances in physics, but one should not assess the value of scientific work merely in the light of future results. In Erwin's own view at the time, it was as exciting and important as any work he had ever done: 'I have found the unitary field equations. They are based only on primitive affine geometry, a way which Weyl opened and Eddington extended, whereupon Albert did the main job in 1923, but missed the goal by a hair's breadth. The result is fascinatingly beautiful. I could not sleep for a fortnight without dreaming of it.'[7]

Albert, however, after twenty-five years of frustrated pursuit of the same goal, was less sanguine: 'Concerning an affine solution of the electrical problem, I have become quite sceptical. This thing, alongside so many others, has been relegated to a pretty spot in the graveyard of my enthusiastic hopes – at the time I found it difficult to separate myself from it . . . One thing is certain. The Lord has not made it easy for us. As long as one is young, one does not notice this so much – luckily.'[8]

A new love

It may be that Erwin was convinced from previous experience that scientific creativity would be promoted and sustained by erotic excitement, but even if he did not consciously decide to act on this principle, the fact is that he had hardly ceased making love with Sheila when he embarked upon a new amorous adventure. In many ways it would be a re-enactment of his affair with Ithi – the distinguished man of the world and the naive virgin. In this case, Erwin dispensed with the pretensions of Arthurian romance – it was to be Valmont and Cecile Volanges rather than the Lancelot and Guinevere he had played with Sheila.

Erwin was introduced to Kate Nolan [a pseudonym] by Hilde March. According to one story, Hilde met Kate through her work with the Red Cross, where she was making up relief parcels to send to Austria through Sweden. They became good friends, even though Erwin later thought there was little psychological basis for a friendship between two such different women. Kate was twenty-six years old, a young woman of limited education who worked full time in a government office in the city. She was a tall, slim, brown-eyed blonde, not so pretty as Sheila and completely lacking her sophistication, but having all the freshness and charm of a strictly brought-up Irish maiden. Her father had died when she was young and she had been raised by her mother and her grandmother in accord with strict Catholic principles.

In the spring of 1945, Erwin took Kate to an *Irish Times* staff dance together with Hilde and some of her theatrical friends. Editor Smylie was very cordial so that the Myles affair must have been forgiven. Erwin, usually so unconcerned with fashion, thought that Kate's gown was far too old for her, and he found her conversation spiritless. He admits that it was not easy to convince himself that he was in love with Kate, but such a conviction was a necessary (and sufficient) condition for a seduction. Kate was at first rather shocked by the news that Erwin loved her, but he explained 'If you can't love me as a lover because I am a married man, just love me as a father.'

Erwin had never devoted much attention to the upbringing of his own daughter Ruth, who was now eleven years old and beginning to be more aware of the world of adults. At about this time she told him in the searching way that children have: 'Ervinilly, you'll never be a grandpa.' 'Why not?' 'Because you have no children.' 'Are you sure?' 'I'd laugh if you had.' 'Well, laugh then!' Not until many years later did he tell Ruth that he was her father. She was about seventeen, and they were on a beach in bathing suits, when he said 'Look at your feet – they are exactly like mine.'

On May 8 came V-E Day, an occasion for relief and celebration even in the Irish Republic – the 'Emergency' was over after almost six years. The event was marked by a serious riot at Trinity College.[9] About 130 students made their way to the roof of a building facing College Green and on a flagpole ran up the Union Jack, the Stars and Stripes, and the Hammer and Sickle, in that order. At sight of the hated British flag all

Erwin with his daughter Ruth (1946).

traffic in College Green came to a halt, and citizens rushed the college gates, assaulting students who tried to bar their way, and smashing everything that was breakable. Police arrived, pushed back the mob and locked the gates. Union Jacks were produced and burned ceremoniously. By now the mob amounted to several thousand and they broke every window facing the street. The rioting went on for several days, but it became difficult to distinguish political statement from drunken revelry.

   Thus it was in an atmosphere of general optimism that Erwin continued to woo Kate. In July he took her to Wicklow, a quaint seaside town south of Dublin, and wrote for her a poem about their outing:

> *On the shore*

> When on the shore of Wicklow
> after the bathe
> we from eachothers mouths
> kissed the cherries from eachothers mouths,
> tell me just what would it mean?
> Is it a pastime for any two?

> When on the shore of Wicklow
> I with my cheek
> 'gainst your bare arm leaning
> went asleep against your bare arm leaning,
> tell me, just what did it mean?
> Is it a pastime for any two?
>
> When on some shore of Wicklow
> I'll once embrace you
> with all strength once will embrace you,
> what is that going to mean:
> that you thenceforth
> from me shall not go, never go.

There must be something infinitely appealing for an inexperienced young woman to have poetry written especially for herself by a love-inspired versifier. Erwin did not hesitate to include even the deity in his appeals:

> I prayed to God, the ever unknown one,
> to him who blossoms from the earth in spring
> and bends the boughs in fall with blessed burden
> to take away from me all joy and mirth,
> leave me an outcast slave, a needy beggar,
> but then be pleased to give me your dear self,
> the only bliss to stay by me for ever . . .

While Kate was gradually falling under the spell of Erwin's charm, Sheila's pregnancy was progressing. Although they were no longer lovers, he often saw Sheila and tried to encourage her when she became despondent about David's lack of interest in the forthcoming addition to his family. She told him, 'If you cannot continue to love me as a lover because I am married, make up for it by loving me as a daughter.' They often met in the mornings on the banks of the Liffey and talked at length until he had to rush away to meet Kate in the city during her lunch hour.

For a time it seemed that Sheila's baby might be born on Ruth's birthday, May 30, but that day came and went, and at last a little girl was born on June 9. She was called Blathnaid Nicolette. Sheila was happy and adored the baby, and even David appeared to welcome it. As Sheila, never at a loss for words, remarked, 'He has always been longing for someone to like him better than anyone else – he has lost his wife and his dog, but now he has the baby.'

Erwin saw Shaw's *Arms and the Man* that night, and commented that 'All great things in the world are worked through love – not only children. It produces everything. Love is not an impediment to great effort but its carrier.' By that time he was convinced that he and Kate

On the banks of the Liffey.

'adored each other'. He brought her to Clontarf for tea and sat adoringly at her feet. Anny was most hospitable and Ruth very friendly, but Hilde was glum and withdrawn.

Erwin's siege of Kate Nolan lasted almost a year before she capitulated. This event was commemorated in another poem, written early in August, 1945.

Zittern

Ich traume von einer Sommernacht
wo du mir in Armen liegst
und ich endlich endlich hab vollbracht
wofür du dich an mich schmiegst.

das Zittern um deine Jungfrauschaft
um Schande Not und Tot
hat uns zur Niedrigkeit erschlaft
hat unsere Liebe verdroht.

wie gestern Nacht dein geliebter Leib
in meiner Umarmung gebebt
das war kein verliebter Zeitvertreib
war Qual – in Wonne erlebt.

Sperlingsgezogene Königin, gib
uns endlich Frieden und Licht
Gib Frieden oder den Todeslieb
es trägt sich langen nicht.

* *Trembling*

I dream of a summer night / when you lie in my arms / and at last at last I have accomplished / that for which you press close to me.

The trembling for your maidenhood / for shame and danger and death / has reduced us to lowliness / has subverted our love.

As last night your beloved body / quivered in my embrace / that was no amorous pastime / 'twas torment – experienced in delight.

O sparrow-circled Queen [Aphrodite] / Give us at last Peace and Light / Give Peace or the death in love / It does not endure for long.

This example of confessional poetry is a fairly explicit account of the difficult initiation of a frightened virgin. Erwin consoled her with a strange statement that recalls the Vienna of Freud and Schnitzler, 'You had no father, now you get one to lie in bed with.' From a conventional viewpoint, the eighteenth-century moral of Laclos might have been appropriate: 'Every woman who consents to receive into her society an immoral man ends by becoming his victim.'[10]

Whereas Kate was worried about becoming pregnant, it is likely that Erwin was consciously motivated by the desire to have a son and the statistical probability 0.5 thereof. The love affair continued until the pregnancy was evident. Kate confided to Lena Lean that 'she did not understand how it ever happened', a statement that would have been more plausible in Ireland than elsewhere.[11] Erwin was afraid that the scenario of the Ithi affair might now be followed too exactly, but to obtain an abortion in Ireland was virtually impossible and wartime restrictions still prevented travel abroad. Kate's mother was indignant and threatened Erwin with public denunciation, but fortunately Stephen Feric, acting as a skilful go-between, was able to arrange a settlement.

The baby was a six and one half pound girl, born June 3, 1946, in the Coolock Nursing Home and christened Linda Mary Therese [Russell], the last name being that of Erwin's English ancestors. The birth was so swift and easy that the nursing sister was quite astonished.

When baby Linda was a few months old, she was taken to the Schrödinger home, where Anny and Lena devoted themselves to her care. Hilde and Ruth meanwhile had returned to the March house in Innsbruck. In her usual unselfish way, Anny offered to give Erwin a divorce if he wished to marry Kate and settle down with their daughter, but neither mother nor father had any interest in such an arrangement. Erwin was delighted with the child and loved it dearly. He wrote about the new member of the family to Hermann Weyl, who replied, 'To Anny and Linda a heartfelt greeting and kiss! In this shattered or shattering world, the natural love of life and gaiety of such a small creature is indeed like a ray of sunshine and a last hope.'[12]

Hiroshima

On August 6, 1945, came news of the first use of the atom bomb – the destruction of Hiroshima. Schrödinger was horrified by the slaughter and the new techniques for mass murder introduced into human affairs. In a letter to Weyl, he wrote 'I find the development of things on this planet so desperate that I close my eyes and don't look around . . . The dangerous enemy is the *State*. The abcess of fascism has been cut out, but the idea lives on in its sworn enemies . . . I shudder at the thought that it can go so far with us, but it has already gone much too far. For example, the atom bomb.'[13]

He saw at once, however, an interesting statistical problem in the chain reactions responsible for nuclear explosions: what is the critical mass below which the occurrence of spontaneous explosion is effectively impossible? In November, he read a paper to the Academy on 'Probability Problems in Nuclear Chemistry'.[14] This paper appeared at a time when publication on such a subject was strictly prohibited in most of the world, and it must have aroused consider-able apprehension as to how many more nuclear secrets might be derived and independently published by maverick Irish physicists. At a colloquium at the Institute, he derived and solved the integral equation now known as the Peierls equation, which allows one to calculate the critical mass of a nuclear explosive.

Unified field theory

Early in 1946, Schrödinger began to attack the unified field theory with fresh enthusiasm. He was encouraged by the renewal of his corres-pondence with Einstein after a hiatus of more than two years. On January 22, Einstein sent him two unpublished papers: 'I am sending them to nobody else, because you are the only person known to me who is not wearing blinkers in regard to the fundamental questions in our science. The attempt depends on an idea that at first seems antiquated and unprofitable, the introduction of a non-symmetrical tensor as the only relevant field quantity . . . Pauli stuck out his tongue at me when I told him about it.'

Schrödinger replied in a long letter on February 19. He said that Einstein's work had made a deep impression and he had studied it intensively for three days. Einstein was astonished that he was able to go so thoroughly into 'my new hobbyhorse' in such a short time. Einstein then set forth clearly the somewhat different approach that he was taking compared with that of Schrödinger:

I can understand it if someone says: I demand from a reasonable theory that the mathematical structure (field variables and expressions derived from these) are restricted through the group only by the demands of economy. The Maxwell equations act somewhat in this way in special relativity theory (if one is prepared to admit a vector – instead of the simple scalar – as field variable). In general relativity theory the requirement is fulfilled (even more completely) by the pure gravitation theory. The new theory, however, does not provide this, because even the non-symmetric tensor is not the most simple structure that is covariant with respect to the group, but decomposes into the indepen-dently transforming parts g_{ik} and $\widehat{g}_{ik}$; the consequence of this is that one can obtain an unsurveyable number of systems of second-order equations.

Thus Einstein wished to impose conditions on the system at this point with which Schrödinger was not in agreement.

Even before receiving this letter, Schrödinger had written two more long letters to Einstein (February 27), to which the latter replied on March 6. The correspondence was now flying back and forth across the Atlantic. Einstein was seeking singularity-free solutions that would perhaps reveal the origin of particles in the structure of space-time. Schrödinger told him: 'You are after big game as I would say in English. You are on a lion hunt, while I am speaking of rabbits . . . If anyone is thinking of the hope that macroscopic electrodynamics, which at first does not appear, can finally be obtained as a con-sequence of a special singularity-free particle structure, then I

Schrödinger held in the hand of God contemplates unified field theory, painting by John Synge.

would not give much for this hope.' No attempt will be made to give the technical details of this spirited correspondence; the essence of all that is of permanent value has been distilled by Schrödinger into the final sections of his lovely little book, *Space-Time Structure* (1950).[15] On April 7, Einstein wrote that 'this correspondence gives me great joy, because you are my closest brother and your brain runs so similarly to mine.'

Schrödinger's paper on 'The General Affine Field Laws' was read before the Academy on April 8, 1946.[16] After his first papers, he decided to go to the general case of a completely asymmetric affine connection. He had then been delayed for two years by the difficulty of finding a solution to a set of linear algebraic equations involving the R_{kl} and derivatives of the affinities Γ_{kl}^i. Still leaving the Lagrangian density unspecified, and using arguments and procedures that seem rather arbitrary, he obtained general equations, which, he contended, present the space-time structure responsible for not only gravitational and electromagnetic fields, but also a classical limit of the meson field similar to the Proca equations. In his summary, he stated: 'In this paper I have at last completed a geometrical field theory on which I began work over two years ago. Whether it is physically right or wrong, that is to say whether it has a direct bearing on the physical fields it purports to account for geometrically, or not, I think it must be called *the* affine field theory, since it rests almost entirely on the assumption that the fundamental connection of space-time is purely affine. This includes, of course, that no metric is envisaged *a priori*.'

On May 1, he wrote to Einstein:

The whole thing is going through my head like a millwheel: to take the Γ alone as primitive variables or the gs and the Γ? To choose the Lagrange function immediately or to leave it undetermined as long as possible? . . . Is the cosmological term (on which Eddington placed so much stress) very important or not? Must a skew field contribute its Maxwell tensor to the gravitational waves or not?

One thing I do know is that my first work [P.R.I.A. 1943] was so imbecilic that it now is repellent to everyone, including you . . . This first work was no advance over 'Einstein 1923', but pretended to be.

Einstein answered on May 20: 'How well I understand your hesitating attitude! I must confess to you that inwardly I am not so certain as I have put it forward.' He is doubtful, however, that one can base a theory on the Γ alone, because it leads to too much arbitrariness. His theory leads to a set of equations for the $g_{ik,l}$, but they seem too complicated to yield useful solutions. 'We have squandered a lot of time on this thing, and the result looks like a gift from the devil's grandmother.' Einstein wonders if it may be necessary to introduce probabilities into field theory instead of trying to specify the 'real situation' of the particles.

Schrödinger replied virtually by return mail (June 13): 'I have not laughed so much for a long time as over the "gift of the devil's grandmother". For in the preceding sentences you described exactly the way of the cross that I also traveled in order to end up with something that is probably even more impossible than your result.'

On July 16, Einstein sent a long letter which set forth the progress that he and his assistant E.G. Straus had made and his present view of the status of the unified field theory. 'Your last letter interested me indescribably. And I was quite moved that you also had paid court to the devil's grandmother with such devotion . . . As long as one cannot express the Γ *in the simplest way* in terms of the $g_{ik,l}$, one has no hope of being able to solve exact problems. Thanks to the truly great skill and persistence of my assistant Straus, we have recently got this far.'

Einstein proceeded to give a point-by-point review of his approach in the hope of convincing Schrödinger of its validity. He actually brought quantum considerations into the field theory:

I come now to the point of our greatest conceptual difference. You reject the g_{ik} as an expression for the electromagnetic field, because the energy components essentially arise as products of the first derivatives of the g_{ik} so that no static forces in the sense of Maxwell's theory can arise and also no energy density in a transverse light wave. It is indeed correct that the energetics occurs quite differently here compared to Maxwell–Poynting. I consider, however, that in this respect the Maxwell theory is really false on account of quantum actualities. In light the energy exists in something like quasi-singularities. The wave field as such should therefore not be the site of energy. One must simply require this and be happy that in the new theory the transverse wave field is indeed present but as such transports no energy.

Such a viewpoint was distinctly different from that of Schrödinger, who was seeking a purely classical wave theory, in which the structure of space-time would yield gravitation, electromagnetism, and even a classical analog of the strong nuclear interactions, but he admitted that in the last case at least, no comparison with experiment was to be expected.

Contacts renewed

As soon as travel became possible after the war, the Dublin Institute invited scientists from overseas. From March 9 to 21, Pauli came from the Institute for Advanced Studies in Princeton to lecture on meson theory, nuclear forces, and elementary particles. His lectures and the resulting discussions brought much new information about unpublished advances in Europe and America. During May and June,

Leon Rosenfeld from Utrecht lectured on nuclear theory, and also gave a general lecture on complementarity. On June 14, Whittaker came from Edinburgh and gave a lecture on cosmology and its religious implications. Many of the Dublin dignitaries attended, including de Valera. Afterwards, Msgr Paddy Browne, in his mischievous way, remarked to Erwin: 'I should have liked to have a nap myself, but the Provost and the Taoiseach were both asleep and, after all, the existence of God was at stake.'

Anny and Erwin had been looking forward to renewing old friendships in their first postwar visit to England and the Continent. Towards the end of July they crossed the Irish Sea for the first time in over six years, going first to Cambridge, where Dirac had arranged a small meeting of physicists. Paul had married Margit Wigner, the sister of Eugene, who had been Erwin's assistant in Berlin and who later was appointed to the position at Princeton that Erwin had declined in 1935. Since they had stood together on the platform in Stockholm to share a prize for quantum mechanics, the research paths of Paul and Erwin had diverged considerably, but their love of mathematical theory provided a lingua franca that allowed spirited discussions.

Erwin was now anxious to renew his love affair with Hansi Bauer. They had corresponded faithfully, but had much to tell each other that could not be put into words. Leaving Anny in Cambridge, he met Hansi in London. They had only five days together, but were soon as ardently in love as ever before. Hansi now had two children, a boy of ten and a younger girl, so that she and Erwin could discuss the problems of parenthood.

The following year Mansi made an Easter visit to Dublin. She met Kate and was not impressed. Apparently she did not meet Sheila, who would have been more of a match for her. From both the intellectual and the purely sexual point of view, Hansi and Sheila were the two foremost women in Erwin's life – they were not only the most clever, they were also the best lovers.

At about this time, Erwin's Aunt Rhoda made her first visit to Dublin. She and Erwin spent many hours recalling their early lives in Vienna. She celebrated her eightieth birthday at Clontarf and Erwin composed and recited an epic poem for the occasion. Another link with the past was provided during visits to the Dunsink observatory, where Hermann Brück resided as Astronomer Royal; he had been a young researcher in Berlin while Schrödinger was there.

Erwin had lectures scheduled in Switzerland; August 1, 1946, he picked up Anny and proceeded to Zürich. Here he gave a nontechnical talk before the Swiss Scientific Research Society on 'Affine Field Theory and Meson'.[17] He began by harking back to 1918 when Weyl at the E.T.H. published his famous book *Raum-Zeit-Materie*, and he reviewed the efforts by Weyl, Einstein and Eddington to discover the laws of electromagnetism in the structure of space-time. He then sketched his affine field theory, and suggested once again that it might be competent to reveal something related to a meson field. It is rather difficult to understand Schrödinger's emphasis on this idea – with its insubstantial logical foundation, it seems to us now as almost based upon nothing more than a fascination by the word 'classical' and a quasimystical feeling that in some sense there must be a primeval physics antecedent to all twentieth-century theories.

It may thus have been appropriate that from Zürich he proceeded to Ascona to take part in an *Eranos Tagung*, one of a series of annual conferences founded by Carl Jung, the famous advocate of ancestral memories.[18] The topic of the meeting was 'The Spirit and Nature', and Erwin's lecture was on 'The Spirit of Science'. He began with an eloquent statement on the true relation of spirit [*Geist*] to science [*Naturwissenschaft*]. Its theme was that 'the spirit is to an eminent degree subject, and thus evades objective examination'. He quoted Śankara's commentary on the Vedanta-sutras: 'Subject and object – the "I" and the "not-I" – are in their essence opposed to each other like light and darkness.' The great Indian philosophers were concerned only with the 'Ego that consists of thought' and its relation to the Godhead. Schrödinger wished to identify this 'Ego that consists of thought' with 'spirit'. Science can examine only the object, the nonself, and 'the spirit, strictly speaking, can never be the object of scientific inquiry, because objective knowledge of the spirit is a contradiction in terms. Yet, on the other hand, all knowledge relates to the spirit, or, more properly, exists in it, and this is the sole reason for our interest in any field of knowledge whatsoever.'

Scientific study can never give us any understanding of the nature of spirit. It would be false to think that even the most exact knowledge of biophysics, physiology, or psychology can ever cause us to regard the spirit as something constrained or 'mechanically determined' by scientific principles. Schrödinger seems to imply that if the dichotomy of mind and matter is to be resolved, it is mind that will survive. It is a pity that the discussions among participants of the conference were not preserved, and thus we have no record of what the great psychologist Carl Jung thought of this contention that his science must be impotent when confronted by the mysteries of the spirit.

The Einstein debacle

Soon after his return to Dublin, Schrödinger resumed his letters to Einstein, keeping him informed of progress with the unified field theory by reports at about fortnightly intervals. Einstein's replies were less frequent. On Christmas Day, he wrote to apprise Einstein that he had carried through the variation procedure for a complex field. On New Year's Day, 1947, Einstein replied that, 'You are a clever rascal [ein raffinierter Gaunerl]'. Erwin was delighted with this epithet: 'No letter of nobility from emperor or king, neither the order of the garter nor the cardinal's red hat could do me greater honor than to be called a clever rascal by you in such circumstances.' In fact he was so pleased that he listed all the reasons why Einstein should move to Ireland. 'One can live in unbelievable peace and tranquility. This is due to the boundless lack of education and intellectual disinterest of the great majority of the population. That is naturally expressed unkindly. One can also say they are a natural, simple people, who do not go in for humbug.' Einstein replied that he could not leave Princeton where they had done so much for him, and besides, a move to Dublin would deprive him of the pleasure of their correspondence. These first days of the new year marked the highest point of Erwin's love and admiration for Einstein, but their collaboration and friendship was about to experience a disastrous misunderstanding.

Schrödinger was scheduled to read his paper on 'The Final Affine Laws' at the meeting of the Royal Irish Academy on January 27, 1947.[19] He had been working on this paper for at least six months, but judging from his letters to Einstein, it had not assumed its final form until the second week in January. The day before the meeting, he wrote a long letter to Einstein to outline what he believed was a major breakthrough:

Today I can report on a real advance. Maybe at first you will grumble frightfully, for you have explained just recently that and why you don't approve of my method. But very soon you will agree with me. A few years ago you pointed out to me (as Pauli never tired of explaining to both of us) that one can *not* take as a basis a non-irreducible form. He is not right about this. And you were right to start from one. For this, as you wrote me a year ago, he stuck out his tongue at you. He will have to draw it back in again.

In brief, the situation in this. If in the affine theory, which I have developed in general form in recent years, one takes the special, the only reasonable Lagrange function, namely the square root of the determinant of the Einstein tensor, then one obtains something fabulously good.

An outline of the paper to be presented the next day at the Academy was then given.

Schrödinger was so entranced by his new theory that he threw caution to the winds, abandoned any pretence of critical analysis, and even though the new theory was scarcely hatched, he presented it to the Academy and to the Irish press as an epoch-making advance.

The nearer one approaches truth, the simpler things become. I have the honour of laying before you today the keystone of the Affine Field Theory and thereby the solution of a 30 year old problem: the competent generalization of Einstein's great theory of 1915. The solution is

$$\delta \int \pounds d\tau = 0 \qquad \text{with} \qquad \pounds = \sqrt{-\det R_{ik}}$$

$$R_{ik} = - \frac{\partial \Gamma^{\varrho}_{ik}}{\partial x_{\sigma}} + \frac{\partial \Gamma^{\varrho}_{i\sigma}}{\partial x_k} + \Gamma^{\varrho}_{i\tau} \Gamma^{\tau}_{\varrho k} - \Gamma^{\varrho}_{\varrho \sigma} \Gamma^{\sigma}_{ik}$$

where Γ is a general affinity of 64 components. That is all. From these three lines my friends would reconstruct the theory, supposing the paper I am handing in got hopelessly lost, and I died on my way home.

I am prepared to see some of my mathematical friends shaking their heads. I should not wonder if some of them thought: The fool, if it is to be as simple as that, why, by God, did he not try that before? Gosh, that is the *simplest*, the most *suggestive* Lagrange function, which *anybody* would, *of course*, try out first.

He points out that something similar was tried by both Einstein and Eddington, but it did not work. 'Why should it work now? Is it the Irish climate? Well, yes, or perhaps the very favourable climate of 64 Merrion Square, where one has time to *think*.'

The reason it failed before was that they tried to use a symmetrical affinity with only 40 (not 64) components. Eddington wanted to proceed in steps, but one cannot always make progress in this way.

I will give you a good simile. A man wants to make a steed take a hurdle. He looks at it and says: 'Poor thing, it has four legs, it will be very difficult for him to control all four of them . . . I'll teach him in successive steps. I'll bind his hind legs together. He'll learn to jump with his fore legs alone. That'll be much simpler. Later on, perhaps, he'll learn it with all four. – This describes the situation perfectly. The poor thing, Γ^i_{kl}, got its hind legs bound together by the symmetry condition, $\Gamma^i_{kl} = \Gamma^i_{lk}$, taking away 24 of its 64 degrees of freedom. The effect was, it could not jump, and it was put away as good for nothing.

. . . I beg my younger fellows of the Institute to take this as a lesson: *never believe in scientific authority*. Even the greatest genius can be wrong.

Alas, it would soon be evident that this remark applied most appropriately to the words of the speaker himself.

I have chatted the time away without telling you much about the theory itself. I have written down the field equations here.

$$\frac{\partial R_{ik}}{\partial x_l} - R_{\sigma k} {}^*\Gamma^{\varrho}_{il} - R_{i\sigma} {}^*\Gamma^{\varrho}_{lk} = 0$$

where,

$$^*\Gamma^i_{kl} = \Gamma^i_{kl} + \tfrac{1}{3} \delta^i_k (\Gamma^{\varrho}_{l\sigma} - \Gamma^{\varrho}_{\sigma l})$$

They will explain a lot. One of the first things they will have to explain is the fact that a rotating mass like the earth is surrounded by a magnetic field. I have no doubt that they will . . . One point that comes out clearly is that the *true* electrodynamics of the world is that of Max Born 1934.

The complete paper was then submitted to the secretary, and Schrödinger was surrounded by reporters from the Dublin papers who had been alerted to the great scientific event.

The *Irish Press* next morning carried the story under the headline 'Einstein Theory of Relativity'.

Twenty persons heard and saw history being made in the world of physics yesterday as they sat in the lecture hall of the Royal Irish Academy, Dublin, and heard Dr Erwin Schrödinger . . . It was later he told me that 'the theory should express everything in Field Physics' . . . The Taoiseach was in the group of professors and students . . . Schrödinger disappeared through the snowy traffic on his veteran bicycle, before he could be questioned further, but later in his Clontarf home, chain smoking, he told me: 'It is practically impossible to reduce the theory to terms that the man in the street can understand. It opens up a new field in the realm of Field Physics. It is the type of thing we scientists should be doing instead of creating atomic bombs.' The reporter asked if he was quite confident in his solution, and Erwin replied, 'This is the generalization. Now the Einstein Theory becomes simply a special case . . . I believe I am right. I shall look an awful fool if I am wrong.'

The story of the great discovery reported in Dublin was picked up by the international wire services and flashed around the world. The science editor of the *New York Times*, William L. Laurence, immediately secured photostats of the original paper and of Schrödinger's remarks to the Academy and sent them to Einstein, Oppenheimer, Wigner and others, asking for their comments.

Einstein would also have had details of the new theory in Schrödinger's letter of January 26. He could hardly believe that such grandiose claims had been made for what was at best a small advance in the work that they had both been pursuing along parallel lines. Schrödinger had adopted the notation of Einstein and Straus, and his final equations were identical with theirs except for the addition of a small 'cosmological constant', which may or may not have been an improvement. It is true that he had derived the equations by a more direct method based upon the general affinity.

Einstein devoted great care to preparing a statement in response to Laurence's request. After explaining the preliminary, formal and purely mathematical state of general field theory, he continued:

Schrödinger's latest effort . . . can be judged only on the basis of its mathematical-formal qualities, but not from the point of view of 'truth' (i.e., agreement with the facts of experience). Even from this point of view I can see no special advantages over the theoretical possibilities known before, rather the opposite. As an incidental remark I want to stress the following. It seems undesirable to me to present such preliminary attempts to the public in any form. It is even worse when the impression is created that one is dealing with definite discoveries concerning physical reality. Such communiqués given in sensational terms give the lay public misleading ideas about the character of research. The reader gets the impression that every five minutes there is a revolution in science, somewhat like the coup d'état in some of the smaller unstable republics. In reality one has in theoretical science a process of development to which the best brains of successive generations add by untiring labor, and so slowly lead to a deeper conception of the laws of nature. Honest reporting should do justice to this character of scientific work.

Einstein's comment also went out over the wires of the international press, together with a quotation of Schrödinger's remark that 'if I am wrong I shall look an awful fool.'

On February 3, before he had seen Einstein's remarks, Erwin sent him a long letter in which he offered an explanation of the overblown newspaper accounts.

I am afraid that in recent days you will have been somewhat pestered to express your view of 'Schrödinger's new theory'. Please don't be angry with me that I have been the cause of this annoyance. I had to indulge in a little hot air in my present somewhat precarious situation. (Explanation of the latter to follow immediately.) I blew myself up quite a bit. I had a newspaper reporter there (from the *Irish Times* with which I am in good standing – it is an outspoken opposition sheet) and de Valera was presiding at the meeting of January 27. And this caused the commotion.

As to the precarious situation: In the first place, our basic salaries have not been increased since 1940. The foremost professors receive only $4800 a year, naturally with heavy taxes. The question of our pensions has not yet been legally settled after six years. For widows there are no pensions at all. (Naturally I cannot return to Austria, at least not yet, although I have been promised a pension there.)

He continued with an outline of the problems of the administration of the Institute, how he had to go to de Valera to settle a question about the duties of the cleaning ladies, and how grateful he was to Heitler for taking over the directorship. He added another far-fetched excuse in an attempt to placate Einstein: 'From the postscript to my little book *What is Life?* and many other things, you know that I don't believe much in "God the Father . . . and Jesus Christ his son" and so forth. Also I have recently published in the socialist party paper here, whose editor [Sheila May] is a good friend of mine, an article on Switzerland, quite unpolitical, and how its fabulously good educational system has raised it to such a high level, which in view of the well known miserable level of the system here, seemed pertinent.' Evidently this article and where it appeared aroused some criticism, and indeed some of the speeches in parliament and editorials in the newspapers were increasingly hostile to the Institute for Advanced Studies. J.M. Dillon, in the Dail, made a 'slashing attack' on the estimates for a School of Cosmic Physics, which he called 'a Machiavellian scheme . . . This thing is being done for the purpose of obtaining cheap and fraudulent publicity for a discredited administration . . . In a year when bread was rationed and milk was a rarity, the establishment of an Institute of Advanced Studies recalled Gulliver's travels to Laputa.'

Schrödinger was no politician – his attempt to exaggerate the success of his researches for political purposes was an absurd miscalculation, and such an excuse could hardly have improved his standing with Einstein. It was bad enough to make a mistake and announce a great discovery in an excess of enthusiasm, but to make the claim so as to bring pressure on the government to raise salaries would have been mere expediency. Actually other evidence indicates that Schrödinger really did believe that he had made a major breakthrough, so that the excuses given to Einstein were dreamed up later. Thus he had written to Hansi Bohm on 14 January 1947:

At my age I had completely abandoned all hope of ever again making a really big important contribution to science. It is a totally unhoped-for gift from God. One could become believing or superstitious [*glaublich oder aberglaublich*], e.g., could think that the Old Gentleman had ordered me specifically to go to Ireland in 1939, the only place in the *world* where a person like me would be able to live comfortably and without any direct obligations, free to follow all his fancies.[20]

He was even thinking of the possibility of receiving a second Nobel prize. In any case, the entire episode reveals a lapse in judgment, and when he actually read Einstein's comment, he was devastated.

Einstein wrote a curt letter on February 2, saying that he believed they had discussed the theory sufficiently, and the time had now come to obtain some rigorous solutions. If anything really was achieved, he would write again to let Schrödinger know. On the technical side, he said:

I was not correct in my objection to your Hamilton-function. But your theory does not really differ from mine, only in the presentation and in the 'cosmological term' which mine lacks. In mine, in the absence of electromagnetic forces (and matter), space is planar, in yours it is a deSitter space (due to the cosmological constant). Not your starting-point but your equations permit a transition to a vanishing cosmological constant, then the content of your theory becomes identical with mine.

Schrödinger wrote once more, on February 7, while sick in bed with the flu. He told Einstein that he was very upset by the newspaper report that 'the eager foxes' got from him. He continued in a wandering letter to describe a book he was reading about a crossing of the Atlantic in a small sailboat. Einstein did not reply and there was no further exchange between the two unified-field theorists for over three years.

Erwin collected the newspaper files and some of the relevant correspondence into a folder labeled *Die Einstein Schweinerei*, an appelation difficult to translate, but which denotes an awful mess.

Probability

In June, 1946, Schrödinger had written to Einstein: 'God knows I am no friend of the probability theory, I have hated it from the first moment when our dear friend Max Born gave it birth. For it could be seen how easy and simple it made everything, in principle, everything ironed out and the true problems concealed. Everybody must jump on this bandwagon [*Ausweg*]. And actually not a year passed before it became an official credo, and it still is.'[21]

At about this time, however, Schrödinger wrote two papers on 'The Foundations of the Theory of Probability'.[22] He had been reading a little book on this subject by Pius Servien, which suggested that an unsatisfactory basis for the concept of probability might cause serious problems for physics.[23] Servien thought that all attempts to define 'probability' beg the question. 'They seize the word and torture it, but can extract from it nothing but itself.' He thought that 'probability' may be a word like 'beauty' rather than a word like 'potential'. Schrödinger therefore tried to set forth an approach to the concept from certain first principles.

One must distinguish the calculus of probabilities, which is derived from set theory, a purely mathematical subject, from the non-mathematical connection which seeks to apply the calculus to physical events, our knowledge about them, or verbal propositions about one or the other.

Most physicists have adopted a simplistic frequency interpretation of probability: if the result A of an experiment has the probability p, it means that if the experiment is repeated many times, the fraction of outcomes that gives result A approximately equals p. Schrödinger said that by such a definition 'we cut ourselves off from ever applying rational probability considerations to a single event'. In particular, the probability calculated from the wavefunction ψ, as in the Born interpretation of wave mechanics, would then refer to a statistical ensemble of systems.

One way to circumvent this objection is the 'propensity' interpretation of probability, according to which probability is determined by the propensity of an individual system to give a certain outcome. For example, a perfect pair of dice have a probability of 1/6 of coming up with a 7. This p is an intrinsic property of the individual system. This interpretation was proposed by Charles Peirce in 1883,[24] and in 1957 Karl Popper adopted it with some modifications.[25] For indistinguishable particles like electrons, the frequency and the propensity descriptions should give identical results. Both these interpretations of probability are objective, in the sense that they refer to properties of actual systems and not to our knowledge or lack of knowledge.

Subjective interpretations of probability have been more popular with philosophers than with scientists, and it is surprising that Schrödinger adopted one: '. . . Given the sum total of our knowledge, the numerical probability p of an event is to be a real number by the indication of which we try . . . to set up a quantitative measure of the strength of our conjecture or anticipation, founded on the said knowledge, that the event comes true.'

With suitable rules and axioms, this definition can be fitted to the mathematical calculus of probabilities. It is not possible to draw too sharp a distinction between intrinsic propensity and our knowledge of the system as the latter becomes more complete. The significant point may be that with one or the other it may be possible to avoid the conclusion that Born probabilities can refer only to statistical ensembles or repeated experiments. The difficulty for any subjective definition of p is to show that it follows the product rule, without which one cannot employ the calculus of probabilities. $[p(A + B) = p(A)p^A(B)]$ where $p^A(B)$ means the probability of B if A is known to be true. At first Schrödinger states the product rule as an axiom, and then in a second paper gives a derivation that is not completely convincing. Any subjective definition of probability, combined with the Born interpretation, would come close to a view that he abhors: the ψ-function does not relate to a system but to our knowledge about a system.

The world view of science

After the Einstein debacle, Schrödinger began to devote more time to philosophy and less to unified field theory. He had reached his sixtieth birthday, a time to put aside storm and stress and to consider the final ends of life. He took up Spinoza, Einstein's favorite philosopher, and began to read again the ancient Greeks in a search for the origins of the dichotomy between scientific and religious thought, but he was not inclined to change his view that physics provides no answers to philosophical questions.

During the autumn of 1947, he finished a long essay, 'On the Characteristics of the World View of Science', which he sent to *Acta*

Physica Austriaca, as a sort of intellectual present to a 'liberated' Austria.[26] He began by quoting with approval the statement of John Burnet that 'science is thinking about the world in the Greek way', and he found that this way was based on two assumptions: (1) comprehensibility – the belief that natural events can be understood and explained; (2) objectivation – the exclusion of the perceiving subject from the world picture that is to be understood and its relegation to the role of an external observer. But what does it mean to *understand* nature? At this time, almost all professional philosophers of science were devoted to a 'received view' somewhat quaintly called 'logical positivism', which was a restatement of the ideas of Mach in more technical terms. Schrödinger had not spent much time reading comtemporary Machians, since he had read every word of their master.

He had never been more than a reluctant disciple of Mach, and in his essay he took a more definite stand against positivism. His first argument was that historical reconstructions provide understanding but they are certainly not merely economical summaries of sense perceptions. Thus positivism cannot provide a general epistemology, although it might still be valid for science. His second argument was based on the structure [*Gestalt*] of fairly complete theories. Many scientific theories, Darwin's theory of evolution for example, cannot be expressed in terms derived from sense perceptions, and thus positivism is not a valid philosophy for all of science. The failure of Mach and his followers to accept the atomic theory in physics and chemistry should in itself raise doubts about the validity of positivism in these fields. The wave-particle duality of quantum mechanics is a crucial case; the Machians say that no meaningful 'picture' is possible, but this conclusion is more like a defeat than a victory for their philosophy. Within a few years, from about 1950, philosophers of science would begin to abandon the 'received view' of logical positivism. Their arguments would be similar to those of Schrödinger, but expressed in more technical language, and with more cogent logical analysis.

To express the absence of mind from the world picture, he made a dramatic comparison with Poe's story, 'The Mask of the Red Death'. When a daring reveler tears the mask and cloak from the dread figure, he finds beneath them – *nothing*. The reason that our perceiving and thinking self is not to be found in the world picture is simply this: *it is the world picture*. It is identical with the whole and thus cannot be found in any part.

How then are we to consider the *apparent* multiplicity of selves? One answer would be a multiplicity of worlds, the 'horrible doctrine of Monads due to Leibniz, each a world to itself, without windows, in agreement through a pre-established harmony'. The opposite alternative is the unification of consciousness. Erwin concludes that the mystical experience of union with God leads regularly to this understanding, and he quotes the words of the Persian poet Aziz Nasafi: 'The world of spirits is a single spirit standing like a light in back of the world of bodies and shining through each individual that comes into existence as through a window. According to the kind and size of the window, more or less light penetrates into the world. But the light always remains the same.'

From his time as a young man in cold and hungry wartime Vienna, when he delved deep into the Upanishads, through his years of great scientific accomplishment, to his situation as a philosopher on the verge of old age, Schrödinger had never deviated from a religious understanding of our mysterious world. His position was captured in a remarkable painting by his Dublin colleague John Synge, who also claimed to be an atheist, which shows Erwin held in the hand of God as he ponders the equations of unified field theory. Yet Erwin could never be a believer in any dogmatic religion – his search for truth could never reach a conclusion. As Hansi once said, 'With him nothing was ever fulfilled. He was always hoping for the ultimate. High expectations by great men cannot be fulfilled.'

A year of changes

As Erwin reviewed the year 1947, he found it to have been one of the happiest in his life, in large measure owing to his love for his baby daughter and her response to him. He had not found any new Irish loves, but had passed some time in flirtation with Betty Dolan, a Dublin *ingénue* whom he suspected might have an interest in a casual *affaire*.

The new year was to bring both sorrows and joys. It began with some improvement in the financial situation of the Institute. The annual budget was increased to £9370, which included three senior professors at £1500 each.

Times were hard in the first years after the war, and there was a widespread feeling that a change from the Fianna Fail government was overdue. As part of the general political skirmishing, the Institute came in for its share of attacks. For example, the *Irish Independent* commented:

There are some people in this country whose minds can never rise from the level of practical finance into the realms of theoretical physics. For all we know there may be some who are so lowbrow and so selfishly concerned with the vulgar problems of dry bread and wet turf that they do not even know what theoretical physics and cosmic physics are . . . Of one thing we feel sure – that Irish is not a compulsory subject in these schools, for any such rule would be a

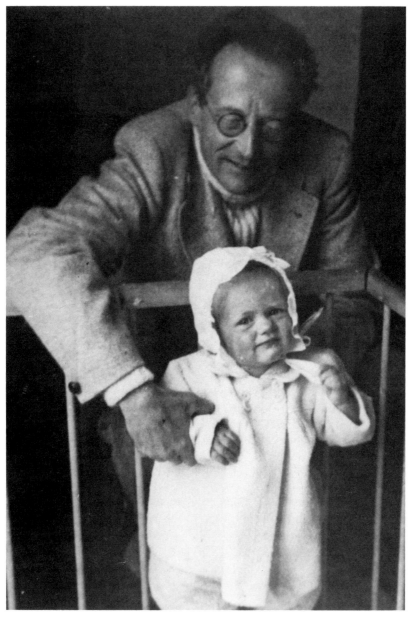

Erwin with daughter Linda.

discourtesy to the miniature League of Nations which labours for dark Rosaleen – at dark Rosaleen's expense of course – in this modest seat of learning.[27]

De Valera called a general election, and the result was that his party was outnumbered in the Dail by a coalition of smaller parties, and on February 18, by a vote of 75 to 70, John Costello was elected Taoiseach. For the first time in sixteen years, Dev took a place on the back benches as leader of the opposition.

On February 17, 1948, Erwin and Anny became citizens of Ireland, in a simple ceremony, which consisted of swearing an oath of loyalty to the Republic before a judge at the Four Courts. It was a fitting gesture of appreciation to the country that had offered him and his family a hospitable refuge during the worst of times. His reasons, however, were more practical, being based on the difficulties of securing travel documents from the Austrian government while it was under the control of the four allied powers. He did not wish to give up Austrian citizenship, for he thought his rights to a pension might depend upon it, but he found that it would be possible to have dual citizenship, so that on becoming Irish, he would remain no less Austrian.

During February, Schrödinger delivered four public lectures at University College, Dublin, on the subject 'Nature and the Greeks', and they were repeated May 24 to 31 as the Shearman Lectures at University College, London.[28] He has found that 'scepticism alone is a cheap and barren affair'. You may get along without philosophy while you are young and everything in life is comfortable and the future looks optimistic. But when you 'grow old and decrepit and begin to face death as a reality', your outlook must change. 'Personal misery, buried hopes, impending disaster, and distrust of the prudence and honesty of the worldly rulers, are apt to make men crave for even a vague hope . . . that the "world" or "life" of experience may be embedded in a context of higher . . . significance.' Schrödinger had the rare ability to say in less than a hundred pages more than most authors can express in five times that many. *Nature and the Greeks* is one of his finest works, probably better in the written version than as heard in lectures, since the density of ideas and the elegance of their expression deserve several readings.

His private view of organized religion had not changed appreciably. Thus he wrote to Benjamin Farrington, the historian of Greek science, in an analysis of the ideas of Lucretius: 'To make the extirpation of religion one's first and principal aim is not very wise anyhow. I believe that a man like Lucretius realizes only half the strength of the enemy he is up against. The hypocrisy of the governing classes who promote superstition as an efficient means of ruling over the dispossessed has a strong ally in the religious desire of many men and women, especially among the poor and unhappy. Lucretius does not seem to have much understanding of this side of human nature. He knows about serious religious longing as little as he knows about the shape of our planet.' Schrödinger believes that the situation today is little changed. 'After a short spell of comparative emancipation, science and learning are again becoming the handmaidens of political rulers, at least in the extremist countries. Great Britain is still something of an oasis.'[29]

After the London lectures, Erwin returned immediately to Dublin, for Anny was suffering from a severe depression. It was probably due mainly to age-related disturbances, but she had also been grieving over the departure of Ruth, whom she dearly loved. After she made a fairly serious attempt at suicide by slashing her wrists, she was admitted to St Patrick's Hospital on June 17. Her psychiatrist was Maurice Drury, the friend of Wittgenstein. He prescribed a short course of electroshock treatments, after which she improved considerably and was able to leave the hospital in a few weeks. For the next eight years, she suffered from recurrent attacks of depression, and when they became too bad, she would get into her car, drive to the hospital, and sign herself in for treatment. She never again attempted suicide and bore her illness with understatement and stoicism. She also had troubles with asthma, although, unlike Erwin, she did not smoke. Asthma is often a psychosomatic illness, but corticosteroids were prescribed for her on a regular basis, as a consequence of which she gained weight and took on the puffy facial appearance caused by these drugs.

Erwin also had a medical problem. For some time cataracts in both eyes had been growing more troublesome, and now he had difficulty in reading even with strong glasses. He consulted Louis Werner, Dublin's most distinguished ophthalmologist, who scheduled the cataract extraction on the right eye for June 29 in the Royal Victoria Eye and Ear Hospital. Dr Werner recalls that 'Erwin approached his operation not with fear but almost with pleasurable anticipation as if we were jointly carrying out an experiment in optics . . . He was a well behaved patient although I suspected that occasionally he disobeyed orders and lifted up his bandage to have a peep around.' The cataract in the left eye was extracted by Dr Werner a year later.

The operations were completely successful and Erwin's vision was restored to full normal with corrective lenses. Erwin wrote to thank Dr Werner: 'The world is full of beauty again, which I owe to your masterful knowledge and skill . . . You have made me a new man who enjoys life once again . . . More than ever before I realize now *what* a human being actually is: a pair of eyes – with something around them to keep them going, and also to turn their givings into "mind-stuff", in

Lucie Rie in her London studio workshop in the late 1940s.

an entirely miraculous and unfathomable manner. All the rest of our sensing, however relevant, is yet only ancillary to sight.'[30] Dr Werner refused to accept any fees for consultations or surgery, and Erwin later dedicated to him his book *Space-Time Structure*.

With Anny not well, the care of baby Linda became primarily the responsibility of Lena Lean, but they also had a maid, Molly, to help. Kate visited frequently, but her relations with Erwin were less than cordial. She often threatened to take the baby away, and might have done so if her mother had not been so fearful of ever being seen with it. The Schrödingers had decided to legally adopt the baby.

In the late summer of 1948, while Erwin and Anny were away in England, Kate met Lena one day walking the baby in its pram; she took it away and there was nothing Lena could do to stop her, since, after all, she was the mother. Erwin was with Hansi in Suffolk when this happened, and the first news came in a telegram brought by an excited messenger, 'Baby abducted from pram. What shall I do?' The messenger was amazed that the tidings were received so calmly. Kate took Baby Linda about as far as she could go from Ireland. Erwin never saw his daughter again and the little girl never heard her father mentioned until many years later, when Ruth established contact with her half-sister. Kate never married, perhaps in obedience to a vow she had made to herself before becoming Erwin's lover. The Schrödingers contributed to the support of Linda, £10 a quarter, later raised to £20 and then to £40. There was also a payment of £1000, which was to be invested for her future needs.

Early in August, Erwin and Hansi set out on a short tour of North Wales. They stayed for a while at the famous seacoast hotel at Portmeirion, and rented a nearby chalet called 'The Blue Dragon'. Bertrand Russell was at the hotel with his eleven-year-old son Conrad. One afternoon there was a garden party at the house of the Welsh philosopher Rupert Crawshay-Williams. Hansi recalls that she wore a remarkable zebra-striped dress. They started drinking tea from mugs and then filled them with whisky to encourage the flow of philosophical ideas. Erwin expounded the absolute identity and hence nonexistence of elementary particles, but most of the time they all listened to Russell who was in sparkling form at the age of seventy-six.

Meanwhile, Erwin had found a new friend in England, Lucie Rie. Lucie was born in Vienna, the daughter of a Jewish doctor. She attended the Vienna Gymnasium and the Kunstgewerbe Schule [School of Applied Arts], where she studied with the famous potter Michael Powolny. In 1938, at the age of thirty-six, already with a considerable reputation in her field, she escaped from Austria to London. She became one of the most influential studio potters of her generation, providing British artists a modern alternative to the folk art and imitative orientalism which had been the prevailing fashions.[31]

Hansi introduced Erwin to Lucie while he was in London for the Shearman lectures and they were immediately fascinated by each other – he saw her almost every day during that visit. When he returned to Dublin, Lucie wrote inviting him to stay with her during his next visit in August. Hansi was not much inclined to jealousy, but as Erwin began to spend more time with Lucie, his relations with Hansi naturally became less intimate. For Erwin, the visual sense was all important, and both these women were professional artists, who had much to teach him. With them he experienced a fusion of romantic and intellectual attractions, so that finally, in his sixties, he must have achieved a considerable degree of psychosexual maturity.

In August, 1948, he wrote a sonnet that must have been for Lucie or Hansi or both, although neither one remembers reading it at the time.

Herbst

Die Nächte langen und die Blätter streuen
Und wehn im Wind, und näher kommt der Tag
Der uns für wehe Wochen trennen mag.
Wie soll ich es ertragen, nun von neuem

Allein zu sein? Mich ohne Dich zu freuen
Wie lern ich? Dort wo helle Sonne lag
Im Finstern sich nicht fürchten, kann man sag
Es lernen und die Einsamkeit nicht scheuen?

Wir werden uns in bangen Wochen quälen
Freudlos am Werk, doch ohne Lust zu feiern.
Sprich, Liebste, wenn macht ärmer was uns Leich macht –
Und müssen wir für alle Zeit bloss stehlen
In seltner Nächte abgedampfter Schleiern
Das glühende Glück, das uns den Göttern gleich macht?*

* *Autumn*

The nights grow longer, and fallen leaves / Swirl and scatter in the wind, and that day nears / On which we may be parted for aching weeks. / How can I endure it – once again

To be alone? How can I learn / To rejoice without you? Where bright sun once lay / How not to fear the darkness? Can one claim / To learn this yet not recoil from solitude?

We shall toil through anxious weeks / Joyless at work, with no heart for celebration. / Speak dearest one, if to be mortal makes us poorer – / Must we then for all time merely steal / In rare nights of transitory revelation / The glowing joy that makes us like to gods.

Philosophical years

Schrödinger attended the Eighth Solvay Congress in Brussels from September 27 to October 2.[32] As in the first conferences after World War I, no scientists from German universities were invited, but many old friends were there, including Bohr, Pauli, Dirac, Kramers, and Lise Meitner. Although the principal topic was 'Elementary Particles', Patrick Blackett gave a lecture on the magnetic field of massive rotating bodies; surprisingly, Schrödinger did not comment on this problem, although his unified field theory had dealt with it explicitly. In fact, his only recorded comment during the meeting was on a paper by de Broglie presented by Mme Tonnelat, where he inquired as to the effect of a small rest mass of 10^{-44} g for the photon. In general his feeling at this meeting must have been that the work he had been doing for the past ten years was far from the concerns of almost all the other theoretical physicists.

After Schrödinger returned to Dublin from the Solvay Congress, his main interest moved even more definitely from mathematical physics to philosophical problems concerning the nature of the physical world and our knowledge of it. In 1949, Heitler accepted a professorship at the University of Zürich, where after some years he turned to studies on the relation of science to religion.

There were a number of new fellows in the School of Theoretical Physics. Walter Thirring was already one of the most promising of the new generation of theoreticians and would eventually succeed his father in the Vienna professorship; his brother Harald had been killed on the Russian front. Ernan McMullin was from Donegal, a recently ordained priest who had studied physics at Maynooth, but was now rather out of touch with the subject. He decided to take an M.A. degree, while supporting himself as a parish priest, and was assigned to work with Janossy. Neville Symonds was an Australian who had taken a Ph.D. in physics in London. Schrödinger advised him to look for problems in theoretical biology, and introduced him to Warren Weaver of the Rockefeller Foundation, who granted him a fellowship a few years later. S.N. Gupta and D. Basu were scholars from India, E. Corinaldesi from Rome, and M. Brdicka from Czechoslovakia – so it was indeed a league of nations.

Erwin was still an enthusiastic hiker. Almost every Saturday a group of ten or twelve would meet at the Institute, pile into cars, and drive about twenty miles into the Wicklow Mountains, where they would set forth on a strenuous walk. They would usually end up late in the afternoon at some small country pub for supper washed down with Guinness or ale. After supper, the talk almost always got around to religion. Erwin would present 'his variety of pantheism', Eva Wills would take the Presbyterian side, and McMullin was expected to defend all of Catholic orthodoxy. The talk went on till long after dark, when they would finally drive back down the mountains. McMullin recalls that at this time Schrödinger was 'far too interested in religion to be called an atheist' – even when he was consulted about physics, the conversation eventually came around to religion.[33]

Towards the end of 1949, Janossy took McMullin aside and told him that he was going to return to Hungary, but did not want it known since he had worked in England on defense projects and an attempt might be made to detain him. He was a dedicated communist and made no secret of his passionate antiamericanism. One day in 1950 he simply disappeared with his family, and only some weeks later did Dublin learn that he had arrived in Budapest to become director of a new cosmic-ray institute. The incident created a sensation, and

Schrödinger was somewhat embarrassed, especially since he was already suspected of socialist leanings because of his articles in the Labour Party paper.

On May 12, 1949, Schrödinger was elected a foreign member of the Royal Society of London. His election had been unduly delayed by complex politics. No German or Austrian citizens were elected from 1938 to 1948, but his actual citizenship was not the problem. Until 1948, the regulations were that foreign nationals living outside Her Majesty's Dominions were eligible for foreign membership, but from 1948, all foreign nationals were eligible. To have elected Schrödinger before 1948, would have been tacitly to admit that Ireland was not a British dominion. The Costello government passed the Republic of Ireland Act in 1948, severing the last links with the British Commonwealth, but not in fact with the Royal Society, since citizens of the Irish Republic are still eligible for election to fellowship.

Gedichte

In 1949, for the first time since 1923, Schrödinger published no scientific articles, but a small book of his poems appeared under the imprint of Helmut Küpper in Godesberg.[34] Most lovers of German poetry have deprecated these poems. As might be expected from someone who had no music in himself, there is often a lack of euphony and a crabbed [*gekrampft*] feeling in the language. Technically they are proficient, especially his mastery of the Petrarchan sonnet. Aside from the love poems, which became almost a standard accessory in his amorous adventures, the poetry expresses two principal themes, his love of nature and a metaphysical desperation often bordering on despair. There is a great sincerity in the poems of this latter kind, in which he abandons the romantic conventions and struggles with the meaning of his existence. Thus one can forgive his imperfections as a poet for the sake of the honesty of his self revelations.

As an epigraph for his *Poems*, Erwin chose a quotation from Goethe's *Egmont*, 'Himmelhoch jauchzend, zum Tode betrübt', ['Exultant unto heaven, dejected unto death']. It is from the song of the heroine Klärchen, which continues 'Glücklich allein, ist die Seele die liebt' [Alone is happy, the soul that loves]. Besides expressing the oscillation of the poems between extremes of joy and sadness, the quotation is interesting in that it calls attention to the character Klärchen, who gains the love of the noble prince Egmont despite her inferior social standing and the opposition of her determined mother, and who chooses to die rather than live without him.

From childhood, when he presented his mother with a little book of his poems, to old age, when he set forth his world view of Vedanta in poetic form, Erwin was often inspired to reveal his inmost doubts and longings in his poetry. As he told Max Born, appreciation of one of his poems gave him much more pleasure than any praise of his scientific papers. He used often to sit at his desk until dawn, making calculations, writing long letters to his friends, and confiding his thoughts to his journal or to the occasional poem. Here is a late Dublin example.

Juni

In jenen kurzen dämmerhellem nächten
wenn dich die sonne aus dem norden grüsst
wenn tag in tag und hell in helle fliesst,
zu einem langen tag sich fast verflechten –

was in dein leben sich vernichtend giesst
die ausgeburt von finstern höllenmächten
vergisst du fast und glaubst an die gerechten
und reinen worte draus erlösung spriesst.

Doch ach die tage neigen sich. Und still
verwintert all dein ungemessnes sehnen
und dunkelheit umschattet dir die seele.

Und all ihr heisses streben wird pasquill.
Nichts bleibt ihr als ein halbersticktes höhnen
ob sie sich auch bis zur vernichtung quäle.*

* *June*

In those short twilight nights / When the sun greets you in the north / And day flows into day and light to light / To near entwine themselves in one long day –

You forget almost the spawn of hellish powers / A flux of self destruction in your life / And once again believe the pure and lawful / Words from which salvation springs.

Yet now alas the days decline / And silent winter chills your boundless longing / As darkness casts its shadow on your soul.

Thus all its ardent striving turns to farce. / With nothing left it but half stifled scorn / Even as it struggles toward annihilation.

Never at rest

The biography of Isaac Newton by Sam Westfall was called *Never at Rest*, which suggests the ceaseless activity of the mind of the great scientist in its perpetual quest for deeper understanding.[35] A similar restless striving motivated Erwin Schrödinger in the later years of his life. In lectures, essays, radio talks, and perhaps most of all in his letters to other scientists, he would select various fundamental ideas and concepts and try to reach some final conclusions about them, or at least to find answers that satisfied both his reason and his intuition.

He wrote a long letter to Sommerfeld for the eightieth birthday of this revered old friend.

I do not believe that we can approach any understanding of the 'mind-matter' problem on a dualistic basis. There is no reason at all for dualism. Matter is a construction from sense impressions and representations [*Vorstellungen*] in a certain combination, and what one calls 'an individual mind' consists of course of the same elements. It is the same material, merely comprehended in a different way. In the first kind of comprehension the world is finally constructed, in the second, the self. The world is certainly no dream, no phantasm, and certainly the self is not merely a 'summation' of connected sensations in a peculiarly fluctuating sequence . . . When I reflect carefully, then two distinct things are not given to me at all. Both come to me uniformly from the same source, I find no inhomogeneity in the flow, no separation of spiritual and material – it is all from the same substance.[36]

In May, 1949, Erwin traveled to London to record in both English and German a talk for the B.B.C. in a series called 'Frontiers of Science'.[37] His subject was 'Free Will and Mechanical Causation Today', but he chose the catchier title 'Can Electrons Think?'. The B.B.C. informed him that 'they were not trying to popularize science', and paid 27½ guineas. The talk dealt with the 'failure of strict causality' at the level of electron physics, and he argued that such a failure has nothing to do with the problem of free will vs determinism. His analysis of this problem was, however, superficial; for example, he does not entirely avoid the logical fallacy of equivalence: because a change cannot be determined [ascertained] it therefore cannot be determined [caused].[38] In any case, since his monistic viewpoint should eliminate the mind-body problem, in which free will vs determinism would appear to be subsumed, it is not clear why he was still pursuing that question. The dualists could contend that 'knowing is mind and being is matter' and the interface between the two is 'consciousness', but Erwin had by now definitely discarded any such view.

The greatest interest in the B.B.C. talk is that a recording has been preserved, in which one can listen to Schrödinger. His voice had a tenor pitch without much timbre, and he spoke clearly, formally, and precisely. Only the very occasional exotic vowel revealed that English was not his native tongue: mootual, pōsitive, revales (reveals).

In August 1950, he recorded for the B.B.C. two talks of about twenty-one minutes each on 'The Future of Understanding'.[39] The fee was £40 and they were broadcast in the European service in English and German. They were not used on the Third Programme, as the manager commented, 'I can imagine them making one very sleepy.' They dealt with his ideas about the evolution of mankind in a society that encourages survival and reproduction of its disadvantaged members.

Max Born had sent Erwin a copy of his important book, *Natural Philosophy of Cause and Chance*,[40] and they had a lively correspondence about some aspects of it, in particular the difficulty of devising a molecular model for the irreversibility of thermodynamics in view of the reversibility of classical mechanics. Born drew clearly the distinction between causality and determinism, and he contended that quantum mechanics had affected the latter but not the former. He believed that 'irreversibility can be understood only by exempting a part of the system from causality. You must violate mechanics in order to obtain a result in obvious contradiction to it.' The violation consists in abandoning the concept that the position and velocity of every particle in the ensemble can in principle be exactly determined (even classically).

Schrödinger did not like this way of resolving the problem, and he published an alternative.[41] He first restated the problem:

The overwhelming majority of all those micro-states that would impress our crude senses as the same observable (=macro-) state do lead to identical, moreover to the actually observed *consequences*. That seems fine. What ails us is only, that we can equally well scan the *antecedents*. And they are – again for an overwhelming majority – entirely wrong, inasmuch as the antecedents are the mirror image in time of the aforesaid consequences; it would thus appear that the system has reached its momentary state by an 'anticipation' of its actual future history in reversed order . . . From this awkward situation, Born, if I understand him aright, proposes the following rescue. If we do not know the actual microstate of the system, we must – and that is where the philosophical loan from quantum mechanics comes in – refrain from drawing inferences from it. We must draw conclusions by averaging over all the microstates that may equally well be at the back of the observed macrostate. That looks splendid. For . . . it is agreed upon by everyone that we thus arrive at the correct prediction of the system's future behaviour. But it would seem to me a rather crude way of killing off the undesirable inference with respect to the opposite direction in time, if one is prohibited any conclusions about the past by saying that our observation of the system in that particular moment is in itself an irreversible process . . .

It is not certain that this restatement of Born's position does it full justice, but Schrödinger then proceeded to give his own solution to the problem. His idea was to reverse the arrow of time in a subsystem temporarily isolated from the rest of the world, if the entropy of that subsystem is decreasing. This subterfuge is even less satisfactory than that of Born, and it led to a lively argument with Karl Popper.[42] It now seems likely that only through cosmology or a new kind of dynamics may we learn to reconcile an irreversible world with reversible mechanical laws (classical or quantal).[43]

Schrödinger's most important work in 1950 was the publication of his 119-page book *Space-Time Structure* by the Cambridge University

Press.[15] Into these pages he condensed all that he had learned about the geometry of space-time and its affine and metric connections. The book has been the *vade-mecum* of two generations of students taking up the subject of general relativity. The book will remain one of the classics of science even though Schrödinger and Einstein, at least as viewed from our present-day perspective, ultimately failed in their efforts to derive electromagnetism from the structure of space-time.

A close call

After Heitler resigned, Schrödinger became Acting Director of the School and actually showed considerable administrative ability in drawing up procedures for its more efficient functioning, despite continuing problems with the old revolutionary McGrianna, who now seemed at times the worse for drink. The budget estimates for 1952–53 were £11 780, including three senior professors at £1800. Cornelius Lanczos, a Hungarian Jew and a great expert in relativity theory, came in 1952 as visiting professor and stayed as a senior professor and later director. He was a small man with an aureole of white hair which made him look like an elderly angel. Like Erwin he was interested in young ladies and in the theater, and he would often invite the waitress who had served his dinner to accompany him to the play. He sat in the front row at lectures and if he fell asleep, Erwin would stealthily approach and suddenly raise his voice, causing Cornelius to awake with a start. Erwin was quite fond of him but this feeling was not reciprocated.

In mid-September, 1952, Schrödinger wrote to John Synge from the Tirol:

I am feeling quite . . . *klein und hässlich* meaning guilty and humble for leaving you alone for so long. Anyway I have been having and am still having a glorious time, and I ought to return filled with so much energy and vigour that it may easily blow up the roof of Old Merrion Square the moment I enter the students' room in the top floor.

This e. & v. . . . is only partly due to the pleasure of adding . . . some contributions to the respective turmoils in two places (Alpbach: Wissenschaft in Imbruch = Science and Learning in Regroupment; Geneva: L'homme devant la science) – to the larger part it is the preceding, interposed and accompanying phenomena as mountains, lakes, good beer, wine, food, last but not least pleasant company, which have marked a rebirth in me, sorely needed for a long time . . .

Please do not judge too severely your most undutiful friend. Being escaped to the Continent is too lovely for ending this state before due necessity compels one to.[44]

The British Society for the History and Philosophy of Science was planning a meeting in London in early December to discuss the interpretation of quantum mechanics. They had invited Bohr, Popper, Born, Schrödinger and others, and everyone was looking forward to exciting confrontations between the upholders of the Copenhagen orthodoxy and their critics. Max Born wrote to Erwin that it would be like the famous 1895 Lübeck conference on atomism, and he did not know whether he would be cast as the matador or the bull, but still he was confident that their friendship would survive the battle. Schrödinger prepared his paper for the conference and sent a copy to Born.

He returned to Dublin the first week in October, feeling in better health and more full of energy than he had been for some years. Three weeks later he was stricken with appendicitis, which he tried to ignore during the first few days. When he was rushed to the hospital, his condition was critical, for the inflamed appendix had burst and peritonitis ensued. An emergency operation was performed at the Bon Secours Hospital by Dr Anthony Burton Clery. Fortunately, antibiotics were by now available and the infection was contained. Erwin began a gradual recovery, but it would be three months before he was fit for serious work. At the age of sixty-five, however, his body lacked resilience, and never again would he feel the almost youthful energy and high spirits of that summer in Austria. The annual attacks of bronchitis now became more wearing. Even when confined to bed with a racking cough, he would fill the room with smoke from his ever present pipe. Perhaps to counteract the resultant winter fatigue and illness, he began to drink more – he had always enjoyed a bottle of good wine, Gewürztraminer being a favorite, but now he fell into the Irish custom of using spirits as an antidote to bad weather.

On December 15, 1952, Willard Quine wrote to invite Schrödinger to come to Harvard for a semester as the William James lecturer in the department of philosophy. He stated that the lectures could be given in the fall term, from roughly October 1 to December 20. Recent holders of the lectureship had been John Dewey, Etienne Gilson, and Bertrand Russell, and Erwin was willing to overcome his aversion to the United States to take part in such a distinguished series. Thus he tentatively accepted the lectureship for the fall term of 1954, repeating the dates as stated by Quine. After he obtained leave of absence from the Institute for the stated period, Quine wrote again to say that he should come about September 25 and expect to stay until January 31 to correct examination papers from students in a seminar. [It is remarkable that a philosopher whose specialty was language and logic could not have written originally a clear statement of the duration of the lectureship.] Schrödinger was willing to undertake the reading of the papers, but after a rather peremptory letter from Quine, he wrote to cancel the Harvard visit, no doubt confirmed in his negative opinion of American customs.

Are there quantum jumps?

The paper that Schrödinger would have given at the London meeting was published later in the *British Journal for Philosophy of Science*.[45] It is a rather rambling account of his ideas on a variety of topics instead of a detailed analysis of the occurrence of discontinuities in physics.

The problem of quantum jumps can be seen most clearly in the theory of spectroscopic transitions. Suppose that a hydrogen atom absorbs a quantum of energy and undergoes a transition between two stationary states with eigenfunctions ψ_1 and ψ_2 having energy eigenvalues E_1 and E_2. Schrödinger attacks the idea that such a transition can occur *instantaneously*. [One current view was that the transition is indeed instantaneous but the time at which it occurs cannot be specified exactly owing to the uncertainty principle in the form $\Delta E \, \Delta t = h/4\pi$.] He was correct in this criticism, and, as Heisenberg pointed out, the atom passes through a series of superposition states, $\psi = c_1 \psi_1 + c_2 \psi_2$, where c_1 and c_2 are continuous functions of t. These superposition states are not eigenfunctions of the energy operator H, but may in certain cases be detectable as eigenfunctions of some other observable. According to Heisenberg, it is the act of observation that precipitates a jump into either state 1 or state 2.[46]

As Heisenberg commented, 'Schrödinger therefore rightly emphasizes that . . . such processes can be conceived of as being more continuous than in the usual picture, but such an interpretation cannot remove the element of discontinuity that is found everywhere in atomic physics: any scintillation screen or Geiger counter demonstrates this element at once. In the usual interpretation of quantum mechanics it is contained in the transition from the possible to the actual. Schrödinger himself makes no counterproposal as to how he intends to introduce this element of discontinuity, everywhere observable, in a different manner from the usual interpretation.'[47] It should be added that quantum mechanics itself, whatever its interpretation, does not account for the transition from the 'possible to the actual'.

The mass of the photon

Schrödinger's aversion to discontinuity may be seen also in the last substantial work that he did in Dublin, in collaboration with Ludvik

Bass, a postdoctoral scholar from Prague and Vienna.[48] In 'Must the Photon Mass be Zero?', they wrote: 'In a reasonable theory we cannot admit even hypothetically that a certain type of modification of Maxwell's equations, however small, would produce . . . grossly discontinuous changes.'

Maxwell's equations are singular in admitting only transverse electromagnetic waves *in vacuo*. In the nineteenth century this was interpreted as incompressibility of the luminiferous aether, in the twentieth as the vanishing of the photon rest-mass. An arbitrarily small but finite rest-mass would permit the existence of longitudinal waves and photons, and hence it would introduce a third independent direction of polarization. In thermal equilibrium, as in blackbody radiation, the three directions of polarization would have equal shares of momentum and energy by equipartition over the degrees of freedom. Radiation pressure and the constants in front of Stefan's and Planck's laws should then have values 3/2 times those actually found. Does this prove that the photon mass must in fact be precisely zero?

The natural generalization of Maxwell's equations for introducing a photon mass is Proca's set of vector equations, in which the four-potential becomes a real physical field coupled to the tensor of field strengths by a characteristic length μ^{-1}, which has the quantum significance of the Compton wavelength of a photon of rest-mass m (divided by 2π): $\mu^{-1} = h/2\pi mc$. Maxwell's equations are recovered as the mass goes to zero: the four-potential is then relegated to its familiar auxiliary role, with gauge invariance restored. Static solutions of the Proca set yield a shielded Coulomb potential $\exp(-\mu r)/r$, so that μ^{-1} could be bounded from the observed accuracy of Gauss' law for the earth's magnetic field. Twelve years earlier, Schrödinger and McConnell had investigated these data in an attempt to estimate μ^{-1} in order to compare it with a prediction from the unified field theory. An upper bound of $\mu^{-1} > 15\,000$ km was inferred, equivalent to $m < 10^{-47}$ g. More recent data, including some from the 1979 *Voyager* fly-past of Jupiter, have reduced this upper bound by several orders of magnitude, incidentally falsifying the 1943 prediction from Schrödinger's unified field theory.

Bass and Schrödinger considered black-body radiation in a cavity made of a perfect conductor. They found that even at the upper limit of $m = 10^{-47}$ g, a cavity with walls as thick as the earth could not confine longitudinal photons, and an appreciable fractional conversion of transverse to longitudinal photons would take longer than the age of the universe. Thus a longitudinal photon must lead a ghostly existence. The hypothetical jump in the properties of black-body radiation does not occur because the relaxation time to equipartition over the three directions of polarization goes to infinity as the photon mass goes to zero. Thus the photon mass is not *required* to be absolutely zero, and the question posed by Bass and Schrödinger continues to attract theoretical interest.[49]

12 Home to Vienna

Soon after the end of the war, Schrödinger's friends in Austria began to explore various avenues for his return to his native land, but it was to prove a long and tedious process. As early as February, 1946, he agreed to accept a professorship at Vienna, provided he could find a suitable successor for his position at D.I.A.S. In April, the Austrian authorities advised that he should apply for a restoration of his professorship at Graz, and then they would arrange a transfer to Vienna, but he was not willing to do this without a more definite understanding.

The main problem was the continuing occupation of Austria by the four allied powers. In January, 1947, General Kurasov told General Mark Clark: 'I expect there will be an Austrian treaty. We have done enough for Austria and it is time we got out.' The winter of 1947 was unusually harsh and there were food riots in Vienna recalling those of 1919. An attempted communist coup in May did not succeed, and in June, Austria was included in the Marshall plan, receiving more dollars per capita during its first year than any other country. In February, 1948, communists took over control of Czechoslovakia, and Stalin showed no willingness to conclude an Austrian peace treaty.[1]

Thus, although Schrödinger told his friends that he wished to return to Austria, he also made it clear that he had no intention of doing so while the threat of a Russian takeover remained. As the peace negotiations dragged on, they became increasingly acrimonious. In October, 1950, there was a second attempt at a communist putsch. Karl Renner, the socialist president, who in 1938 had advised his followers to vote for Hitler, protested to the British government: 'You sacrificed the Austro-Hungarian Empire in 1919, and in 1938 you abandoned Austria for Czechoslovakia, which you then abandoned also. These are bad precedents.'

The situation at the Vienna Physics Institute was rather surprising.[2] As the Russians approached in 1945, the Nazi boss of the Institute, Georg Stetter, fled to Egypt with his assistant Ortner, and Hans Thirring was recalled to his old position. By now he was more concerned with the peace movement than with physics. Przibram who had escaped to Belgium at the Anschluss returned to his old chair but soon retired. After a few years, Thirring brought Stetter back, and soon Ortner was also recalled. Thirring extended his pacifist philosophy even to the former Nazis, and believed that old enmities should be buried and a fresh start made, a feeling widely shared by Austrian politicians and intellectuals. Nevertheless, the old political divisions between the socialists and the clericals were soon revived, but with 90% of Austrian Jews either exiled or exterminated, the blacks were bereft of some of their favorite slogans. Both parties were striving to attract the former Nazis who had comprised the majority of the population, and even the reds were not anxious to encourage any return of Jewish exiles.

Sabbatical in Innsbruck

The University of Innsbruck, at the suggestion of Arthur March, invited Schrödinger to spend the winter term, through March 1951, as a visiting professor, and he gladly accepted. Innsbruck was in the French occupation zone and so presented no political hazards. He was also invited by the Education Ministry in Vienna to deliver a course of four lectures on 'General Relativity Theory and Field Theory'. At sixty-three, Erwin was still an enthusiastic walker, but he sometimes became short of breath, his lungs beginning to falter after years of abuse. Lack of practice had brought him back to a beginner's status on skis. He was happy to be again in the place he loved best in the world, the mountains of Tirol, and with March's assistant Ferdinand Cap and other young colleagues he made frequent excursions to the Stubaital and Seefeld, favorite spots in the environs of Innsbruck.[3] Not the least of the pleasures of Innsbruck was reunion with Hilde and with his daughter Ruth, now a young lady of sixteen.

While in Innsbruck, he was asked by the university authorities whether he would consider favorably a permanent appointment there, and his response was positive. After considerable correspondence with the University of Innsbruck and with the Education Ministry in Vienna, in which agreement appeared to have been reached concerning the conditions of an appointment, he was surprised to receive in April, 1953, a letter from the Cabinet Director, Wilhelm Klastersky, which stated that no appointment in Innsbruck would be possible because that university had no openings in physics.[4] He replied that this was for him 'a hard disillusionment whereby I must bury the hope I had cherished for many a year, that the mountain air of Tirol would provide a stimulus for the last decades of my work, which declines with advancing age.' He expressed his astonishment that the grounds cited for the refusal must already have existed at the time he was given the invitation. He then pointed out that one of the perquisites of his professorship at Graz had been a widow's pension, and he asked if this might be reinstated since there was no such provision in Dublin. It would be detrimental to the reputation of the Austrian Republic if his widow, like the widow of Heinrich Hertz, were to become an object of international charity.[5] The Vienna bureaucracy was adamant – no such pension was possible unless he was an Austrian professor. They pointed out that Vienna was the most important Austrian university and had first call on available funds for reconstruction. Would he accept a professorship there? Schrödinger's answer was: not under the present political conditions.

He had good reason for his fears. Vienna was surrounded by the Russian zone and although it was easy to enter, it was not so easy to leave. On one occasion Erwin and Anny were hauled off a bus at the Russian check-point at Semmering, and only the immediate protest of Hans Thirring to the Russian commandant in Vienna prevented a most unpleasant incident.[6]

Alpbach

After his sabbatical at Innsbruck, Erwin spent as much time in Austria as possible. He became an active participant in the summer conferences at Alpbach [*Hochschulwochen*], which brought together intellectuals from different fields for discussions of interdisciplinary topics. Arthur Köstler has given a malicious description of one of these conferences in his novel *The Call Girls*.[7] Alpbach is a beautiful Tirolean village at an altitude of about 1000 m, situated in a sunny valley between the mighty Zillertal and Kitzbühel Alps. It became for Erwin his favorite place on earth, and he was increasingly reluctant to leave for rainy Dublin at the end of the holidays. He wrote that 'sunshine is a lovely thing. You cannot really feel that you have enough of it before you feel that you have had too much of it.'

As Erwin neared his sixty-seventh birthday, his physical health was far from satisfactory. In retrospect one can see that his condition in Alpbach during the summer of 1954 was similar to that of his father in Millstatt 35 years earlier, a severe atherosclerosis complicated by chronic respiratory troubles. Lise Meitner came for a visit and they took her for a five-day tour of the neighboring mountains. At seventy-five she was tireless; Anny would drop back with her asthma after about 150 metres, while Erwin would plod slowly behind them both, hardly able to get up a small hill. They had planned to make a tour of Sicily in early September, but this had to be canceled.

When they returned to Alpbach, he became seriously ill, but refused to see a doctor. As Anny reported: 'he imbibed more freely of the good Austrian wine than was good for him and in between he drank ice cold beer and you know he never said no.' In early September she finally persuaded him to see Dr Ursinn, who diagnosed a severe emphesema, bronchitis, high blood pressure and a weakened heart. He advised him to get back home as quickly as possible and to go to bed with absolute rest for at least ten days with no smoking and no alcohol. They arrived back in Dublin on September 15, having stopped for two days in Innsbruck with the Marches and two days in Zürich with the Weyls.[8]

After the prescribed ten-day bed rest, Erwin improved remarkably. He could go to the Institute for a short time each day; Anny drove him there and picked him up, the bus was strictly forbidden. He drank only a mixture of milk and soda water and went to bed at 9 p.m., but he was unable to give up smoking completely. 'The future prognosis is not too good, but he doesn't think about that at the moment, only that he can no longer climb mountains, but he is happy that he can work again.'

Farewell to Dublin

Despite his miserable health, Erwin continued to work with a concentration and efficacy that would have done credit to a man in his prime. From 1954 to 1956 he published fourteen papers, several of them of a popular nature, and he completed another of his elegant short monographs, *Expanding Universes*.[9] He was delighted with a new group of young scholars at the Institute, especially with Bruno Bertotti from Pavia, with whom he formed an enduring friendship, for the lively young Italian shared his interests not only in relativity theory but also in philosophy and poetry.

Anny's visits to Lucan for electroshock therapy had now become routine. Her attacks of depression always seemed to occur early in December, aggravated perhaps by the Advent season and mourning for a distant Catholic childhood when the message of salvation was more easily decipherable. Her stays at the sanitorium, however, never coincided with the times when Erwin was seriously ill and required her attention. Typically she would return from Lucan just in time to put him to bed. While Anny's poor health was mainly psychosomatic in origin, Erwin's was due to heredity and tobacco. He regarded Christmas, and anything to do with the Christian religion, with outward tolerance but inward antipathy. He believed that everything beautiful in life and in art is a consequence of sex, and he could not abide the degradation of this vital force by a puritanical clergy.

Once Christmas was over, they both tended to revive and to look forward to springtime and then, in early summer, the Tirol. The day after New Year, 1955, Erwin was up at 7 o'clock, ready for a large breakfast of ham and eggs, and a full day's work. As he became more energetic, Anny suffered a relapse and spent another ten days in Lucan at the end of March. With the last of the Irish daffodils in April, it was Erwin's turn to be downhearted, but he had a logical reason since he now realized that his unified field theory was no longer tenable. In a paper in *Physical Review*, I. Callaway[10] had shown that his equations did not yield the electrical Lorentz force at the same approximation as the gravitational force, and in another paper the same year [1953] C.P. Johnson[11] gave an apparently convincing argument against any unification of gravity and electromagnetism by the Schrödinger–Einstein method. Einstein dealt with the latter objection but Schrödinger was still shaken by it. His correspondence with Einstein had resumed in 1950 and continued until the death of the latter in 1955, but both unified-field theorists were now quite discouraged.

Schrödinger was pleased, however, with progress in plans for his return to Vienna. After the death of Stalin in March, 1953, Khrushchev had decided to resolve the Austrian situation, and in February, 1955, he instructed Molotov to come to an agreement that would guarantee the neutrality of Austria after the Swiss model. He wrote to Raab, the Austrian Chancellor: 'Follow my example and turn communist, but if I really can't convince you, for God's sake stay as you are.' The final treaty was more favorable to the Austrians than earlier versions rejected by the Russians, and the war guilt clause was removed. On May 15, 1955, the treaty was ceremoniously signed in the Belvedere, and on November 5, the last foreign soldier left Austria. After almost ten years, the way was at last clear for Erwin Schrödinger to return to a professorship in his native land.

In early June, Erwin and Anny traveled to Italy, where he gave a paper on the photon mass at a meeting of the Italian Physical Society in Pisa. They remained a week in Tuscany, visiting Lucca, Sienna, Orvieto, and Firenze. Then the heat became so fearful that they fled to the Alps. In Innsbruck they picked up a new car, a Fiat 2000, and drove to the South Tirol. They had a lovely stay at Neustift, where in mid-July Erwin received official word of his appointment in Vienna. They created for him an *Ordinarius Extra-Status* from January 1, 1956, with full pension rights upon retirement. He accepted immediately with great joy.

After a difficult drive over the Bermina Pass, with both Erwin and the car complaining in their different ways, they arrived at Lake Garda where they enjoyed an idyllic two weeks. Hence back over the

Brenner to Alpbach. This was almost the last time that they could both enjoy touring through their favorite landscapes, all their financial worries over, and their health sufficiently good to breathe freely in the good mountain air.

Soon after their return to Dublin, Anny went to Lucan for ten days. Ruth came to stay for four months to help them with the move to Vienna. Erwin was busy with his final arrangements at the Institute; after Christmas he would have no more formal duties there. He was scheduled to go to Cambridge in January to deliver the Tarner lectures at Trinity College but an attack of phlebitis made it necessary to postpone the lectures till October. He had barely recovered when Anny had to enter Lucan for further treatment. Ruth had meanwhile gone back to Innsbruck. As soon as Anny returned to Clontarf, Erwin had an unusually severe bronchitis. He was so despairing and depressed that on Sunday night, February 11, he took four different kinds of sleeping pills washed down with whisky. On Monday morning, Anny had difficulty in rousing him and summoned Dr Dempsey, who said: 'I've never seen him like that. He nearly killed himself.' Dempsey wanted to call in one of the psychiatrists from Lucan for a consultation but Erwin vehemently refused. All that day he was lethargic and exhausted, with a rapid and irregular pulse, and that evening Dempsey gave him a heavy injection so that he slept about twelve hours. The next morning he was normal. He was allowed up for the first time on Saturday and Anny reported that he was pathetically grateful to her for taking care of him.[12]

Home again

By the time they boarded the cross-channel steamer, Erwin was fairly exhausted by the farewell parties in Dublin, lunch with de Valera, lunch with the President, and a large party with colleagues from the Institute. They finally set out on March 23, and Dev came down to the pier to see them off. Some of Erwin's Dublin friends thought that after seventeen years Ireland should have remained his homeland for the rest of his life, but his heart and soul were still in Austria, and even had this not been so, practical considerations would have required his return, since he could not have lived on his Irish pension of £50 a month, barely enough to buy a few books from Blackwell's and to pay the allowance for his youngest daughter.

They stayed two days in London with an old friend Elisabeth Ullmann and then traveled to Innsbruck, where they arrived on March 28. Arthur March was seriously ill and Hilde was upset and distraught, but Ruth was managing to cope with the situation. The drive to Vienna was splendid, alpine spring flowers in Kitzbühel, where they stayed a day with the Thirrings, and then Easter Sunday in Salzburg, and on Monday an overnight stop at Dürnstein, on the Danube not far from Vienna.

They reached Vienna at 4 o'clock and at 5 the avalanche of welcomes began. They were staying at the Pension Atlanta very near the Physics Institute. Their rooms were filled with flowers, including a large purple azalea with a card reading: 'The Academic Senate warmly sends greetings to the great son of Austria upon his return home.' There was a mountain of letters and telegrams of welcome and congratulation. They were besieged by reporters and photographers, and stories and interviews with Austria's greatest scientist were on the front pages of all the newspapers. Anny reported to Elisabeth that 'Erwin behaved like a lamb – fabulously. I never in my wildest dreams thought that so many people would be so happy at his return.'

At noon on April 13, Schrödinger delivered his inaugural address in the Auditorium Maximum of the University. The platform was adorned by civic and academic dignitaries and the great hall including the gallery was overflowing. Among the invited guests were Dr Mayer, director of his old school, the *Akademisches Gymnasium*, Erwin's cousin Dora Halpern, with whom he used to play and sometimes fight as a little boy, and his early great love, Felicie Bianchi. After a few words of greeting from Dean Laska of the philosophical faculty, Erwin approached the lectern, and as he picked up the spray of narcissus that was lying there, he was greeted by thunderous applause, which continued for several minutes. It was an historic occasion and his pleasure and emotion were evident, but they did not detract from his urbane style as he delivered one of his most polished lectures on 'The Crisis of the Atomic Concept'. He said that the materialistic picture of the world has been severely shaken and is today more uncertain than ever before, while modern physics is undergoing a revolution, the duration of which cannot be foreseen. Experiments have shown that waves are real and even fundamental entities, which can be directly and even easily observed, which can be said for only a few things in our world. Many physicists today are prepared to consider matter as simultaneously a wave and a particle phenomenon, but one must be wary of a physics that is too philosophical, since it may lead to specious solutions and prevent a deeper understanding. At the conclusion of his inaugural address, he was given a standing ovation.[13]

On May 5, Schrödinger was awarded the Prize of the City of Vienna at a ceremony in the Festival Room of the Rathaus, which was followed by an elaborate lunch hosted by the Bürgermeister. Erwin enjoyed such occasions although he was supposed to be on a strict diet. Other festivities occurred in June when Lise Meitner came to

Inaugural lecture in Vienna, April, 1956.

Vienna for a celebration of the fiftieth anniversary of her doctorate from the University.

It was not easy to find a suitable apartment in Vienna but the Schrödingers finally located one at Pasteurgasse 4, about a kilometre from the Physics Institute. There were five rooms on the third floor, and the building had a lift. It cost £2100 to secure and renovate the apartment, and the monthly maintenance charge was £3.10s.0d. They were glad to be leaving the Atlanta, for Anny thought the cost of £2.10s.0d. a day half pension for the two of them was 'sinfully expensive', and Erwin had once been imprisoned for three days on the third floor when the lift failed. They planned to move to their new dwelling after spending the summer in Alpbach.

Vienna had never seemed so beautiful as during those first months after their return home; the Burgtheater, the Opera, the Ringstrasse, Kärntnerstrasse, Graben, the splendid monuments and the elegant shops, all bathed in the warmth and sunshine of early summer. They went often to the theater; once the Minister of Education sent his tickets in the first row of the parterre circle for a play by the old favorite Grillparzer, and as they drove there in their pretty car, Erwin exclaimed 'Na – so fein war ich in meinem ganzen Leben noch nie!'

But intimations of mortality were never far distant. As they were walking in the courtyard of the university, looking at the sculptures of its famous professors from bygone times, Erwin suddenly said, 'Let's get away from here – I feel as if I were walking in my future cemetery.'

He lectured twice a week on General Relativity and Expanding Universes, subjects that had received little attention in Vienna for many years. His teaching was more a discussion than a formal lecture course; he invited the rather small number of auditors to interrupt him at any point that was not perfectly clear, and he would then elucidate with careful detail. There was also a weekly seminar but he thought that 'this was more like a higher kindergarten, not like the Dublin Seminar.'[14]

Mind and matter

As soon as the summer term at the university ended, Erwin and Anny moved to Alpbach, and after a few weeks of mountain air, they were both feeling remarkably well. In early September, however, there was a recurrence of his heart weakness, and he was forced to cancel his planned visit to Cambridge in October to deliver the Tarner lectures. Fortunately, John Wisdom, the professor of philosophy, agreed to read the lectures from Erwin's manuscript.[15] The book *Mind and Matter*[16] was published in 1958, being dedicated to 'my famous and beloved friend Hans Hoff in deep devotion'. [cf. p. 466] This book is a fitting sequel to *What is Life?*, but it has not yet had such a far-reaching influence, perhaps because psychobiology is still at such an early stage of development, compared, for example, with biochemistry.

The first lecture was on 'The Physical Basis of Consciousness'. 'The world is a construct of our sensations, perceptions, memories. It is convenient to regard it as existing objectively on its own. But it certainly does not become manifest by its mere existence.' He is thus led to ask whether one can believe that the existence of the brains of higher animals is a 'necessary condition for the world to flash up to itself in the light of consciousness. Would it otherwise have remained a play before empty benches, not existing for anybody, thus quite properly speaking not existing? This would seem to me the bankruptcy of a world picture.'

He seeks a solution to this problem based upon the concept that consciousness is associated with *learning* by a living substance. The *knowing how* of life is consciousness, whether it occurs in an animal nervous sytem, a plant or an amoeba. Living processes in general are associated with consciousness whenever they represent the emergence of something new. This daring view recalls the hylozoism of the Ionians, the pantheism of Spinoza, and the plant-souls of

Fechner, but Schrödinger suggests further that such consciousness is intimately associated with organic evolution.

A similar relation of consciousness to evolution can be found in the realm of ethics. Man is still undergoing an evolution towards a social animal, and the ethical law 'be unselfish' reflects this process. Every individual life is only an insignificant chisel stroke in the never finished sculpture of our species. 'The resistance of our primitive will is the psychical correlate of the resistance of the existing shape to the transforming chisel.' He does not contend, however, that the ethical law, the Kantian imperative, is based upon evolution, on the contrary, it is simply *there* and may provide a direction for the future evolution of mankind.

The second lecture was essentially the same as his 1950 B.B.C. talk on 'The Future of Understanding'. The main point is that behavior can influence the process of natural selection, while education and environment can influence behavior, so that a darwinian mechanism of evolution can result in a sort of apparent lamarckism.

The third lecture was 'The Principle of Objectivation', in which he restated in the clearest possible terms a basic principle that has often been misunderstood by biologists and psychologists. The false conclusion that I am part of a real world that itself arises from my consciousness lets loose a 'pandemonium of disastrous logical consequences'. One of these is the 'fruitless quest for the place where mind acts on matter or vice-versa'. The material world can be constructed only at the price of excluding the self, the mind, from it. Spinoza, 'the greatest philosopher of the seventeenth century', was absolutely clear on this point: 'The body cannot determine the mind to think, neither can the mind determine the body to motion or rest or any state different from these (if such there be).' [*Ethics*, III]. Mind has constructed the outside world of the natural philosopher out of its own substance, but it could not cope with this gigantic task except by paying the high price of withdrawing itself from its own creation.

Schrödinger quotes with approval a statement made by Carl Jung at the 1946 Eranos Meeting: 'All science is a function of the soul, in which all knowledge is rooted. The soul is the greatest of all cosmic wonders, it is the *conditio sine qua non* of the world as an object. It is remarkable that Western mankind, with a few exceptions, has accorded so little value to this fact. The flood of external objects of cognizance has made the subject of all cognizance withdraw into the background, and often into apparent nonexistence.'

Schrödinger agrees with Jung that science must be made over again, but great care will be needed in changing foundations that have endured for at least two thousand years. Some interpretations of quantum mechanics have suggested that the distinction between subject and object has broken down even as a basis for physics. He refuses to say that the barrier between subject and object has 'broken down' since this barrier does not exist.

The fourth lecture was devoted to 'The Oneness of Mind'. The reason why our feeling, perceiving, thinking Self occurs nowhere in our scientific world picture can easily be expressed in six words: It is itself that world picture. The 'arithmetical paradox' is that out of many conscious Selves only one world is concocted. Schrödinger sees only one solution: the unity of all Selves in one consciousness. He rejects any dualist interpretation of the world, yet it would not be correct to say that he rejects 'realism'. The reality, however, is not in some material world that is outside consciousness, but is consciousness itself. This is the same view of the world that he wrote down in Part One of *Meine Weltansicht* in 1925, at this time not yet published, and it does not differ appreciably from the conclusions he reached as a young researcher in Vienna in 1920. From youth to old age, there has been no essential change in his philosophy. In this lecture he paid tribute by frequent quotations to the work of Charles Sherrington, *Man on His Nature*, and thus may be said to have joined the society of latter-day gnostics for whom this book has become a basic text.[17,18]

The fifth lecture, 'Science and Religion', was devoted to questions concerning 'the other world' and 'life after death'. He at once disavowed any attempt to answer such questions, and wished only to examine what might be relevant concerning them in science and philosophy. There are truths that do not depend on space and time, for example, certain mathematical theorems. Kant showed the ideality of space and time; they are categories imposed (in its present situation anyhow) by the mind. But Kant also taught that this *thing*, mind or world, could in principle be capable of other forms that do not depend on space or time. Probably Schopenhauer was the first to understand Kant in this way. This declaration of freedom leaves open the way to religious beliefs, while preserving them from contradictions arising from either science or philosophy. A mode of existence without time would of course render the concept of 'after' meaningless, and 'life after death' would not be a meaningful expression. The work of Einstein has even set limits to the concepts of 'before' and 'after' in the physical world, and this mutability of time is consistent with the idea that the timescale of human life is of little importance. Schrödinger thought that the statistical interpretation of time, as elaborated by Gibbs and Boltzmann, is even more significant than relativity theory in this context, for it shows that the direction of time is dependent upon a sequence of events that we observe. What we construct in our mind cannot have a dictatorial power over our mind itself, neither the power to call us into life nor the power to destroy us. Thus the statistical theory 'frees us from the tyranny of Father Kronos', and we see that Mind cannot be destroyed by Time.

A hard winter

Old age is itself a burden, old age and poor health are worse, and old age, ill health and poverty are the worst of all. The Schrödingers were spared only the last of these afflictions. As he entered his seventieth year, Erwin wrote that 'the only thing I have enough of now is money'.

When they returned to Vienna for the beginning of the winter term 1956/57, he was in no condition to withstand the cold, wet weather with his already weakened heart and lungs. In retrospect, Anny at least realized that they should have fled to a warm climate, but there was only one more year to go before retirement and Erwin resolved to see it through. There is little doubt that he would have received an adequate pension in any case, but it was not in his nature to surrender when faced with a challenge to his mind and willpower.

They had excellent medical attention, the famous psychiatrist Hans Hoff, the heart specialist Professor Kurt Polzer,[19] and a physician in general practice, Joseph Schneeweiss, who was always prepared to make house calls.[20] Anny's worst fear was that the arteriosclerosis would affect Erwin's brain. She would drive him to the Physics Institute for his lectures and collect him when he had finished; after a lecture he was sometimes drowsy and incoherent, but the evidence of his letters and other writings at this time shows that his intellectual clarity was not affected by these episodes. Polzer was able to effect a marked improvement in the cerebral symptoms although the underlying disease continued its relentless course, and despite all efforts, Erwin could give only about half his scheduled classes.

Every Saturday morning Polzer and Schneeweiss came, and after the examination and discussion. Erwin would protest, 'Herr Colleague, I cannot accept the fact that you devote so much of your valuable time to me', and Polzer would reply, 'Herr Professor, you must allow this, that we do everything for you that we have done for our dear Bundespresident and will do much more.' This was not overly encouraging, since President Korner who had suffered a stroke in July died on January 4.

Even when confined at home, Erwin loved to have visitors, and he soon gathered a large circle of interesting people from the theatrical and literary worlds; they were almost always on the left politically, i.e., social democrats or Austrian 'reds'. The Russian response to the Hungarian revolt in November brought an influx of refugees to Vienna, and for a few weeks caused apprehension that Soviet tanks might roll on into Austria. Erwin said that in such an event he would stay put, being too old to move again.

Although Schrödinger was officially treated well at the university, he was not so pleased with the physics department. His friend Hans Thirring retired, and there ensued the usual political manoeuvering about his replacement. The various factions were agreed only on one thing: to prevent Schrödinger from using his enormous prestige to influence the decision. This academic politics did not bother him so much, however, as did the failure of the department to provide adequate office facilities.

There were compensations in his family life. In May, Ruth had married Arnulf Braunizer, a member of a strongly anti-Nazi family, which for a long time had hoped for a restoration of the monarchy. When Hitler entered Austria, only the communists on the far left and the monarchists on the far right had seriously opposed the Anschluss. Ruth came to visit during December, with the news that Arthur March was mortally ill with cancer of the throat. Hilde was unable to cope with the situation and Ruth and Arnulf had to take the management responsibility, while Anny provided financial help as needed. Ruth's first child was expected in February.[21]

Despite her worries, Anny was entranced by the Christmas cheer in Vienna. As usual while Erwin needed nursing, her depressions were banished and her spirits soared. On New Year's Day, she wrote to Elisabeth:

What an experience when I came on December 29 to Fritz the Baker's. When one has lived twenty years in the British Isles one thinks such bounty is impossible, a little shop with about 20 serving girls and the Chef whom they treat like a lord . . . It is like a beehive and quite expensive and every variety of baked goods can be found, croissants, rolls, salt rolls, kümmel buns, sour bread, sandwich bread and different kinds of black bread, brioches, milk sticks, pretzels and bosniaks, not to forget six different kinds of tarts. In the Operngasse, the stalls were filled with specialties from all the provinces – wonderful.

They had a dinner for eleven people on the 30th and that was the beginning of an attack of acute bronchitis for Erwin, fortunately brought under control with aureomycin.

On February 19, Ruth came to Vienna to await the arrival of her baby. Arnulf was anxious to get her away from the March house, where Arthur was dying and Hilde was in a state of nervous exhaustion. Another factor in Ruth's coming to Vienna was the orthodox Catholic policy of the Innsbruck Hospital to sacrifice the mother if necessary to deliver a baby safely. Anny was happily busy getting things ready and Erwin was looking forward to his first grandchild. On the 27th, Anny and Erwin drove Ruth to the University Clinic, and then spent a sleepless night making periodic telephone inquiries. Labor began at 6 a.m. and at 12:35 p.m. a strong boy came into the world. He was called Andreas. Arnulf arrived the next day. Erwin was delighted with his grandson but unfortunately was not well enough to enjoy him as much as he had hoped. He had to struggle to get through his lectures for the second term, and although it was a mild winter, he seldom left the house except to go to the Physics Institute.

Arthur died on April 17, after a last-ditch trial of special radiation treatment in Bern. When Erwin wrote to Hilde he began by quoting a verse from Rilke:

> Wir wissen nichts von diesem Hingehn, das
> nicht mit uns teilt. Wir haben keinen Grund,
> Bewunderung und Liebe oder Hass
> dem Tod zu zeigen, den ein Maskenmund
> tragischer Klage wunderlich entstellt.*

* We know nothing of this passing, which / sends us no message. We have no reason, / to show admiration or love or hatred / towards Death, which a face mask / of tragic sorrow strangely distorts.

Erwin wrote that one should try to fill the last days of a dying man with beauty rather than simply trying to prolong his life for a few days more in a hateful hospital bed. We ought to be allowed to carry him into a favorite spot under a canopy of cherry blossoms. Then one would give him a *Heuriger* [last year's wine] in two or three *Viertels*, as much as he wishes. Then we should begin to add more and more opium, until with a radiant look he happily goes to sleep in that passing of which we know nothing. Unfortunately we must conquer Religion before such a passing will be possible, for as Lucretius said, 'Tantum religio potuit suadere malorum.' Erwin spoke with special feeling for, as he told Hilde, he did not think that his own half-health could last much longer, three or four years at the most. 'But I am glad of the few years, for the world is very, very beautiful.'[22]

The second week in May, Anny's asthma became so bad that she left Erwin in Vienna and undertook a new treatment in the sanitorium at Solbad Hall near Innsbruck. She had been there about ten days when Arnulf came with the March car. 'Don't be alarmed, Anny, but you must go immediately to Wien. Erwin is suddenly worse and you must get back.' She rushed to catch the express train at Jenbach, and by 11 p.m. was at the Westbahnhof where Thirring and his wife met her and took her to the apartment. Dr Schneeweiss was there; he had not left Erwin's bedside since 9 a.m. A case of grippe had developed into pneumonia in the right lung. Penicillin and streptomycin had no effect, and the left lung was also attacked. A trial of terramycin and magnamycin was begun. By the evening of May 29, Erwin's body was close to exhaustion, and Polzer offered little hope, but Schneeweiss kept watch at the bedside for twenty hours, administering stimulants as needed. Next day, however, Ruth's birthday, the fever began to fall, and Polzer now thought there was some hope. Hoff came and was definitely optimistic – the patient must be allowed to sleep, sleep and more sleep. There was one nursing sister by day and one by night, and Erwin began to complain about the day sister – an encouraging sign. During his fever he had talked a lot of physics and a lot of nonsense, but sometimes he made sense: 'How good that our house could be brought together here, and all our books that we waited for.' A friend from Graz had saved many of his books and it had been a joy to see them again after twenty years.

Some of the books were those he had kept from his father's library: the collected works of Grillparzer in ten volumes, the poems of Kleist, Eichendorff's *Von Wald und Welt*, the novels of Tolstoy. Among the many books that he had bought himself were volumes of Schweitzer, Darwin, Freud, Rilke, Hofmannsthal, Dante, Schopenhauer, Nietzsche, Russell, Popper-Lynkeus, and a great variety of texts and commentaries on eastern religions.

On May 31, the patient was out of immediate danger. This was the day that the order 'Pour le Mérite' was officially conferred in Bonn on Schrödinger and on Lise Meitner. Lise was the second woman to receive the honor, the first had been Käthe Kollwitz. Erwin had been greatly pleased when he heard from Otto Hahn the news of this highest nonmilitary decoration of the German government, restricted to thirty living persons. The fact that the two distinguished Jewish women were willing to accept any award from the Germans would have seemed strange to Einstein, who had said that any official contact with Germany made him feel unclean. He had refused all invitations to participate in German organizations, for example, writing to the Max Planck Gesellschaft: 'The crime of the Germans is truly the most abominable ever to be recorded . . . The conduct of the German intellectuals – seen as a group – was no better than that of the mob. And even now there is no indication of any regret or any real desire to repair whatever little may be left to restore after the enormous murders. In view of these circumstances I feel an irrepressible aversion to participating in anything that represents any aspect of public life in Germany.'[23]

Schrödinger did not share this reaction, but as he so often said, he was an unpolitical person. Only his intellect and not his emotions were affected by the crimes of Germany. In an article for his seventieth birthday in the Socialist paper *Arbeiter Zeitung* he was called an enemy of the Nazis [*Nazigegner*]. He objected to this designation: 'I am of course *very unpolitical*. Nevertheless I am, for example, an enemy of assassins, rapists, sadists, and incendiaries.' He admitted, however, that he felt closer to the socialists than to all other parties.

Schrödinger's contempt for politics would have been approved by Voltaire:

True greatness consists in having received from heaven a powerful genius and in having used it to enlighten himself and others. A man such as Newton, the like of whom is scarcely to be found in ten centuries, is the truly great man, and the politicians and conquerors, in which no period has been lacking, are usually nothing more than illustrious criminals. It is to the man that rules over minds by the power of truth, not to those who enslave men by violence, it is to the man who understands the universe and not to those who disfigure it, that we owe our respect.[24]

Vienna anecdotes

Schrödinger's work as a professor in Vienna was greatly facilitated by a devoted assistant, Leopold Halpern.[25] As a teenager in 1938, he had escaped from Vienna to Palestine, but unlike most of the Jewish exiles, he decided to return home after the war, even though such returns were not encouraged by the Austrian authorities.

Leopold was rather naive about scientific research at that time, and he believed that in order to publish a paper, one must have made a considerable discovery. After a year, Schrödinger asked him how many publications were expected from his work in progress. 'So far I do not have any', said Leopold. 'Well that's not very many', commented Schrödinger. Leopold explained that he was working on a very significant problem which would be worth the time spent. 'Well, maybe', said Schrödinger. 'That reminds me of a colleague who went on a tiger hunt in India, and did not kill a single tiger. On his return somebody remarked this was not very many, and he replied, "But after all, they were tigers"' Schrödinger then admitted, however, that he thought better of those who published little than of those who published too much.

When news of the untimely death of Pauli was received, Schrödinger remarked, 'I don't understand why he was never elected to membership in the Austrian Academy – I wonder if the reason might perhaps be sought in antisemitism?' Halpern mentioned that Pauli had been openly critical of Austrian physicists, especially during and after the war. Schrödinger began to pull vigorously on his pipe, a clear sign that a definitive statement would be forthcoming. At last he said, 'Pauli was a very *honest* person [*ehrlicher Mensch*].'

Last professorial year

Schrödinger's final academic year at the university was 1957/58, but he was relieved of formal teaching duties. His last scientific lecture was given on March 26 in the Main Hall of Physics Institute II to a joint meeting of the Austrian Physical Society and the Chemical-Physics Society. His subject was 'Is the Energy Principle also perhaps only Meaningful Statistically?' The 'also' referred to entropy which was known to have a statistical basis. In this lecture he returned to the subject of his inaugural lecture in Zürich of thirty-six years ago, and now brought forth some new arguments: (1) In classical mechanics, the specification of energy depends upon an integration constant which is only an approximation that breaks down in the limit of very small systems. (2) The so called 'stationary' energy levels, the only ones admitted in quantum mechanics, can be so densely packed that they cannot be safely distinguished on account of the uncertainty principle. (3) An energy eigenfunction is not an adequate description of the state of an isolated system with sharply defined energy, because there are many such states which are nonequilibrium states. Schrödinger calls an energy eigenfunction 'dead', since *nothing can happen* in an isolated system whose wavefunction is an energy eigenfunction. In the Zürich lecture, he had been concerned with energy exchanges in the interactions between atoms and radiation, but now he was interested in developing an analogy between energy and entropy for large-scale systems.[26]

Bruno Bertotti attended this meeting and stayed in Vienna for several weeks. Erwin was delighted at this opportunity for discussions with the young scientist whom he hoped might extend some of his ideas. Anny gave a midday dinner party for Bertotti and Victor Weisskopf (who had come from Geneva). She served soup with dumplings [Griesnockerlsuppe], roast loin of veal [Kalbsnierenbraten] with rice and vegetables, Sachertorte with whipped cream, and a special Mocha coffee. The physicists of three generations, aged 71, 49 and 27, are said to have had a lively discussion even after this filling repast.

Bertotti had scarcely left Vienna, however, when Erwin suffered another attack of phlebitis in the left leg, which spread into the abdominal cavity. He was confined to bed, at first in hospital and then at home for about two months, but he made a good recovery and was able to get to Alpbach in July. He wrote to Bertotti to resume the discussion about sharp energy levels. Weisskopf had objected to his idea that an emphasis on frequencies instead of energies would help. Schrödinger commented 'To this one must reply that, of course, nothing depends on the *names* (energy – frequency). But in one case (sharp frequencies) one means the physical *nature* of the object in question, in the other (sharp energies) one means the *state* of the physical object. Or, to use my old *analogy*: a vibrating string has a sharp discrete series of proper modes, but this does not mean that it usually, or ever, vibrates with just one of these eigen-frequencies.'[26]

On September 30, Schrödinger became Professor Emeritus of the University of Vienna. After the summer in Alpbach, he was feeling remarkably well, although, as he wrote to Bertotti,[27] he was

too old for working out any new suggestions . . . But do not take this 'too old' to mean that I am feeling a weak old man. Not at all. I am feeling quite happy and merry, and so is my wife. But I do not think that I shall write much more papers on physics. My little library, regained after about twenty years, keeps me going on all sorts of subjects, that I can now follow up with the complete leisure of an Emeritus.

Professor Emeritus

During the summer and fall of 1959, Erwin and Anny spent four happy months in the Tirol. They stayed in Alpbach until almost all the summer visitors had departed. Lise Meitner was there for part of their stay and they found a congenial new friend in an old German nobleman, Max Trauttmansdorff. Erwin wrote to John Synge that 'I think you know us well enough to tell that neither Anny nor I are admirers . . . of aristocracy. But this man has a peculiar charm . . . We agreed so well on everything as I never thought I could agree with a man of that description.'[28]

About the middle of October they drove to Bolzano and the stay there was a 'lovely dream' with excursions to nearby beauty spots that 'beggar description'. When the weather grew colder, however, Erwin's old bronchitis recurred and he sometimes had trouble getting to sleep at night. He would then get up and write long letters to friends, such as the one to Synge which included a lively statement of his ideas about the philosophy of science.[28]

With very few exceptions (such as Einstein and Laue) all the rest of the theoretical physicists were unadulterated asses and I was the only sane person left . . . The one great dilemma that ails us . . . day and night is the wave-particle dilemma. In the last decade I have written quite a lot about it and have almost tired of doing so: just in my case the effect is null . . . because most of my friendly (truly friendly) nearer colleagues (. . . theoretical physicists) . . . have formed the opinion that I am – naturally enough – in love with 'my' great success in life (*viz.*, wave mechanics) reaped at the time I still had all my wits at my command (1926 at the age of 39) and therefore, so they say, I insist upon the view that 'all is waves'. Old-age dotage closes my eyes towards the marvelous discovery of 'complementarity' [Niels Bohr]. So unable is the good average theoretical physicist to believe that any sound person could refuse to accept the Kopenhagen oracle . . .

Discussion with Lise Meitner at Alpach.

He asked Synge to read Section 2 of his paper on energy as a statistical concept, which gives the most concise summary of his 'philosophical complaint against the great Dane'. In any criticism of Bohr's philosophy, one can never be sure that one has deciphered the meaning of the statements that one wishes to controvert, but Schrödinger does provide an effective criticism of at least one version, the idea that the wavefunction does not represent the state of a physical object, but rather the relation of the object to the subject, and hence to the knowledge that the subject has acquired about it. 'The discrimination between the ego and the world outside appears to me to be based on an epistemology out of date for some time . . . The *being shared by everybody*, this community, is the one and only hall-mark of physical reality.'

Complementarity, he says, is a 'thoughtless slogan'. 'If I were not thoroughly convinced that the man [Bohr] is honest and really believes in the relevance of his – I do not say theory but – sounding word, I should call it intellectually wicked.

> Denn eben wo Begriffe fehlen
> Da stellt ein *Wort* zu rechte Zeit sich ein. (Faust I)*

---

* For just where concepts and logic are at the end of their tether / You are sure to hit on a *word* to help you in your troubles.

Austrian banknote and first-day cover for 100th anniversary, 1987.

After leaving Alpbach in the fall, Erwin and Anny traveled for three weeks in Italy, visiting Mantova, Cremona, Pacenza, Parma, Verona and Venice. 'All very beautiful if one did not become tired.' One wonders why they did not spend the next few months in some sunny and warm southern place, but probably the need for excellent medical care made them return to the refuge of their small flat in wintry Vienna.

Erwin began to work intensively on a second section of his book *My World View*. He maintained an extensive correspondence, and his letters were always full of good spirits and new ideas. Occasionally he would comment on problems due to his ill health, but always in an objective and noncomplaining way. He bore the burdens of the flesh with optimistic resignation.

He found consolation in the ancient Indian 'This Thou Art'. As he explained to Bertotti,[29] it 'is of course not a physical but rather a metaphysical statement. It is so simple that it is impossible to explain it. It cannot be grasped by the intellect, but it may spring up in you on some occasion like a spark, and then it is there and will never really leave you, even though it is not a practical maxim to use every hour of your life.' The poet of the Bhagavadgita sees himself 'not only in his dear friend, wife or son, but also in the lion that attacks him furiously, the snake that is about to give his child the deadly blow . . . finally even in the KZ [concentration camp] officer who inflicts devilish torture on his prisoners'. Thus at first there is *nothing moral* in the *Tat-twam-asi*, it is something between intellectual and intuitional. If it were shared by the mighty (at least), then the world would be a paradise, but to the Brahmin this is only a very welcome by-effect. The main point is to have recognized the truth. It will comfort him in the hour of death, just as the 'viaticum' comforts the Roman Catholic. Einstein, when he once (a long time ago, in his fifties) was dangerously ill, was asked by a friend: Are you not afraid of death? No, said he, I feel so intimately connected with the whole of the world, that I cannot all of a sudden drop out of it.

Schrödinger maintained an interest in many things besides science and philosophy, including even the Vienna Physics Institute. He wrote to Born early in 1960: 'The ship of physics has been leaderless [here] ever since Hasenöhrl was killed. For 30 years the School was directed by my friend Thirring. Next semester his son takes over and all hell breaks loose. Hans was not gifted for theoretical physics. Walther is an excellent specialist.'[30]

In the spring of 1960, Erwin's respiratory symptoms seemed difficult to interpret and an extensive series of tests was carried out. To the surprise of everyone, evidence was found of a recurrence of the pulmonary tuberculosis of forty years ago. He was ordered to begin a *Liegekur* with a course of tuberculostatic drugs. Thus in the late spring he found himself lying on a *chaise longue* on a sunny verandah in Alpbach with a view of snowcapped mountains, an almost exact replay of his situation in Arosa in 1921.

What is real?

During his *Liegekur* in Alpbach, Schrödinger completed the second part of *My World View*, which he called *What is Real?*, [*Was ist Wirklich?*].[31] In sixty pages he gave a final statement of his philosophy – he knew that his life was drawing to an end. The book was written in an informal style, serious but undogmatic, with a number of homely examples to illustrate difficult points. It was not meant to be a philosophical treatise with logical and weighty arguments, but simply a translucent statement of his personal beliefs. Indeed, as he admitted, it is not possible to give any rigorous demonstration of what he called these mystical-metaphysical propositions.

As he had done in the first part of *Meine Weltansicht* thirty-five years previously, he rejected any dualistic view of our world. His first chapter is called 'The Reasons for the Abandonment of the Dualism of Thinking and Being or of Mind and Matter'. The most cogent reason, he says, is that we are not able to explain how a material world can act upon an immaterial mind, or *vice versa*. To eliminate this intractable problem, we accept monism and reject dualism. Such a choice is also more economical of hypotheses and is thus encouraged by the threat of Occam's razor.

Having opted for monism, we must now choose between matter and mind. One of the earliest exponents of a purely material world was Democritus of Abdera, who taught that the soul is made of atoms of an especially fine, smooth, spherical, readily mobile kind. A noted advocate of the primacy of mind was Spinoza, who said that the all was one substance, with two attributes, extension and thought, and this unity he called God. Bertrand Russell in *The Analysis of Mind* (1933) thought that psychic states and material bodies are composed of elements of the same kind related in different ways, but in later writings he reverted to the concept of a real external world. Schrödinger admits that a *representation* of the real world is essential to everyday life, but we do not need the existence of an object that is represented, indeed, we do not know what 'to exist' would mean for such an object. He says that if we are to have only one realm, it must be the psychic since the psychic certainly exists (*cogitat – est*).

The theme of Chapter 2 is that we become aware of the commonality of the world only through language, which includes both verbal and nonverbal communication. This shared experience is a basic fact. The hypothesis of a common external world does not 'explain' this fact – it merely restates it in different words. 'Reality, existence, and the like, are empty words.' Even if the hypothesis of a common external world is plausible, it still does not explain how we recognize its commonality. The next chapter explores 'The Imperfection of Understanding': no science is ever exact or is ever completed. There is an interesting discussion of color vision and color blindness, and a comparison of the senses of sight and sound.

After these preliminaries Schrödinger restates the 'Identity Doctrine', the teaching of Vedanta that the underlying reality is a unity of Mind. In this last statement of his religious beliefs, he gave up any reliance on logical argument. 'It is now necessary to admit that the considerations of this chapter are *logically* not meant quite so seriously

as all that was previously discussed, but *ethically* much more seriously
. . . Now I shall not keep free of metaphysics, nor even of mysticism;
they play a role in all that follows.' In brief, the meaning of Vedanta is
that 'we living beings all belong to one another, that we are all actually
members or aspects of a single being, which we may in western
terminology call God, while in the Upanishads it is called Brahman.'

The Indian thinkers have derived two consequences from this
'Identity Doctrine' [*Identitätslehre*], one ethical, the other eschatological. He agrees willingly with the first, but rejects the second. The
ethical doctrine is to be found in this verse from the Vedas, as given by
Schopenhauer:

> Die eine höchste Gottheit
> In allen Wesen stehend
> Und lebend wenn sie sterben,
> Wer diese sieht, ist sehend.
> Denn welcher allerorts den höchsten Gott gefunden,
> Der Mann wird durch sich selbst sich selber nicht verwunden.*

* The one highest Godhead / In all beings existent / And when they die yet living /
Who sees this is seeing. / For the one who has found the highest God everywhere /
That man will not harm himself through himself.

'These beautiful words need no commentary. Here mercy and
goodness towards all living things (not merely fellow human beings)
are glorified as the highest attainable goal – almost in the sense of
Albert Schweitzer's reverence for life.'

The second lesson of Vedanta, the eschatological, is summarized in
the verses:

> Im Geiste sollen merken sie:
> Nichts ist hier Vielheit irgendwie;
> Von Tod zu Tode wird verstrickt,
> Wer eine Vielheit hier erblickt.*

* In spirit you are obliged to take note: / Nothing here is in any way multiplicity; /
From death to death will be ensnared / He who sees a multiplicity here.

Only the inner vision and realization of the unity in Brahman will
free the soul from the recurrent *Samsara* of reincarnation, worldly life,
and death. 'One must make the mystical doctrine one's own, understand it with the entire soul, and not only with lip-service.'
Schrödinger calls this a 'salvation through knowledge', which can be
achieved by only a few persons since it requires intelligence and
leisure for contemplation. Like Schopenhauer, he had no personal
interest in the pathway of ascetic practices followed by most Indian
mystics. Also, he rejects the eschatological teaching of Vedanta, and
even advises that anyone else who accepts the Vedantic world view
should leave out the transmigration of souls.

The concluding chapter is called 'Two Causes for Astonishment. Ersatz Ethics'. Schrödinger believes that the distinction between two remarkable findings is the contribution that is new in his study. The first is that despite the hermetic separation of my sphere of consciousness from every other one, a common language has developed, which has yielded a wide-ranging structural agreement in that part of our experiences called external, or, in other words, has led to the conclusion that we all live in the same world. Since we all become accustomed to this as growing children, the astonishment that it deserves has been lost through familiarity.

Although the origin of the agreement can be explained by language and development, these do not explain the fact of the agreement. This requires one of two metaphysical assumptions: (a) the existence of a real world, or (b) the assumption that we are all different aspects of a Unity. 'I will not argue with anyone who contends that these are actually the same thing. That is pantheism, and the Unity is called God-Nature.' But even the recognition of the metaphysical character of the first form of the hypothesis (a real external world) is far removed from vulgar materialism. The ethical consequences are more easily derived from the second formulation.

He admits that the identity doctrine still *appears* mystical and metaphysical, and the very different degrees of commonality of experience are not easily understood in its light. The hypothesis of a real world explains some of these differences more readily since it includes the reality of space-time. A real world can also provide a basis for a utilitarian morality, which he calls ersatz-ethics. Nevertheless Schrödinger prefers the Identity Doctrine because of its superior ethical content and the deep religious consolation [*Trost*] it provides in our ephemeral life.

He alludes briefly to the possible disparity between beliefs and actions, saying that as a guide to living a good life, Sancho Panza may be preferable to Schopenhauer. One was a decent and kindly rascal while the other was an irascible intellectual. He mentions the incident when the philosopher threw an old cleaning lady down the stairs and injured her back so that he had to pay her a monthly allowance for the rest of her life. When she died, he entered in his diary no word of regret, but the laconic notice, '*obiit anus, abiit onus*, [the old woman has died, the burden has gone].' Erwin also had a bad temper, but he never injured anyone physically; the worst report is that to punish it for refusing to deliver its salt he once threw a salt cellar through a glass door at Clontarf.

His studies of Vedanta must have made him aware that it prescribes a way of life and not only a system of beliefs, yet in all his writings, including this last testament, he remains curiously aloof and impersonal. It is not known that his belief in Vedanta ever influenced his actions as distinct from his philosophical writings, and he also kept this belief scrupulously separated from his work in theoretical physics, even from his interpretation of wave mechanics. He ends *Meine Weltansicht* with a warning that superstition and silly nonsense may be fostered by even the most subtle and abstract religious beliefs.

The connection between pantheism and the golden rule may not be so obvious as Schrödinger contended. The *Identitätslehre* is a static, timeless, nonhistorical doctrine, and its acceptance may be inconsistent with the practical politics of people working with people to achieve viable compromises for actions in space and time. Also, considering the tendency of many persons to damage themselves, the equation of the self with the other may not provide an ethical panacea.

During his *Liegekur* Schrödinger also wrote a short (forty-page) autobiography *Mein Leben*.[32] It gave most emphasis to his childhood and student days. The time of famine in Vienna after World War I had left indelible traces in his memory and, although he would not have liked the word, in his conscience. Even now, in the last year of his life, he blamed himself for not doing more for his mother and father when they had lost all their money. Actually he was financially helpless at that time, and from this impotence came his lifelong dislike for bankers and profiteers.

He wrote letters particularly to Bertotti, Born, Synge and the poet and novelist Franz Csokor. His letters ranged over many subjects, from physics to literature and even politics (there was tension between Austria and Italy over South Tirol, and he wrote to assure Bertotti that he understood both sides of the question).

To Born he wrote: 'Maxel, you know I love you and nothing can change that. But I do need to give you once a thorough head washing. So stand still. The impudence with which you assert time and again that the Copenhagen interpretation is practically universally accepted, assert it without reservations, even before an audience of the laity – who are completely at your mercy – it's at the limit of the estimable . . . Have you no anxiety about the verdict of history? Are you so convinced that the human race will succumb before long to your own folly?'[33]

In his last letter to Born, on October 24, he mentioned that he had twice nominated for a Nobel prize the Viennese physicist Marietta Blau, who had discovered the 'stars' in photographic plates caused by massive high-energy particles in cosmic-ray showers.[34] Blau was one of the few Jewish refugees who returned to Austria after the War. She lived in poverty since, in accord with the usual Austrian policy, she was offered no reparations. She was not even paid a salary for work that she did at the Radium Institute. One administrator told her quite frankly, 'You know, you are a woman and Jew, and the two together are simply too much.' Erwin managed to obtain for her the Schrödinger prize, which had been established by the Academy of Sciences, but this provided only a temporary support.[35]

He had ample time for reading, both books from his old library and some new ones of current interest, *Christ Stopped at Eboli* and *The Leopard* [in Italian], Selma Lagerlöf, Plutarch in a bilingual edition, and as always Goethe and some Shakespeare (which he preferred in German translation). Perhaps he read the words of Prospero, so close to his own way of thinking:

> Our revels now are ended . . . These our actors,
> As I foretold you, were all spirits, and
> Are melted into air, into thin air,
> And, like the baseless fabric of this vision,
> The cloud-capped towers, the gorgeous palaces,
> The solemn temples, the great globe itself,
> Yea, all which it inherit, shall dissolve,
> And, like this insubstantial pageant faded,
> Leave not a rack behind: we are such stuff
> As dreams are made on; and our little life
> Is rounded with a sleep.

Last days[36]

On October 20, Anny narrowly escaped death in the worst asthma attack she had ever experienced. By good luck there was a Red Cross doctor nearby, who brought oxygen to save her from asphyxiation. She was rushed to the hospital at Solbad-Hall. She was destined to survive Erwin for over four years.

On November 9, Erwin decided to return to Vienna. A young friend in Alpbach drove him back in Anny's car. There was no maid at the apartment and Erwin had considerable difficulty managing for himself. He was helped by Anny's sister Irmgard and by the couple who were caretakers of the building. He drew up a schedule for himself 'For the Time When Anny is not Here', which ended with the words, 'If I become really *sick*, naturally I must enter the hospital, that is clear.'

Anny worried about their separation as 'joy and sorrow has bound us so closely together in the past 41 years that we don't want to be separated during the few remaining years of our lives.' They wrote loving letters to each other every few days.

Erwin became too weak to take care of himself and on December 2, he was brought to the Lainz Hospital where Polzer intended to make an examination of his lungs. He had a fine corner room with a view of the

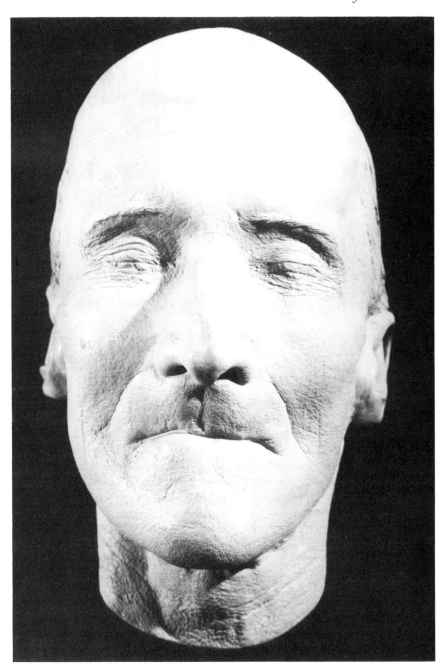

Death mask.

Lainz Zoo, but he protested so furiously that he was transferred to the university psychiatric-neurological clinic. When he calmed down, he was much amused by this, exclaiming, 'Now at last *I* am with Hoff.' (Anny had spent many weeks under the care of the famous psychiatrist.)

In early December, Anny was well enough to return to Vienna and she was able to spend several hours a day with Erwin. He asked to be brought home, saying, 'I was born at home and I'll die at home even if it shortens my life. At the age of 73, I am not going to put up with forced nursing.' In the week before Christmas, his condition became critical, with a urological problem complicating the heart disease. Anny held his hand for hours a day, leaving the bedside only for meals and sleep. He often said, 'Oh since I have you again, everything is good again.' Hoff and Polzer were very kind, and they assured Anny that seeing her again had greatly lightened his end.

They finally agreed that he should be brought home in accord with his wishes, and he was put into his bed there at 10 a.m. on January 3. He was still conscious but he could no longer breathe without an oxygen apparatus. At 4 o'clock in the morning he became unconscious, but a doctor who immediately came gave him an injection which brought him back to consciousness. Anny sat by his bedside. He could no longer take nourishment. In the afternoon she gave him a sip of orange juice, and then heard his last words. 'Annichen, Du bleibst bei mir, – auf das ich nicht hinunter stürze' [Annikin, stay with me – so that I don't crash]. Then he became still and seemed to sleep. Around 5 o'clock the doctor came and said that he could last only a few hours. At about 6:55 p.m. his pulse stopped – he was dead. Anny kissed him, stroked his hair, and took her departure from him.

Hans Thirring came immediately and took over everything. The coroner arrived and certified the official death and the body was taken to the Medicolegal Institute where two death masks were taken and also a casting of his hand. Dr Holczabek, dean of the medical faculty, performed an autopsy. The cause of death was reported to be general ageing of the heart and the arteries, with no specific pathology.

The body was transferred to Alpbach where it arrived exactly as the bell of the little church struck noon on Saturday January 9, 1961. There was some problem about burial in the churchyard since Erwin was not a Catholic, but the priest relented when informed that he was a member in good standing of the Papal Academy, and a plot was made available at the edge of the *Friedhof*. The ceremony at the graveside on Sunday was simple but impressive. Hans Thirring gave a talk and Father Metzner, a Jesuit, said the Lord's Prayer. Almost all the villagers of Alpbach were gathered around. The next day the gravemound was completely covered by snow.[31]

References

Introduction

1 Schrödinger, E. (1961) *Meine Weltansicht*. Wien: Paul Zsolnay.
2 Born, Max (1968) *My Life and Views*, p. 48, New York: Scribner.
3 Hermann, Armin (1963) 'Erwin Schrödinger – eine Biographie' in *Die Wellenmechanik, Dokumente der Naturwissenschaft*, vol. 3, pp. 173–92. Stuttgart: Ernst Battenberg.
4 Kuhn, Thomas S. (1970) *The Structure of Scientific Revolutions*, 2nd edn. Chicago: Chicago University Press.
5 Schrödinger, E. *Mein Leben*, Ms, Schrödinger Archive, Alpbach. Later published in 1985 in *Mein Leben, Meine Weltansicht*. Wien: Paul Zsolnay.
6 Edel, Leon (1984) *Writing Lives*. New York: W.W. Norton.
7 Quoted in Micheál MacLiammoir (1952) *Put Money in Thy Purse*. London: Methuen.
8 Musil, Robert (1930–1949) *Der Mann ohne Eigenschaften*. [Translated by Eithne Wilkins & Ernest Kaiser (1960) *The Man Without Qualities*. London: Secker & Warburg.]
9 Thirring, Hans (1972) 'Erwin Schrödinger' in *Neue Österreichischer Biographie ab 1815* **18**, 63–8. Wien: Amalthea.
10 Kretschmer, Ernst (1929) *Körperbau und Charakter*, 7th edn. Berlin: Julius Springer.
11 Müllern-Schönhausen, Johannes (1959) *Die Lösung des Rätsels Adolf Hitler*. Vienna: Verlag zür Forderung Wissenschaftlicher Forschung.
12 Cassirer, Ernst (1981) *Kant's Life and Thought*, (translated by J. Haden). New Haven CT: Yale University Press.
13 Schrödinger, E. (1949) *Gedichte*, Godesberg: Helmut Kupper.
14 Grillparzer, Franz (1969) *Plays on Classic Themes* (translated by Samuel Solomon), New York: Random House.
15 Born, Max (1978) *My Life, Recollections of a Nobel Laureate*. London: Taylor & Francis.

Chapter 1

1 Arzberger, Rhoda. Family History of Schrödingers, unpublished Ms, Schrödinger Archive, Alpbach.
2 Leamington Spa, Town Directories.

3 Strunz, Franz (1909) 'Alexander Bauer und die Geschichte der Chemie Oesterreichs', *Chem.-Z.* **88**, 669–70, 679–80.
4 Lechner, Natalie Bauer (1980) *Recollections of Gustav Mahler* (translated by D. Newlin). London: Faber & Faber.
5 Gay, Peter (1984) *The Bourgeois Experience*, vol. 1, New York: Oxford University Press.
6 Bamberger, Emily. Recollections of Erwin Schrödinger as a Child, unpublished Ms, Schrödinger Archive, Alpbach.
7 Schrödinger, E. (1935) in *Les Prix Nobel en 1933*. Stockholm: Norstedt & Soener, pp. 86–8.
8 Musil, Robert (1930–1949) *Der Mann ohne Eigenschaften*. [Translated by Eithne Wilkins & Ernst Kaiser (1960) *The Man Without Qualities*, London: Secker & Warburg.]
9 Schrödinger, E. (1961) *Meine Weltansicht*. Wien: Paul Zsolnay.
10 Barnes, Djuna (1936) *Nightwood*. London: Faber & Faber.
11 Paulsen, Friedrich (1889) *Das Realgymnasium und die Humanistische Bildung*. Berlin: W. Hertz.
12 Laue, Max (1961) *Gesammelte Schriften und Vorträge*, vol. 3. Braunschweig: Fr. Vieweg & Sohn.
13 Halporn, Robert (1985) Personal recollections, Bloomington, Indiana.
14 Schrödinger, E. *Mein Leben*, Ms, Schrödinger Archive, Alpbach. Later published in 1985 in *Mein Leben, Meine Weltansicht*. Wien: Paul Zsolnay.
15 Boltzmann, Ludwig (1905) *Populäre Schriften*. Leipzig: J.A. Barth.
16 Zweig, Stefan (1944) *Die Welt von Gestern*. Stockholm: Bermann-Fischer.
17 Hermann, Armin (1963) 'Erwin Schrödinger – eine Biographie' in *Die Wellenmechanik, Dokumente der Naturwissenschaft*, vol. 3. Stuttgart: Ernst Battenberg, pp. 173–92.
18 Rozenblit, Martha (1985) *The Jews in Vienna*. Ithaca, NY: SUNY Press.
19 Rella, Mario. Personal communication and interview, Vienna 1986.
20 Schrödinger, E. (1930) 'Verabsäumte Pflichten' in *Reclams Universum* **47**, 263.
21 Schnitzler, Arthur (1968) *Jugend in Wien*. Wien: Molden.
22 Keil-Budoschowsky, Verena (1983) *Die Theater Wiens*. Wien: Paul Zsolnay.
23 Freud, Sigmund (1955) 'Die Briefe von Sigmund Freud an Arthur Schnitzler', *Die Neue Rundschau* **66**, No. 1.
24 Grillparzer, Franz (1969) *Plays on Classic Themes* (translated by Samuel Solomon). New York: Random House.
25 Schrödinger, E. *Theaternotizbuch*, unpublished Ms, Schrödinger Archiv, Zentralbibliothek für Physik, Wien.
26 Schorske, Carl (1980) *Fin de Siècle Vienna: Politics and Culture*. New York: Alfred Knopf.
27 Schrödinger Archive, The Bohr Library, American Institute of Physics, New York.

Chapter 2

1 Du Moulin Eckhart, Richard (1929) *Geschichte der deutscher Universitäten*. Stuttgart: Ferdinand Enke.

2 Bittner, Lotte (1949) *Geschichte des Studienfaches Physik an der Wiener Universität in den letzten hundert Jahren.* Dissertation, Universität Wien.

3 Karlik, Berta & Schmid, Erich (1982) *Franz S. Exner und seiner Kreis.* Wien: Verlag Österreichischen Akademie der Wissenschaften.

4 Meister, Richard (1947) *Geschichte der Akademie der Wissenschaften in Wien. 1847–1947.* Wien: Holzhausen.

5 Broda, Engelbert (1955) *Ludwig Boltzmann.* Wien: Deuticke.

6 Schrödinger, E. (1929) *Sitzber. Preuss. Akad. Wiss. Phys. – Math. Kl.,* C–CII.

7 Thirring, Hans (1959) 'Friedrich Hasenöhrl' in *Neue Österreichische Biographie ab 1815* **13**, 192–200.

8 Schrödinger, E. Letter to Arthur Eddington, 1940.3.22, A.H.Q.P., Bohr Library, American Institute of Physics, New York, Microfilm 37.

9 Blackmore, John D. (1972) *Ernst Mach, His Life, Work and Influence.* Berkeley, CA: University of California Press.

10 Mach, Ernst (1906) *Die Analyse der Empfindungen und das Verhältnis des Physischen zum Psychischen,* 5th edn, Jena: G. Fischer.

11 Fechner, Gustav (1860) *Elemente der Psychophysik.* Leipzig: Breitkopf & Hartel.

12 Mach, Ernst (1923) *Populär-wissenschaftliche Vorlesungen,* 5th edn. J.A. Barth. Leipzig. [Lecture before Academy of Sciences, Vienna, May 25, 1882.]

13 Avenarius, Richard (1888) *Kritik der reiner Erfahrung.* Leipzig: Fues.

14 Avenarius, Richard (1912) *Der Menschliche Weltbegriff,* 3rd edn. Leipzig: O.R. Reisland.

15 Skalnik, Kurt (1957) 'Karl Lueger' in *Neue Österreichische Biographie ab 1815* **12**, 107–16. Wien: Amalthea.

16 Johnston, William M. (1976) *Austrian Mind: An Intellectual and Social History 1848–1938.* Berkeley, CA: University of California Press.

17 Wirtinger, Wilhelm. Letter to E. Mach, 1910.7.28, quoted in Ref. (9), p. 162.

18 Thirring, Hans (1972) 'Erwin Schrödinger' in *Neue Österreichischer Biographie ab 1815* **18**, 63–8. Wien: Amalthea.

19 Ref. (14), Ch. 1, p. 16.

20 Semon, Richard (1904) *Die Mneme als erhaltendes Prinzip.* Leipzig: Engelmann.

21 Schatzmann, Jurg (1968) *Richard Semon und seine Mnemetheorie.* Zürcher Mediziner Abhandlungen No. 58, Zürich: Jans.

22 Haeckel, Ernst (1876) *Die Perigenesis der Plastidule oder die Wellenzeugung der Lebensteilchen.* Berlin: Greimer.

23 Butler, Samuel (1880) *Unconscious Memory.* London: David Bogue.

24 Einstein, Albert (1916) *Phys. Z.* **17**, 101–4.

25 Goethe, Wolfgang Johann (1810) *Zur Farbenlehre.* Pt 5, para. 739. Tübingen: J.G. Cottaschen Buchhandlung.

26 Schrödinger, E. Ref. (14), Ch. 1, p. 15.

27 Benndorf, Hans (1927) *Phys. Z.* **28**, 397–409.

28 Weber, H. (1900) *Die Partial Differentialgleichungen der mathematischen Physik,* Braunschweig: F. Vieweg & Sohn.

29 Schweidler, Egon (1905) Über die Schwankungen der Radioaktiven Umwandlung.' Liège: 1° Congrès International pour l'Étude de la Radiologie et de l'Ionisation.

30 Salpeter, Jakob. Biographical Notes, unpublished Ms, Cornell University Library.

31 Archiv der Universität Wien.

32 Schrödinger, E. (1910) 'Über die Leitung der Elektrizität auf der Oberfläche von Isolatoren am feuchten Luft', *Sitzber. k. Akad. Wiss. Wien, Math. – Naturwiss. Kl.* **119**, 1215–22.

33 Rothenberg, Gunther E. (1976) *The Army of Francis Joseph*. W. Lafayette, IN: Purdue University Press.

34 Schrödinger, E. (1912) 'Zur kinetischen Theorie der Magnetismus', *Sitzber. k. Akad. Wiss. Wien, Math. – Naturwiss. Kl.* **121**, 1305–28.

35 Bohr, Niels (1911) *Studier over Metallernes Elektrontheori*. Kobenhavn: Thanning & Appel. [English translation Nielsen, Rud, J. (1972) *Collected Works*, vol. 1. Amsterdam: North Holland.]

36 Schrödinger, E. (1912) *Sitzber. k. Akad. Wiss. Wien, Math. – Naturwiss. Kl.* **121**, 1937–72.

37 Bianchi, Johanna. Interviews 1986, 1987, Vienna.

38 Schrödinger Archiv, Zentralbibliothek für Physik, Wien.

39 Schrödinger, E. (1912) *Sitzber. k. Akad. Wiss. Wien, Math. – Naturwiss. Kl.* **121**, 2391–406.

40 Hess, Victor Franz (1928) *The Electrical Conductivity of the Atmosphere and its Causes*. New York: D. van Nostrand.

41 Schrödinger, E. (1913) *Sitzber. k. Akad. Wiss. Wien, Math. – Naturwiss. Kl.* **122**, 2023–67.

42 Schrödinger, Annemarie. A.H.Q.P. Interview, 1963.4.5, The Bohr Library, American Institute of Physics, New York.

43 Ref. (17), Ch. 1.

44 (1913) *Phys. Z.* **14**, 1073–5.

45 Frank, Philipp (1948) *Einstein, His Life and Times*. London: J. Cape.

46 Schrödinger, E. (1913) *Verh. Deut. Physik. Ges.* **15**, 1167–72.

47 Schrödinger, E. (1918) 'Dielektrizität', *Handbuch der Elektrizitat und des Magnetismus*, Band I, 157–231. Leipzig: J.A. Barth. [written 1914]

48 Schrödinger, E. (1914) *Ann. Phys.* **44** (4), 916–34.

49 Boltzmann, Ludwig (1905) *Populäre Schriften*. Leipzig: J.A. Barth.

Chapter 3

1 Crankshaw, Edward (1974) *The Fall of the House of Habsburg*. London: Cardinal.

2 Schnitzler, Arthur (1968) *Jugend in Wien*. Wien: Molden.

3 Trotsky, Leon (1930) *My Life, An Attempt at an Autobiography*. New York: Scribner.

4 Brook-Shepherd, Gordon (1968) *The Last Habsburg*. New York: Weybright & Talley.

5 Ref. (14), Ch. 1.

6 Schrödinger, E., War Diary, Schrödinger Archiv, Zentralbibliothek für Physik, Wien.
7 Schrödinger, E. (1915) *Ann. Phys.* **46** (4), 413–18.
8 Ref. 33, Ch. 2.
9 Schrödinger, E. (1915) *Phys. Z.* **16**, 289–95.
10 Millikan, Robert A. (1922) *The Electron.* Chicago: Chicago University Press.
11 Klossowski de Rola, Stanislas (1983) *Balthus.* New York: Harper & Row.
12 Imperial War Archive, Vienna.
13 Thirring, Hans (1916) *Vierteljahrsberichte Wiener Vereins zur Forderung phys. chem. Unterrichts* No. 1.
14 Mark, Hermann. Interview, 1984. Brooklyn, N.Y.
15 Schrödinger, E. (1917) *Phys. Z.* **18**, 445–53; (1917) Nachtrag, 567.
16 Schrödinger, E. (1917) *Naturwiss.* **5**, 561–7.
17 Teske, Andrzej (1975) 'Marian Smoluchowski' in *Dictionary of Scientific Biography* (ed. C.C. Gillespie). New York: Scribners.
18 Smoluchowski, Marian (1904) *Ann. Phys.* **21** (4), 756–80.
19 Ref. (29), Ch. 2.
20 Schrödinger, E. (1918) *Sitz. k. Akad. Wiss. Wien, Math. – Naturwiss. Kl.* **127**, 237–62.
21 Schrödinger, E. (1919) *Sitz. k. Akad. Wiss. Wien, Math. – Naturwiss. Kl.* **128**, 177–237.
22 Schrödinger, E. (1918) *Phys. Z.* **19**, 4–7.
23 Schrödinger, E. (1918) *Phys. Z.* **19**, 20–2.
24 Einstein, Albert (1918) *Phys. Z.* **19**, 115–16. 165–6.
25 Bullard, Robert Lee (1925) *Personalities and Reminiscences of the War*, Garden City, NY: Doubleday Page.
26 Eisenmenger, Anna (1932) *Blockade: The Diary of an Austrian Middle-class Woman 1914–1924.* New York: Long & Smith.
27 Ref. (14), Ch. 1.

Chapter 4

1 Magee, Bryan (1983) *The Philosophy of Schopenhauer.* New York: Oxford University Press.
2 Schrödinger, E. Vedanta Notebook, Schrödinger Archiv, Zentralbibliothek für Physik, Wien.
3 Hearn, Lafcadio (1927) *Gleanings in Buddha Fields.* London: J. Cape. [Reprint.]
4 Ref. (20), Ch. 1.
5 Handlirsch, Anton (1920) *Verh. Zool.-Botan. Ges. Wien* **70**, 35–6.
6 Ref. (14), Ch. 1.
7 Planck, Max (1950) *A Scientific Autobiography.* London: Williams & Norgate.
8 Planck, Max (1900) *Verh. Deut. Phys. Ges.* **2**, 237–45.
9 Einstein, Albert (1905) *Ann. Phys.* **17**, 132–48.
10 Schrödinger, E. (1919) *Ann. Phys.* **61** (4), 69–86.
11 Committee on Colorimetry, Optical Society of America (1953) *The Science of Color.* New York: Thomas Y. Crowell.

12 Exner, Franz (1919) *Vorlesungen über die physikalischen Grundlagen der Naturwissenschaften*. Wien: Deuticke.

13 Macadam, David (ed.) (1970) *Sources of Color Science*. Cambridge MA: M.I.T. Press.

14 Born, Max (1963) *Naturwiss*. **50**, 29–39.

15 Schrödinger, E. (1920) *Ann. Phys.* **62** (4), 603–22.

16 Schrödinger, E. (1920) *Ann. Phys.* **63** (4), 397–426; 427–56; 481–520.

17 Schrödinger, E. (1924) *Naturwiss*. **12**, 925–9.

18 Schrödinger, E. (1925) *Naturwiss*. **13**, 373–6.

19 Schrödinger, E. (1925) *Sitzber. k. Akad. Wiss. Wien, Math. – Naturwiss. Kl.* **134**, 471–90.

20 Schrödinger, E. (1926) 'Die Gesichtsempfindungen' in *Muller-Pouillets Lehrbuch der Physik 2/1*, 11th edn, pp. 456–560. Braunschweig: Vieweg.

21 Urban, Paul (1980) *Gedächtnisprotokoll*, Zentralbibliothek für Physik, Wien.

22 Ref. (2), Ch. 2.

23 Ref. (18), Ch. 2.

24 Steiger, Gunter (1971) *Ich wurde doch nach Jena gehen*. Jena: Auslieferung Max Kessler.

25 Eucken, Rudolf (1922) *Lebens Erinnerungen*. Leipzig: K.F. Koehler.

26 Rathgeber-Regener, Erika. Interview, 1986, Sydney.

27 Schrödinger, E. Letters to Stefan Meyer and to Hans Thirring, 1920.10.20 and 1920.10.[23]. Schrödinger Archiv, Zentralbibliothek für Physik, Wien.

28 Sommerfeld, Arnold (1919) *Atombau und Spektrallinien*. Braunschweig: Friedr. Vieweg & Sohn.

29 Ref. (7), Ch. 1.

30 Hevesy, Georg Letter to Niels Bohr, 1913.9.23, Niels Bohr Archive, Copenhagen.

31 Benz, Ulrich (1975) *Arnold Sommerfeld*. Stuttgart: Wissenschaftliche Verlagsgesellschaft MBH.

32 Schrödinger, E. (1921) *Z. Phys.* **4**, 347–54.

33 Ref. (1), Ch. 2.

34 Rasche, G. & Staub, H.H. (1979) 'Physik und Physiker an der Universität Zürich', *Vierteljahrschrift Naturfor. Ges. Zürich* **124**, 205–20.

35 University of Zürich Archive. [All documents cited concerning Schrödinger's appointment are located here.]

Chapter 5

1 Schrödinger, E. Letter to W. Pauli, 1922.11.8. [Published in Hermann, A., Meyenn, K.v. & Weisskopf, V.F. (eds) (1979). *Wolfgang Pauli, Wissenschaftlicher Briefwechsel, Band I*. New York: Springer Verlag.]

2 Ewald, Ella. Interview, 1984, Ithaca, N.Y.

3 Schrödinger, E. (1922) *Z. Phys* **12**, 13–23.

4 Weyl, Hermann (1918) *Raum, Zeit, Materie*, 4th edn. Berlin: J. Springer.

5 Yang, Chen Ning (1987) 'Square Root of Minus One, Complex Phases and Erwin Schrödinger' in Kilmister, C.W. (ed.) *Schrödinger: Centenary Celebration of a Polymath*. Cambridge: Cambridge University Press.

6 London, Fritz. Letter to Schrödinger, 1926.12.7, A.H.Q.P., Bohr Library, American Institute of Physics, New York, Microfilm 41.

7 Muralt, Alexander v. Personal Letter, 1985.

8 Debye, Pieter. Interview A.H.Q.P., Bohr Library, American Institute of Physics, New York.

9 Ball, Hugo (1946) *Die Flucht aus der Zeit*. Luzern: Stocker.

10 Dumont, Hervé (1973) *Das Zürcher Schauspielhaus von 1921 bis 1938*. Lausanne: Editions Publi.

11 Hückel, Erich (1975) *Ein Gelehrtenleben*. Weinheim: Verlag Chemie.

12 Schrödinger, E. (1923) *Neue Zürcher Zeitung*, February 3.

13 Schrödinger, E. (1929) *Naturwiss.* **17**, 9–11.

14 Schrödinger, E. Letter to W. Pauli, 1922.11.8. (cf. Ref. 1).

15 Exner, Franz (1908) Über Gesetze in Naturwissenschaft und Humanistik'. Vienna: Selbstverlag der k.k. Universität. Cf. Hanle, Paul A. (1975) 'Erwin Schrödinger's Statistical Mechanics, 1912–1925', Ph.D. Thesis, Yale University.

16 Ref. (12), Ch. 4.

17 Cassirer, Ernst (1956) *Determinism and Indeterminism in Modern Physics*. New Haven: Yale University Press.

18 Hamel, George (1904) *Z. Math. Phys.* **50**, 1. Cf. Eugene Wigner (1964) *Proc. Nat. Acad. Sci.* **51**, No. 5.

19 Raman, V.V. & Forman, Paul (1969) *Historical Studies in the Physical Sciences* **1**, 291–314.

20 Spengler, Oswald (1962) *The Decline of the West*. New York: Alfred Knopf.

21 Weyl, Hermann (1919) 'Das Verhaltnis der kausalen zur statistischen Betrachtungsweise in der Physik'. Published in 1968 in *Gesammelte Werke*, Band II, 113–23. Berlin: Springer.

22 Mehra, Jagdish (1975) *The Solvay Conferences in Physics*. Dordrecht: D. Reidel.

23 Einstein, Albert & Besso, Michele (1979) in Speziali, Pierre (ed.) *Correspondance 1903–1955*. Paris: Hermann.

24 Joffe, A.F. (1967) *Begegnungen mit Physiker*. Leipzig: Teubner.

25 Schrödinger, E. (1922) *Phys. Z.* **23**, 301–3.

26 Slater, John C. (1977) *Solid State and Molecular Theory: A Scientific Autobiography*. New York: John Wiley.

27 Bohr, Niels (1923) *Naturwiss.* **11**, 606–24.

28 Bohr, Niels, Kramers, H.A. and Slater, J.C. (1924) *Phil. Mag.* **47**, 785–802.

29 Bothe, W. & Geiger, H. (1925) *Z. Phys.* **32**, 639–63.

30 Schrödinger, E. Letter to Niels Bohr, 1924.5.24. Published in Rudinger, E. (ed.) (1984) *Collected Works*, vol. 5, p. 490. Amsterdam: North Holland.

31 Schrödinger, E. (1924) *Naturwiss.* **12**, 720–4.

32 *Verh. Deut. Phys. Ges.* **5** (3), 21. *Naturwiss.* 1924.11.21 and *Phys. Z.* 1924.

33 Cap, Ferdinand (1958) 'Arthur March zum Gedenken', *Acta Phys. Austriaca*, 11, 289–93.

34 Cap, Ferdinand (1986) 'Erwin Schrödinger and Tirol', *Jahrbuch Überblicke Math*, 211–16.

35 Schrödinger, E. Letter to Arnold Sommerfeld, 1925.7.21, Sommerfeld Archiv, Deutsches Museum, Munich.

36 Schrödinger, E. Letter to Sommerfeld, 1925.3.7.
37 Wien, Wilhelm. Letter to Schrödinger, 1925.9.16, Schrödinger Archiv, Zentralbibliothek für Physik, Wien.
38 Schrödinger, E. Letter to Sommerfeld, 1926.1.29.
39 Miller, D.C. (1925) *Proc. Nat. Acad. Sci (U.S.A.)* **11**, 306–14.
40 Wien Archiv, Deutsches Museum, Munich.
41 Schrödinger, E. Letter to Wilhelm Wien, 1925.9.17.
42 Schrödinger, E. Letter to W. Wien, 1925.12.27.
43 Ref. (9), Ch. 1.
44 Muller, F. Max (1890) *The Sacred Books of the East, vol. 34. The Vedanta Sutras with commentary of Śankara* (translated by George Thibaut). Oxford: Oxford University Press.
45 Armstrong, A.H. (1960) 'Plotinus' in *The Concise Encyclopaedia of Western Philosophy and Philosophers*. London: Hutchinson.
46 Schopenhauer, Arthur (1958) *The World as Will and Representation* (translated by E.F.J. Payne) vol. 2, p. 172. Indian Hills, CO: Falcon's Wing Press.
47 Sackur, O. (1912) *Ann. Phys.* **40** (4), 87–106; Tetrode, H. (1912) *Ann. Phys.* **38** (4), 434–42.
48 Schrödinger, E. (1924) *Phys. Z.* **25**, 41–5.
49 Hanle, Paul A. (1977) *ISIS* **68**, 606–9; cf. Hanle Thesis, Ref. (15).
50 Schrödinger, E. (1925) *Sitzber. Preuss. Akad. Wiss. Phys. – Math. Kl.* **134**, 434–41.
51 Bose, S.N. (1924) *Z. Phys.* **26**, 178–81.
52 Schrödinger, E. Letter to Einstein, 1925.11.5, Einstein Archive, Princeton NJ.
53 Broglie, Louis de (1924) *Recherche sur la Theorie des Quanta*, Thèses presentées à la Faculté des Sciences de l'Université de Paris. Paris: Masson & Cie.
54 Broglie, Louis de & Andrade e Silva, J.L. (1971) *La Réinterprétation de la Mécanique undulatoire*. Paris: Gauthier-Villars.
55 Schrödinger, E. (1926) *Phys. Z.* **27**, 95–101.

Chapter 6

1 Weyl, Hermann, quoted in Abraham Pais (1986) *Inward Bound*. Oxford: Oxford University Press.
2 Bloch, Felix (1976) *Physics Today* **29** (December), 23–7.
3 Schrödinger, E. Letter to Einstein, 1925.11.3. Einstein Archive. Princeton Univ.
4 Schrödinger, E. Letter to A. Landé, 1925.11.16, Bohr Library, American Institute of Physics, New York.
5 Fues, Erwin, interview and correspondence with Thomas Kuhn, Bohr Library, American Institute of Physics, New York.
6 Schrödinger, E. A.H.Q.P., Microfilm 40, Bohr Library, American Institute of Physics, New York.
7 Ref. (18), Ch. 2.
8 Fischer, Ernst Peter (1984) *Social Research* **51**, 809–35.

9 Schrödinger, E. Letter to W. Wien, 1925.12.27.
10 Yourgrau, W. & Mandelstam, S. (1968) *Variational Principles in Dynamics and Quantum Theory* 3rd edn, London: Isaac Pitman & Sons.
11 Kragh, Helge (1979) *On the History of Early Wave Mechanics*. Roskilde: Roskilde Universitetscenter. (1982) *Centaurus* **26**, 154–97.
12 Schlesinger, Ludwig (1900) *Einführung in die Theorie der Differential gleichungen mit einer unabhängigen Variable*. Leipzig: Goschen.
13 Courant, Richard & Hilbert, David (1924) *Methoden der Mathematischen Physik*. Berlin: Springer.
14 Frank, Philipp & Mises, Richard v. (1925) *Differential und Integralgleichungen der Mechanik und Physik*. Braunschweig: Friedr. Vieweg & Sohn.
15 Schrödinger, E. (1926) *Abhandlungen zur Wellenmechanik*. Leipzig: J.A. Barth.
16 Junger, Itha. Personal interviews, 1985, London.
17 Schrödinger, E. (1926) *Ann. Phys.* **79**, 361–76.
18 Dirac, P.A.M. (1929) *Proc. Roy. Soc. London* **A123**, 713–33.
19 Oppenheimer, J. Robert (1957) *Engineering & Technology* March.
20 Schrödinger, E. (1926) *Ann. Phys.* **79** (4), 489–527.
21 Planck, Max to Schrödinger, 1926.4.2, in Karl Prizbram (ed.) (1963) *Briefe zur Wellenmechanik*. Wien: Springer.
22 Ref. (21).
23 Ref. (21).
24 Ehrenfest, Paul. letter to Schrödinger, 1926.5.29, Rijksmuseum voor de Geschiedenis der Natuurwetenschappen, Leiden.
25 Wien, Wilhelm. Letter to Schrödinger, 1926.2.6. Schrödinger Archiv, Zentralbibliothek für Physik, Wien.
26 Heisenberg, Werner (1925) *Z. Phys.* **33**, 879–93; Born, Max & Jordan, Pascual (1925) *Z. Phys.* **34**, 858–88; Born, Max, Heisenberg, Werner & Jordan, Pascual (1926) *Z. Phys.* **36**, 557–615.
27 Mehra, Jagdish & Rechenberg, Helmut (1982) *Historical Development of Quantum Theory*. vol. 4. New York: Springer Verlag.
28 Eckart, Carl H. (1926) *Phys. Rev.* **28**, 711–26.
29 Schrödinger, E. (1926) *Ann. Phys.* **79**, 734–56.
30 Schrödinger, E. *Ann. Phys.* **80** (4), 437–90.
31 Lorentz, Hendrik, correspondence with Schrödinger, 1926 (in Ref. 21).
32 Wessels, Linda (1975) 'Schrödinger's Interpretations of Wave Mechanics', Thesis, Indiana University.
33 Schrödinger, E. (1926) *Ann. Phys.* **81**, 109–39.
34 Ref. (5), Ch. 5.
35 Dirac, P.A.M. (1972) *Fields and Quanta* **3**, 139–64.
36 Born, Max (1926) *Z. Phys.* **37**, 863–7.
37 Rasche, Gunther (1987) *Neue Zürcher Zeitung*, August 12, 1987.
38 Schrödinger, E. Letter to Lorentz, 1926.6.6, Ref. (21).
39 Heisenberg, W. Letter to W. Pauli, 1926.6.8. Ref. (1), Ch. 5.
40 Pauli, W. Letter to Schrödinger, 1926.11.22. ibid.
41 Schrödinger, E. Letter to Pauli, 1926.12.15. ibid.
42 Heisenberg, W. Letter to Pauli, 1926.7.28 [also editorial notes] ibid.

43 Sommerfeld, A. Letter to Pauli, 1926.7.26. ibid.
44 Ref. (16).
45 Schrödinger, E. Letter to W. Wien, 1926.8.26. Ch. 5 Ref. (40).
46 Heisenberg, Werner (1969) *Der Teil und das Ganze*. Munich: Piper Verlag.
47 Schrödinger, E. Letter to W. Wien, 1926.10.21. Ch. 5 Ref. (40).
48 Galton, Francis (1883) *Inquiries into Human Faculty and its Development*. New York: Macmillan.
49 Schrödinger, E. Letter to Niels Bohr, 1926.10.23, in Kalckar, J. (ed.) (1985) *Bohr, Collected Works* vol. 6. Amsterdam: North Holland.
50 Bohr, Niels. Letter to Schrödinger, 1926.12.2 in Ref. (49).
51 Klein, O. (1927) *Z. Phys.* **41**, 407–22.

Chapter 7

 1 University of Wisconsin Archive.
 2 A.H.Q.P., Microfilm P–13, Bohr Library, American Institute of Physics, New York.
 3 Meyenn, Karl v. (1982) *Gesnerus* **39**, 261–76.
 4 Schrödinger, Annemarie, 'Amerikanische Reise', Schrödinger Archiv, Zentralbibliothek für Physik, Wien.
 5 Ref. (24), Ch. 5.
 6 Johns Hopkins University Archive.
 7 (1973) *Die Humboldt Universität zu Berlin*. Berlin: VEB Deutsche Verlag der Wissenschaften.
 8 Humboldt University of Berlin, Archive.
 9 Schrödinger Archiv, Alpbach. [Cf. Ref. (17), Ch. 1, p. 186.]
10 Flamm, Ludwig (1961) 'Erwin Schrödinger', *Almanach Osterr. Akad. Wiss. Wien* **104**, 402–11.
11 (1928) *Electrons et Photons, Rapports de VIe Conseil de Physique Solvay*. Paris: Gauthier-Villars.
12 Schilpp, Paul A. (ed.) (1962) *Albert Einstein, Philosopher Scientist*. New York: Alfred Knopf.
13 *Phys. Z.* vol. 28 (1928) to vol. 34 (1933) list physics courses at German universities.
14 Ledermann, Walter. Interview 1986, Brighton.
15 (1975) *Die Berliner Akademie der Wissenschaften in der Zeit des Imperialismus*. Berlin: Akademie Verlag.
16 (1929) *Sitzbr. Preuss. Akad. Wissen. Phys. – Math. Kl.* C–Cii.
17 (1941) *Jahrbuch der Preussischen Akademie der Wissenschaften 1940*. Berlin: Verlag der Akademie der Wissenschaft.
18 Schrödinger, E. (1928) *Four Lectures on Wave Mechanics*. London & Glasgow: Blackie & Sons.
19 Pauli, W. Letter to Bohr, 1929.4.25. Ref. (1). Ch. 5.
20 Schrödinger, E. (1932) *Über Indeterminismus in der Physik*. Leipzig: J.A. Barth.
21 Holmsten, Georg (1984) *Die Berlin-Chronik*. Dusseldorf: Droste.
22 Ref. (45), Ch. 2.

23 Elsasser, Walter (1978) *Memoirs of a Physicist in the Atomic Age*. New York: Science History Publications.

24 Schrödinger, E. 'Die Wandlung des physikalischen Weltbegriffs'. Published in 1984 in *Gesammelte Abhandlungen* vol. 4, 600–8. Wien: Verlag Österreichischen Akademie der Wissenschaften.

25 Sullivan, J.W.N. (1931) 'Interviews with Great Scientists', *The Observer* (London) January 11.

26 Dirac, P.A.M. (1930) *The Principles of Quantum Mechanics*. Oxford: Oxford University Press.

27 Schrödinger, E. (1931) *Sitzber. Preuss. Akad. Wiss. Phys. – Math. Kl.* 63–72; (1932), 105–28.

28 Schrödinger, E. (1931) *Sitzber. Preuss. Akad. Wiss. Phys. – Math. Kl.* 144–53.

29 Cramer, John G. (1986) *Rev. Mod. Phys.* **58**, 647–87.

30 Castellan, Georges (1969) *L'Allemagne de Weimar 1918–1933*. Paris: Armand Colin; Eyck, Erich (1960) *A History of the Weimar Republic*, Cambridge: Cambridge University Press.

31 Weisskopf, Viktor. Interview, 1986, Cambridge, MA.

32 Köstler, Arthur (1969) *Arrow in the Blue*. London: Hutchinson.

33 Pelzer, Heinz. Personal communications, 1985.

34 Cremer, Erika. Interview, 1986, Innsbruck.

35 Casimir, Hendrik (1983) *Haphazard Reality*. New York: Harper & Row.

36 Hoffman, Dieter (1984) *Erwin Schrödinger*. Leipzig: B.G. Teubner.

37 Bleuel, Hans Peter (1968) *Deutschlands Bekenner*. Bern: Scherz.

38 Schrödinger, E. (1933) *Forschungen und Fortschritte* **9**, 125–6.

39 Schrödinger, E. (1933) *Sitzber. Preuss. Akad. Wiss. Phys. – Math. Kl.* 165.

40 Schrödinger, Annemarie. Interview with Thomas Kuhn, A.H.Q.P., 1963.4.5, Bohr Library, American Institute of Physics, New York.

41 Beyerchen, Alan D. (1977) *Scientists under Hitler*, New Haven: Yale University Press.

42 Heilbron, J.L. (1986) *Dilemmas of an Upright Man*. Berkeley, CA: University of California Press.

43 Lüders, Marie-Elisabeth (1963) *Fürchte Dich nicht: Persönliches und Politisches aus mehr als 80 Jahren 1878–1962*. Köln: Westdeutscher Verlag.

44 Haberer, Joseph (1969) *Politics and the Community of Science*. New York: D. van Nostrand.

45 Laue, Max. Letter to Schrödinger, 1936.3.4, A.H.Q.P., Microfilm P–13, Sect. 1, Pt 7. Bohr Library, American Institute of Physics, New York.

46 Heisenberg, Werner. Letter to Max Born, 1933.6.2.

47 Krieck, Ernst (1937) *Nationalpolitische Erziehung*. Leipzig: Azuranen Verlag.

48 Harrod, Roy F. (1959) *The Prof – A Personal Memoir of Lord Cherwell*. London: Macmillan.

49 The Cherwell Archive, Oxford.

50 Hermann, Armin, Ref. (17), Ch. 1.

51 Schrödinger, Annemarie. 'Was ein Kleines BMW Erzählen Kann', Schrödinger Archiv, Zentralbibliothek für Physik, Wien.

52 Bargmann, Valentine. Personal communication, 1987.

53 March, Arthur. Letter to Schrödinger, 1934.2.1, Schrödinger Archiv, Alpbach.
54 Schrödinger, E. Letter to Einstein, 1933.8.12. [Princeton University]
55 Born, Max (1978) *My Life, Recollections of a Nobel Laureate.* London: Taylor & Francis.

Chapter 8

1 (1934) *Structure et Propriétés des Noyaux Atomiques: Rapports et Discussions du VIIeme Conseil de Physique Solvay.* Paris: Gauthier-Villars.
2 Ref. (24), Ch. 5.
3 Imperial Chemical Company Archives.
4 Ref. (17), Ch. 1.
5 Schrödinger, E. Letter to Max Born, 1943.1.13, Staatsbibliothek Preussischer Kulturbesitz, Berlin, Born Nachlass 704.
6 Halasz, A.N. (1960) *Nobel: A Biography.* London: Hale.
7 Friedman, R.M. (1981) 'The Nobel Prizes in Physics', *Nature* **292**, 293–8.
8 Crawford, Elisabeth (1984) *The Beginnings of the Nobel Institution: The Science Prizes 1901–1915.* Cambridge: Cambridge University Press.
9 Nobel Archives, Stockholm.
10 Dirac, P.A.M. (1963) *Scientific American* **208** (May) 45–53.
11 (1935) *Les Prix Nobel en 1933.* Stockholm: Norstedt & Soener.
12 Schrödinger, Annemarie. 'Stockholm 1933', A.H.Q.P., Bohr Library, American Institute of Physics, New York.
13 Princeton University Archives.
14 Institute for Advanced Study Archive, Princeton NJ.
15 Laue, Max. Letter to Erwin Freundlich. June 1934. Library, University of St. Andrews.
16 Schrödinger, Erwin. Letter to Bruno Bertotti, 1960.5.8. [Bertotti Collection, Pavia.]
17 Ayer, A.J. (1977) *Part of My Life.* London: Collins.
18 Ref. (55), Ch. 7.
19 Schrödinger, E. (1935) *Cursos de la Universidad Internacional de Verano en Santander* **1**, 1–73. Madrid: Signo.
20 *The Listener*, 1935.6.5, pp. 952–3.
21 Einstein, Albert, Podolsky, Boris & Rosen, Nathan (1935) *Phys. Rev.* **47**, 777–800.
22 All the Einstein–Schrödinger correspondence is in the Einstein Archive at Princeton University.
23 Rosenfeld, Leon in Rozental, S. (ed.) (1968) *Niels Bohr. His Life and Work.* Amsterdam: North Holland.
24 Bohr, Niels (1935) *Phys. Rev.* **48**, 696–702.
25 Schrödinger, E. Letter to W. Pauli, 1935.7.[1]. Ref. (1), Ch. 5.
26 Pauli, Wolfgang. letter to Schrödinger, 1935.7.9. ibid.
27 Bohm, David (1952) *Phys. Rev.* **85**, 166–79; 180–93.
28 Schrödinger, E. (1935) *Naturwiss.* **23**, 807–12; 823–8; 844–9.
29 Neumann, John v. (1932) *Mathematische Grundlagen der Quantenmechanik.* Berlin: Springer.

30 Greenberger, Daniel M. (1953) *Rev. Mod. Phys.* **55**, 875–905.
31 Schrödinger, E. (1935) *Proc. Camb. Phil. Soc.* **31**, 555–63; (1936) **32**, 446–52.
32 Bell, John (1964) *Physics* **1**, 195–200.
33 Clauser, John & Shimony, A. (1978) *Rep. Prog. Phys.* **41**, 1881–1927.
34 Aspect, A., Grangier, P. & Roger, G. (1981) *Phys. Rev. Lett.* **47**, 460–3.
35 Espagnat, Bernard d' (1979) *Scientific American* **242**, 158–81; (1979) *A La Recherche du Réal.* Paris: Gauthier-Villars; (1985) *Une incertaine Réalité*, Paris: Gauthier-Villars.
36 Schrödinger, E. Letter to (1935.10.13) and from (1935.10.26) Niels Bohr, Schrödinger Archiv, Alpbach.
37 Schrödinger, E. Letter to Einstein, 1936.3.23.
38 Simon, Franz. Royal Society Archive, London.
39 Schrödinger, E. (1936) *Nature* **138**, 13–14.
40 University of Edinburgh Archive.
41 Thirring, Hans. Letter to Schrödinger 1936.2.20, Schrödinger Archiv, Zentralbibliothek für Physik, Wien.

Chapter 9

1 Stadler, Karl R. (1971) *Austria.* New York: Prager.
2 Gedye, George (1939) *Betrayal in Central Europe: Austria and Czechoslovakia the Fallen Bastions.* New York: Harper.
3 Pontifical Academy of Sciences Archive.
4 Schrödinger Archiv, Alpbach.
5 Imperial Chemical Industries Archives.
6 Kilmister, Clive W. (1966) *Sir Arthur Eddington.* Oxford: Oxford University Press. Douglas, Allie V. (1957) *The Life of Arthur Stanley Eddington.* London: Nelson.
7 Eddington, Arthur S.(1923) *Mathematical Theory of Relativity.* Cambridge: Cambridge University Press.
8 Eddington, Arthur S. (1936) *The Nature of the Physical World.* Cambridge: Cambridge University Press.
9 Schrödinger, E. (1929) *Naturwiss.* **17**, 695.
10 Eddington, Arthur S. (1936) *Relativity Theory of Protons and Electrons.* Cambridge: Cambridge University Press.
11 Schrödinger, E. (1937) *Nature* **140**, 742–4.
12 Whittaker, E.T. (1937) *The Observatory* **60**, 14–23.
13 (1937) *Rendiconto de la Celebrazione Galvaniana e Atti del Congresso di Fisica* Bologna: Societa Italiana de Fisica.
14 Schrödinger, E. (1940) *Wiss. natuurkundig. Tijdschrift* **10**, 2–9.
15 Schrödinger, E. (1937) *Commentationes Pontificiae Acad. Scient.* **2**, 321–64.
16 Muller, Robert (1939) *Phys. Z.* **40**, 366–84.
17 Schrödinger, E. (1938) *Ann. Phys.* **32** (5), 49–55.
18 Mark, Hermann. Interview, 1985, Brooklyn, N.Y.
19 Zückmayer, Carl (1966) *Als Wars ein Stück von Mir.* Wien: S. Fischer.
20 *Der Wiener Tag*, 1938.2.19.
21 Clare, George (1980) *Last Waltz in Vienna.* London: Macmillan. [This is one of many similar accounts of Nazi brutalities in Vienna.]

22 *Grenzfeste Deutscher Wissenschaft*. Wien: Verlag fur Gesellschaftkritik, 1985.
23 Cherwell Archive, Christ Church College, Oxford.
24 *Graz Tagespost*, 1938.3.30.
25 *Nature*. 1938.5.21.
26 Cherwell Archive, Christ Church College, Oxford.
27 Schrödinger Archiv, Alpbach.
28 Dublin Institute for Advanced Studies, Archive.
29 Simon Archive, Royal Society, London.
30 Schrödinger, Annemarie. Unpublished Ms, Dublin Institute for Advanced Studies.
31 Lady Charlotte Simon. Interview, 1985.
32 Fondation Francqui, Brussels, Archive.
33 Schrödinger, E. (1939) *Physica*, **6**, 899–912.
34 Schrödinger, E. (1939) *Nature*, **144**, 593.
35 Universitet zu Gent, Archive.

Chapter 10

1 Longford, Frank P. (1970) *Eamon de Valera*. London: Hutchinson.
2 Flexner, Abraham (1940) *I Remember: The Autobiography of Abraham Flexner*. New York: Simon and Schuster.
3 Dublin Institute for Advanced Studies, Archive.
4 *Eamon de Valera Centenary* (1983) Institiúid Ard-Léinn Bhaile Átha Cliath.
5 Dublin Institute for Advanced Studies, School of Theoretical Physics, *Fifteen Year Report*, Dublin, 1961.
6 Longford, Christine (1936) *A Biography of Dublin*. London: Methuen.
7 McConnell, Albert J. Interview, 1985, Dublin.
8 *Irish Catholic* 1942.11.29.
9 Ryan, John (1975) *Remembering How We Stood*. New York: Toplinger.
10 *Irish Times* 1939.11.15.
11 Schrödinger, E. (1940) *Proc. Roy. Irish Acad.* **46**A, 9–16.
12 Schrödinger, E. (1941) *Proc. Roy. Irish Acad.* **46**A, 183–206.
13 Schrödinger, E. Letter to Max Born, 1941.1.29. Staatsibliothek Preussischer Kulturbesitz, Berlin, Born Nachlass 704.
14 Royal Irish Academy, Dublin, Archives.
15 Schrödinger, E. (1946) *The Bell* **8**, 405–6.
16 Maire Cruise O'Brien. Personal communication, 1985.
17 O'Kennedy, Sean. Personal communications, 1985–86.
18 O'Keeffe, Timothy (ed.) (1973) *Myles, Portraits of Brian O'Nolan*. London: Martin, Brian & O'Keeffe.
19 Trinity College Dublin, Archives.
20 Scaife, Brendan. Interview, 1986, Dublin.
21 School of Theoretical Physics, Dublin Institute for Advanced Studies, Minutes of Governing Board.
22 Feric, Stephen. Personal memoir, 1986.
23 Schulhof, Alfred. Personal communications, 1987.
24 Macliammoir, Micheál (1946) *All for Hecuba*. London: Methuen.

25 Acton, Charles. Personal discussion, 1985, Dublin.

26 Kavanagh, Patrick (1964) *Self Portrait*. Dublin: Dolmen Press.

27 Anderson, Ronald. Interview, 1985, Dublin.

28 Lieven, Alexander. Interview, 1986, London.

29 Mac Cana, Proinsias (1983) 'David Greene (1915–1981)', *Erin* **34**, 1–10.

30 *Irish Press* 1940.11.1.

31 Tinney, Sheila Power. Interview, 1985, Dublin.

32 Schrödinger, E. Letter to H. Weyl, 1941.4.23, University of Zürich Archive.

33 Maugham, W. Somerset (1940) *The Summing Up*. London: Heinemann.

34 Schrödinger, E. *Ephemeridae*, 1941.6.22, Schrödinger Archiv, Alpbach.

35 Schrödinger, E. Letter to Max Born, 1941.10.5. Staatsbibliothek Preussischer Kulturbesitz, Berlin, Born Nachlass 704.

36 *Irish Times* 1941.11.6.

37 *Irish Times* 1942.4.10.

38 Minutes of the Council, Dublin Institute for Advanced Studies.

39 Schrödinger, E. (1948) *Acta Physica Austriaca* **1**, 245.

40 Eddington, Arthur S. (1942) *Comm. Dublin Inst. Adv. Studies* No. 2.

41 Dirac, P.A.M. (1941) *Comm. Dublin Inst. Adv. Studies* No. 1.

42 Schrödinger, E. Letter to Max Born, 1942.7.27. Staatsbibliothek Preussischer Kulturbesitz, Berlin, Born Nachlass 704.

43 Born, Max & Infeld, Leopold (1934) *Proc. Roy. Soc.* (London) A **144**, 425–51.

44 Pais, Abraham (1972) 'The Early History of the Theory of the Electron, 1897–1947' in Salam, A. & Wigner, E. (eds) *Aspects of Quantum Theory*. Cambridge: Cambridge University Press.

45 Schrödinger, E. (1935) *Proc. Roy. Soc.* (London) A **150**, 465–77.

46 Dirac, Paul (1977) in *History of Twentieth Century Physics* Proc. Int. School Phys. Enrico Fermi, course LVII, ed. C. Wiener. New York: Academic Press.

47 Schrödinger, E. (1942) *Proc. Roy. Irish Acad.* **47** A, 72–117.

48 Schrödinger, E. Letter to Max Born, 1942.5.3. Staatsbibliothek Preussischer Kulturbesitz, Berlin, Born Nachlass 704.

49 Bertotti, Bruno (1985) *Stud. Hist. Phil. Sci.* **16**, 83–100.

50 Barthes, Roland (1980) in Goldsmith, M., Mackay, A. & Woudhuysen, J. (eds) *Einstein: The First Hundred Years*. Oxford: Pergamon Press.

51 Feynman, Richard (1963) 'Lectures on Gravitation 1962–63' (mimeographed) California Institute of Technology.

52 Schrödinger, E. (1944) *Nature* **153**, 572–5.

53 Einstein, Albert (1951) *The Meaning of Relativity*, 5th edn. London: Methuen.

54 Schutz, Bernard (1985) *A First Course in General Relativity*. Cambridge: Cambridge University Press.

55 Misner, Charles W., Thorne, Kip S. & Wheeler, John A. (1973) *Gravitation*. San Francisco: Freeman.

56 Schrödinger, E. (1943) *Proc. Roy. Irish Acad.* **49** A, 43–58.

57 Ref. (10), Ch. 6.

58 Delbrück, Max, Timofeef-Ressovsky, N.W. & Zimmer, K.G. (1935) *Nachr. Biol. Ges. Wiss. Göttingen* **1**, 189–245.

59 Schrödinger, E. (1944) *What is Life?* Cambridge: Cambridge University Press.

60 Carlson, Elof Axel (1981) *Genes, Radiation and Society: The Life and Work of H.J. Muller*. Ithaca: Cornell University Press.

61 Delbrück, Max in E.A. Carlson (1971) *J. Hist. Biol.* **4**, 149–70.

62 Clancy, Basil. Personal communication, 1986.

63 Muller, Hermann (1946) *J. Heredity* **37**, 90–2.

64 Stent, Gunther (1966) 'Waiting for the Paradox' in *Phage and the Origins of Molecular Biology*, ed. John Cairns, G.S. Stent, & J.D. Watson. Cold Spring Harbor, N.Y.: Laboratory of Quantitative Biology.

65 Watson, James D. Lecture at Indiana University, 1984.10.4.

66 Symonds, Neville (1986) *Quart. Rev. Biol.* **61**, 221–6.

67 May, Sheila (1943) *Dublin Magazine* **18**, 73–4.

68 May, Sheila (1944) 'Two Dublin Slums', *The Bell* **7**, 351–6.

69 Mann, Thomas (1920) *Betrachtungen eines Unpolitischen*. Berlin: Fischer.

70 Schrödinger, E. (1949) *Gedichte*. Godesberg: Helmut Kupper.

71 Schopenhauer, Arthur. Ref. (46), Ch. 5, vol. 2, p. 536.

72 Schrödinger Archiv, Alpbach.

Chapter 11

1 Schrödinger, E. (1946) *Statistical Thermodynamics*. Cambridge: Cambridge University Press.

2 Schrödinger, E. Letter to Max Born, 1944.8.19. Staatsbibliothek Preussischer Kulturbesitz, Berlin, Born Nachlass 704.

3 Kirdly, Peter & Ziegler, Maria Narayne (1987) *Janossy Lajos (1912–1978) es Schrödinger Erwin (1887–1961) Levelzesse*. Budapest: Hungarian Academy of Sciences.

4 Peng, H.W. Personal communication, 1985.

5 Schrödinger, E. (1944) *Proc. Roy. Irish Acad.* **49**A, 225–35; 237–44; 275–87.

6 Schrödinger, E. & McConnell, James (1944) *Proc. Roy. Irish Acad.* **49** A, 259–73.

7 Quoted by Hittmair, 'Schrödinger's Unified Field Theory Seen 40 Years Later', Ref. (5), Ch. 5, p. 168.

8 Einstein, Albert. Letter to Schrödinger, 1943.4.26.

9 Kavanagh, Peter (1977) *Beyond Affection*. New York: The Peter K. Hand Press.

10 Laclos, Pierre Choderlos de (1782) *Les Liaisons Dangereuses*, Paris, Durand.

11 McManus, Lena Lean. Interview, 1987, London.

12 Weyl, Hermann. Letter to Schrödinger, 1947.9.4.

13 Schrödinger, E. Letter to Weyl, 1945.10.11, E.T.H. Archiv, Zürich.

14 Schrödinger, E. (1945) *Proc. Roy. Irish Acad.* **51**A, 1–8.

15 Schrödinger, E. (1950) *Space-Time Structure*. Cambridge: Cambridge University Press.

16 Schrödinger, E. (1946) *Proc. Roy. Irish Acad.* **51**A, 41–50.

17 Schrödinger, E. (1946) *Verh. Schweiz. Naturforsch. Gesell.* 126, 53–61.

18 Schrödinger, E. (1947) *Eranos Jahrbuch 1946* **14**, 491–520, Zürich: Rhein.

19 Schrödinger, E. 1947) *Proc. Roy. Irish Acad.* **51**A, 163–71.

20 Schrödinger, E. to Bohm, Hansi, 1947.1.14. Bohm collection.
21 Schrödinger, E. Letter to Einstein, 1946.6.13.
22 Schrödinger, E. (1947) *Proc. Roy. Irish Acad.* **51**A, 51–66; 141–6.
23 Servien, Pius (1942) *Probabilité*. Paris: Hermann & Cie.
24 Peirce, Charles S. (1932) *Collected Papers* vol. 2. Cambridge MA: Harvard University Press.
25 Popper, Karl (1957) in *Observation and Interpretation*, ed. S. Korner. London: Constable; (1959) *Brit. J. Phil. Sci.* **10**, 25–42.
26 Schrödinger, E. (1948) *Acta Physica Austriaca* **1**, 201–45.
27 *Irish Independent*, 1947.5.15.
28 Schrödinger, E. (1954) *Nature and the Greeks*. Cambridge: Cambridge University Press.
29 Schrödinger, E. Letter to Benjamin Farrington, A.H.Q.P., Bohr Library, Microfilm 37, 1949.4.19.
30 Schrödinger, E. Letter to Louis Werner, 1948.8.20.
31 Houston, John (ed.) (1981) *Lucie Rie: A Survey of her Life and Work*. London: Crafts Council.
32 *Les Particules Elémentaires: Rapports et Discussions de 8me Conseil de Physique*. Bruxelles: R. Stoops, 1950.
33 McMullin, Ernan. Interview, 1986, Bloomington, IN.
34 Schrödinger, E. (1949) *Gedichte*. Godesberg: Helmut Kupper.
35 Westfall, R.S. (1980) *Never at Rest: the Life of Isaac Newton*. Cambridge: Cambridge University Press.
36 Schrödinger, E. Letter to Arnold Sommerfeld, 1949.2.13, Bohr Library, New York.
37 Schrödinger, E. BBC Written Archives, Reading, U.K.
38 Turner, J.E. (1930) *Nature* 126, 995.
39 Schrödinger, E. 'The Future of Understanding', BBC European Service, 1950.9.13 and 20.
40 Born, Max (1949) *Natural Philosophy of Cause and Chance*. Oxford: Oxford University Press.
41 Schrödinger, E. *Proc. Roy. Irish Acad.* **53**A, 189–95.
42 Popper, Karl (1976) *Unended Quest*. Glasgow: Fontana/Collins.
43 Davies, P.C.W. (1977) *Space and Time in the Modern Universe*, Ch. 3. Cambridge: Cambridge University Press.
44 Schrödinger, E. Letter to John Synge, 1952.9.[12], Dublin Institute for Advanced Study, Archive.
45 Schrödinger, E. (1952) *Brit. J. Phil. Sci.* **3**, 109–23; 233–142.
46 Macomber, James D. (1973) *Dynamics of Spectroscopic Transitions*. New York: John Wiley & Sons.
47 Heisenberg, W. (1958) *Physics and Philosophy*. London: Allen & Unwin.
48 Bass, Ludvik & Schrödinger, Erwin (1955) *Proc. Roy. Soc.* (London) A **232**, 435–77; (1956) *Il Nuovo Cimento* Suppl. (10) 825–6.
49 Goldhaber, Alfred S. & Nieto, Michael M. (1971) *Rev. Mod. Phys.* **43**, 277–96.

Chapter 12

1 Cronin, Audrey K. (1986) *Great Power Politics and the Struggle Over Austria 1945–1955*. Ithaca NY: Cornell University Press.
2 Halpern, Leopold. Interviews 1986, 1987, Pasadena CA.
3 Cap, Ferdinand, Ref. (34), Ch. 5.
4 Klastersky, Wilhelm. Letter to Schrödinger, 1953.4.16, A.H.Q.P., Bohr Library, New York.
5 Schrödinger, E. Letter to W. Klastersky, 1953.4.26, Bohr Library, New York.
6 Bass, Ludvik, Interview 1985, Brisbane.
7 Koestler, Arthur (1972) *The Call Girls*. London: Hutchinson.
8 Schrödinger, Annemarie. Letter to Elisabeth Ullmann, 1956.3.1. Ullmann collection.
9 Schrödinger, E. (1956) *Expanding Universes*. Cambridge: Cambridge University Press.
10 Callaway, I. (1953) *Phys. Rev.* **92**, 1567–70.
11 Johnson, C.P. (1953) *Phys. Rev.* **89**, 320–21.
12 Schrödinger, Annemarie. Letter to Elisabeth Ullmann, 1956.4.15.
13 *Neues Österreich*, 1956.4.14.
14 Schrödinger, E. Letter to B. Bertotti, 1957.12.27.
15 Wisdom, John. Personal communication 1987.12.11.
16 Schrödinger, E. (1959) *Mind and Matter*. Cambridge: Cambridge University Press.
17 Sherrington, Charles (1940) *Man on His Nature*. Cambridge: Cambridge University Press.
18 Ruyer, Raymond (1974) *Le Gnose de Princeton*. Paris: Fayard.
19 Polzer, Kurt. Interview 1985, Vienna.
20 Schneeweiss, Joseph (1986) *Keine Führer, Keine Götter*. Wien: Junius.
21 Schrödinger, Annemarie. Letter to Elisabeth Ullmann, 1957.1.1.
22 Schrödinger, E. Letter to Hilde March, 1957.4.28. Schrödinger Archiv, Alphach.
23 Einstein, Albert (1972) Quoted in *Albert Einstein Creator and Rebel*, by Banesh Hoffmann and Helen Dukas, New York: Viking, p. 237.
24 Voltaire (1980) *Letters on England* (translated by Leonard Tancock) Harmondsworth UK: Penguin Books.
25 Halpern, Leopold 'Reminiscences of Erwin Schrödinger', unpublished Ms, 1987.
26 Schrödinger, E. Letter to Bruno Bertotti, 1958.1.11.
27 Schrödinger, E. Letter to Bruno Bertotti, 1959.1.7.
28 Schrödinger, E. Letter to John Synge, 1959.11.9.
29 Schrödinger, E. Letter to Bruno Bertotti, 1959.4.17.
30 Schrödinger, E. Letter to Max Born, 1960.1.8. Staatsbibliothek Preussischer Kulturbesitz, Berlin, Born Nachlass 704.
31 Ref. (9), Ch. 1.
32 Ref. (14), Ch. 1.
33 Schrödinger, E. Letter to Max Born, 1960.10.10. Staatsbibliothek Preussischer Kulturbesitz, Berlin, Born Nachlass 704.

34 Schrödinger, E. Letter to Max Born, 1960.10.24. Staatsbibliothek Preussischer Kulturbesitz, Berlin, Born Nachlass 704.
35 Halpern, Leopold. 'The Life of Marietta Blau', unpublished Ms., 1987.
36 Schrödinger, Annemarie. Letter to Elisabeth Ullmann, 1961.1.15.

Name index

Acton, Charles 371
Adler, Viktor 44, 53
Anderson, Carl 45, 257, 336
Anderson, Ronald 371
Arrhenius, Svante 282
Aspect, Alain 311
Aston, Francis 276, 328
Auerbach, Felix 132
Avernarius, Richard 1, 19, 43, 170
Ayer, Alfred 298

Ball, Hugo 150
Bär, Richard 148, 158, 165, 247, 341–2, 344
Barkla, Charles 73
Barnes, Djuna 20
Barthes, Roland 386
Bass, Ludvik 452–4
Bauer, Alexander 8–12, 16, 26, 41, 121
Bauer, Fritz 116, 333
Bauer-Artzberger, Rhoda (Aunt Rhoda) 7, 9, 11
Bauer-Bamberger, Emily (Aunt Emmy) 10, 14, 17
Bauer-Böhm, Hansi 131, 275, 318–19, 333, 341, 346, 407, 428, 434, 443
Bauer-Schrödinger, Georgine 9–19, 17, 63, 108, 133, 256
Baumgartner, Andreas 33
Bell, John 311
Benda, Julien 376
Benndorf, Hans 53
Benzer, Seymour 404
Berliner, Arnold 97, 325
Bertel-Schrödinger, Annemarie: American travels 230–1; asthma 480; Berlin 236–7, 249, 262; B.M.W. 270, 400, 438; depressions 441, 458; divorce considered 175–6, 191, 255, 429; Dublin 369, 364, 382, 440; engagement 116; Erwin S. & 69–70, 81, 94, 116, 132, 226, 338; Ewalds & 395, 410; Felicie & 66; Graz 321, 344; Italy 474; Junger family 223, 227; Oxford 295, 297; Russian incident 457; Spain 300; Stockholm 290;

Sylvester Abend 323; Vienna 321, 333, 461–2, 466, 471, 482; wedding 131; Weyl & 275, 299, 323; Zürich 145, 149, 195
Bertotti, Bruno 386, 458, 471–2, 475, 479
Bethe, Hans 269
Birkhoff, George 354
Bjerrum, Niels 98, 135
Blackett, Patrick 416, 444
Blackmore, John 42
Blau, Marietta 479
Bloch, Eugene 284
Bloch, Felix 192
Bogner, Maria 11
Bohm, David 306, 311
Böhme, Jakob 173
Bohr, Niels: biology 394, 398–9; complementarity 228, 472; Copenhagen School 98, 152, 159, 210; Danish proverb 320; Dirac & 290; E. P. R. reply 305–6, 312–14; father 7; Galvani Congress 328–9; Heisenberg & 210; hydrogen-atom theory 3, 135–6, 220; – Kramers–Slater paper 160–1; London conference (1920) 450; metal theory 60; Nobel 282, 286, 288–90; Papal Academy 322; Schrödinger & 226, 266–9, 251; Solvay Conferences 157, 302, 279, 444; Zürich chair 141
Boldt, Erika 249, 296
Boltzmann, Ludwig 4, 20, 34, 37–40, 75, 105, 118, 130, 153, 177, 226, 245, 465
Born, Max: Berlin chair 234; Cambridge 276, 295, 311; Edinburgh chair 319; Göttingen 159; Germany, escape 270; Heisenberg & 266; irreversibility 258, 449; Lindemann & 268; London conference (1952) 451; matrix mechanics 208, 210; *Natural Philosophy of Cause and Chance* 449; Nobel 285; optics, nonlinear 383–4; philosophy and physics 1; probability waves 220, 225, 240, 435–6; Schrödinger &: escape from Graz 342, 345–6, – letters from 278, 280, 345, 361, 368, 376, 381, 446,

Subject index

Academic Assistance Council 268
affine connection (affinity) 391–2
affine geometry 391
affine theory 426, 430
affine transformation 123
Akademisches Gymnasium 20
Alpbach 450, 457–8, 462, 472–5, 480, 482
Altamira 300
angular momentum 307
Ann Arbor, Michigan 232
Annalen der Physik 120, 200
Anschluss 333–7
antisemitism 24, 44, 80, 263, 320, 404
aperiodic crystal 398, 404
Arosa 145–7, 194–7
atmospheric electricity 53, 67–70
atomism 38–9
Austria 329–2, 334–6; politics 320, 455
Austria–Hungary 5, 107

Berlin 209, 222, 247–50, 260–3
Berlin Colloquium 242
Berlin University 233–7, 241–2, 271, 343
Bhagavad-Gita 114
black-body radiation 34, 118, 210, 453
Bohr–Kramers–Slater theory 161–3, 206
Born–Infeld electrodynamics 382–5, 393, 432
Bose–Einstein statistics 180–1, 380
Brahman 171–2, 401, 477
Breslau University 134
British Broadcasting Corporation (B.B.C.) 301, 448
British Society History and Philosophy of Science 450
Brixen (Bressanone) 273
Broglie equation 183–7
Brownian motion 83, 99

California Institute of Technology 232, 289
Cambridge University Press 402, 415, 458
categorical imperative 252
causality 112, 155, 245–6, 299, 359–60, 378, 448–9

chemical bond 398, 400, 403
Christoffel symbols 389, 391
chromosome 396
Clontarf 356, 369, 373, 377
coherence time 309
color theory 36, 41, 120–9
complementarity 228, 473
Compton effect 160
conductivity 157
configuration space 208, 239–41
Connemara 362
consciousness 174, 252–3, 409, 448, 463
conservation laws 152–4, 162, 393
Copenhagen interpretation 226, 228, 241, 250, 260, 303, 314, 451, 479
Copenhagen physics 163
correspondence principle 159, 161, 279
cosmic rays 68, 479
cosmological constant 390, 426, 434
cosmology 106–7, 349–50, 428, 449
curvature tensor 389
Czernowitz University 1, 109

Dalmatia 247
degeneracy (gas) 179–81
determinism 317, 400, 418
dielectrics 61, 75
differential equations 196, 199, 204
differential geometry 123
diffusion equation 100–5, 258–9
dipole moment 61
Dirac equation 256
dispersion, anomalous 74
dispersion formula 218
Doppler effect 33, 160
dualism 251, 448, 476
Dublin 355, 358, 373; bombing 373, 375; slums 405
Dublin Colloquia 368, 375, 379–81
Dublin Institute for Advanced Studies 340, 342, 347, 353–5, 362, 365, 374, 378, 433
Dublin University Metaphysical Society 359
Dunquin 363–4, 407